AF388679

Bacillus thuringiensis Toxins: Functional Characterization and Mechanism of Action

Bacillus thuringiensis Toxins: Functional Characterization and Mechanism of Action

Editors

Yolanda Bel
Juan Ferré
Patricia Hernández-Martínez

MDPI • Basel • Beijing • Wuhan • Barcelona • Belgrade • Manchester • Tokyo • Cluj • Tianjin

Editors
Yolanda Bel
Universitat de València
Spain

Juan Ferré
Universitat de València
Spain

Patricia Hernández-Martínez
Universitat de València
Spain

Editorial Office
MDPI
St. Alban-Anlage 66
4052 Basel, Switzerland

This is a reprint of articles from the Special Issue published online in the open access journal *Toxins* (ISSN 2072-6651) (available at: https://www.mdpi.com/journal/toxins/special_issues/Bt_toxins).

For citation purposes, cite each article independently as indicated on the article page online and as indicated below:

LastName, A.A.; LastName, B.B.; LastName, C.C. Article Title. *Journal Name* **Year**, *Volume Number*, Page Range.

ISBN 978-3-0365-2049-0 (Hbk)
ISBN 978-3-0365-2050-6 (PDF)

Cover image courtesy of Yolanda Bel, Juan Ferré and Patricia Hernández-Martínez.

Contents

About the Editors

Yolanda Bel (Dr.) studied Biology (major in Biochemistry) and received her Ph.D. in Biology from the University of Valencia (UV), Spain. Her Ph.D. was undertaken in the Department of Genetics of the UV and in the Biology Division of the Oak Ridge National Laboratory, Oak Ridge (Tennessee, USA). She undertook her postdoctoral studies (1992–1993) in Sandoz, Basel, Switzerland. After a contract at the University of Valencia (1993–1996) to work on *Bacillus thuringiensis* (Bt), she completed a "Master in Industrial Wastewater Treatment", and worked in water microbiology in the private industry. She returned to the UV in 2003, where she currently works as a Research Associate in the Institute of Biotechnology and Biomedicine. Her current research interests include the screening and study of new Bt pesticidal proteins, the binding of Bt toxins to pests' midguts, study of their mode of action, and the biochemical and genetic bases of insect resistance.

Juan Ferré (Prof.) received his Ph.D. degree in Chemistry from the University of Valencia (UV), Spain, with a study which he carried out in both the Department of Genetics of the UV and the Biology Division of the Oak Ridge National Laboratory, Oak Ridge (Tennessee, USA). He conducted his postdoctoral studies in the Department of Reproductive Genetics of the Magee Womens Hospital (Pittsburgh, Pennsylvania, USA). He became a Professor of Genetics in 2000, and served as the Head of the Department of Genetics of the UV for 7 years. He is currently Director of the Institute of Biotechnology and Biomedicine of the UV (Biotecmed). His current research interests, ongoing since 1990, are (i) to understand the biochemical and genetic bases of insect resistance to *Bacillus thuringiensis* (Bt) toxins; (ii) to study the mode of action of Bt toxins; and (iii) to find novel Bt strains and insecticidal protein genes for the development of Bt-based insecticides to control agricultural insect pests.

Patricia Hernández-Martínez (Dr.) studied Biology (1998–2003) at the University of Valencia (UV), Spain. She received her Ph.D. in Biology from the UV in 2009, with the work "Study of susceptibility and resistance towards *Bacillus thuringiensis* proteins on different *Spodoptera exigua* colonies", which she conducted in the Department of Genetics of the UV. After her Ph.D., her research career continued at the Institute of Biotechnology and Biomedicine (Biotecmed), where she still works as a Research Fellow. Her current research interests are (i) to study the mode of action of *B. thuringiensis* (Bt) proteins; (ii) to study the response mechanisms of different hosts after Bt protein challenge; and (iii) to understand the bases of insect resistance to Bt proteins.

 *toxins*

Editorial

Bacillus thuringiensis Toxins: Functional Characterization and Mechanism of Action

Yolanda Bel [1,2,*], Juan Ferré [1,2,*] and Patricia Hernández-Martínez [1,2,*]

1   Instituto de Biotecnología y Biomedicina (BIOTECMED), Universitat de València, 46100 Burjassot, Spain
2   Department of Genetics, Universitat de València, 46100 Burjassot, Spain
*   Correspondence: yolanda.bel@uv.es (Y.B.); juan.ferre@uv.es (J.F.); patricia.hernandez@uv.es (P.H.-M.)

Received: 30 November 2020; Accepted: 8 December 2020; Published: 10 December 2020

Bacillus thuringiensis (*Bt*)-based products are the most successful microbial insecticides to date. This entomopathogenic bacterium produces different kinds of proteins whose specific toxicity has been shown against a wide range of insect orders, nematodes, mites, protozoa, and human cancer cells. Some of these proteins are accumulated in parasporal crystals during the sporulation phase (Cry and Cyt proteins), whereas other proteins are secreted in the vegetative phase of growth (Vip and Sip toxins). Currently, insecticidal proteins belonging to different groups (Cry and Vip3 proteins) are widely used to control insect pests and vectors both in formulated sprays and in transgenic crops (the so-called Bt crops). Despite the extensive use of these proteins in insect pest control, especially Cry and Vip3, their mode of action is not completely understood.

The aim of this Special Issue was to gather information that could summarize (in the form of review papers) or expand (research papers) the knowledge of the structure and function of Bt proteins, as well as shed light on their mode of action, especially regarding the insect receptors. This subject has generated great interest, and this interest has been materialized into the 18 papers published in this issue.

This Special Issue, "*Bacillus thuringiensis* Toxins: Functional Characterization and Mechanism of Action", includes five review papers and 13 research papers. The review papers bring up to date important aspects of Bt pathogenicity, such as its interaction with the intestinal microbiota and the immune system of the insect [1]. The current knowledge about Vip proteins has also been reviewed [2], as has the contribution that the use of toxin mutants has made to the knowledge of the mode of action of the three-domain Cry proteins [3]. On the other hand, two more review papers recapitulate the information on the cytocidal activity of Bt proteins [4] or the insecticidal activity of Bt proteins against coleopteran pests [5]. All these review papers are of high value, allowing readers to stay updated on the different aspects of the Bt field described here.

The Special Issue also gathers information that could expand the knowledge of the structure and function of Bt proteins and sheds light on their mode of action, especially regarding the insect receptors. Publishing papers focusing on the steps that remain blurred within the mode of action of all Bt insecticidal proteins, including the three-domain Cry proteins, was one of the main goals. The role of receptors such as cadherin, ABCC2, and ABCA2 on the toxicity of Bt proteins in different lepidopterans has been investigated in three different papers [6–8]. In addition, other steps in the mode of action (that comprises protein solubilization, activation, binding, oligomerization, and pore formation) have also been addressed. Examples of these steps include the involvement of a novel trypsin protein for toxin activation in *Plutella xylostella*, discovered after studying a Cry1Ac resistant strain [9], and the promotion of oligomerization of the activated Cry1Ia with insect brush border midgut vesicles, in vitro [10]. The toxicity-promoting effect of a Bt chitin-binding protein that binds to the insect peritrophic matrix has also been studied [11]. Moreover, the Special Issue includes a paper highlighting the synergistic mosquitocidal activity of the parasporal Cry and Cyt proteins present in *B. thuringiensis* ser. *israelensis* [12], and it also includes a manuscript focused on deciphering the amino acid residues important for the interaction of Cyt2A protein with membrane lipids, a binding step necessary to exert its cytolytic action [13].

The vegetative insecticidal proteins (Vip3) secreted by *Bacillus thuringiensis* are nowadays considered as the new generation of insecticidal Bt toxins because of their different structural and molecular properties regarding the classical Bt 3-D Cry proteins. Vip3 toxins have been already introduced in Bt-crops to control lepidopteran pests. However, little is known about their mode of action. In the Special Issue, five papers analyze different aspects of its biology. They cover aspects ranging from its crystal structure [14] and structural–functional domain analyses [15] to different aspects in the mode of action, such as a study of a possible receptor (the alkaline phosphatase) in a resistant strain [16], the role of oligomerization in toxicity [17], and the study of intracellular events promoted by Vip3A intoxication in *Spodoptera frugiperda* Sf9 cells [18].

In summary, the Special Issue brings together papers of important scientific value in the field of Bt. The review and research papers included will help keep readers up to date on the topic and, at the same time, will contribute to increasing the vast knowledge of Bt and its insecticidal proteins. These studies will help to provide useful information for the development of new strategies to fight against pest insects, in the least aggressive and harmful but better environmental scenario.

Funding: This research received no external funding.

Acknowledgments: The editors are grateful to all of the authors who contributed to this Special Issue "*Bacillus thuringiensis* Toxins: Functional Characterization and Mechanism of Action". Special thanks to the peer reviewers for their expertise evaluations, which have contributed to increasing the quality of the research works and reviews compiled in this Special Issue. Finally, we thank the MDPI management team and staff for their valuable contributions, organization, and editorial support.

Conflicts of Interest: The authors declare no conflict of interest.

References

1. Li, S.; De Mandal, S.; Xu, X.; Jin, F. The Tripartite interaction of host immunity–*Bacillus thuringiensis* infection–gut microbiota. *Toxins* **2020**, *12*, 514. [CrossRef] [PubMed]
2. Syed, T.; Askari, M.; Meng, Z.; Li, Y.; Abid, M.A.; Wei, Y.; Guo, S.; Liang, C.; Zhang, R. Current insights on vegetative insecticidal proteins (Vip) as next generation pest killers. *Toxins* **2020**, *12*, 522. [CrossRef] [PubMed]
3. Vílchez, S. Making 3D-Cry toxin mutants: Much more than a tool of understanding toxins mechanism of action. *Toxins* **2020**, *12*, 600. [CrossRef] [PubMed]
4. Mendoza-Almanza, G.; Esparza-Ibarra, E.L.; Ayala-Luján, J.L.; Mercado-Reyes, M.; Godina-González, S.; Hernández-Barrales, M.; Olmos-Soto, J. The Cytocidal spectrum of *Bacillus thuringiensis* toxins: From insects to human cancer cells. *Toxins* **2020**, *12*, 301. [CrossRef]
5. Domínguez-Arrizabalaga, M.; Villanueva, M.; Escriche, B.; Ancín-Azpilicueta, C.; Caballero, P. Insecticidal activity of *Bacillus thuringiensis* proteins against Coleopteran pests. *Toxins* **2020**, *12*, 430. [CrossRef]
6. Zhang, J.; Jin, M.; Yang, Y.; Liu, L.; Yang, Y.; Gómez, I.; Bravo, A.; Soberón, M.; Xiao, Y.; Liu, K. The cadherin protein is not involved in susceptibility to *Bacillus thuringiensis* Cry1Ab or Cry1Fa toxins in *Spodoptera frugiperda*. *Toxins* **2020**, *12*, 375. [CrossRef] [PubMed]
7. Wang, X.; Xu, Y.; Huang, J.; Jin, W.; Yang, Y.; Wu, Y. CRISPR-mediated knockout of the *ABCC2* gene in *Ostrinia furnacalis* confers high-level resistance to the *Bacillus thuringiensis* Cry1Fa toxin. *Toxins* **2020**, *12*, 246. [CrossRef] [PubMed]
8. Li, X.; Miyamoto, K.; Takasu, Y.; Wada, S.; Iizuka, T.; Adegawa, S.; Sato, R.; Watanabe, K. ATP-binding cassette subfamily a member 2 is a functional receptor for *Bacillus thuringiensis* Cry2A toxins in *Bombyx mori*, but not for Cry1A, Cry1C, Cry1D, Cry1F, or Cry9A toxins. *Toxins* **2020**, *12*, 104. [CrossRef] [PubMed]
9. Gong, L.; Kang, S.; Zhou, J.; Sun, D.; Guo, L.; Qin, J.; Zhu, L.; Bai, Y.; Ye, F.; Akami, M.; et al. Reduced expression of a novel midgut trypsin gene involved in protoxin activation correlates with Cry1Ac resistance in a laboratory-selected strain of *Plutella xylostella* (L.). *Toxins* **2020**, *12*, 76. [CrossRef] [PubMed]

10. Khorramnejad, A.; Domínguez-Arrizabalaga, M.; Caballero, P.; Escriche, B.; Bel, Y. Study of the *Bacillus thuringiensis* Cry1Ia protein oligomerization promoted by midgut brush border membrane vesicles of lepidopteran and coleopteran insects, or cultured insect cells. *Toxins* **2020**, *12*, 133. [CrossRef] [PubMed]

11. Qin, J.; Tong, Z.; Zhan, Y.; Buisson, C.; Song, F.; He, K.; Nielsen-LeRoux, C.; Guo, S. A *Bacillus thuringiensis* chitin-binding protein is involved in insect peritrophic matrix adhesion and takes part in the infection process. *Toxins* **2020**, *12*, 252. [CrossRef] [PubMed]

12. Valtierra-de-Luis, D.; Villanueva, M.; Lai, L.; Williams, T.; Caballero, P. Potential of Cry10Aa and Cyt2Ba, Two minority δ-endotoxins produced by *Bacillus thuringiensis* ser. *israelensis*, for the control of *Aedes aegypti* larvae. *Toxins* **2020**, *12*, 355. [CrossRef] [PubMed]

13. Tharad, S.; Promdonkoy, B.; Toca-Herrera, J.L. Protein-Lipid Interaction of Cytolytic Toxin Cyt2Aa2 on Model lipid bilayers of erythrocyte cell membrane. *Toxins* **2020**, *12*, 226. [CrossRef] [PubMed]

14. Jiang, K.; Zhang, Y.; Chen, Z.; Wu, D.; Cai, J.; Gao, X. Structural and Functional Insights into the C-terminal Fragment of insecticidal Vip3A toxin of *Bacillus thuringiensis*. *Toxins* **2020**, *12*, 438. [CrossRef] [PubMed]

15. Gomis-Cebolla, J.; Ferreira dos Santos, R.; Wang, Y.; Caballero, J.; Caballero, P.; He, K.; Jurat-Fuentes, J.L.; Ferré, J. Domain shuffling between Vip3Aa and Vip3Ca: Chimera stability and insecticidal activity against European, American, African, and Asian pests. *Toxins* **2020**, *12*, 99. [CrossRef] [PubMed]

16. Pinos, D.; Chakroun, M.; Millán-Leiva, A.; Jurat-Fuentes, J.L.; Wright, D.J.; Hernández-Martínez, P.; Ferré, J. Reduced membrane-bound alkaline phosphatase does not affect binding of Vip3Aa in a *Heliothis virescens* resistant colony. *Toxins* **2020**, *12*, 409. [CrossRef] [PubMed]

17. Shao, E.; Zhang, A.; Yan, Y.; Wang, Y.; Jia, X.; Sha, L.; Guan, X.; Wang, P.; Huang, Z. Oligomer formation and insecticidal activity of *Bacillus thuringiensis* Vip3Aa toxin. *Toxins* **2020**, *12*, 274. [CrossRef] [PubMed]

18. Hou, X.; Han, L.; An, B.; Zhang, Y.; Cao, Z.; Zhan, Y.; Cai, X.; Yan, B.; Cai, J. Mitochondria and lysosomes participate in Vip3Aa-induced *Spodoptera frugiperda* Sf9 cell apoptosis. *Toxins* **2020**, *12*, 116. [CrossRef] [PubMed]

Publisher's Note: MDPI stays neutral with regard to jurisdictional claims in published maps and institutional affiliations.

Obituary

A Tribute to a *Bacillus thuringiensis* Master: Professor David J. Ellar

Susana Vílchez

Institute of Biotechnology, Department of Biochemistry and Molecular Biology I, Faculty of Science, University of Granada, 18071 Granada, Spain; svt@ugr.es

Received: 27 November 2020; Accepted: 29 November 2020; Published: 3 December 2020

This Special Issue, on *Bacillus thuringiensis* and its toxins, seems to be the right place to pay tribute to one of the most influential scientists in the field of research into this peculiar bacterium. Professor David J. Ellar passed away at his home in Cambridge, UK, on 21 May 2020, aged 80, as a result of pneumonia, after a long and debilitating illness. Everyone who knew him and had the opportunity to work hand in hand with him remain devastated by his loss.

Professor Ellar developed a very relevant scientific career in the Department of Biochemistry at the University of Cambridge. Through his laboratory, commonly known as Skylab, passed literally hundreds of postdoctoral researchers, visiting professors, PhD students, master's students, and undergraduate students. We all received the best training possible on *B. thuringiensis*, Cry toxins, their receptors, and above all on how cutting-edge and quality science is conducted. Professor Ellar has been a referent in the field of *B. thuringiensis* for all researchers working with this special bacterium and its entomopathogenic toxins. Thanks to his more than 160 research papers, published in the best scientific journals, we all know today a little bit more about this microorganism, which has proved to be extremely useful in the area of biotechnology.

Professor Ellar was a pioneer in *B. thuringiensis* research. Thanks to him, we were able to "see" for the first time the three-dimensional structure of a Cry toxin [1]. He was also the first to show us what a Cyt toxin, another entomotoxin produced by *B. thuringiensis*, looked like [2]. Thanks to his research we learned about the existence of the first Cry toxin receptor described [3], present on the enterocyte membrane of an insect, further identified as the well-known APN receptor [4]. This created an opportunity to begin understanding the mechanism of action of Cry toxins. In addition, he was the first to relate the Domain II Loops of a Cry toxin to specificity [5], explaining why some Cry toxins are active against some insects and not against others. He was the first person who managed to successfully display a functional Cry toxin on the surface of a phage [6], opening the possibility of using phage display technology for the in vitro evolution of Cry toxins.

However, David not only stood out for being a brilliant professional. On a personal level, he was simply a great person. He was always willing to help to anyone in need. He was extremely generous, and he had an outstanding sense of humour. He also had the remarkable ability to keep a huge research group motivated, in which each of its members worked with the precision of a Swiss watch.

I have the pleasure to say that David was a real master for me: the person that I have as a model in life and the person I want to become when I grow older. With this humble letter, and on behalf of the *B. thuringiensis* research community, I would like to give you the most sincere thanks for all your work and knowledge, and for your human greatness. Rest in peace.

Funding: This research received no external funding.

Acknowledgments: I would like to thank the Guest Editors Yolanda Bel, Juan Ferré and Patricia Hernández-Martínez for letting me include this tribute in this Special Issue.

Conflicts of Interest: The author declares no conflict of interest.

References

1. Li, J.; Carroll, J.; Ellar, D. Crystal structure of insecticidal delta-endotoxin from *Bacillus thuringiensis* at 2.5 A resolution. *Nature* **1991**, *353*, 815–821. [CrossRef]
2. Li, J.; Koni, P.A.; Ellar, D. Structure of the mosquitocidal δ-endotoxin CytB from *Bacillus thuringiensis* sp. *kyushuensis* and implications for membrane pore formation. *J. Mol. Biol.* **1996**, *257*, 129–152. [CrossRef] [PubMed]
3. Knowles, B.H.; Knight, P.J.K.; Ellar, D.J. N-acetyl galactosamine is part of the receptor in insect gut epithelia that recognizes an insecticidal protein from *Bacillus thuringiensis*. *Proc. Biol. Sci.* **1991**, *245*, 31–35. [CrossRef] [PubMed]
4. Knight, P.J.K.; Crickmore, N.; Ellar, D.J. The receptor for *Bacillus thuringiensis* CryIA (c) delta-endotoxin in the brush border membrane of the lepidopteran *Manduca sexta* is aminopeptidase N. *Mol. Microbiol.* **1994**, *11*, 429–436. [CrossRef]
5. Smith, G.P.; Ellar, D.J. Mutagenesis of two surface-exposed loops of the *Bacillus thuringiensis* CryIC δ-endotoxin affects insecticidal specificity. *Biochem. J.* **1994**, *302*, 611–616. [CrossRef]
6. Vilchez, S.; Jacoby, J.; Ellar, D.J. Display of biologically functional insecticidal toxin on the surface of lambda phage. *Appl. Environ. Microbiol.* **2004**, *70*, 6587–6594. [CrossRef] [PubMed]

Publisher's Note: MDPI stays neutral with regard to jurisdictional claims in published maps and institutional affiliations.

Review

The Tripartite Interaction of Host Immunity–*Bacillus thuringiensis* Infection–Gut Microbiota

Shuzhong Li, Surajit De Mandal, Xiaoxia Xu and Fengliang Jin *

Laboratory of Bio-Pesticide Innovation and Application of Guangdong Province, College of Agriculture, South China Agricultural University, Guangzhou 510642, China; shuzhongli@stu.scau.edu.cn (S.L.); surajit_micro@scau.edu.cn (S.D.M.); xuxiaoxia111@scau.edu.cn (X.X.)
* Correspondence: jflbang@scau.edu.cn; Tel.: +86-20-85280203; Fax: +86-20-85280293

Received: 28 June 2020; Accepted: 7 August 2020; Published: 12 August 2020

Abstract: *Bacillus thuringiensis* (Bt) is an important cosmopolitan bacterial entomopathogen, which produces various protein toxins that have been expressed in transgenic crops. The evolved molecular interaction between the insect immune system and gut microbiota is changed during the Bt infection process. The host immune response, such as the expression of induced antimicrobial peptides (AMPs), the melanization response, and the production of reactive oxygen species (ROS), varies with different doses of Bt infection. Moreover, *B. thuringiensis* infection changes the abundance and structural composition of the intestinal bacteria community. The activated immune response, together with dysbiosis of the gut microbiota, also has an important effect on Bt pathogenicity and insect resistance to Bt. In this review, we attempt to clarify this tripartite interaction of host immunity, Bt infection, and gut microbiota, especially the important role of key immune regulators and symbiotic bacteria in the Bt killing activity. Increasing the effectiveness of biocontrol agents by interfering with insect resistance and controlling symbiotic bacteria can be important steps for the successful application of microbial biopesticides.

Keywords: *Bacillus thuringiensis*; antimicrobial peptide; gut microbiota

Key Contribution: This review focused on describing the tripartite interaction of host immunity, Bt infection, and gut microbiota.

---

1. Introduction

The Gram-positive bacterium *Bacillus thuringiensis* (Bt) and its toxins are used to control several orders of insects, including agricultural pests and pathogen vectors [1,2]. Due to their selective insecticidal activity, *B. thuringiensis* toxins have become the most widely used commercial biopesticide worldwide [3,4]. Besides, the isolated Bt toxin genes have also been expressed in several transgenic Bt crops, and these strategies have reduced reliance on chemical pesticides [5–7]. The most common virulence factors of Bt are the crystal (Cry) toxin proteins produced during the sporulation phase of its growth cycle when ingested by susceptible insect larvae. The Cry toxins solubilize in the gut and are further activated by the host gut protease. The active fragments cross the peritrophic membrane and bind to the protein receptor located on the brush border membrane of midgut epithelial cells and create pores that induce osmotic cell lysis and subsequent death [8–10].

The widespread use of Bt spray products in high-value horticulture and the large-scale cultivation of Bt transgenic cotton and maize has resulted in cases of field resistance in several lepidopteran pest species and the western corn rootworm, *Diabrotica virgifera virgifera* [11,12]. The most common factors associated with Bt resistance are alteration of the Bt toxin receptors' binding site, mutations, and altered expressions of the midgut receptor genes [12–14]. Several Bt Cry toxin receptors, such as aminopeptidase-N (APN), alkaline phosphatase (ALP), cadherin, and ATP-binding cassette transporter (ABC transporter), have been identified and characterized in the midgut membrane of the insects [9,15,16]. *B. thuringiensis* resistance has also been linked to several other factors, such as inactivation of the midgut protease required for processing the Bt protoxins [17], gut stem cell proliferation, and differentiation [18]. However, the host immune response and the function of the gut microbiota during Bt infection, which are important aspects of Bt research, has still been inadequately studied and remains controversial [19–22].

The insect's innate immune system consists of both humoral and cellular immune responses, which depend on the non-self recognition of microbes and the subsequent production of immune effectors [23]. The humoral immune response of insects includes the induction of antimicrobial peptides (AMPs), lysozymes, and the rapidly activated phenoloxidase (PO) cascade-mediated melanization [24]. AMPs are produced by two major immune pathways, Toll and IMD (immune deficiency). These pathways produce and regulate the expression of AMPs that are specific to either Gram-positive bacterial/fungal and Gram-negative bacterial infection, respectively [25,26]. The insect cellular immune process consists of encapsulation, nodulation, and phagocytosis, which is primarily driven by the hemocyte [27]. Recognition by pattern recognition receptors (PRRs) triggers immune signal transduction, and results in the activation of the Toll, Imd, Janus kinase/signal transducer and activator of transcription (JAK/STAT), c-Jun N-terminal kinase (JNK), and prophenoloxidase (PPO) pathways [28–30]. Furthermore, insects possess both midgut-specific and systemic immune responses to combat the infection, and reactive oxygen species (ROS) production mediated by dual oxidase (DUOX) is another inducible immune defense mechanism of insects [23].

The insect gut microbiota includes not only the bacterial community but also fungi, protists, and archaea, although the bacterial species dominate in the gut microbial community [31], and plays a vital role in insect development, nutrition, immunity, metabolism, and colonization resistance against pathogens [32–37]. Several factors, such as environmental habitat, host, developmental stage, and diet, play a significant role in the structure and function of the insect gut microbiota [38,39]. It has also been reported that Bt toxicity is often associated with the abundance of the gut microbiota. The lepidopteran pest *Spodoptera exigua* can tolerate the action of Bt toxin when it contains an increased midgut microbiota load [40]. Native gut microbiota can also stimulate the host immune system. It has been reported that the native gut microbiota of bee is associated with the upregulated expression of AMPs, such as apidaecin and hymenoptaecin [41]. This indicates that gut microbiota could help the host to maintain an appropriate immune level, and play a vital role in the survival against Bt toxicity. In this review, we describe the tripartite interaction between host immunity, Bt infection, and gut microbiota (Figure 1). We discuss the effects of Bt infection on the host immune response and intestinal microbes, and how gut microbiota responds to Bt toxicity or causes Bt resistance, as well as the mechanism used by the host to limit Bt infection while maintaining intestinal homeostasis.

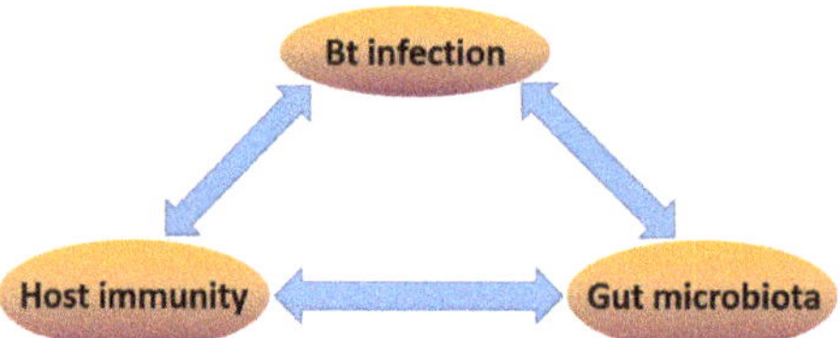

Figure 1. The tripartite interaction model between host immunity, Bt (*Bacillus thuringiensis*) infection, and gut microbiota.

2. The Interaction between Host Immunity and Gut Microbiota

2.1. Native Gut Microbiota-Induced Host Immune Response

It is well known that the host insect immune system is stimulated upon immune challenge by invading pathogens [23]. The local gut immunity plays a vital role in maintaining gut homeostasis by inhibiting or removing invading pathogens and limiting the growth of the symbionts [42]. Native gut microbiota participates in various symbiotic interactions and also affects host immunity, but the relationship between native gut microbiota and host immune function has so far been less well studied. Kwong et al. (2007) reported that the expression of the antimicrobial peptides (AMPs) is highly upregulated in gut tissue having normal gut microbiota in comparison with the gut tissues deficient in the microbiota. They suggest that the native microbiota may induce an immune response in the host [41]. Intriguingly, colonization of one specific intestinal bacteria, *Snodgrassella alvi,* to the host alone does not induce AMPs expression. However, colonization of another gut symbiont *Frischella perrara* results in a strong host immune response involving the upregulation of AMPs and the genes associated with the melanization cascade [43]. It suggests that different microbial species may have a different regulatory function in the host. Similar to the above results, the gut commensal microbiota of Red palm weevil (RPW), *Rhynchophorus ferrugineus* Olivier can help to protect against pathogenic infection by priming the immune system [44], and the colonization of gut commensal microbiota could enhance the immunocompetence of the host. Futo et al. (2016) reported that *Tribolium castaneum* larvae with less microbiota load showed a decrease in survival rate upon immune challenge by Bt [45], which indicates that gut microbiota is essential for immune priming.

Another important aspect is the messengers of immune priming between hosts and microbes, which explains why different gut commensal bacteria showed a different effect on host immunity and physiology. It has been reported that peptidoglycan and uracil, which are released from intestinal commensal bacteria, can induce AMPs gene expression and ROS production to maintain the gut homeostasis [46–49]. Growing evidence revealed that messenger molecules not only involve peptidoglycan and uracil but also contain numerous bioactive compounds, such as short-chain fatty acids (SCFAs), choline metabolites, and lipids [50–52]. Furthermore, it has been confirmed that the axenic population of *D. melanogaster* has altered lipid metabolism and insulin signaling, but the host physiology can be restored after the administration of the gut microbial metabolite acetate [53]. A recent study found that gut microbiota may also affect the systemic immune response apart from gut immunity in Red palm weevil (RPW) *Rhynchophorus ferrugineus* Olivier larvae [44]. They might derive some metabolites, which can cross the gut epithelium and enter the host hemolymph. However, more studies are required to decipher this mechanism.

In addition to immune priming to defend microbial pathogen infection, gut commensal bacteria-mediated immune responses are also crucial for efficient arboviral acquisition in mosquitoes. Intestinal symbiotic bacterium *Serratia* Y1 has been shown to inhibit successful establishment of the *Plasmodium* through direct activation of the mosquito immune response [54], and gut microbiota could elicit a protective immune response against the *Plasmodium* transmission [55]. In contrast, *Serratia* J1, another *Serratia* strain isolated from field-caught mosquito, has no impact on *Plasmodium* development [54]. Likewise, different strains of the same bacterial species have a different effect on *Plasmodium* infections in the *Anopheles* mosquito midgut [56]. It further reminds us that the interaction between host and gut commensal bacteria is complex and may involve strain-specific outcomes according to the corresponding metabolites.

2.2. Multiple Immune Reactions Help to Maintain Gut Homeostasis

The insect immune system not only protects the host against pathogen infection but also regulates the colonization of symbiotic microorganisms in the gut to maintain host homeostasis [57]. Several interesting mechanisms contribute to the proper maintenance of the microbiota by balancing the complex interaction between the host and the microbiota, which is mainly under the control of Toll and IMD pathways, and dual oxidase (DUOX) pathways, respectively (Figure 2) [23,58]. However, the functions of the Toll pathway are not consistent in different insect species, e.g., the Toll pathway is not found to be associated with the regulation of local gut immunity in *D. melanogaster* [59]. Recent reports by Abrar et al. found that the Spatzle-mediated Toll-like signaling pathway could regulate the homeostasis of gut microbiota by mediating the synthesis of AMPs in Red palm weevil, *Rhynchophorus ferrugineus* Olivier [60]. Royet et al. (2011) reported that the Toll signaling pathway could also be activated in the midgut of *P. xylostella* larvae by oral ingestion of pathogenic microbes. They also found that several essential elements for the Toll signaling pathway, including Spatzle, Toll receptor, tube, pelle, cactus, and dorsal, were expressed in *P. xylostella* midgut after the infection (Figure 2) [61]. Both Toll and IMD pathways can be activated following the detection of peptidoglycan (PGN) released from bacteria by different peptidoglycan recognition proteins (PGRPs). The family of PGRP is one of the key modulators in this process, which coordinates between the host immune response with the gut commensal bacteria. Similar to invading pathogens, gut commensal microorganisms can produce many immune-activating compounds (such as peptidoglycan) during growth and proliferation. A total of seven PGRPs were identified in *D. melanogaster* that can degrade peptidoglycan into non-immunostimulatory muropeptides. With the help of amidase activity, the peptidoglycan released from intestinal bacteria was maintained at a low basal level, so that the host can avoid the overactivation of the Toll and IMD pathway by gut microbiota (Figure 2) [62]. It has also been revealed that the low levels of peptidoglycan were limited by the PGRPs with amidase activity to transfer across the epithelial barrier and reach into hemolymph to stimulate the systematic immune response [63].

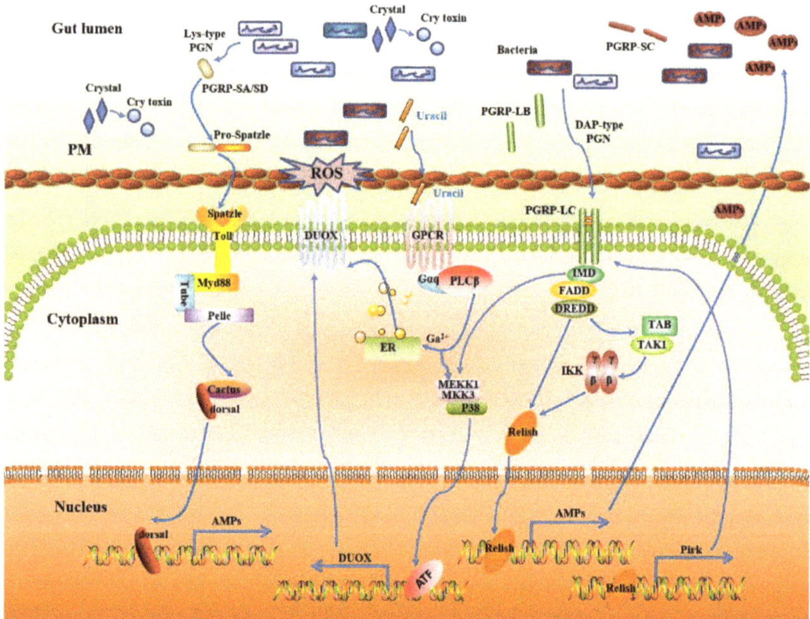

Figure 2. Insect gut immunity protects against infections and maintains gut microbiota homeostasis. DAP-type peptidoglycan (PGN) from intestinal bacteria is sensed by PGRP-LC, which triggers the IMD-dependent MEKK1-MKK3-p38 DUOX-expression pathway. Uracil also activates MEKK1-MKK3-p38 in a PLCβ-dependent manner; the activation of p38 enhances the transactivating function of ATF, which in turn activates the transcription of dual oxidase (DUOX). On the other hand, PLCβ-calcium signaling is responsible for the induction of DUOX enzymatic activity. Both contribute to the production of reactive oxygen species (ROS) in the gut lumen, where they control endogenous and infectious bacteria [64]. DAP-type PGN recognition by PGRP-LC also triggers the IMD pathway through the translocation of the nuclear factor-κB (NF-κB) family member Relish, which then induces increased transcription of antimicrobial peptides (AMPs) genes [23]. Besides, the IMD pathway has established a negative feedback loop to prevent overactivation. One is the members of the PGRP family gene (PGRP-LB or PGRP-SC) with amidase activity can cleave PGN and therefore blocks the activation of the

IMD pathway. Another is Pirk, which interferes with the plasma membrane localization of PGRP-LC [62]. In some insect species, the Toll signaling pathway is activated with the Lys-type PGN recognition by PGRP-SA or PGRP-SD after microbial infection. This initiates a proteolytic cascade that ultimately cleaves pro-Spatzle into an active ligand for Toll, leading to the activation of the NF-κB-like transcription factors dorsal and then translocation into the nucleus to induce increased transcription of the AMP gene. Finally, these immune regulatory networks cooperatively help to maintain gut homeostasis.

Similarly, the PGRP-LB homolog with amidase activity also acts as a negative modulator in the immunity of Red palm weevil, *Rhynchophorus ferrugineus* Olivier, in which abnormal expression alters the abundance and community structure of gut microbiota [65]. The intracellular protein Pirk can prevent PGRP-LC from recognizing extracellular peptidoglycan, thereby preventing hyperactivation of the gut immune response in flies [66]. Besides, it has been shown that peritrophic membrane (PM) integrity is related to the gut microbiota homeostasis in *A. stephensi* [67] and that PGRP-LD can help the PM to maintain structural integrity by preventing overactivation of the gut immune response, in turn limiting *P. berghei* infection. The knockdown of PGRP-LD can increase gut immunity and alters the gut microbial spatial distribution, which results in the dysbiosis of the gut microbiota. It suggests that PGRP-LD acts as a negative regulator of the immune signal pathway.

Research was also conducted to study other immune pathway regulators for maintaining gut homeostasis. Relish is an important regulator gene of the IMD pathway. Silencing the expression of Relish in the model insect *G. mellonella* results in a significant increase in the concentration of gut bacteria and decreases in the expression of AMPs [68]. Similar findings were also reported in Red palm weevil, which showed a compromised ability of pathogen clearance and increased gut bacterial load after silencing the Relish expression [69]. Moreover, a change in the gut commensal microbiota was observed after the inhibition of Caudal, a transcription repressor of NF-κB-mediated expression of AMPs [70]. The elimination of gut microbiota through antibiotics results in the downregulation of the IMD pathway and AMP gene expression [68]. Collectively, these results indicate that the IMD pathway plays a vital role in maintaining gut microbiota homeostasis.

The production of reactive oxygen species (ROS) is another inducible defense mechanism in the gut in addition to AMPs production (Figure 2) [23]. ROS are produced by the DUOX protein with an N-terminal extracellular peroxidase domain, which can convert H_2O_2 into HOCl in the presence of chloride, and thereby are detoxified in the presence of IRC catalase [58,71]. Unlike the gut IMD pathway, it is the uracil nucleobase, not peptidoglycan (PGN), that acts as an agonist to induce DUOX-dependent ROS production [48]. However, DUOX cannot be activated by most of the symbiotic bacteria under natural conditions, which suggest that symbiotic bacteria may block their uracil secretion pathway under natural conditions, and initiate it under specific dysregulated gut environments [48]. DUOX is also involved in the regulation of gut permeability in *Anopheles gambiae* [72]. The knockdown of DUOX increases the overall bacterial load in the oriental fruit fly *Bactrocera dorsalis*; however, the relative abundance of the bacterial symbionts *Enterobacteriaceae* is decreased in the gut [73].

3. The Host Immune System in Response to Bt Infection

Insects can initiate humoral and cellular immune responses to reduce the damage caused by Bt infection [74–80]. It has been reported that Bt tolerance in the flour moth, *Ephestia kuehniella*, can be achieved by the preexposure of low-concentrated Bt endotoxins (Syngenta, North Ryde, NSW, Australia) [81]. This phenomenon is mostly denoted as immune priming, which implies that the primary exposure to pathogen activates the basic immune response result in an improved immune response upon second exposure [82–84]. Therefore, the mechanism of Bt toxicity does not only depend on the host receptor but is also associated with the elevated immune response of the host. It has also been reported that Bt endotoxin-tolerant *E. kuehniella* larvae can increase the lipid carrier lipophorin in the gut lumen, which inactivates Bt toxins through the aggregation of lipophorin particles to break down toxins into coagulation products [85]. A soluble toxin-binding glycoprotein is also found in the intestinal lumen of the Bt (Cry1Ac)-resistant larvae of the lepidopteran pest *Helicoverpa armigera*, which can bind to Cry1Ac and GalNAc-specific lectins and forms an insoluble aggregate [78]. An LC5 dose of Bt ssp *galleria* strain 69-6 can trigger phagocytic activity in the larvae of *Galleria melonella*, whereas an LC15 dose of Bt increases the encapsulation rate in the hemolymph during infection [86]. Similarly, both the LC15 dose and LC50 dose of Bt resulted in elevated hemolymph phenoloxidase, and lysozyme-like activity in Bt-infected *Galleria mellonella* larvae. However, the difference is that low doses of Bt can increase the humoral and cellular immune response, involve an increased encapsulation response, and enhance the phagocytic activity of hemocytes. However, a higher dose decreases cellular reactions, involving the coagulation index and activity of phenoloxidase in hemocytes [77]. The host's immune response to Bt is likely dose dependent, as a sublethal dose of Bt damages the mid-intestinal epithelial cells, but it can be repaired by stem cell proliferation, and the enhanced immune response of the hemocytes can help to limit further infection and prevent septicemia [87]. At the LC50 dose for Bt, the situation is generally different, as the symbiotic bacteria and destroyed intestinal cells lead to dysfunctional humoral and cellular immune reactions. This indicates that Bt infection not only stimulates the local immune response in the gut but also induces the systematic immune response, where the radiation of immune pathway activation begins at the site of initial infection and radiates out, reaching the hemolymph. We found the LC50 dose of Bt infection could suppress the humoral immune response in the third instar larvae of *Plutella xylostella* [80]. Growing evidence suggests that Bt-induced immunity is a dose-dependent effect [88–91].

Comparatively, less information is available on the intestinal melanization response during Bt infection. It can be assumed that hemocytes can be recruited to seal perforations in the site of intestinal damage, and melanization may play a key role in this process. It has been reported that plasma phenoloxidase (PO) activity can be induced by both low and high concentrations of Bt in *G. melonella* and *E. kuehniella* larvae [77,81]. The prophenoloxidase (PPO) of insects comes from different sources and performs diverse functions, such as wound repairing, protection against pathogens, catalyzing, and detoxifying phenolics in the diet [92–94]. Several studies have shown that the PO may come from the hemolymph of the adult mosquito midguts [95,96]. However, other studies have shown that PPO is secreted into the foregut and can be activated by gut proteinase to detoxify phenolic present in the diet of Lepidoptera [92,93]. A recent study revealed that the PPO cascade is triggered after the infection of the Bt strain (Bt8010) in the midgut of *P. xylostella* larvae, which involves pattern recognition receptors (PRRs) and genes encoding proteases and protease inhibitors in the PPO cascade [94]. PPO can also be secreted into the hindgut to clean fecal bacteria by induced melanization of feces [92]. Similarly, the melanization response was reported in the hindgut of *Drosophila* mutant species [97]. PO-mediated melanization in the midgut might prevent symbiotic bacteria from escaping into the hemocoel through damaged midgut epithelial cells. However, despite various scientific investigations, the origin of the PO remains controversial in insects [98].

B. thuringiensis also synthesize another insecticidal protein (Vip) during the vegetative growth phase [99], and Vip3A, Cry1, and Cry2 genes have pyramided in cotton and maize to control lepidopteran insects [100]. The resistance to Vip3A has been selected in several lepidopteran species under laboratory conditions [101,102], while little is known of the biochemical mechanisms of resistance to Vip3A. Studies have shown this toxin does not share binding sites with Cry1 or Cry2 toxins [103,104], and in a laboratory-selected population of *Heliothis virescens*, resistance to Vip3A was shown to confer little cross-resistance to Cry1Ab and no cross-resistance to Cry1Ac [101]. The genome-wide analysis showed that most of the immune response genes, including AMPs, were upregulated, and genes involved in the metabolism and digestion process were downregulated in *Spodoptera exigua* larvae in response to Vip3 insecticidal challenge [105,106]. Similar to Bt and Bt Cry toxins, Vip3A toxin also triggers the PPO cascade and upregulates most of the genes involved in the midgut melanization process of *S. litura* and *S. exigua* [74,105]. Vip proteins also have a dose-dependent effect on the host. An increasing concentration increases the number of upregulated genes involved in the immune system and hormone modulation, and the downregulated genes involved in peritrophic membrane stability and the digestion process [106]. The genome-wide microarray analysis of Vip3Aa toxin-treated beet armyworm, *Spodoptera exigua*, showed that the upregulated enriched genes are involved in innate immune response, such as AMPs and *repat* genes [105]. This information helps to understand the host insect immune response after Bt Vip protein toxin challenge.

It is well known that the interaction between Cry toxin and toxin receptors from the host midgut brush border membrane vesicles (BBMVs) is the initial step in the insecticidal activity of the Cry protein toxins [107–110]. Several researchers also showed the interaction between midgut immune-related proteins and Cry toxin [111–113]. There is evidence showing that immune-related protein like Dorsal and peroxidase C in the midgut juice of *P. xylostella* and *S. exigua* can bind to the Cry1Ab1 protein toxin [111]. The protein Dorsal plays a significant role in the insect immune system, especially in the Toll pathway; therefore, a possible insecticidal mechanism of Bt Cry1Ab1 mediated by the midgut immune-related protein can be proposed.

Similarly, it has been reported that C-type lectin-20 (CTL-20) in *Aesdes aegyptii* has the potential to bind to both toxin receptors and Cry toxins to affect the interactions between Cry toxins and toxin receptors to reduce Cry toxicity in *Aedes aegypti* [112]. Similarly, another study on immune-related peptidoglycan recognition protein (PGRP) gene expression and PO activities in Cry1Ac-susceptible and -resistant *P. xylostella* found that the resistant strain of *P. xylostella* had higher PO activity compared with the susceptible strain [114]. Moreover, among three different *P. xylostella* strains, a Cry1Ac-susceptible, a Cry1Ac–resistant strain, and a field strain, both PGRP1 (belong to PGRP-SA family) and PGRP3 (PGRP-LF) showed higher expression levels in the gut of susceptible strains compared to the resistant strain and field strains, and PGRP2 (PGRP-LB) showed the highest expression levels in the gut of resistant strains [114]. It has been found that Cry1Ah toxins can bind directly to the PPO proteins in *Ostrinia furnacalis* [115]. The interaction between Cry protein toxins and the host midgut immune-related proteins requires further investigation as the study progresses.

The above studies have greatly enriched our knowledge of the host immune response after Bt or Bt toxin infection, and it is now confirmed that Bt or Bt toxin protein can affect the host's immune system in a dose-dependent manner. It is known that the insect gut harbors a diverse indigenous microbiota, and the host immune system plays an important role in maintaining the gut homeostasis [42]. In the next section, we discuss the intestinal microbiota and its functions when the immune system of the insect host has been compromised by Bt or Bt toxins.

4. The Interaction between Bt and Host Gut Microbiota

4.1. B. thuringiensis Infection Altered Host Insect Gut Microbiota

In general, insects maintain a balanced local intestinal microbial community that plays a vital role in their host, including host development, nutrition, and tolerance against pathogens [116,117]. It has been shown that gut microbiota is also associated with the resistance against Bt SV2 in mosquito and Bt HD-1 in Indian meal moth, *Plodia interpunctella* (Hübner) [118,119]. The diversity and richness of gut microbiota are changed by pathogen infection. However, relatively few studies have been published on the effects of Bt or Bt toxin infection on host gut microbiota. An investigation into mosquito larvae exposed to time increasing doses of Bt showed that the lowest diversity of gut microbiota comes from the most tolerant mosquito larvae [88]. Interestingly, the same study also found that the most tolerant larvae had the highest inter-individual difference. Similarly, *B. thuringiensis* infection significantly reduced the diversity and abundance of the gut microbiota in the Bt-resistant line of *G. mellonella* [120]. However, honeybees feeding on transgenic Cry1Ah maize pollen did not result in significant changes in the gut microbiota community composition under laboratory conditions [121]. Intestinal epithelial cells act as a barrier to separate the microbiota of midgut and hemocoel, and the microbial composition differs between these two tissues under normal conditions. However, the bacterial profile between the gut and hemocoel has been reported to be similar following treatment of *Spodoptera littoralis* larvae with an LC50 dose of Bt Cry1Ca toxin [122]. This indicates that gut bacteria cross the intestinal barrier to hemolymph as a result of the Bt toxin infection and then reproduce in the insect hemolymph. However, further research is needed to reveal the action mechanism of Bt on host gut microbiota.

4.2. The Function of Gut Microbiota in Response to Bt Infection

The effect of insect gut microbiota in Bt toxicity has long been controversial. In 2006, Broderick et al. reported that midgut microbiota is required for Bt subspecies *kurstaki* insecticidal activity in the larvae of the gypsy moth, *Lymantria dispar*. They also found that Bt *kurstaki* was unable to multiply in insect hemolymph in vitro, indicating that intestinal bacteria cause septicemia and contribute to Bt toxicity, but without Bt, intestinal bacteria cannot induce death [21]. However, several studies showed contrasting results. In 2009, Johnston reported that intestinal bacteria were not responsible for Bt HD73 strain toxicity in the tobacco hornworm, *Manduca Sexta* [22]. Interestingly, Bt HD73 Cry$^-$ cells can grow rapidly in plasma after intra-hemocoelic inoculation in many species. The same year, another study also confirmed that midgut microbiota is not required for the pathogenicity of Bt HD-73 and Bt HD-1 strains in the larvae of *P. xylostella* [123]. Work on the same insect host found that inoculation with one isolated gut bacteria *Enterobacter* sp. Mn2 has a different effect on Bt HD-1 and Bt HD-73 strain pathogenicity [123]. The contrasting results of different Bt strain pathogenicity after inoculation with the same gut bacteria to host insects indicate that we need to have an in-depth knowledge of different Bt strains before designing the experiment. From such studies, it is clear we need to pay attention to the host insect gut bacterial community, whether different diets and environments cause different gut bacteria communities between species or diverse populations within one species, and how it influences the interaction between Bt and host gut microbiota.

The interaction between Bt and gut microbiota can be competitive. *B. thuringiensis* can produce bacteriocin to inhibit the growth of gut bacteria [124]; on the other hand, insect gut microbiota can inhibit Bt multiplication, growth, and alteration of its toxins [118,125,126]: This is a kind of competition relationship. Conversely, Bt and host gut microbiota also show beneficial interactions to a certain extent; for example, some intestinal bacteria species can produce proteases that help solubilize Bt protoxins to their active form [127]. Furthermore, *B. thuringiensis* infection can promote translocation of gut-opportunistic pathogenic bacteria to hemocoel, which relies on gut epithelial damage caused by Bt toxins or some other factors, and then rapidly reproduce in the hemocoel and participate in host septicemia, finally leading to the death of the host [20,128]. One study showed that hemolymph microbiota are changed dramatically and the change is dominated by *Serratia* and *Clostridium* species upon Bt infection in *Spodoptera littoralis* larvae, which switch from asymptomatic gut symbionts to hemocoelic pathogens [122]. This translocation phenomenon agrees with the hypothesis discussed earlier in Section 4.1 of the present review.

Many gut symbiotic bacteria have been isolated and characterized; some of them showed a beneficial effect on the host and can be called probiotic. Such bacteria are widely used as animal feed additives in food production [129]. *Enterococcus mundtii* bacteria isolated from the feces of *Ephestia kuehniella* have the function of protecting the flour beetle, *Tribolium castaneum*, against Bt infection [130]. The surface properties test showed that this isolate has intense levels of auto-aggregation, which is related to the formation of colonies in the host insect's gut [131]. Moreover, bacteria cell wall compounds, such as lipopolysaccharide (LPS) and peptidoglycan (PGN), have been well studied, and can stimulate the host immune response [132,133]. However, *Tribolium castaneum* larvae exposed to the corresponding supernatant can also increase the resistance to Bt infection [130]. This perhaps suggests that the protective function of probiotic bacteria is based on the secreted proteins or some small peptides, which may act directly against Bt infection or through the triggering of immune priming.

Similarly, another *E. mundtii* strain isolated from *S. littoralis* also showed a protective function for the host insect, which can directly inhibit competitors and suppress pathogens' growth through its antimicrobial activity [134]. Previous studies confirmed that *E. mundtii* cells accumulate on the surface of the intestinal epithelium and form a biofilm-like structure, which helps to keep its predominant colonization status in the host insect's gut [135]. Additionally, after removing the dominant bacteria from the gut, this resulted in increased susceptibility of the spruce budworm larvae to Bt infection [20]. Although growing evidence confirmed that intestinal bacteria play important roles in the host defense response [136,137], the molecular function mechanism requires more research. It has been reported that *E. mundtii* SL can secrete a kind of bacteriocin, which strongly inhibits some of the competing organisms and can impair pathogen colonization in vivo [134]. Through this, we can speculate that Bt may also shown inhibited growth and has limited activity in the insect gut lumen, and the battle between Bt and probiotic *E. mundtii* may depend on the dose effect. It is also noteworthy that the secreted bacteriocin showed a selective antibacterial activity and has no influence on other intestinal bacteria, and as a result, the gut microbiota can develop normally. There is another study report that normal gut microbiota mediated pathogen clearance from the gut lumen [138]; this suggests that the gut microbiota can act as another form of protection response in organisms, or at least an important complement to host gut immune protection.

Insect gut microbiota also play an important role in Bt infection indirectly through intestinal epithelium cell regeneration. The Lepidoptera larvae intestinal epithelium mainly includes four kinds of cell types: Columnar cells, goblet cells, enteroendocrine cells, and intestinal stem cells. Each cell type has a specific role and helps maintain normal gut functions. The intestinal stem cell is the only type capable of division, which mediates epithelial renewal and the healing response [139]. *B. thuringiensis* infection can disrupt the gut epithelial cells by producing toxins, and insects mount a series of defensive responses, which involves melanization, AMP-mediated antimicrobial activity, and gut stem cell proliferation and differentiation in response to gut damage [18,94]. It has been shown that REPAT and MAPK p38 signaling pathways may be involved in the regulation of the gut defensive response to Bt toxins [18,140]. The REPAT gene was also predicted to be associated with a regenerative response in Bt-resistant insect species and showed constitutively increased expression in a Bt-resistant *S. exigua* [141]. Moreover, it showed that Cry1Ac resistance is related to an enhanced midgut healing response in the tobacco budworm [142,143]. These results suggest that intestinal stem cell activity is associated with Bt resistance because bacteria cannot get through the healthy midgut epithelial cells. This allows the host to quickly repair the damage, resulting in a limited number of invaders into the hemocoel.

It has been reported that indigenous gut microbiota can modulate the activity of intestinal stem cells in *Drosophila*, which correlates with the activated JAK/STAT pathway and epidermal growth factor receptor (EGFR) pathways [33,144]. Both the rate of epithelium renewal and the number of dividing intestinal stem cells were reduced after removing all the intestinal bacteria. The abundance of gut microbiota can also be increased after host immune suppression [145]. Interestingly, the number of mitotic intestinal stem cells is increased after blocking the *Drosophila* IMD pathway, which caused an abnormal intestinal microbiota, and many genes related to stem cell proliferation and differentiation were also upregulated by the induced gut microbiota [144]. It indicated that the intestinal stem cell proliferation could be stimulated by increased gut microbiota. In summary, these results demonstrate that insect gut microbiota can affect Bt resistance by mediated intestinal stem cell activity.

5. Conclusions

B. thuringiensis infection induces a variety of host immune responses, and interferes with the gut microbiota of the host. The resulting dysbiosis, in turn, stimulates both the expression of AMPs and the production of ROS by different ligand molecules. DUOX also plays a key role in regulating the gut microbial homeostasis. The interaction between the host immune system and gut symbionts is more cooperative rather than antagonistic. However, *B. thuringiensis* or other pathogenic infections can cause dysregulated gut environments in insects, which makes it possible to convert some symbiotic bacteria into pathobiont, known as an opportunistic pathogen. The above interaction relationship has an important effect on Bt pathogenicity or toxicity. Most studies have focused on the interaction between Bt and the host immune system or the interaction between Bt and microbiota, which significantly expands our knowledge about the dynamic Bt infection process. However, a few important aspects are still unanswered and need to be explored, i.e., (i) How does Bt trigger the immune signaling pathway? (ii) Do the membranes of the intestinal lumen or intestinal epithelial cells have any toxin recognition receptors attached to the Toll and IMD pathways? (iii) Why do different intestinal bacteria have a different effect on the host during Bt infection? and (iv) which bacterial metabolite plays a significant role in Bt toxicity and host immunocompetence?

Author Contributions: Conceptualization, F.J. and S.L.; software, S.L.; writing—original draft preparation, S.L.; S.D.M.; F.J. and X.X.; writing—review and editing, S.L.; F.J.; S.D.M. and X.X.; supervision, F.J.; funding acquisition, F.J. All authors have read and agreed to the published version of the manuscript.

Funding: This work was supported by a grant from the National Natural Science Foundation of China (31972345), Natural Science Foundation of Guangdong, China (2019A1515011221,2020A1515010300), Provincial Agricultural Science and technology innovation and Extension project of Guangdong Province (2019KJ147).

Acknowledgments: We would like to thank anonymous referees for their valuable comments and suggestion.

Conflicts of Interest: The authors declare no conflict of interest.

References

1. Bravo, A.; Likitvivatanavong, S.; Gill, S.S.; Soberon, M. *Bacillus thuringiensis*: A story of a successful bioinsecticide. *Insect Biochem. Mol. Biol.* **2011**, *41*, 423–431. [CrossRef] [PubMed]
2. Raymond, B.; Johnston, P.R.; Nielsen-LeRoux, C.; Lereclus, D.; Crickmore, N. *Bacillus thuringiensis*: An impotent pathogen? *Trends Microbiol.* **2010**, *18*, 189–194. [CrossRef] [PubMed]
3. Ibrahim, M.A.; Griko, N.; Junker, M.; Bulla, L.A. *Bacillus thuringiensis*: A genomics and proteomics perspective. *Bioeng. Bugs* **2010**, *1*, 31–50. [CrossRef] [PubMed]
4. Van Rie, J. *Bacillus thuringiensis* and its use in transgenic insect control technologies. *Int. J. Med. Microbiol.* **2000**, *290*, 463–469. [CrossRef]
5. Lu, Y.; Wu, K.; Jiang, Y.; Guo, Y.; Desneux, N. Widespread adoption of Bt cotton and insecticide decrease promotes biocontrol services. *Nature* **2012**, *487*, 362–365. [CrossRef]
6. Sanahuja, G.; Banakar, R.; Twyman, R.M.; Capell, T.; Christou, P. *Bacillus thuringiensis*: A century of research, development and commercial applications. *Plant Biotechnol. J.* **2011**, *9*, 283–300. [CrossRef]
7. Kumar, S.; Chandra, A.; Pandey, K.C. *Bacillus thuringiensis* (Bt) transgenic crop: An environment friendly insect-pest management strategy. *J. Environ. Biol.* **2008**, *29*, 641–653.
8. Pardo-Lopez, L.; Soberon, M.; Bravo, A. *Bacillus thuringiensis* insecticidal three-domain Cry toxins: Mode of action, insect resistance and consequences for crop protection. *FEMS Microbiol. Rev.* **2013**, *37*, 3–22. [CrossRef]
9. Pigott, C.R.; Ellar, D.J. Role of receptors in *Bacillus thuringiensis* crystal toxin activity. *Microbiol. Mol. Biol. Rev.* **2007**, *71*, 255–281. [CrossRef]
10. De Maagd, R.A.; Bravo, A.; Berry, C.; Crickmore, N.; Schnepf, H.E. Structure, diversity, and evolution of protein toxins from spore-forming entomopathogenic bacteria. *Annu. Rev. Genet.* **2003**, *37*, 409–433. [CrossRef]
11. Tabashnik, B.E.; Carrière, Y. Surge in insect resistance to transgenic crops and prospects for sustainability. *Nat. Biotechnol.* **2017**, *35*, 926–935. [CrossRef] [PubMed]
12. Peterson, B.; Bezuidenhout, C.C.; Van den Berg, J. An overview of mechanisms of cry toxin resistance in lepidopteran insects. *J. Econ. Entomol.* **2017**, *110*, 362–377. [CrossRef] [PubMed]
13. Jurat-Fuentes, J.L.; Karumbaiah, L.; Jakka, S.R.; Ning, C.; Liu, C.; Wu, K.; Jackson, J.; Gould, F.; Blanco, C.; Portilla, M.; et al. Reduced levels of membrane-bound alkaline phosphatase are common to lepidopteran strains resistant to Cry toxins from *Bacillus thuringiensis*. *PLoS ONE* **2011**, *6*, e17606. [CrossRef] [PubMed]
14. Adang, M.J.; Crickmore, N.; Jurat-Fuentes, J.L. Diversity of *Bacillus thuringiensis* Crystal Toxins and Mechanism of Action. *Adv. Insect Physiol.* **2014**, *47*, 39–87. [CrossRef]
15. Heckel, D.G. Learning the ABCs of Bt: ABC transporters and insect resistance to *Bacillus thuringiensis* provide clues to a crucial step in toxin mode of action. *Pestic. Biochem. Physiol.* **2012**, *104*, 103–110. [CrossRef]
16. Guo, Z.J.; Sun, D.; Kang, S.; Zhou, J.L.; Gong, L.J.; Qin, J.Y.; Guo, L.; Zhu, L.H.; Bai, Y.; Luo, L.; et al. CRISPR/Cas9-mediated knockout of both the PxABCC2 and PxABCC3 genes confers high-level resistance to *Bacillus thuringiensis* Cry1Ac toxin in the diamondback moth, *Plutella xylostella* (L.). *Insect Biochem. Mol. Biol.* **2019**, *107*, 31–38. [CrossRef]

17. Oppert, B.; Kramer, K.J.; Johnson, D.E.; MacIntosh, S.C.; McGaughey, W.H. Altered protoxin activation by midgut enzymes from a *Bacillus thuringiensis* resistant strain of *Plodia interpunctella*. *Biochem. Biophys. Res. Commun.* **1994**, *198*, 940–947. [CrossRef]

18. Castagnola, A.; Jurat-Fuentes, J.L. Intestinal regeneration as an insect resistance mechanism to entomopathogenic bacteria. *Curr. Opin. Insect Sci.* **2016**, *15*, 104–110. [CrossRef]

19. Khan, I.; Prakash, A.; Agashe, D. Experimental evolution of insect immune memory versus pathogen resistance. *Proc. Biol. Sci.* **2017**, *284*, 20171583. [CrossRef]

20. Van Frankenhuyzen, K.; Liu, Y.H.; Tonon, A. Interactions between *Bacillus thuringiensis* subsp kurstaki HD-1 and midgut bacteria in larvae of gypsy moth and spruce budworm. *J. Invertebr. Pathol.* **2010**, *103*, 124–131. [CrossRef]

21. Broderick, N.A.; Raffa, K.F.; Handelsman, J. Midgut bacteria required for *Bacillus thuringiensis* insecticidal activity. *Proc. Natl. Acad. Sci. USA* **2006**, *103*, 15196–15199. [CrossRef] [PubMed]

22. Johnston, P.R.; Crickmore, N. Gut bacteria are not required for the insecticidal activity of *Bacillus thuringiensis* toward the tobacco hornworm, *Manduca sexta*. *Appl. Environ. Microbiol.* **2009**, *75*, 5094–5099. [CrossRef] [PubMed]

23. Lemaitre, B.; Hoffmann, J. The host defense of *Drosophila melanogaster*. *Annu. Rev. Immunol.* **2007**, *25*, 697–743. [CrossRef] [PubMed]

24. Sheehan, G.; Garvey, A.; Croke, M.; Kavanagh, K. Innate humoral immune defences in mammals and insects: The same, with differences ? *Virulence* **2018**, *9*, 1625–1639. [CrossRef]

25. Silverman, N.; Paquette, N.; Aggarwal, K. Specificity and signaling in the *Drosophila* immune response. *Invert. Surviv. J. ISJ* **2009**, *6*, 163–174.

26. Tanji, T.; Ip, Y.T. Regulators of the Toll and Imd pathways in the *Drosophila* innate immune response. *Trends Immunol.* **2005**, *26*, 193–198. [CrossRef]

27. Strand, M.R. The insect cellular immune response. *Insect Sci.* **2008**, *15*, 1–14. [CrossRef]

28. Kim, C.H.; Park, J.W.; Ha, N.C.; Kang, H.J.; Lee, B.L. Innate immune response in insects: Recognition of bacterial peptidoglycan and amplification of its recognition signal. *BMB Rep.* **2008**, *41*, 93–101. [CrossRef]

29. Tsakas, S.; Marmaras, V.J. Insect immunity and its signalling: An overview. *Invert. Surviv. J. ISJ* **2010**, *7*, 228–238.

30. Lu, Y.Z.; Su, F.H.; Li, Q.L.; Zhang, J.; Li, Y.J.; Tang, T.; Hu, Q.H.; Yu, X.Q. Pattern recognition receptors in *Drosophila* immune responses. *Dev. Comp. Immunol.* **2020**, *102*, 103468. [CrossRef]

31. Engel, P.; Moran, N.A. The gut microbiota of insects-diversity in structure and function. *FEMS Microbiol. Rev.* **2013**, *37*, 699–735. [CrossRef] [PubMed]

32. Storelli, G.; Defaye, A.; Erkosar, B.; Hols, P.; Royet, J.; Leulier, F. *Lactobacillus plantarum* promotes *Drosophila* systemic growth by modulating hormonal signals through TOR-dependent nutrient sensing. *Cell Metab.* **2011**, *14*, 403–414. [CrossRef] [PubMed]

33. Buchon, N.; Broderick, N.A.; Chakrabarti, S.; Lemaitre, B. Invasive and indigenous microbiota impact intestinal stem cell activity through multiple pathways in *Drosophila*. *Genes Dev.* **2009**, *23*, 2333–2344. [CrossRef] [PubMed]

34. Sabree, Z.L.; Kambhampati, S.; Moran, N.A. Nitrogen recycling and nutritional provisioning by *Blattabacterium*, the cockroach endosymbiont. *Proc. Natl. Acad. Sci. USA* **2009**, *106*, 19521–19526. [CrossRef]

35. Koch, H.; Schmid-Hempel, P. Socially transmitted gut microbiota protect bumble bees against an intestinal parasite. *Proc. Natl. Acad. Sci. USA* **2011**, *108*, 19288–19292. [CrossRef] [PubMed]

36. Gonzalez-Ceron, L.; Santillan, F.; Rodriguez, M.H.; Mendez, D.; Hernandez-Avila, J.E. Bacteria in midguts of field-collected Anopheles albimanus block Plasmodium vivax sporogonic development. *J. Med. Entomol.* **2003**, *40*, 371–374. [CrossRef]

37. Combe, B.E.; Defaye, A.; Bozonnet, N.; Puthier, D.; Royet, J.; Leulier, F. *Drosophila* microbiota modulates host metabolic gene expression via IMD/NF-kappa B signaling. *PLoS ONE* **2014**, *9*, e94729. [CrossRef]

38. Yun, J.-H.; Roh, S.W.; Whon, T.W.; Jung, M.-J.; Kim, M.-S.; Park, D.-S.; Yoon, C.; Nam, Y.-D.; Kim, Y.-J.; Choi, J.-H.; et al. Insect gut bacterial diversity determined by environmental habitat, diet, developmental stage, and phylogeny of host. *Appl. Environ. Microbiol.* **2014**, *80*, 5254–5264. [CrossRef]

39. Dillon, R.J.; Webster, G.; Weightman, A.J.; Charnley, A.K. Diversity of gut microbiota increases with aging and starvation in the desert locust. *Antonie Van Leeuwenhoek* **2010**, *97*, 69–77. [CrossRef]

40. Hernandez-Martinez, P.; Naseri, B.; Navarro-Cerrillo, G.; Escriche, B.; Ferre, J.; Herrero, S. Increase in midgut microbiota load induces an apparent immune priming and increases tolerance to *Bacillus thuringiensis*. *Environ. Microbiol.* **2010**, *12*, 2730–2737. [CrossRef]

41. Kwong, W.K.; Mancenido, A.L.; Moran, N.A. Immune system stimulation by the native gut microbiota of honey bees. *R. Soc. Open Sci.* **2017**, *4*, 170003. [CrossRef] [PubMed]

42. Buchon, N.; Broderick, N.A.; Lemaitre, B. Gut homeostasis in a microbial world: Insights from *Drosophila melanogaster*. *Nat. Rev. Microbiol.* **2013**, *11*, 615–626. [CrossRef] [PubMed]

43. Emery, O.; Schmidt, K.; Engel, P. Immune system stimulation by the gut symbiont Frischella perrara in the honey bee (*Apis mellifera*). *Mol. Ecol.* **2017**, *26*, 2576–2590. [CrossRef] [PubMed]

44. Muhammad, A.; Habineza, P.; Ji, T.; Hou, Y.; Shi, Z. Intestinal microbiota confer protection by priming the immune system of red palm weevil *Rhynchophorus ferrugineus* olivier (Coleoptera: Dryophthoridae). *Front. Physiol.* **2019**, *10*, 1303. [CrossRef]

45. Futo, M.; Armitage, S.A.; Kurtz, J. Microbiota plays a role in oral immune priming in *Tribolium castaneum*. *Front. Microbiol.* **2015**, *6*, 1383. [CrossRef]

46. You, H.; Lee, W.J.; Lee, W.-J. Homeostasis between gut-associated microorganisms and the immune system in *Drosophila*. *Curr. Opin. Immunol.* **2014**, *30*, 48–53. [CrossRef]

47. Ryu, J.-H.; Ha, E.-M.; Lee, W.-J. Innate immunity and gut-microbe mutualism in *Drosophila*. *Dev. Comp. Immunol.* **2010**, *34*, 369–376. [CrossRef]

48. Lee, K.-A.; Kim, S.-H.; Kim, E.-K.; Ha, E.-M.; You, H.; Kim, B.; Kim, M.-J.; Kwon, Y.; Ryu, J.-H.; Lee, W.-J. Bacterial-derived uracil as a modulator of mucosal immunity and gut-microbe homeostasis in *Drosophila*. *Cell* **2013**, *153*, 797–811. [CrossRef]

49. Lee, K.-A.; Kim, B.; Bhin, J.; Kim, D.H.; You, H.; Kim, E.-K.; Kim, S.-H.; Ryu, J.-H.; Hwang, D.; Lee, W.-J. Bacterial uracil modulates *Drosophila* DUOX-dependent gut immunity via hedgehog-induced signaling endosomes. *Cell Host Microbe* **2015**, *17*, 191–204. [CrossRef]

50. Canfora, E.E.; Jocken, J.W.; Blaak, E.E. Short-chain fatty acids in control of body weight and insulin sensitivity. *Nat. Rev. Endocrinol.* **2015**, *11*, 577–591. [CrossRef]

51. Lee, W.J.; Hase, K. Gut microbiota-generated metabolites in animal health and disease. *Nat. Chem. Biol.* **2014**, *10*, 416–424. [CrossRef] [PubMed]

52. Nicholson, J.K.; Holmes, E.; Kinross, J.; Burcelin, R.; Gibson, G.; Jia, W.; Pettersson, S. Host-gut microbiota metabolic interactions. *Science* **2012**, *336*, 1262–1267. [CrossRef] [PubMed]

53. Hang, S.; Purdy, A.E.; Robins, W.P.; Wang, Z.; Mandal, M.; Chang, S.; Mekalanos, J.J.; Watnick, P.I. The acetate switch of an intestinal pathogen disrupts host insulin signaling and lipid metabolism. *Cell Host Microbe* **2014**, *16*, 592–604. [CrossRef]

54. Bai, L.; Wang, L.; Vega-Rodriguez, J.; Wang, G.; Wang, S. A gut symbiotic bacterium serratia marcescens renders mosquito resistance to plasmodium infection through activation of mosquito immune responses. *Front. Microbiol.* **2019**, *10*, 1580. [CrossRef] [PubMed]

55. Yilmaz, B.; Portugal, S.; Tran, T.M.; Gozzelino, R.; Ramos, S.; Gomes, J.; Regalado, A.; Cowan, P.J.; d'Apice, A.J.F.; Chong, A.S.; et al. Gut microbiota elicits a protective immune response against malaria transmission. *Cell* **2014**, *159*, 1277–1289. [CrossRef]

56. Bando, H.; Okado, K.; Guelbeogo, W.M.; Badolo, A.; Aonuma, H.; Nelson, B.; Fukumoto, S.; Xuan, X.; Sagnon, N.; Kanuka, H. Intra-specific diversity of *Serratia marcescens* in *Anopheles mosquito* midgut defines plasmodium transmission capacity. *Sci. Rep.* **2013**, *3*, 1641. [CrossRef]

57. Cerf-Bensussan, N.; Gaboriau-Routhiau, V. The immune system and the gut microbiota: Friends or foes? *Nat. Rev. Immunol.* **2010**, *10*, 735–744. [CrossRef]

58. Ha, E.M.; Oh, C.T.; Bae, Y.S.; Lee, W.J. A direct role for dual oxidase in *Drosophila* gut immunity. *Science* **2005**, *310*, 847–850. [CrossRef]

59. Davis, M.M.; Engstrom, Y. Immune response in the barrier epithelia: Lessons from the fruit fly *Drosophila melanogaster*. *J. Innate Immun.* **2012**, *4*, 273–283. [CrossRef]

60. Muhammad, A.; Habineza, P.; Wang, X.; Xiao, R.; Ji, T.; Hou, Y.; Shi, Z. Spatzle homolog-mediated toll-like pathway regulates innate immune responses to maintain the homeostasis of gut microbiota in the red palm weevil, *Rhynchophorus ferrugineus* olivier (Coleoptera: Dryophthoridae). *Front. Microbiol.* **2020**, *11*, 846. [CrossRef]

61. Lin, J.; Xia, X.; Yu, X.Q.; Shen, J.; Li, Y.; Lin, H.; Tang, S.; Vasseur, L.; You, M. Gene expression profiling provides insights into the immune mechanism of *Plutella xylostella* midgut to microbial infection. *Gene* **2018**, *647*, 21–30. [CrossRef]

62. Royet, J.; Gupta, D.; Dziarski, R. Peptidoglycan recognition proteins: Modulators of the microbiome and inflammation. *Nat. Rev. Immunol.* **2011**, *11*, 837–851. [CrossRef] [PubMed]

63. Gendrin, M.; Welchman, D.P.; Poidevin, M.; Herve, M.; Lemaitre, B. Long-Range Activation of systemic immunity through peptidoglycan diffusion in *Drosophila*. *PLoS Pathog.* **2009**, *5*, e1000694. [CrossRef] [PubMed]

64. Kim, S.-H.; Lee, W.-J. Role of DUOX in gut inflammation: Lessons from *Drosophila* model of gut-microbiota interactions. *Front. Cell. Infect. Microbiol.* **2014**, *3*. [CrossRef] [PubMed]

65. Dawadi, B.; Wang, X.; Xiao, R.; Muhammad, A.; Hou, Y.; Shi, Z. PGRP-LB homolog acts as a negative modulator of immunity in maintaining the gut-microbe symbiosis of red palm weevil, *Rhynchophorus ferrugineus* Olivier. *Dev. Comp. Immunol.* **2018**, *86*, 65–77. [CrossRef] [PubMed]

66. Kleino, A.; Myllymaki, H.; Kallio, J.; Vanha-aho, L.M.; Oksanen, K.; Ulvila, J.; Hultmark, D.; Valanne, S.; Ramet, M. Pirk is a negative regulator of the *Drosophila* Imd pathway. *J. Immunol.* **2008**, *180*, 5413–5422. [CrossRef] [PubMed]

67. Song, X.; Wang, M.; Dong, L.; Zhu, H.; Wang, J. PGRP-LD mediates *A. stephensi* vector competency by regulating homeostasis of microbiota-induced peritrophic matrix synthesis. *PLoS Pathog.* **2018**, *14*, e1006899. [CrossRef]

68. Sarvari, M.; Mikani, A.; Mehrabadi, M. The innate immune gene *Relish* and *Caudal* jointly contribute to the gut immune homeostasis by regulating antimicrobial peptides in *Galleria mellonella*. *Dev. Comp. Immunol.* **2020**, *110*, 103732. [CrossRef]

69. Xiao, R.; Wang, X.; Xie, E.; Ji, T.; Li, X.; Muhammad, A.; Yin, X.; Hou, Y.; Shi, Z. An IMD-like pathway mediates the intestinal immunity to modulate the homeostasis of gut microbiota in *Rhynchophorus ferrugineus* Olivier (Coleoptera: Dryophthoridae). *Dev. Comp. Immunol.* **2019**, *97*, 20–27. [CrossRef]

70. Ryu, J.H.; Kim, S.H.; Lee, H.Y.; Bai, J.Y.; Nam, Y.D.; Bae, J.W.; Lee, D.G.; Shin, S.C.; Ha, E.M.; Lee, W.J. Innate immune homeostasis by the homeobox gene caudal and commensal-gut mutualism in *Drosophila*. *Science* **2008**, *319*, 777–782. [CrossRef]

71. Ritsick, D.R.; Edens, W.A.; McCoy, J.W.; Lambeth, J.D. The use of model systems to study biological functions of Nox/Duox enzymes. *Biochem. Soc. Symp.* **2004**, *71*, 85–96.

72. Kumar, S.; Molina-Cruz, A.; Gupta, L.; Rodrigues, J.; Barillas-Mury, C. A peroxidase/dual oxidase system modulates midgut epithelial immunity in *Anopheles gambiae*. *Science* **2010**, *327*, 1644–1648. [CrossRef] [PubMed]

73. Yao, Z.; Wang, A.; Li, Y.; Cai, Z.; Lemaitre, B.; Zhang, H. The dual oxidase gene BdDuox regulates the intestinal bacterial community homeostasis of *Bactrocera dorsalis*. *ISME J.* **2016**, *10*, 1037–1050. [CrossRef] [PubMed]

74. Song, F.; Chen, C.; Wu, S.; Shao, E.; Li, M.; Guan, X.; Huang, Z. Transcriptional profiling analysis of *Spodoptera litura* larvae challenged with Vip3Aa toxin and possible involvement of trypsin in the toxin activation. *Sci. Rep.* **2016**, *6*, 23861. [CrossRef] [PubMed]

75. Crava, C.M.; Jakubowska, A.K.; Escriche, B.; Herrero, S.; Bel, Y. Dissimilar regulation of antimicrobial proteins in the midgut of *Spodoptera exigua* larvae challenged with *Bacillus thuringiensis* toxins or baculovirus. *PLoS ONE* **2015**, *10*, e0125991. [CrossRef]

76. Lei, Y.; Zhu, X.; Xie, W.; Wu, Q.; Wang, S.; Guo, Z.; Xu, B.; Li, X.; Zhou, X.; Zhang, Y. Midgut transcriptome response to a Cry toxin in the diamondback moth, *Plutella xylostella* (Lepidoptera: Plutellidae). *Gene* **2014**, *533*, 180–187. [CrossRef]

77. Grizanova, E.V.; Dubovskiy, I.M.; Whitten, M.M.; Glupov, V.V. Contributions of cellular and humoral immunity of *Galleria mellonella* larvae in defence against oral infection by *Bacillus thuringiensis*. *J. Invertebr. Pathol.* **2014**, *119*, 40–46. [CrossRef]

78. Ma, G.; Roberts, H.; Sarjan, M.; Featherstone, N.; Lahnstein, J.; Akhurst, R.; Schmidt, O. Is the mature endotoxin Cry1Ac from *Bacillus thuringiensis* inactivated by a coagulation reaction in the gut lumen of resistant *Helicoverpa armigera* larvae? *Insect Biochem. Mol. Biol.* **2005**, *35*, 729–739. [CrossRef]

79. Contreras, E.; Benito-Jardon, M.; Lopez-Galiano, M.J.; Real, M.D.; Rausell, C. *Tribolium castaneum* immune defense genes are differentially expressed in response to *Bacillus thuringiensis* toxins sharing common receptor molecules and exhibiting disparate toxicity. *Dev. Comp. Immunol.* **2015**, *50*, 139–145. [CrossRef]

80. Li, S.; Xu, X.; Shakeel, M.; Xu, J.; Zheng, Z.; Zheng, J.; Yu, X.; Zhao, Q.; Jin, F. *Bacillus thuringiensis* suppresses the humoral immune system to overcome defense mechanism of *Plutella xylostella*. *Front. Physiol.* **2018**, *9*, 1478. [CrossRef]

81. Rahman, M.M.; Roberts, H.L.; Sarjan, M.; Asgari, S.; Schmidt, O. Induction and transmission of *Bacillus thuringiensis* tolerance in the flour moth *Ephestia kuehniella*. *Proc. Natl. Acad. Sci. USA* **2004**, *101*, 2696–2699. [CrossRef] [PubMed]

82. Roth, O.; Sadd, B.M.; Schmid-Hempel, P.; Kurtz, J. Strain-specific priming of resistance in the red flour beetle, *Tribolium castaneum*. *Proc. R. Soc. B Biol. Sci.* **2009**, *276*, 145–151. [CrossRef] [PubMed]

83. Pham, L.N.; Dionne, M.S.; Shirasu-Hiza, M.; Schneider, D.S. A specific primed immune response in *Drosophila* is dependent on phagocytes. *PLoS Pathog.* **2007**, *3*, e26. [CrossRef] [PubMed]

84. Little, T.J.; Kraaijeveld, A.R. Ecological and evolutionary implications of immunological priming in invertebrates. *Trends Ecol. Evol.* **2004**, *19*, 58–60. [CrossRef] [PubMed]

85. Mahbubur Rahman, M.; Roberts, H.L.; Schmidt, O. Tolerance to *Bacillus thuringiensis* endotoxin in immune-suppressed larvae of the flour moth *Ephestia kuehniella*. *J. Invertebr. Pathol.* **2007**, *96*, 125–132. [CrossRef]

86. Dubovskiy, I.M.; Krukova, N.A.; Glupov, V.V. Phagocytic activity and encapsulation rate of *Galleria mellonella* larval haemocytes during bacterial infection by *Bacillus thuringiensis*. *J. Invertebr. Pathol.* **2008**, *98*, 360–362. [CrossRef]

87. Buchon, N.; Broderick, N.A.; Kuraishi, T.; Lemaitre, B. *Drosophila* EGFR pathway coordinates stem cell proliferation and gut remodeling following infection. *BMC Biol.* **2010**, *8*, 152. [CrossRef]

88. Tetreau, G.; Grizard, S.; Patil, C.D.; Tran, F.H.; Van, V.T.; Stalinski, R.; Laporte, F.; Mavingui, P.; Despres, L.; Moro, C.V. Bacterial microbiota of *Aedes aegypti* mosquito larvae is altered by intoxication with *Bacillus thuringiensis* israelensis. *Parasites Vectors* **2018**, *11*, 121. [CrossRef]

89. Rabelo, M.M.; Matos, J.M.L.; Santos-Amaya, O.F.; Franca, J.C.; Goncalves, J.; Paula-Moraes, S.V.; Guedes, R.N.C.; Pereira, E.J.G. Bt-toxin susceptibility and hormesis-like response in the invasive southern armyworm (*Spodoptera eridania*). *Crop Prot.* **2020**, *132*, 105129. [CrossRef]

90. Liu, J.; Wang, L.; Zhou, G.; Gao, S.; Sun, T.; Liu, J.; Gao, B. Midgut transcriptome analysis of *Clostera anachoreta* treated with lethal and sublethal Cry1Ac protoxin. *Arch. Insect Biochem. Physiol.* **2020**, *103*, e21638. [CrossRef]

91. Steijven, K.; Steffan-Dewenter, I.; Haertel, S. Testing dose-dependent effects of stacked Bt maize pollen on in vitro-reared honey bee larvae. *Apidologie* **2016**, *47*, 216–226. [CrossRef]

92. Shao, Q.M.; Yang, B.; Xu, Q.Y.; Li, X.Q.; Lu, Z.Q.; Wang, C.S.; Huang, Y.P.; Soderhall, K.; Ling, E.J. Hindgut innate immunity and regulation of fecal microbiota through melanization in insects. *J. Biol. Chem.* **2012**, *287*, 14270–14279. [CrossRef] [PubMed]

93. Wu, K.; Zhang, J.; Zhang, Q.L.; Zhu, S.L.; Shao, Q.M.; Clark, K.D.; Liu, Y.N.; Ling, E.J. Plant phenolics are detoxified by prophenoloxidase in the insect gut. *Sci. Rep.* **2015**, *5*, 16823. [CrossRef] [PubMed]

94. Lin, J.; Yu, X.Q.; Wang, Q.; Tao, X.; Li, J.; Zhang, S.; Xia, X.; You, M. Immune responses to *Bacillus thuringiensis* in the midgut of the diamondback moth, *Plutella xylostella*. *Dev. Comp. Immunol.* **2020**, *107*, 103661. [CrossRef] [PubMed]

95. Osta, M.A.; Christophides, G.K.; Vlachou, D.; Kafatos, F.C. Innate immunity in the malaria vector *Anopheles gambiae*: Comparative and functional genomics. *J. Exp. Biol.* **2004**, *207*, 2551–2563. [CrossRef]

96. Whitten, M.M.; Shiao, S.H.; Levashina, E.A. Mosquito midguts and malaria: Cell biology, compartmentalization and immunology. *Parasite Immunol.* **2006**, *28*, 121–130. [CrossRef]

97. Seisenbacher, G.; Hafen, E.; Stocker, H. MK2-Dependent p38b Signalling protects *Drosophila* hindgut enterocytes against JNK-induced apoptosis under chronic stress. *PLoS Genet.* **2011**, *7*, e1002168. [CrossRef]

98. Wu, K.; Yang, B.; Huang, W.R.; Dobens, L.; Song, H.S.; Ling, E.J. Gut immunity in Lepidopteran insects. *Dev. Comp. Immunol.* **2016**, *64*, 65–74. [CrossRef]

99. Palma, L.; Munoz, D.; Berry, C.; Murillo, J.; Caballero, P. *Bacillus thuringiensis* toxins: An overview of their biocidal activity. *Toxins* **2014**, *6*, 3296–3325. [CrossRef]

100. Yang, F.; Kerns, D.L.; Head, G.P.; Price, P.; Huang, F. Cross-resistance to purified Bt proteins, Bt corn and Bt cotton in a Cry2Ab2-corn resistant strain of *Spodoptera frugiperda*. *Pest Manag. Sci.* **2017**, *73*, 2495–2503. [CrossRef]

101. Pickett, B.R.; Gulzar, A.; Ferré, J.; Wright, D.J. *Bacillus thuringiensis* Vip3Aa Toxin Resistance in *Heliothis virescens* (Lepidoptera: Noctuidae). *Appl. Environ. Microbiol.* **2017**, *83*. [CrossRef] [PubMed]

102. Mahon, R.J.; Downes, S.J.; James, B. Vip3A resistance alleles exist at high levels in Australian targets before release of cotton expressing this toxin. *PLoS ONE* **2012**, *7*, e39192. [CrossRef] [PubMed]

103. Lee, M.K.; Miles, P.; Chen, J.S. Brush border membrane binding properties of *Bacillus thuringiensis* Vip3A toxin to *Heliothis virescens* and *Helicoverpa zea* midguts. *Biochem. Biophys. Res. Commun.* **2006**, *339*, 1043–1047. [CrossRef] [PubMed]

104. Chakroun, M.; Ferré, J. In vivo and in vitro binding of Vip3Aa to *Spodoptera frugiperda* midgut and characterization of binding sites by (125)I radiolabeling. *Appl. Environ. Microbiol.* **2014**, *80*, 6258–6265. [CrossRef] [PubMed]

105. Bel, Y.; Jakubowska, A.K.; Costa, J.; Herrero, S.; Escriche, B. Comprehensive analysis of gene expression profiles of the beet armyworm *Spodoptera exigua* larvae challenged with *Bacillus thuringiensis* Vip3Aa toxin. *PLoS ONE* **2013**, *8*, e81927. [CrossRef]

106. Hernandez-Martinez, P.; Gomis-Cebolla, J.; Ferre, J.; Escriche, B. Changes in gene expression and apoptotic response in *Spodoptera exigua* larvae exposed to sublethal concentrations of Vip3 insecticidal proteins. *Sci. Rep.* **2017**, *7*, 16245. [CrossRef]

107. Wang, S.; Kain, W.; Wang, P. *Bacillus thuringiensis* Cry1A toxins exert toxicity by multiple pathways in insects. *Insect Biochem. Mol. Biol.* **2018**, *102*, 59–66. [CrossRef]

108. Melo, A.L.; Soccol, V.T.; Soccol, C.R. *Bacillus thuringiensis*: Mechanism of action, resistance, and new applications: A review. *Crit. Rev. Biotechnol.* **2016**, *36*, 317–326. [CrossRef]

109. Ma, Y.; Zhang, J.; Xiao, Y.; Yang, Y.; Liu, C.; Peng, R.; Yang, Y.; Bravo, A.; Soberón, M.; Liu, K. The cadherin Cry1Ac binding-region is necessary for the cooperative effect with ABCC2 transporter enhancing insecticidal activity of *Bacillus thuringiensis* Cry1Ac toxin. *Toxins* **2019**, *11*, 538. [CrossRef]

110. Hu, X.; Zhang, X.; Zhong, J.; Liu, Y.; Zhang, C.; Xie, Y.; Lin, M.; Xu, C.; Lu, L.; Zhu, Q.; et al. Expression of Cry1Ac toxin-binding region in *Plutella xyllostella* cadherin-like receptor and studying their interaction mode by molecular docking and site-directed mutagenesis. *Int. J. Biol. Macromol.* **2018**, *111*, 822–831. [CrossRef]

111. Lu, K.; Gu, Y.; Liu, X.; Lin, Y.; Yu, X.Q. Possible insecticidal mechanisms mediated by immune-response-related cry-binding proteins in the midgut juice of *Plutella xylostella* and *Spodoptera exigua*. *J. Agric. Food Chem.* **2017**, *65*, 2048–2055. [CrossRef] [PubMed]

112. Batool, K.; Alam, I.; Zhao, G.; Wang, J.; Xu, J.; Yu, X.; Huang, E.; Guan, X.; Zhang, L. C-Type lectin-20 interacts with ALP1 receptor to reduce cry toxicity in *Aedes aegypti*. *Toxins* **2018**, *10*, 390. [CrossRef] [PubMed]

113. Batool, K.; Alam, I.; Jin, L.; Xu, J.; Wu, C.; Wang, J.; Huang, E.; Guan, X.; Yu, X.Q.; Zhang, L. CTLGA9 interacts with ALP1 and APN receptors to modulate Cry11Aa toxicity in *Aedes aegypti*. *J. Agric. Food Chem.* **2019**, *67*, 8896–8904. [CrossRef] [PubMed]

114. Liu, A.; Huang, X.; Gong, L.; Guo, Z.; Zhang, Y.; Yang, Z. Characterization of immune-related PGRP gene expression and phenoloxidase activity in Cry1Ac-susceptible and -resistant *Plutella xylostella* (L.). *Pestic Biochem. Physiol.* **2019**, *160*, 79–86. [CrossRef] [PubMed]

115. Shabbir, M.Z.; Zhang, T.; Prabu, S.; Wang, Y.; Wang, Z.; Bravo, A.; Soberon, M.; He, K. Identification of Cry1Ah-binding proteins through pull down and gene expression analysis in Cry1Ah-resistant and susceptible strains of *Ostrinia furnacalis*. *Pestic Biochem. Physiol.* **2020**, *163*, 200–208. [CrossRef] [PubMed]

116. Blum, J.E.; Fischer, C.N.; Miles, J.; Handelsman, J. Frequent replenishment sustains the beneficial microbiome of *Drosophila melanogaster*. *MBio* **2013**, *4*, e00860-13. [CrossRef]

117. Russell, J.A.; Dubilier, N.; Rudgers, J.A. Nature's microbiome: Introduction. *Mol. Ecol.* **2014**, *23*, 1225–1237. [CrossRef]

118. Patil, C.D.; Borase, H.P.; Salunke, B.K.; Patil, S.V. Alteration in *Bacillus thuringiensis* toxicity by curing gut flora: Novel approach for mosquito resistance management. *Parasitol. Res.* **2013**, *112*, 3283–3288. [CrossRef]

119. Orozco-Flores, A.A.; Valadez-Lira, J.A.; Oppert, B.; Gomez-Flores, R.; Tamez-Guerra, R.; Rodriguez-Padilla, C.; Tamez-Guerra, P. Regulation by gut bacteria of immune response, *Bacillus thuringiensis* susceptibility and hemolin expression in *Plodia interpunctella*. *J. Insect Physiol.* **2017**, *98*, 275–283. [CrossRef]

120. Dubovskiy, I.M.; Grizanova, E.V.; Whitten, M.M.; Mukherjee, K.; Greig, C.; Alikina, T.; Kabilov, M.; Vilcinskas, A.; Glupov, V.V.; Butt, T.M. Immuno-physiological adaptations confer wax moth *Galleria mellonella* resistance to *Bacillus thuringiensis*. *Virulence* **2016**, *7*, 860–870. [CrossRef]
121. Jiang, W.Y.; Geng, L.L.; Dai, P.L.; Lang, Z.H.; Shu, C.L.; Lin, Y.; Zhou, T.; Song, F.P.; Zhang, J. The Influence of Bt-transgenic maize pollen on the bacterial diversity in the midgut of Chinese honeybees, *Apis cerana cerana*. *J. Integr. Agric.* **2013**, *12*, 474–482. [CrossRef]
122. Caccia, S.; Di Lelio, I.; La Storia, A.; Marinelli, A.; Varricchio, P.; Franzetti, E.; Banyuls, N.; Tettamanti, G.; Casartelli, M.; Giordana, B.; et al. Midgut microbiota and host immunocompetence underlie *Bacillus thuringiensis* killing mechanism. *Proc. Natl. Acad. Sci. USA* **2016**, *113*, 9486–9491. [CrossRef] [PubMed]
123. Raymond, B.; Johnston, P.R.; Wright, D.J.; Ellis, R.J.; Crickmore, N.; Bonsall, M.B. A mid-gut microbiota is not required for the pathogenicity of *Bacillus thuringiensis* to diamondback moth larvae. *Environ. Microbiol.* **2009**, *11*, 2556–2563. [CrossRef] [PubMed]
124. Cherif, A.; Rezgui, W.; Raddadi, N.; Daffonchio, D.; Boudabous, A. Characterization and partial purification of entomocin 110, a newly identified bacteriocin from *Bacillus thuringiensis* subsp. Entomocidus HD110. *Microbiol. Res.* **2008**, *163*, 684–692. [CrossRef]
125. Shan, Y.; Shu, C.; Crickmore, N.; Liu, C.; Xiang, W.; Song, F.; Zhang, J. Cultivable gut bacteria of scarabs (Coleoptera: Scarabaeidae) inhibit *Bacillus thuringiensis* multiplication. *Environ. Entomol.* **2014**, *43*, 612–616. [CrossRef]
126. Takatsuka, J.; Kunimi, Y. Intestinal bacteria affect growth of *Bacillus thuringiensis* in larvae of the oriental tea tortrix, *Homona magnanima* diakonoff (Lepidoptera: Tortricidae). *J. Invertebr. Pathol.* **2000**, *76*, 222–226. [CrossRef]
127. Regode, V.; Kuruba, S.; Mohammad, A.S.; Sharma, H.C. Isolation and characterization of gut bacterial proteases involved in inducing pathogenicity of *Bacillus thuringiensis* toxin in cotton bollworm, *Helicoverpa armigera*. *Front. Microbiol.* **2016**, *7*, 1567. [CrossRef]
128. Mason, K.L.; Stepien, T.A.; Blum, J.E.; Holt, J.F.; Labbe, N.H.; Rush, J.S.; Raffa, K.F.; Handelsman, J. From Commensal to Pathogen: Translocation of *Enterococcus faecalis* from the midgut to the Hemocoel of *Manduca sexta*. *MBio* **2011**, *2*, e00065-11. [CrossRef]
129. Sha, Y.; Wang, L.; Liu, M.; Jiang, K.; Xin, F.; Wang, B. Effects of lactic acid bacteria and the corresponding supernatant on the survival, growth performance, immune response and disease resistance of *Litopenaeus vannamei*. *Aquaculture* **2016**, *452*, 28–36. [CrossRef]
130. Grau, T.; Vilcinskas, A.; Joop, G. Probiotic *Enterococcus mundtii* isolate protects the model insect *Tribolium castaneum* against *Bacillus thuringiensis*. *Front. Microbiol.* **2017**, *8*, 1261. [CrossRef]
131. Kos, B.; Suskovic, J.; Vukovic, S.; Simpraga, M.; Frece, J.; Matosic, S. Adhesion and aggregation ability of probiotic strain *Lactobacillus acidophilus* M92. *J. Appl. Microbiol.* **2003**, *94*, 981–987. [CrossRef] [PubMed]
132. Rao, X.J.; Yu, X.Q. Lipoteichoic acid and lipopolysaccharide can activate antimicrobial peptide expression in the tobacco hornworm *Manduca sexta*. *Dev. Comp. Immunol.* **2010**, *34*, 1119–1128. [CrossRef] [PubMed]
133. Wang, Q.; Ren, M.; Liu, X.; Xia, H.; Chen, K. Peptidoglycan recognition proteins in insect immunity. *Mol. Immunol.* **2019**, *106*, 69–76. [CrossRef] [PubMed]
134. Shao, Y.; Chen, B.; Sun, C.; Ishida, K.; Hertweck, C.; Boland, W. Symbiont-derived antimicrobials contribute to the control of the lepidopteran gut microbiota. *Cell Chem. Biol.* **2017**, *24*, 66–75. [CrossRef]
135. Shao, Y.Q.; Arias-Cordero, E.; Guo, H.J.; Bartram, S.; Boland, W. In Vivo Pyro-SIP Assessing Active Gut Microbiota of the Cotton Leafworm, Spodoptera littoralis. *PLoS ONE* **2014**, *9*, e85948. [CrossRef]
136. Scott, J.J.; Oh, D.C.; Yuceer, M.C.; Klepzig, K.D.; Clardy, J.; Currie, C.R. Bacterial protection of beetle-fungus mutualism. *Science* **2008**, *322*, 63. [CrossRef]
137. Forsgren, E.; Olofsson, T.C.; Vasquez, A.; Fries, I. Novel lactic acid bacteria inhibiting *Paenibacillus larvae* in honey bee larvae. *Apidologie* **2010**, *41*, 99–108. [CrossRef]
138. Endt, K.; Stecher, B.; Chaffron, S.; Slack, E.; Tchitchek, N.; Benecke, A.; Van Maele, L.; Sirard, J.C.; Mueller, A.J.; Heikenwalder, M.; et al. The microbiota mediates pathogen clearance from the gut lumen after non-typhoidal *Salmonella diarrhea*. *PLoS Pathog.* **2010**, *6*, e1001097. [CrossRef]
139. Baldwin, K.M.; Hakim, R.S. Growth and differentiation of the larval midgut epithelium during molting in the moth, *Manduca sexta*. *Tissue Cell* **1991**, *23*, 411–422. [CrossRef]

140. Guo, Z.; Kang, S.; Chen, D.; Wu, Q.; Wang, S.; Xie, W.; Zhu, X.; Baxter, S.W.; Zhou, X.; Jurat-Fuentes, J.L.; et al. MAPK signaling pathway alters expression of midgut ALP and ABCC genes and causes resistance to *Bacillus thuringiensis* Cry1Ac toxin in diamondback moth. *PLoS Genet.* **2015**, *11*, e1005124. [CrossRef]
141. Hernandez-Martinez, P.; Navarro-Cerrillo, G.; Caccia, S.; de Maagd, R.A.; Moar, W.J.; Ferre, J.; Escriche, B.; Herrero, S. Constitutive activation of the midgut response to *Bacillus thuringiensis* in Bt-resistant *Spodoptera exigua*. *PLoS ONE* **2010**, *5*, e12795. [CrossRef] [PubMed]
142. Forcada, C.; Alcacer, E.; Garcera, M.D.; Tato, A.; Martinez, R. Resistance to *Bacillus thuringiensis* Cry1Ac toxin in three strains of heliothis virescens: Proteolytic and SEM study of the larval midgut. *Arch. Insect Biochem. Physiol.* **1999**, *42*, 51–63. [CrossRef]
143. Martinez-Ramirez, A.C.; Gould, F.; Ferre, J. Histopathological effects and growth reduction in a susceptible and a resistant strain of *Heliothis virescens* (Lepidoptera: Noctuidae) caused by sublethal doses of pure Cry1A crystal proteins from *Bacillus thuringiensis*. *Biocontrol. Sci. Technol.* **1999**, *9*, 239–246. [CrossRef]
144. Broderick, N.A.; Buchon, N.; Lemaitre, B. Microbiota-induced changes in *Drosophila melanogaster* host gene expression and gut morphology. *MBio* **2014**, *5*, e01117-14. [CrossRef] [PubMed]
145. Jakubowska, A.K.; Vogel, H.; Herrero, S. Increase in gut microbiota after immune suppression in baculovirus-infected larvae. *PLoS Pathog.* **2013**, *9*, e1003379. [CrossRef]

 *toxins*

Review

Current Insights on Vegetative Insecticidal Proteins (Vip) as Next Generation Pest Killers

Tahira Syed [†], Muhammad Askari [†], Zhigang Meng, Yanyan Li, Muhammad Ali Abid, Yunxiao Wei, Sandui Guo, Chengzhen Liang * and Rui Zhang *

Biotechnology Research Institute, Chinese Academy of Agricultural Sciences, Beijing 100081, China; syedtahira98@gmail.com (T.S.); 2017Y90100082@caas.cn (M.A.); mengzhigang@caas.cn (Z.M.); liyanyan01@caas.cn (Y.L.); abid@caas.cn (M.A.A.); weiyunxiao@caas.cn (Y.W.); guosandui@caas.cn (S.G.)
* Correspondence: liangchengzhen@caas.cn (C.L.); zhangrui@caas.cn (R.Z.); Tel.: +86-10-82106127 (R.Z.)
† These authors contributed equally.

Received: 7 July 2020; Accepted: 11 August 2020; Published: 14 August 2020

Abstract: *Bacillus thuringiensis* (Bt) is a Gram positive soil bacterium. This bacterium secretes various proteins during different growth phases with an insecticidal potential against many economically important crop pests. One of the important families of Bt proteins is vegetative insecticidal proteins (Vip), which are secreted into the growth medium during vegetative growth. There are three subfamilies of Vip proteins. Vip1 and Vip2 heterodimer toxins have an insecticidal activity against many Coleopteran and Hemipteran pests. Vip3, the most extensively studied family of Vip toxins, is effective against Lepidopteron. Vip proteins do not share homology in sequence and binding sites with Cry proteins, but share similarities at some points in their mechanism of action. Vip3 proteins are expressed as pyramids alongside Cry proteins in crops like maize and cotton, so as to control resistant pests and delay the evolution of resistance. Biotechnological- and in silico-based analyses are promising for the generation of mutant Vip proteins with an enhanced insecticidal activity and broader spectrum of target insects.

Keywords: *Bacillus thuringiensis*; vegetative insecticidal proteins; insecticidal activity; resistance; pyramids

Key Contribution: This review addressed the recent advances in the structure, function, and mode of action of vegetative insecticidal proteins (Vip), and explored the strategies applied so far to modify the Vip proteins for an enhanced insecticidal activity.

---

1. Introduction

Common soil bacterium *Bacillus thuringiensis* (Bt) is a Gram positive motile bacterium that gained significant popularity over the past decades because of its role as an invertebrate pest killer [1]. *B. thuringiensis* has been extensively studied for its ability to produce immense arsenal of toxins with an insecticidal potential against insect vectors of human diseases, agricultural pests, nematodes, fungi, gastropods, and protozoans [1–5]. Bt proteins, as an active ingredient of biopesticides, are a valuable eco-friendly approach to replacing chemical insecticides. Bt genes are engineered in many crops like maize, cotton, soybean, and rice, offering a sustainable solution to control insect pests [6]. A Bt-transformed cotton plant expressing Cry1Ac was first introduced in 1996 in Australia and the United States of America [7]. The agriculture land covered by transgenic Bt crops reached 98.5 million hectors in 2016 [8]. Nevertheless, the rise of insect resistance is becoming a major hurdle in the commercialization of Bt transgenic crops [9].

B. thuringiensis produce Crystal (Cry) and Cytolytic (Cyt) toxins during sporulation, which are stored in parasporal crystalline inclusions and released after the disintegration of the cell wall in a culture medium. However, vegetative cells produce non-crystalline toxins, such as vegetative insecticidal proteins (Vip) and secreted insecticidal proteins (Sip), secreted as soluble proteins in a growth medium. Vip proteinaceous toxins were isolated from the culture medium of both *Bascillus cereus* and *B. thuringiensis* after screening [10,11]. The Vip toxin family is classified into four subfamilies, namely Vip1, Vip2, Vip3, and Vip4, based on their amino acid similarity, which also guides Vip proteins' nomenclature [10,12]. The Bt Toxin Nomenclature Committee assigned a four-rank name to each toxin—primary rank (two toxins with 45% similarity), denoted by an Arabic number; secondary rank (<78% similarity), denoted by an uppercase letter; tertiary rank (95% similarity), denoted by a lower case letter; and quaternary rank (>95% similarity), denoted by the final number (Figure 1).

To date, 15 Vip1 proteins, 20 Vip2 proteins, 111 Vip3 proteins, and 5 Vip4 proteins have been reported and named [13]. (Vip1 and Vip2 heterodimer toxins are effective against insects from Coleopteran and Hemipteran orders [14]. The largest family, Vip3, is effective against many species of Lepidoptera, and crops like cotton and maize have been successfully transformed with various Vip3 toxins [15]. Interestingly, Vip proteins have no sequence homology with Cry proteins, and do not share common binding sites in target insects [16–19]. This makes them ideal toxins to be used in combination with Cry proteins in insect resistance management (IRM) programs.

Significant knowledge about the structure and mode of action is available for Cry proteins, but this information is still rudimentary for Vip toxins. This review discusses recent insights on the structure and mechanism of action of Vip toxins. Detailed knowledge about the structure and functional characterization of Vip toxins will lead to the development of new strategies for designing improved toxins against insects that have developed resistance. The Vip family of Bt toxins is a potential candidate against these resistant pests. This study also focused on the natural and in vitro evolution of Vip toxins, and the strategies developed so far that improve the insecticidal activity of these toxins at a molecular level.

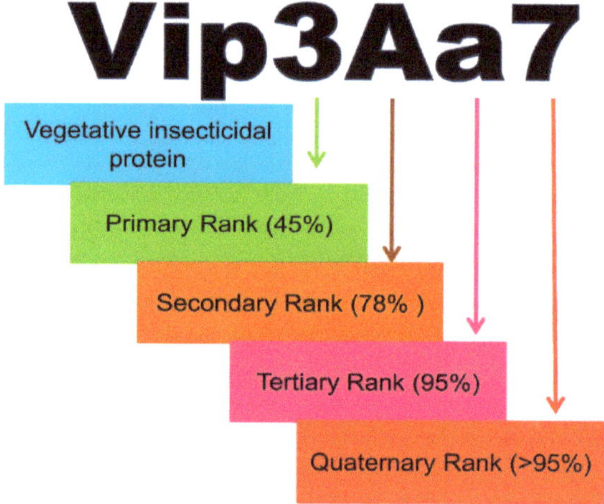

Figure 1. Schematic representation of vegetative insecticidal proteins (Vip) proteins' nomenclature system. Each protein is assigned a four ranked name—primary rank, given to proteins sharing less than 45% homology in amino acid sequences; secondary and tertiary ranks, with less than 78% and 95% similarity, respectively; and finally, the quaternary rank is given with more than 95% identical proteins.

2. Structure and Function of Vip Toxins

Vip proteins are widely distributed among Bacillus species. These proteins are not produced in parasporal crystalline inclusions, but are instead secreted into the culture medium. Like other Bt toxins, Vip proteins are also inactive in their native form, and are activated after being secreted in the membranes of insect midgut cells through the action of enzymes [20]. Detailed structural information of the Vip proteins is not yet elucidated; therefore, the current structural and functional predictions are based on in silico and mutagenic studies. A comparison between the Vip family of proteins is summarized in Table 1.

Table 1. Comparison of Vip family proteins.

Traits	Vip1/Vip2	Vip3	Vip4
Number of proteins	15 Vip1 20 Vip2	111	5
Gene size	~4 kb	~2.4 kb	2895 bp
Number of amino acids	881 Vip1 462 Vip2	787 to 789	965
Protein size	Vip1 80 kDa Vip2 45 kDa	89 kDa	~108 kDa
Target insects	Coleopteran and Hemipteran	Lepidopteron	Not available
Mode of action	ADP ribosyltransferase Activity/cytoskeleton abnormalities	Apoptotic cell death/pore formation	Not available
Commercialized crops	None	Cotton and maize	None

2.1. Structure of Vip1/Vip2 Binary Toxins

Vip1 and Vip2 act as binary toxins possessing an ADP ribosyltransferase activity [21]. The genes of the Vip1 and Vip2 proteins are located in a ~4 kb single operon with different reading frames [12,22]. To date, 35 *Vip1* and *Vip2* genes have been listed in the Bt nomenclature database. Sequence analysis found that Vip1 is synthesized as a protoxin of 100 kDa, with an N terminal signal peptide sequence of 35 amino acids. Similarly, Vip2 proteins in protoxin form are 52 kDa in size, with an N terminal signal peptide of 50 amino acids [22,23]. After their modification at the N terminal signal peptide, Vip1 and Vip2 are transformed into a mature proteins of 80 kDa and 45 kDa, respectively [24]. The structural analysis of Vip2 has confirmed two domains, N-terminal (Nt) and an NAD-binding (Nicotinamide adenine dinucleotide) C-terminal (Ct) domain [15,25]. Moreover, X-ray crystallography revealed homology between the N and C terminal domains of the Vip2 protein, and both domains are formed by the perpendicular packing of five mixed β sheets, with one flanking α helix and three anti parallel β sheets with three flanking α helices [25].

Research based on their sequence homology suggests that Vip1 and Vip2 act as binary toxins of A + B type, with similarities to many mammalian toxins. Vip1 has very little structural similarity with the *Clostridium spiroforme* toxin, protective antigen of *Bacillus anthracis*, and CdtB toxin of *Clostridium dificile*. Vip2 has a structural similarity to the active domain of CdtA toxin produced by *C. difcile* [26]. Both Vip1 and Vip2 toxins have a similarity to the C2 toxin of *Clostridium botulinum* and the domain Ia of the iota toxin produced by *Clostridium perfringens* [27]. Altogether, this predicts Vip2 as an ADP-ribosyltransferase toxin inhibiting the polymerization of actin filaments, causing cytoskeleton abnormalities and insect cell death [28]. Vip1 is inferred to act as a B toxin (Binding domain) responsible for the translocation of Vip2 inside insect midgut cells. Vip2 is a cytotoxic A toxin with a binary toxin response, showing no toxicity to insects when applied alone [28].

2.2. Structure and Function of Vip3 Proteins

Vip3 toxin is also produced during the vegetative growth phase of Bt, and Vip3A is the most widely studied Vip toxin so far. Vip3 is a diverse family of toxins with 95% similarity between its members, and it shares no primary sequence homology to any other Vip families or Bt toxins [29]. These proteins show a strong inhibition of insect larval growth at a low concentration [29,30], and the structural differences among the Vip3 members predict a broader mode of action against a wide spectrum of insects.

The Vip3 gene encodes a protein of 89 kDa having 787 to 789 amino acids [29]. Functional characterization of the *vip3Aa16* gene revealed that its −35 and −10 promoter regions have homology to the *Bacillus subtilis* promoters, which suggests that the *Vip3* gene promoter is under the control of the σ^{35} holoenzyme [31]. It shares no sequence homology with any other toxins produced by *B. thuringiensis* [1]. Phylogenetic analysis showed that the Vip3 protein belongs to distant clade rather than to Cry toxins.

A comprehensive structure of Vip3A has not yet been resolved, and is only derived through in silico modeling [32,33]. Notably, the Vip3A signal peptide located at the N-terminal is responsible for the translocation of protein. The N-terminal region is highly conserved and performs important regulatory insecticidal functions [34], while the C-terminal region undergoes various modifications and has the ability to target insect specificity. At present, the role of both domains in insecticidal activity is mainly perceived by mutagenic studies. The deletion or addition of amino acids at the N- and C-terminals greatly affects the entomocidal property of a particular protein. For example, the deletion of amino acids at the Nt region causes a negative effect on the insecticidal activity of the Vip3A protein [35,36]. Substitutions at position T167 or G168 at the N-terminal of Vip3Af with alanine leads to a decreased insecticidal activity [32]. In contrast, the deletion of 200 amino acids from the N-terminal enhanced the toxicity of Vip3Aa against various Lepidoptera pests, and the deletion of 200 amino acids from the C-terminal region abolished the insecticidal activity of Vip3BR. Also, the deletion of 127 amino acids at the C-terminal maintained a low level of insecticidal activity against *Agrotis ipsilon* (Lepidoptera: Noctuidae) and *Helicoverpa armigera* (Lepidoptera: Noctuidae) [37], which suggests the important roles of the N- and C-terminal parts in the insecticidal activity of Vip3 proteins. It is also possible that the interactions of other plant or insect proteins with Vip3 C- and N-terminal domains could enhance their insecticidal activity. Hence, further studies are needed to explore the exact mechanism of action.

Each C-terminal amino acid plays an important role in the target specificity and toxicity against many Lepidopteron pests. Meanwhile, Vip3A11 mutants are generated after replacing nine residues at the C-terminus with Vip3A39 residues by site-targeted mutagenesis. Here, the cysteine residue CYS784 of the C-terminal region is found to be a crucial trypsin cleave site for bioactivity and toxicity [38]. However, the Vip3 C-terminal region alone does not possess an insecticidal activity, as the expression and purification of the C-terminal region of Vip3Ab1 and Vip3Bc1 cause no harm to the insects. Therefore, in contrast with the Cry proteins, both the C- and N-terminal regions are important for oligomerization and proteolytic stability, with a significant contribution to the toxicity of the Vip3 proteins [39].

Site directed mutagenesis anticipates putative trypsin cleave sites Lys195, Lys197, and Lys198 inside Vip3Aa. The mutants generated by replacing these three Lysine residues with alanine lose sensitivity to trypsin or midgut juices (MJ), and also show toxicity against *Spodoptera exigua* (Lepidoptera: Noctuidae) [40]. In the same manner, the substitution of cysteine with serine at the C-terminal also reduces the Vip3A7 protein insecticidal activity against *Plutella xylostella* (Lepidoptera: Plutellidae), possibly due to the disruption of disulfide bonds between cysteine residues [41]. An alanine scanning analysis of 588 residues unveiled a five-domain structure of the Vip3Af1 protein and its role in toxicity. This approach revealed 50 residues with a significant impact on Vip3Aa structural conformation and toxicity. Among them, two clusters of 19 substitutions, located near the N-terminus region between Leu167–Tyr27 or on the C-terminus between Gly689–Phe741, abolished toxicity to *Agrotis segetum* (Lepidoptera: Noctuidae). Another 19 substitutions also reduced toxicity to *Spodoptera frugiperda* (Lepidoptera: Noctuidae). Hence, it is evident that each amino acid within the Vip3 protein plays a diverse role in protein stability and toxicity [32]. Transmission electron microscopy and single particle analysis of Vip3Ag4 have revealed the surface topology of its tetramers. After trypsin treatment, the protein forms an octamer containing tetramers of 65 kDa and 22 kDa fragments. In addition, the tetrameric form and main topology are retained even after trypsin treatment [42].

Although no clear evidence about the Vip3 3D structure is available, in silico analyses point to the presence of five domains in the Vip3Af protein. The trypsin fragmentation of alanine mutants depicted five domains in the Vip3Af proteins structure. Domain I spans from amino acids 12 to 198, domain II 199 to 313, domain III 314 to 526, domain IV 527 to 668, and domain V 669 to 788 amino acids. Domains I to III are necessary for tetramerization, however not domain V. In addition, the role of domain IV remains unclear [43]. This evidence is further supported by the 2D structure of the Vip3B protein predicted by X-ray diffraction studies. According to this model, Vip3B is composed of five domains; two domains carrying α helices (DI and DII) at the N-terminus and three β sheet containing domains (DIII, DIV, and DV) on the C-terminal region. Domain III shows a slight resemblance to domain II of the Cry4Aa and Cry1Ac proteins. Domain IV and V share homology with carbohydrate binding modules (CBM), indicative of glycosylated receptor-binding inside the midgut [44]. A carbohydrate binding motif (CBM; CBM_4_9) has also been identified at the C-terminal region in all of the Vip3 protein members, except for Vip3Ba. It is then possible that the C-terminal region plays a crucial role in recognition and binding to midgut receptors [45].

A recent report sheds more light on the structure of all five domains of Vip3A and their related function. Domain II has two highly conserved hydrophobic α helices, predicting their role in membrane insertion and pore formation inside the insect midgut. Domain III comprises three β sheets potent for cell binding, along with domain II, persistent with previous results. In Vip3A, two CBM domains with different glycan binding pockets are found in C-terminal region, which forms domain IV and domain V [46]. This indicates a specificity in their binding capacity with glycan on the targeted cell surface. The cryo-EM (cryogenic electron microscopy) structural analysis solves the structure of Vip3A and showed how the toxin forms pores in the insect midgut. The protein architecture has five distinct domains in protoxin form, with domains I (coiled α 1–α 4) ending at the primary protease cleavage site, and domain II having five α helices mainly producing the core before trypsin digestion. Antiparallel β sheets of domain III form a β prism fold analogous to the Cry toxins. In the Cry toxin, this fold assists in receptor recognition. Similar to previous results, two CBM folds form the last two domains. After trypsinization, all four monomers stay connected, and no conformational change takes place in domains II to V. Three N-terminal α helices form a parallel four helix coiled coil, which forms a long dipole to lodge the ions in its cavity. Its dimension has the ability to form pores in the lipid bilayer [47].

2.3. Vip4 Toxins

Vip4 is the least characterized toxin of the Vpb class. Only five Vip4 proteins have been identified to date. The first reported Vip4 toxin was Vip4Aa1 (now named Vpb4Aa1), isolated from Bt strain Sbt009, with no insecticidal activity against any pests. Its molecular mass is ~108 kDa and it possesses 965 amino acids [13]. The main region spanning from 47 to 77 amino acids is a PA14 domain, and the region from 218–631 residues is named the bacterial Binary_ToxB domain. This novel protein shares 34% identity with the Vip1Aa1 protein and 65% with the Ia domain of the Iota toxin of *B. cereus*, specifically to the B component of the binary toxin [26].

3. Modern Classification of Vip Proteins

Recently, the classification and names of the Vip family of toxins were modified by Crickmore et al. [48]. Accordingly, all Vip toxins are placed in three different classes, namely Vip3, Vpa, and Vpb. The Vip3 family mnemonic remains the same. In class Vpa, all of the Vip2 proteins are placed. Class Vpb contains a binary toxin component of Vip2, such as Vip1, and its structural analogues, previously known as Vip4 [13].

4. Mechanism of Action

The mode of action of the Vip toxins inside insect guts is not yet clear and needs further investigation. Differences in the structure of the Vip and Cry toxins determine the different target sites in the insect midgut, making them suitable candidates for insects-resistant control [16].

4.1. Mechanism of Action of Vip1/Vip2 Binary Toxin

There is no clear mechanism of action for the Vip1/Vip2 binary toxin. Each protein alone is not enough to cause toxicity, but rather they act in combination as A + B binary toxins [28]. This multistep process begins with toxin entry at the insect midgut and ends with larvae death. There are many proposed models on this mechanism. Toxins first enter the midgut and are digested by trypsin-like proteases. Enzymatic action by trypsin or midgut juices (MJ) cleaves Vip1Ac into its activated form before entering into the brush border membrane (BBM). After enzymatic activation, Vip1Ac oligomerizes to form seven monomers containing a multimeric structure [49], which binds the BBMVs of cotton aphids with a high target specificity [14]. Similar results were observed with the activated form of the Vip1Ad and Vip2Ag binary toxin activity. Vip2Ag binding, insecticidal activity, and toxicity is increased tremendously in the presence of Vip1Ad in *Holotrichia parallela* (Coleoptera: Scarabaeidae). This shows strong evidence for Vip1 as a potential receptor of Vip2 after trypsin activation [50].

After internalization, Vip2 transfers the ADP ribose group to the actin filaments, inhibiting its polymerization. This ultimately leads to abnormal microfilament formation and cell death [15,23]. The insecticidal activity of Vip2Ag and Vip1Ad is characterized by the vacuolization and destruction of BBMVs and microvilli depletion, similar to what has been observed in *H. armigera* fed on Cry2Ab and Vip3AcAa/Cry1Ac binary toxins [51,52]. In a nutshell, the histopathological effects of Vip1/Vip2 binary toxins are comparable to those of Cry proteins.

4.2. Mechanism of Action of Vip3 Toxin

The Vip3 toxins mechanism of action has some similarity to the Cry toxin-like protease activation, binding with midgut cells and pore formation. The complex multistep process is not fully elucidated yet. At first, all Vip3 toxins are activated by midgut juices. Proteolytic analysis of Vip3A toxin has revealed several fragments (62–66 kDa, 45 kDa, 33 kDa, and 19–22 kDa) with a similar pattern after trypsin treatment or insect midgut proteases. The main cleavage product from the C-terminus region is a 62–66 kDa core peptide, generally considered to be the main part of the activated toxin [19,37,53,54]. However, a 19–22 kDa fragment comprising the 199 amino acids at the N-terminus, together with a 62–66 kDa fragment, are crucial for lethality, as shown by bioactivity assays after purification [43]. Furthermore, 45 kDa and 33 kDa fragments are formed after cleavage of a 62–66 kDa fragment [53]. Other research on protease cleavage inferred that these smaller fragments formed after the digestion of 65 kDa fragment could be the result of denaturing conditions of SDS-Page, as the C-terminal (65 kDa fragment) domain remains intact under native conditions [55]. In contrast, the proteolytic digestion of Vip3Ca produces a 70 kDa fragment [56]. Variations in the insecticidal activity of the Vip3 toxin seem to depend on the hydrolysis pattern inside insects' midgut [54], however, the proteolytic digestion pattern of Vip3Ab1 and Vip3Bc1 in a non-susceptible insect, *Ostrinia nubilalis* (Lepidoptera: Crambidae), was identical to the susceptible insects, and shed light on the fact that resistance might be unrelated to protein cleavage [39].

After the trypsin processing of Vip3A, two of its fragments, 19–22 kDa of the N-terminal region and 62–66 kDa fragment of the C-terminal region join to form a ~360 kDa homo-tetramer, which cannot be degraded by proteases [57]. Moreover, another study proposed the formation of a >240 kDa complex and identified a novel site, S164, crucial for the formation and stability of this complex. Mutations at this site lead to a loss of insecticidal activity [58]. In addition, the two fragments are eluted together in gel permeation chromatography, emphasizing the fact that they may remain together after cleavage [55]. The interaction between the 22 kDa and 65 kDa fragment is necessary for their stability and toxicity [39].

The next phase is the binding of the activated toxin with the BBMVs inside the midgut of susceptible insects. The Vip3 protein binding sites do not overlap with the Cry proteins. A clear mechanism regarding the binding of the Vip3 toxin is still not available, and only a few studies have addressed the recognition of binding molecules inside the insect midgut cells. So far, only a few Vip3 binding proteins have been identified, one of which is ribosome S2 protein, identified by yeast hybrid assay, and confirmed by vitro pull-down assays in sf21 cells of *S. frugiperda*. A knock down of the ribosomal S2 gene in the Sf21 cells and the larvae of *Spodoptera litura* (Lepidoptera: Noctuidae) resulted in a reduced larvicidal activity, considering that S2 is one of the proteins involved in the Vip3 insecticidal mechanism [59]. Another potential protein is a 48 kDa tenascin-like glycoprotein, which strongly binds to Vip3Aa in BBMVs from black cutworm [36,60].

Many current studies are focused on the recognition of Vip3 potent receptors in order to understand the mechanism of action. A novel receptor, fibroblast growth factor receptor-like protein (Sf-FGFR), has been identified on the membrane of Sf9 cells, as a result of its binding affinity toVip3, confirmed by in vitro analysis. Silencing of the *Sf-FGFR* gene resulted in a reduced toxicity of Vip3Aa to Sf9 cells. The localization of Sf-FGFR and Vip3Aa on the surface and then inside the cytoplasm suggests that binding takes place on the surface, leading to internalization [61]. Similarly, Vip3Aa has shown a strong interaction with scavenger receptor class C like protein (Sf-SR-C) in both in vivo and in vitro analysis with Sf9 cells of *S. frugiperda*. Knocking down the expression of these receptor genes results in a reduced mortality of Sf9 cells and *S. exigua* larvae to Vip3Aa [62]. Further studies are required to fully clarify the specificities of the Vip3 toxin receptor binding.

Proteolytic activation and receptor mediated binding of Vip3 toxins leads to pore formation and cell death. After feeding Vip3 toxins, the insect midgut is damaged, which is proposed as the main target site for the Vip3 toxin. Histopathological analyses have revealed similar symptoms to Cry toxins, like swollen or lysed midguts and pore formation [19,63]. However, clear mechanisms on toxin intercellular localization and pore formation in BBM are not available. The localization mechanism of active the Vip3Aa protein inside Sf9 cells, by laser scanning confocal microscopy with fluorescently labeled Vip3Aa (Alexa488-actVip3Aa), has demonstrated that Vip3Aa is not internalized by the endocytic- or clathrin-dependent endocytic pathway. Instead, this seems to happen through receptor mediated endocytosis, after which the Vip3Aa protein interacts with various cytosolic proteins (e.g., ribosomal S2 protein) [64].

Voltage clamping and planar lipid bilayer experiments predict the ability of a toxin to form discrete ion channels without involving receptors, which differs from Cry1Ab [16]. Pore formation in lipid bilayers by an activated toxin inside the midgut has been reported in H. armigera using the florescent quenching method [65]. Additionally, the maximum potential of the activated Vip3Aa toxin to form pores has been seen at specific pHs during in vitro analyses, showing that pore formation only happens at acidic or neutral pH [57]. The mechanism of Vip3 pore formation and virulence is generally the most accepted (Figure 2).

Contrary to the pore formation model, is another mechanism in which the Vip3 toxin induces apoptotic cell death in insects. The intercellular localization of Vip3 causes abnormalities of cell division and leads to the apoptosis of insect midgut cells. Vip3A treated Sf9 cells undergo arrest at the G2/M phase and the disruption of the mitochondrial membrane potential ($\Delta\Psi m$), leading to apoptotic cell death via the sf-caspase-I mediated pathway [62]. Another study has evidenced the involvement of regulatory proteins and lysosomes in apoptosis. Furthermore, symptoms of apoptosis and mitochondrial collapse are prevalent in sf9 cells when administered Vip3Aa, such as the accumulation of reactive oxygen species (ROS), caspases (caspase 3,9), and cytochrome c [66] (Figure 3). Because of the lack of sound information, more investigation is needed to clarify the downstream mechanism of Vip3 induced apoptosis and cell death.

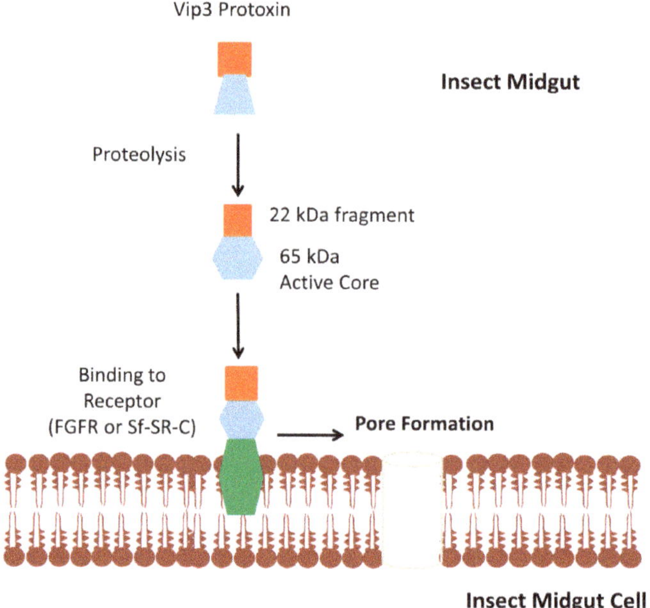

Figure 2. Proposed mechanism of pore formation by the Vip3 toxin. The Vip3A toxin is activated by proteolysis inside the insect midgut. In the next step, activated toxins, including 22 kDa and 65 kDa fragments, bind with receptors, leading to pore formation in the insect midgut cells and, ultimately, to the death of the insects.

To understand the mechanism of insect response to toxins, the transcriptomic and proteomic characterization of genes and proteins is of great interest. The gene expression profiles of toxic-dose-treated larvae of *S. exigua* and *S. litura* have been analyzed in two independent studies. From the analysis of the transcriptome profile in *S. exigua* larvae, >29,000 unigenes were obtained, in which the S2 and tenascin-like protein gene expression, was stable. The up regulated genes were mostly related to immune reposes and defense mechanism while down regulated genes were mainly metabolic ones [67]. Similarly, the genes coding for lysosomes and antimicrobial peptides have been found to be up-regulated in *S. exigua* [68]. In the gene expression profile of the Vip3 toxin treated larvae of *S. exigua*, immune response genes are up-regulated and the genes involved in the digestion process are down regulated. The up regulation of initiator and effector caspases genes and antimicrobial effectors provides strong evidence for the apoptosis of insect cells, similar to previous reports [69]. In another analysis, 56,498 unigenes were identified in *S. litura* larvae. The transcription levels of the trypsin related genes increased in this case after toxin induction, which supports the role of trypsin in the metabolism of the Vip3Aa toxin [70]. However, further investigation is necessary to elucidate how Vip3 toxins cause apoptotic cell death.

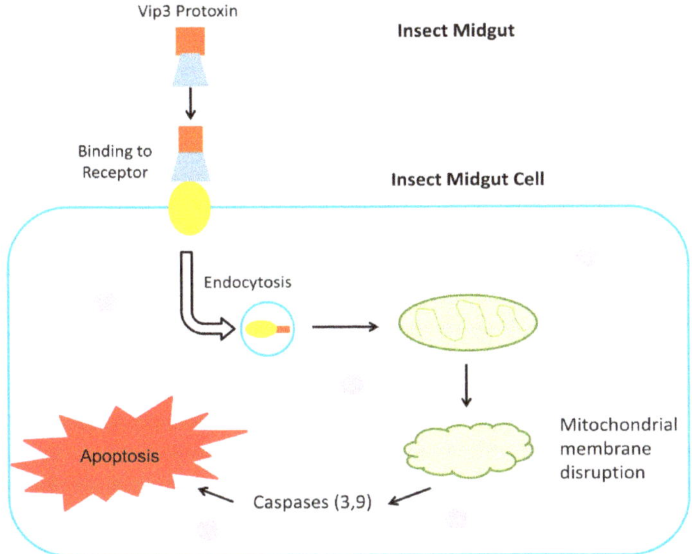

Figure 3. Schematic representation of the mechanism of Vip3 toxin induced apoptotic cell death of insect midgut cells. Vip3A protoxin binds with receptors, and the receptor mediated internalization of toxin takes place. Toxin internalization leads to changes like DNA damage, mitochondrial membrane disruption, and the activation of caspases (caspase 3 or 9), in turn promoting apoptotic cell death.

5. Insecticidal Activity of Vip Proteins

After research on various insect species, it has been found that Vip1/Vip2 has an insecticidal activity against some pests of Coleopteran and Hemipteran orders [14]. Vip1 and Vip2 act in combination, and none of the toxins have an insecticidal activity when administered alone. The combination of Vip1Aa/Vip2Aa or Vip2Ab has been found to be affective against *Diabrotica* spp [12]. The Vip1Ad/Vip2Ag toxins, when combined and expressed as a binary toxin, show toxicity against *H. parallela, H. oblita,* and *Anomala corpulenta* (Coleoptera: Scarabaeidae) [24].

The most extensively studied Vip3 protein family is Vip3Aa, which is widely known for their insecticidal activity against many species of Lepidopteron and pests like *S. exigua, H. armigera, S. frugiperda, Heliothis virescens* (Lepidoptera: Noctuidae), *Helicoverpa zea* (Lepidoptera: Noctuidae), and *A. ipsilon*. Although they have very minor differences in their sequences, Vip proteins exhibit great variability in their targeted insects. For instance, one of the most recently discovered members of the Vip3 family protein, Vip3Ca, has 70% homology with Vip3Aa and has been found to be toxic against *Chrysodeixis chalcites* (Lepidoptera: Noctuidae), *Mamestra brassicae* (Lepidoptera: Noctuidae), and *Trichoplusia ni* (Lepidoptera: Noctuidae). However, the Vip3Ca toxin shows a moderate insecticidal activity against *Cydia pomonella* (Lepidoptera: Tortricidae; non-susceptible to the Cry toxin) and *O. nubilalis* (susceptible to the Cry toxin) [71]. Vip3Ca is the most potent toxin against *Mythimna separate* (Lepidoptera: Noctuidae), with an LC$_{50}$ value 3.4 µg/g. This toxin could be used in future maize crop protection to control Oriental armyworm [72]. Vip3Ca is also more toxic to *Ostrinia furnacalis* (Lepidoptera: Crambidae), which is more similar to Cry1Ab than Vip3A, and can be an effective candidate against Cry1Ab-resistant colonies of *O. furnacalis* [73].

Likewise, Vip3Ae and Vip3Af are toxic to *Spodoptera littoralis* (Lepidoptera: Noctuidae), *M. brassicae,* and *Lobesia botrana* (Lepidoptera: Tortricidae), and Vip3Ab is lethal against *A. ipsilon*. Nonetheless, Vip3Ad exerts no toxicity to insects like *H. armigera, M. brassicae, S. frugiperda, S. exigua, S. littoralis,* and *A. ipsilon* [29]. The insecticidal activity of Vip3Aa59 is significantly higher than Vip3Aa58 towards

Dendrolimus pini (Lepidoptera: Lasiocampidae) [20], and Vip3Aa45 shows a 40-fold higher toxicity than Cry proteins against *S. exigua* [26].

In some reports, the toxicity of the Vip3 protoxin was found to be more than active toxin, e.g., Vip3Ae protoxin is more insecticidal than the active toxin when compared with Vip3Aa [54]. These could result from differences in the protocols for protein isolation and purification, bioassay conditions, and quantification methods. For example, metal chelate chromatography clearly affects the insecticidal activity of Vip3 [20,54].

6. Evolution of Resistance and Cross Resistance to Vip3 Toxins

Since the application of Vip3 transformed Bt crops, only few cases of practical resistance have been reported. The reported cases of resistance are from both laboratory and field selected insects. For example, in laboratory conditions, insects of *S. litura* show a 280-fold resistant to Vip3A after 12 generations of selection. This is probably due to the lower protease activity in those insects [74]. Similarly, the laboratory selection of *H. virescens* with Vip3Aa for 12 generations results in a >2040-fold resistance. This resistance is polygenic and decreases after 13 to 28 generations, without toxin administration. A lack of cross resistance is also seen against Cry toxins [75].

In a laboratory evolved resistance study, conducted with transgenic maize expressing Vip3Aa20 in Brazil, the target pest, *S. frugiperda,* showed a >3200-fold resistance. The pattern of resistance inheritance was autosomal recessive and monogenic, with a very low frequency of resistant alleles (0.00009 estimated by the F_2 screening method) [76]. A Vip3A resistant strain of *S. frugiperda* evolved >632-fold resistance to Vip3A, with minor cross resistance to Cry1F, Cry2Ab2, or Cry2Ae toxins [77].

Despite the fact that Vip3A expressing crops are not yet commercialized in Australia, a high frequency of resistant alleles has been observed in various studies conducted on *H. armigera* and *Helicoverpa punctigera* (Lepidoptera: Noctuidae). In a study screening for resistance alleles in *H. armigera* and *H. punctigera* using the F_2 method, a natural polymorphism and very high baseline frequency of 0.027 and 0.008, respectively, were observed, with no cross resistance to Cry1Ac or Cry2Ab. The presence of both resistance alleles on the same locus confirmed resistance to be recessive [78]. Another biochemical study also found a high frequency of resistant alleles in the same insect, *H. armigera,* with no significant change in the binding of Vip3 toxins to BBMVs compared with susceptible insects. Instead, a low proteolytic activity was found in resistant insects. [79].

Despite the intrinsic specificities of Vip3 toxicity, no cross resistance to Vip3C has been observed in insects of different species, previously found to be resistant to Cry1A, Cry2Ab, Dipel (Mixture of Cry1 and Cry2). Vip3C shows cross resistance to *H. armigera* colonies resistant to Vip3Aa or Vip3Aa/Cry2Ab, and toxicity against *O. furnacalis,* which is nonsusceptible to Vip3A [73]. The biochemical basis of resistance could not be established by the down regulation of membrane bound alkaline phosphatase (mALP) isoform HvmALP1, observed in Vip3 resistant insects and results does not support it to be the functional receptor of Vip3. In addition, mALP can be used as a marker for the detection of Vip3A resistance [80]. Moreover, Cry1F and Vip3A do not share common binding sites in *S. frugiperda* [17], and also lack cross resistance [54].

7. Identification of Bt Isolates Containing *Vip Genes*

Over the past few decades, researchers have been trying to find new *B. thuringiensis* isolates from different geographical regions and diverse environments, to develop new toxins with a high insecticidal potential and to cope with resistance. The discovery of new isolates not only helped in the production of new pesticides of a wide insecticidal spectrum, but also in overcoming insecticidal resistance [81]. *B. thuringiensis* strains were isolated from diverse habitats, like milk and mossy pine cone [82], soil, leaf [83] and insect cadavers [84], and goat gut [85].

After the characterization of native *B. thuringiensis* isolates isolated from soil, and fig leaves and fruits from a Turkish collection, a new *B. thuringiensis* isolate, 6A, was identified carrying a high expression of Vip3Aa. The identified protein, named Vip3Aa65, has a similar insecticidal activity against *Grapholita molesta* (Lepidoptera: Tortricidae) and *H. armigera*, but is less toxic to Spodoptera spp. compared with Vip3Aa16 [86]. In another study, two Bt strains, BnBt and MnD, were found to produce Vip proteins in isolates of great potential with a high toxicity (LC_{50} = 41.860 ng/μL of BnBt and 55.154 ng/μL of MnD) in the second instar larvae of *S. littoralis* [87]. Lone and co-workers isolated and expressed a novel Vip3Aa61 gene in *Escherichia coli* from isolate *B. thuringiensis* JK37. Nucleotide analysis found differences in many amino acids compared with Vip3A. Because of its high insecticidal activity (LC_{50} = 169.63 ng/cm^2) against second instar larvae of *H. armigera*, the Vip3Aa61 toxin is a potential candidate for transgenic crop production and pest protection [88].

Bacterial 16S ribosomal RNA (rRNA) sequencing is frequently used to characterize microbes at a species rank. In the past, this method was not successful because of the close resemblance of *B. thuringiensis* to other strains, making it ambiguous for phylogenetic and diversity analysis [89]. A new and more reliable method of phylogenetic analysis is multiple locus sequence typing (MLST). MLST-based analysis of *B. thuringiensis* kurstaki isolates from Assam, India, confirmed the presence of Vip1 (53.3%), Vip2 (46.6%), and Vip3 (40%) genes [90]. The results were in contrast to a previous study, where the Vip3 gene was more abundant than Vip1 and Vip2 [91]. Recently, the characterization of an indigenous Bt strain, found Vip3 gene in this strain, and that the spore crystal mixture of this isolate had a high mortality rate against *S. frugiperda* [92].

8. Transgenic Crops Expressing Vip Proteins

Despite the high toxicity of Vip1/Vip2 toxins against corn rootworms, Bt maize crops containing these binary toxins cannot be developed because of the cytotoxicity of Vip2 proteins [93].

Vip3 proteins are introduced in crops like cotton and maize. The *Vip3Aa19* gene was first introduced into Bollgord cotton (COT102), expressed as a single insecticidal protein (VIPCOT commercialized in 2008 in Unite States of America). This toxin provided protection against three major cotton pests, *H. virescens*, cotton bollworm *H. zea*, and *Pectinophora gossypiella* (Lepidoptera: Gelechiidae). Later, it was pyramided with Cry genes (modified Cry1Ab) for insect resistance management. Vip3Aa20 was introduced (MIR162) in corn and was commercialized in 2009 in the United States. Later, Vip3 proteins were pyramided with Cry genes (Cry1Ab + Vip3Aa20) and (Cry1Ab, Vip3Aa20, and mCry3A) in corn.

Other than commercialized crops, *Vip3* genes are successfully transformed and the proteins have been expressed in transgenic crops in laboratory-based studies. Transgenic sugar cane lines expressing Vip3A showed a high mortality rate against sugar borer *Chilo infuscatellus* (Lepidoptera: Pyraloidea) [94]. Cow pea is an important food crop in many African countries, and is harmed by Lepidopteron pest, *Maruca vitrata* (Family: Crambidae). The *Vip3Ba1* gene, isolated from Australian Bt isolates, was transformed and expressed in cowpea to provide protection against legume pod borer (*M. vitrata*), by strongly inhibiting larvae growth [95]. Similarly, a transgenic corn event (C008 and C010) expressing Vip3Aa19 has been found to be highly toxic against black cut worm [96]. When a tobacco plant was transformed with the N-terminal deletion mutant of the Vip3BR protein (Ndv200), it acquired full protection against *S. littoralis*, *A. ipsilon*, and *H. armigera* [97]. These results are helpful for future Bt-derived mutant protein transformation in crops.

9. Biotechnological Strategies to Improve Toxicity and Insecticidal Spectrum of Vip3 Proteins

In vitro directed evolution to increase the insecticidal potential of Vip proteins can be employed with success. The fact that Vip3 toxins share no sequence homology with Cry toxins makes them an ideal candidate for insect resistance management (IRM) programs. Vip3 proteins are found to be toxic against insects that are less susceptible to Cry toxins, such as Lepidopteron [10]. Various strategies can be employed to increase the toxicity and insecticidal spectrum of Vip proteins. For example, sequence or domain swapping to form chimeric Vip toxins successfully enhances toxicity against susceptible and resistant insects.

Genetic Engineering of Vip3A Genes to Form Chimeric Proteins

To increase the insecticidal spectrum and activity of the Vip3 proteins, these were genetically modified. In this process, the genes were swapped for the construction of protein chimeras, where one protein expresses the sequence or domains of another protein. These novel chimeras have shown great toxicity against resistant and non-susceptible insect pests. The successful implication of a domain swapping method to generate chimeric Cry1 proteins (Cry1Ba/Cry1Ia hybrid) with enhanced toxicity against Colorado potato beetle has already been reported [98]. However, fewer reports on Vip3 chimera formation are available. A chimera of two Vip3 proteins, Vip3AcAa generated by sequence shuffling of Vip3Aa and Vip3Ac, not only had enhanced toxicity against the fall armyworm, but also to European corn borer, against which Vip3Ac was not toxic, even at high concentrations. These new chimeric proteins caused growth retardation in a Vip3A non-susceptible insect *O. nubilalis* [99].

Using domain shuffling, six chimeric proteins were generated by joining fragments of N-terminal, C-terminal, and the central part of the core protein from Vip3Aa and Vip3Ca. Two of these chimeras, in which only Nt domain (Vip3C having Nt domain of Vip3Aa) was shuffled, had shown no effect over stability and solubility. The exchange of the Ct domain in four chimeric proteins resulted in proteins that were insoluble and unstable to trypsin, except for one soluble and stable chimera. Compared with the parental proteins, one chimeric protein formed by shuffling the Nt domain of Vip3C with Vip3Aa showed an enhanced insecticidal activity against *S. frugiperda* (Table 2). Another chimeric protein was highly unstable and formed after shuffling Ct domain of Vip3Aa with Vip3Ca *O. furnacalis*, which is susceptible to Vip3C and non-susceptible to Vip3Aa, and also showed vulnerability against those chimeras containing the Ct domain of Vip3C. Considering the above mentioned observations, it is evident that the Ct domain is involved in the specificity of Vip3 proteins for targeted insects [100].

The administration of a dual toxin with non-homologous mechanisms of action is an effective way to circumvent resistance. For this purpose, chimeric proteins were formed by fusing *Vip* and *Cry* genes sequences. By combining the sequence of the full-length Vip3Aa16 toxin gene with the Nt region of the Cry1Ac activated core, a Vip3A16-Cry1Ac chimeric protein of 150 kDa was generated. The resulting fusion protein toxicity was triggered against the first-instar larva of *Ephestia kuehniella* (Lepidoptera: Pyralidae) in contrast with parental toxin Vip3Aa [33]. Similarly, another successful chimeric protein was made by fusing the nucleotide sequence of Vip3Aa7 and the Nt region of a synthetic toxin Cry9Ca, with an enhanced insecticidal activity (compared with the single parent proteins or a mixture of both) against *P. xylostella* [41].

Even if domain shuffling and sequence swapping are successfully implemented to form new toxin combination with improved insecticidal activity against new or resistant pests, site directed mutations and in silico analyses still provide crucial information necessary to understand protein toxicity. With the help of ever-evolving bioinformatics, it will be possible to better understand the effect of mutations on the mechanism of action of particular toxins against target insects (Table 2).

Table 2. Chimeric Vip3 proteins and their toxicity profiles.

Toxin	Chimera Type [1]	Insecticidal Activity [2]	Reference
Vip3AcAa	Chimera of Vip3Ac N terminus + Vip3Aa C-terminus	Toxic to *O. nubilalis*, insecticidal activity against *S. frugiperda*, *H. zea*, and *Bombyx mori*	[99]
Vip3AaAc	Chimera of Vip3Aa N-terminus + Vip3Ac C-terminus	RA against *S. frugiperda* and *H. zea*; LA against *B. mori*	
Vip3Aa14	Chimera of Vip3Aa14 and Cry1Ac	Toxic against *H. armigera* and *P. xylostella*; RA against *S. litura* than Vip3Aa	[101]
Vip3Aa7	Chimera of Cry1C promoter + Vip3Aa7 (Nt 39 aa deleted) + Cry1C C-terminus	RA against *P. xylostella*, *H. armigera*, and *S. exigua*	[102]
Vip3Aa7	Chimera of Vip3Aa7 + Cry9Ca N-terminus	Increased activity against *P. xylostella*	[41]
Vip3Aa16	Chimera of Vip3Aa16 + Cry1Ac N-terminus (48-609 aa)	Increased toxicity against *E. kuehniella*	[33]
Vip3Aa and Vip3Ca	Chimera of Vip3Aa N-terminus + Vip3Aa central domain + Vip3Ca C-terminus	Not active against *A. gemmatalis*, *M. brassicae*, *O. furnacalis*, *S. frugiperda*, *S. littoralis*, *H. armigera*, and *S. exigua*	[100]
	Chimera of Vip3Aa N-terminus + Vip3Ca central domain + Vip3Ca C-terminus	RA against *Anticarsia gemmatalis*, *M. brassicae*, and *O. furnacalis*; strong activity against *S. frugiperda* than Vip3Ca	
	Chimera of Vip3Aa N-terminus + central domain of Vip3Ca + Vip3Aa C-terminus	Insoluble protein	
	Chimera of Vip3Ca N-terminus + central domain of Vip3Aa + Vip3Ca C-terminus	LA against *S. littoralis*, *S. frugiperda*, *H. armigera*, *M. brassicae*, *S. exigua*, and *A. gemmatalis*; except *O. furnacalis*	
	Chimera of Vip3Ca N-terminus + central domain of Vip3Ca + Vip3Aa C-terminus	Insoluble protein	
	Chimera of Vip3Ca N-terminus + central domain of Vip3Aa + Vip3Aa C-terminus	RA against *S. littoralis*, *S. frugiperda*, *H. armigera*, and *M. brassicae*; except *S. exigua* compared with Vip3Aa. LA against *A. gemmatalis* and *O. furnacalis*	

[1] aa = amino acids; [2] LA = lost activity; RA = reduced activity.

10. In Silico Analyses for Generation of Mutagenic Vip3 Proteins

Over the past two decades, researchers have been trying to produce mutagenic proteins with an enhanced insecticidal activity against specific pests. By using computational methodology the analysis of the chimeric protein, Vip3Aa-Cry1Ac (formed by the fusion of the functional regions of Cry1Ac and Vip3Aa), unveiled its enhanced toxicity and broad-spectrum insect control. Molecular docking analysis was performed with five Lepidopteron insect receptors, forming a strong interaction. This new protein is proposed to be the potential toxin for future crop protection against Lepidopteron pests [103].

Effect of Amino Acid Modifications on Toxicity of Mutant Vip3A

Mutagenic analyses have been widely utilized to explore the amino acids present at specific sites critical for toxicity. For this purpose, Vip3A11 mutants were generated by replacing nine residues at N-terminus with Vip3A39 residues, using site targeted mutagenesis. An approximately two-fold increase in toxicity for three mutants (S9N, S193T, and S194L) was seen against *H. armigera* larvae compared with Vip3A11. Furthermore, the N-terminal amino acids also played a great role in toxicity and insect specificity against Lepidopteron pests [104].

Similarly, the docking and binding site prediction analysis identified the amino acids Y616, H618, Y619, W552, K557, E627, and Q652 in the Ct region as crucial sites for Vip3Aa toxin binding and insecticidal activity. The insecticidal activity of only one mutant, Y619A, was increased against *H. armigera* and *S. exigua* [38]. In another case, the Vip3Aa protein substitutions at site S164 with alanine or proline resulted in a loss of oligomer formation, and an ablation of the insecticidal potential against *S. litura*. Notably, substitution with threonine resulted in only a 35% reduction in toxicity [58].

A cysteine residue at the C-terminal region, CYS784, is a crucial site for trypsin cleavage and for the formation of the active core for toxicity. Hence, both the C- and N-terminal regions are necessary elements for toxicity. Cysteine to serine substitutions at the C-terminal also reduced the Vip3A7 protein insecticidal activity against *P. xylostella*, likely due to the disruption of the disulfide bonds between the cysteine residues [41]. A modified Vip3Ca protein, ARP150v02, with amendments at eight locations near the N-terminus region, was expressed and purified in *E. coli*. The ARP150v02 protein showed an insecticidal activity against many insects, but a high insecticidal effect against *S. frugiperda* (LC_{50} = 450 ng/cm^2). In contrast, this protein was ineffective against *H. armigera*, even at a high dose. The binding assays revealed that the ARP150v02 protein competes for binding with Vip3Aa in *S. frugiperda* [105]. More studies based on site directed mutagenesis are necessary in order to overcome pest resistance.

11. Synergism and Antagonism in Vip3 and Cry Proteins

Various studies reported the presence of synergism in Vip3 and other Bt toxin (Cry and Cyt). The coexpression of Vip3Aa and Cyt2Aa in *E.coli* lead to synergism in *S. exigua* and *Chilo suppressalis* (Lepidoptera: Crambidae) [106]. Similar synergism was observed between Vip3A and Cry1Ia in *S. frugiperda*, *Spodoptera albula*, and *Spodoptera cosmioides* (Lepidoptera: Noctuidae) [107]. Another synergistic combination was found between Cry9Aa and Vip3Aa, possibly due to the binding mechanism between two toxins with BBMVs in *C. suppressalis* and *O. furnacalis*. Interestingly, the synergism between Vip3Aa and Cry9Aa mutants was disturbed moderately in *C. suppressalis*, and severely in and *O. furnacalis*. Synergism resulted in an improved toxicity of Vip3Aa and Cry9Aa in *C. suppressalis*, which is a great threat to rice crops in China [108]. Strong synergism was also seen in the Cry1Ab/Vip3Ca protein combination. These can be useful in *M. separate* and *O. furnacalis* control in future pyramided gene stacking [72]. A high rate of synergism was identified in the Vip3Aa and Cry1Ab combination against the neonatal larvae of *S. frugiperda*, without competing for binding sites [109]. In the same study, several other Bt protein combinations, like Vip3Aa/Cry2Ab, Cry1Ab/Cry2Ab, Cry1Ab/Cry2Ab/Vip3Aa, Cry1Ea/Cry1Ca, and Vip3Ca/Cry1Ea, also showed synergism against *S. frugiperda*.

Finally, it is worth mentioning that some toxin combinations show antagonism as well. For example, Vip3Aa showed slight antagonism with Cyt2Aa in *Culex quinquefasciatus* (Diptera: Culicidae) [106], and with Cry1Ia in *Spodoptera eridania* (Lepidoptera: Noctuidae) [107]. These antagonisms could result from direct competition between the CRY and Vip3A toxin for the same binding sites in insect BBMVs [110]. Another study identified many antagonist combinations in various Cry and Vip3 protein pairs, such as Vip3A/Cry1A or Cry1Ca, Cry1Ca/Vip3Aa, Cry1Ca/Vip3Ae, Cry1Ca/Vip3Af, Vip3Af/Cry1Aa, or Cry1A. Vip3A and Cry1Ca showed more antagonism in *S. frugiperda* at LC$_{90}$ [111]. This knowledge can be helpful in the future for stacking genes in pyramids in order to broaden the insect spectrum, and for managing the evolution of insect resistance.

Efficacy of Pyramided Vip3 and Cry Proteins

The past decade marked an increased use of Vip3A with Cry proteins in pyramided crops for broader insecticidal activity and in insect resistance management [93]. The pyramiding of Cry1A and Cry2A with Vip3A is a promising strategy in IRM programs. No cases of cross resistance have been reported yet in pyramided Bt crops. Meanwhile, registered varieties of pyramided Bt cotton and maize containing Vip3Aa19 and Vip3Aa20 are commercialized worldwide [112].

In order to form pyramided Bt rice, a fusion gene (C1V3) was formed by combining truncated Cry1Ab and the full-length Vip3A by a linker, to generate a chimeric protein that could be digested efficiently by trypsin. After digestion into activated fragments, both toxins function just like an individual toxin of Cry1Ab and Vip3A. Transgenic rice with the fusion gene (C1V3) showed insecticidal activity against two major rice pests, *C. suppressalis* and *Cnaphalocrocis medinalis* (Lepidoptera: Crambidae). A high toxin content was seen after two generations in fields, along with disease spots. No difference in phenotypes was seen in the transgenic (A1L3) and control rice plants. Further investigations will clarify the implications of this strategy [113]. Cry1Ac and Vip3Aa are potential candidates for sugarcane protection against *Diatraea flavipennella* (Lepidoptera: Crambidae) and *Elasmopalpus lignosellus* (Lepidoptera: Pyralidae), through pyramided transgenic Bt sugarcane production and commercialization [114].

The main threat to the commercialization of these crops is the existence of resistance to Cry1A or Cry2A, which may spark the evolution of resistance to these pyramided crops [112]. To overcome this barrier, other strategies could be implemented, alone or by pyramiding them with Bt toxins, in order to control resistant pests. For this purpose, post transcriptional gene silencing of insects specific genes involved in various physiological functions could be effective to inhibit insect growth and development. In this method, double stranded RNAs (dsRNA) are designed to target essential insect genes, disrupting their expression by RNA interference (RNAi). Short sequences of dsRNA are incorporated into insects through diet and are also transformed in plants [115,116]. The chances of cross resistance are very low in this case, as both pathways have diverse and independent mechanisms of action.

Pyramided Cry toxins and RNAi corn plants targeting *D. v. virgifera* have already been developed [117]. Another pyramid formed by the combination of Bt toxins, Cry1Ac and dsRNA, designed to target the metabolism of juvenile hormone (JH) in *H. armigera*, was introduced in cotton. Two types of cotton plants, JHA (targeting JH acid methyltransferase) and JHB (JH transporter protein), showed a high activity against resistant *H. armigera* [118]. For the safety and high efficiency of this strategy, dsRNA can be transformed into plastids to be expressed with plastid genomes rather than a nuclear genome. The chloroplast transformation of dsRNA also increases the protein content in the cell, as RNAi machinery is absent in the chloroplast compartment. Introducing dsRNA into potato plastids targeted the β-actin gene of the deadly potato pest, Colorado potato beetle, and protected the crop against this notorious pest [119] (Table 3).

Table 3. Commercialized pyramided Bt crops with the Vip3A protein [120,121].

Plant	Event	Pyramid	Target Insects	Country
Maize	BT11/GA21	Cry1Ab, Vip3Aa20	*A. ipsilon, O. nubilalis, H. zea,* and *S. frugiperda*	Canada (2005), South Korea (2006/2008), Japan, Mexico, Philippines (2007), Argentina, Brazil (2009), Uruguay (2011), and Colombia (2012)
	BTT11/GA21/MIR162	Cry1Ab, Vip3Aa20	*H. zea, S. frugiperda,* and *A. ipsilon*	Brazil (2011) and Colombia (2012)
	BT11/MR162	Cry1Ab, Vip3Aa20	*A. ipsilon, O. nubilalis, H. zea, S. frugiperda,* and *Spodoptera albicosta*	United States (2009)
	BT11/MIR162/MIR604	Cry1Ab, mCry3A, Vip3Aa20	*O. nubilalis, Diatraea crambidoides, S. frugiperda, Pseudaletia unipunctata, S. exigua, A. ipsilon, Striacosta albicosta, Diatraea saccharalis, Diabrotica virgifera, Diabrotica barberi,* and *Papaipema nebris*	United States (2009)
	BT11/MIR162/MIR604/GA21	Cry1Ab, mCry3A, Vip3a20	*Diabrotica spp., H. zea, O. nubilalis, S. frugiperda,* and *A. ipsilon*	Colombia (2012)
	BT11 × MIR162 × TC1507 × GA21	Cry1Ab + Cry1Fa + Vip3Aa	Lepidopteran	United States (2011)
	BT11 × MIR162 × MIR604 × TC1507 × 5307	Cry1Ab + mCry3A + Vip3A + chimeric (Cry3A-Cry1Ab)	Lepidopteron and Coleopteran	Brazil (2019)
Cotton	COT102 × COT67B	mCry1Ab + Vip3Aa19	*H. virescens, H. zea, P. gossypiella, S. frugiperda, S. exigua,* and *T. ni*	United States (2008)
	COT102 × COT67B × MON88913	Cry1Ab + Vip3A	Lepidopteron	Costa Rica (2009)
		Cry1Ac + Cry1Fa + Vip3Aa	Lepidopteron	United States (2013)
	COT102 × MON15985	Cry1Ac + Cry2Ab + Vip3A	Lepidopteron	Japan, Australia, and Mexico (2014)
	COT102 × MON15985 × MON88913	Cry1Ac + Cry2Ab + Vip3A	Lepidopteron	Japan, Australia (2014), Brazil, Taiwan (2016), and South Korea (2015),

Table 3. *Cont.*

Plant	Event	Pyramid	Target Insects	Country
	COT102 (IR102)	Vip3A	Lepidopteron	Australia, New Zealand, United States (2005), Canada (2011), Costa Rica (2017), South Korea (2014), Mexico (2010), Japan (2012), Taiwan, Philippines China (2015), and Columbia (2016)
	3006-210-23 × 281-24-236 × MON88913 × COT102	Cry1Ac + Cry1Fa + Vip3Aa	Lepidopteron	Japan (2016), Mexico, and South Korea (2014)
	281-24-236 × 3006-210-23 × COT102 × 81910	Cry1Ac + Cry1Fa + Vip3Aa	Lepidopteron	Japan (2016) and Brazil (2019)
	281-24-236 × 3006-210-23 × COT102	Cry1Ac + Cry1F + Vip3A	Lepidopteron	Brazil (2018)
	T304-40 × GHB119 × COT102	Cry1Ab + Cry2Ae + Vip3A	Lepidopteron	Brazil (2018)
	GHB811 × T304-40 × GHB119 × COT102	Cry1Ae + Cry2Ae + Vip3A	Lepidopteron	Brazil (2019)
	GHB614 × T304-40 × GHB119 × COT102	Cry1Ab + Cry2Ae + VIP3A	Lepidopteron	Philippines (2020)

12. Future Perspectives

There is an increasing need for new IRM strategies. The effective control of resistant pests and the delay of adaptive evolution to resistance in insects will not be achieved solely with pyramid strategies. In depth knowledge of the mechanism of action of Bt proteins and the mechanisms of insect resistance is crucial for prolonged benefits of Bt toxins in pest control. For effective resistance management, it is necessary to develop novel toxins, and to combine more than one strategy in pest control. Vip toxins are a promising new generation of insecticides to be used in spray formulation and transgenic crops, because of their broad spectrum of insect targets. Researchers are focusing on the structure and function of Vip proteins and are attempting to find new Vip proteins from already identified and novel Bt strains. Finding new proteins could be a strategy of choice to manage resistance to Bt toxins. Next generation sequencing (NGS) can accelerate the discovery of novel proteins through the complete sequencing of novel genomes and already known Bt strains.

Author Contributions: Conceptualization, T.S., C.L., and R.Z.; writing (original draft preparation), T.S., M.A., Z.M., M.A.A., Y.W., Y.L., S.G., C.L., and R.Z. All authors have read and agreed to the published version of the manuscript.

Funding: This work was supported by grants from the Ministry of Agriculture of China (grant nos. 2016ZX08005004 and 2016ZX08009003-003-004).

Acknowledgments: We would like to apologize to all the investigators whose research could not be cited appropriately, owing to space limitations.

References

1.	De Maagd, R.A.; Bravo, A.; Berry, C.; Crickmore, N.; Schnepf, H.E. Structure, diversity, and evolution of protein toxins from spore-forming entomopathogenic bacteria. *Ann. Rev. Genet.* **2003**, *37*, 409–433. [CrossRef]

2.	Zheng, J.; Gao, Q.; Liu, L.; Liu, H.; Wang, Y.; Peng, D.; Ruan, L.; Raymond, B.; Sun, M. Comparative genomics of Bacillus thuringiensis reveals a path to specialized exploitation of multiple invertebrate hosts. *MBio* **2017**, *8*. [CrossRef] [PubMed]

3.	Gomaa, E.Z. Chitinase production by Bacillus thuringiensis and Bacillus licheniformis: Their potential in antifungal biocontrol. *J. Microbiol.* **2012**, *50*, 103–111. [CrossRef] [PubMed]

4.	Abd El-Ghany, A.M.; Abd El-Ghany, N.M. Molluscicidal activity of Bacillus thuringiensis strains against Biomphalaria alexandrina snails. *Beni Suef Univ. J. Basic Appl. Sci.* **2017**, *6*, 391–393. [CrossRef]

5.	Kondo, S.; Mizuki, E.; Akao, T.; Ohba, M. Antitrichomonal strains of Bacillus thuringiensis. *Parasitol. Res.* **2002**, *88*, 1090–1092. [CrossRef]

6.	Sanchis, V. From microbial sprays to insect-resistant transgenic plants: History of the biospesticide Bacillus thuringiensis: A review. *Agron. Sustain. Dev.* **2011**, *31*, 217–231. [CrossRef]

7.	Tabashnik, B.E.; Brévault, T.; Carrière, Y. Insect resistance to Bt crops: Lessons from the first billion acres. *Nat. Biotechnol.* **2013**, *31*, 510–521. [CrossRef]

8.	Briefs, I. *Global Status of Commercialized Biotech/GM Crops in 2017: Biotech Crop Adoption Surges as Economic Benefits Accumulate in 22 Years*; The International Service for the Acquisition of Agri-biotech Applications: Ithaca, NY, USA, 2017.

9.	Tabashnik, B.E.; Carrière, Y. Surge in insect resistance to transgenic crops and prospects for sustainability. *Nat. Biotechnol.* **2017**, *35*, 9262017. [CrossRef]

10.	Estruch, J.J.; Warren, G.W.; Mullins, M.A.; Nye, G.J.; Craig, J.A.; Koziel, M.G. Vip3A, a novel Bacillus thuringiensis vegetative insecticidal protein with a wide spectrum of activities against lepidopteran insects. *Proc. Natl. Acad. Sci. USA* **1996**, *93*, 5389–5394. [CrossRef]

11.	Warren, G.W.; Koziel, M.G.; Mullins, M.A.; Nye, G.J.; Carr, B.; Desai, N.M.; Kostichka, K.; Duck, N.B.; Estruch, J.J. Auxiliary Proteins for Enhancing the Insecticidal Activity of Pesticidal Proteins. U.S. Patent 5,770,696, 23 June 1998.

12.	Warren, G.W. *Vegetative Insecticidal Proteins: Novel Proteins for Control of Corn Pests*; Taylor and Francis: London, UK, 1997; pp. 109–121.

13. Crickmore, N.; Berry, C.; Panneerselvam, S.; Mishra, R.; Connor, T.R.; Bonning, B.C. Bacterial Pesticidal Protein Resource Center. 2020. Available online: https://www.bpprc.org/database/ (accessed on 8 August 2020).

14. Sampurna, S.; Maiti, M.K. Molecular characterization of a novel vegetative insecticidal protein from Bacillus thuringiensis effective against sap-sucking insect pest. *J. Microbiol. Biotechnol.* **2011**, *21*, 937–946.

15. Jucovic, M.; Walters, F.S.; Warren, G.W.; Palekar, N.V.; Chen, J.S. From enzyme to zymogen: Engineering Vip2, an ADP-ribosyltransferase from Bacillus cereus, for conditional toxicity. *Protein Eng. Des. Sel.* **2008**, *21*, 631–638. [CrossRef]

16. Lee, M.K.; Walters, F.S.; Hart, H.; Palekar, N.; Chen, J.-S. The mode of action of the Bacillus thuringiensis vegetative insecticidal protein Vip3A differs from that of Cry1Ab δ-endotoxin. *Appl. Environ. Microbiol.* **2003**, *69*, 4648–4657. [CrossRef]

17. Sena, J.A.; Hernández-Rodríguez, C.S.; Ferré, J. Interaction of Bacillus thuringiensis Cry1 and Vip3A proteins with Spodoptera frugiperda midgut binding sites. *Appl. Environ. Microbiol.* **2009**, *75*, 2236–2237. [CrossRef] [PubMed]

18. Gouffon, C.V.; Van Vliet, A.; Van Rie, J.; Jansens, S.; Jurat-Fuentes, J. Binding sites for Bacillus thuringiensis Cry2Ae toxin on heliothine brush border membrane vesicles are not shared with Cry1A, Cry1F, or Vip3A toxin. *Appl. Environ. Microbiol.* **2011**, *77*, 3182–3188. [CrossRef] [PubMed]

19. Chakroun, M.; Ferré, J. In vivo and in vitro binding of Vip3Aa to Spodoptera frugiperda midgut and characterization of binding sites by 125I radiolabeling. *Appl. Environ. Microbiol.* **2014**, *80*, 6258–6265. [CrossRef] [PubMed]

20. Baranek, J.; Kaznowski, A.; Konecka, E.; Naimov, S. Activity of vegetative insecticidal proteins Vip3Aa58 and Vip3Aa59 of Bacillus thuringiensis against lepidopteran pests. *J. Invertebr. Pathol.* **2015**, *130*, 72–81. [CrossRef] [PubMed]

21. Barth, H.; Hofmann, F.; Olenik, C.; Just, I.; Aktories, K. The N-Terminal Part of the Enzyme Component (C2I) of the BinaryClostridium botulinum C2 Toxin Interacts with the Binding Component C2II and Functions as a Carrier System for a Rho ADP-Ribosylating C3-Like Fusion Toxin. *Infect. Immun.* **1998**, *66*, 1364–1369. [CrossRef] [PubMed]

22. Shi, Y.; Ma, W.; Yuan, M.; Sun, F.; Pang, Y. Cloning of vip1/vip2 genes and expression of Vip1Ca/Vip2Ac proteins in Bacillus thuringiensis. *World J. Microbiol. Biotechnol.* **2007**, *23*, 501. [CrossRef]

23. Shi, Y.; Xu, W.; Yuan, M.; Tang, M.; Chen, J.; Pang, Y. Expression of vip1/vip2 genes in Escherichia coli and Bacillus thuringiensis and the analysis of their signal peptides. *J. Appl. Microbiol.* **2004**, *97*, 757–765. [CrossRef]

24. Bi, Y.; Zhang, Y.; Shu, C.; Crickmore, N.; Wang, Q.; Du, L.; Song, F.; Zhang, J. Genomic sequencing identifies novel Bacillus thuringiensis Vip1/Vip2 binary and Cry8 toxins that have high toxicity to Scarabaeoidea larvae. *Appl. Microbiol. Biotechnol.* **2015**, *99*, 753–760. [CrossRef]

25. Han, S.; Craig, J.A.; Putnam, C.D.; Carozzi, N.B.; Tainer, J.A. Evolution and mechanism from structures of an ADP-ribosylating toxin and NAD complex. *Nat. Struct. Biol.* **1999**, *6*, 932–936. [PubMed]

26. Palma, L.; Muñoz, D.; Berry, C.; Murillo, J.; Caballero, P. Bacillus thuringiensis toxins: An overview of their biocidal activity. *Toxins* **2014**, *6*, 3296–3325. [CrossRef] [PubMed]

27. Knapp, O.; Benz, R.; Popoff, M.R. Pore-forming activity of clostridial binary toxins. *Biochim. Biophys. Acta Biomembr.* **2016**, *1858*, 512–525. [CrossRef] [PubMed]

28. Barth, H.; Aktories, K.; Popoff, M.R.; Stiles, B.G. Binary bacterial toxins: Biochemistry, biology, and applications of common Clostridium and Bacillus proteins. *Microbiol. Mol. Biol. Rev.* **2004**, *68*, 373–402. [CrossRef] [PubMed]

29. De Escudero, I.R.; Banyuls, N.; Bel, Y.; Maeztu, M.; Escriche, B.; Muñoz, D.; Caballero, P.; Ferré, J. A screening of five Bacillus thuringiensis Vip3A proteins for their activity against lepidopteran pests. *J. Invertebr. Pathol.* **2014**, *117*, 51–55. [CrossRef]

30. Chakroun, M.; Bel, Y.; Caccia, S.; Abdelkefi-Mesrati, L.; Escriche, B.; Ferré, J. Susceptibility of Spodoptera frugiperda and S. exigua to Bacillus thuringiensis Vip3Aa insecticidal protein. *J. Invertebr. Pathol.* **2012**, *110*, 334–339. [CrossRef]

31. Mesrati, L.A.; Tounsi, S.; Jaoua, S. Characterization of a novel vip3-type gene from Bacillus thuringiensis and evidence of its presence on a large plasmid. *FEMS Microbiol. Lett.* **2005**, *244*, 353–358. [CrossRef]

32. Banyuls, N.; Hernández-Rodríguez, C.; Van Rie, J.; Ferré, J. Critical amino acids for the insecticidal activity of Vip3Af from Bacillus thuringiensis: Inference on structural aspects. *Sci. Rep.* **2018**, *8*, 1–14. [CrossRef]

33. Sellami, S.; Jemli, S.; Abdelmalek, N.; Cherif, M.; Abdelkefi-Mesrati, L.; Tounsi, S.; Jamoussi, K. A novel Vip3Aa16-Cry1Ac chimera toxin: Enhancement of toxicity against Ephestia kuehniella, structural study and molecular docking. *Int. J. Biol. Macromol.* **2018**, *117*, 752–761. [CrossRef] [PubMed]

34. Rang, C.; Gil, P.; Neisner, N.; Van Rie, J.; Frutos, R. Novel Vip3-related protein from Bacillus thuringiensis. *Appl. Environ. Microbiol.* **2005**, *71*, 6276–6281. [CrossRef] [PubMed]

35. Li, C.; Xu, N.; Huang, X.; Wang, W.; Cheng, J.; Wu, K.; Shen, Z. Bacillus thuringiensis Vip3 mutant proteins: Insecticidal activity and trypsin sensitivity. *Biocontrol Sci. Technol.* **2007**, *17*, 699–708. [CrossRef]

36. Estruch, J.J.; Yu, C.-G. Plant Pest Control. U.S. Patent 6,291,156, 18 September 2001.

37. Gayen, S.; Hossain, M.A.; Sen, S.K. Identification of the bioactive core component of the insecticidal Vip3A toxin peptide of Bacillus thuringiensis. *J. Plant Biochem. Biotechnol.* **2012**, *21*, 128–135. [CrossRef]

38. Chi, B.; Luo, G.; Zhang, J.; Sha, J.; Liu, R.; Li, H.; Gao, J. Effect of C-terminus site-directed mutations on the toxicity and sensitivity of Bacillus thuringiensis Vip3Aa11 protein against three lepidopteran pests. *Biocontrol Sci. Technol.* **2017**, *27*, 1363–1372. [CrossRef]

39. Zack, M.D.; Sopko, M.S.; Frey, M.L.; Wang, X.; Tan, S.Y.; Arruda, J.M.; Letherer, T.T.; Narva, K.E. Functional characterization of Vip3Ab1 and Vip3Bc1: Two novel insecticidal proteins with differential activity against lepidopteran pests. *Sci. Rep.* **2017**, *7*, 1–12. [CrossRef]

40. Zhang, J.; Pan, Z.-Z.; Xu, L.; Liu, B.; Chen, Z.; Li, J.; Niu, L.-Y.; Zhu, Y.-J.; Chen, Q.-X. Proteolytic activation of Bacillus thuringiensis Vip3Aa protein by Spodoptera exigua midgut protease. *Int. J. Biol. Macromol.* **2018**, *107*, 1220–1226. [CrossRef]

41. Dong, F.; Shi, R.; Zhang, S.; Zhan, T.; Wu, G.; Shen, J.; Liu, Z. Fusing the vegetative insecticidal protein Vip3Aa7 and the N terminus of Cry9Ca improves toxicity against Plutella xylostella larvae. *Appl. Microbiol. Biotechnol.* **2012**, *96*, 921–929. [CrossRef]

42. Palma, L.; Scott, D.J.; Harris, G.; Din, S.-U.; Williams, T.L.; Roberts, O.J.; Young, M.T.; Caballero, P.; Berry, C. The Vip3Ag4 insecticidal protoxin from Bacillus thuringiensis adopts a tetrameric configuration that is maintained on proteolysis. *Toxins* **2017**, *9*, 165. [CrossRef]

43. Quan, Y.; Ferré, J. Structural Domains of the Bacillus thuringiensis Vip3Af protein unraveled by tryptic digestion of alanine mutants. *Toxins* **2019**, *11*, 368. [CrossRef]

44. Zheng, M.; Evdokimov, A.G.; Moshiri, F.; Lowder, C.; Haas, J. Crystal structure of a Vip3B family insecticidal protein reveals a new fold and a unique tetrameric assembly. *Protein Sci.* **2020**, *29*, 824–829. [CrossRef]

45. Mushtaq, R.; Shakoori, A.R.; Jurat-Fuentes, J.L. Domain III of Cry1Ac is critical to binding and toxicity against soybean looper (Chrysodeixis includens) but not to velvetbean caterpillar (Anticarsia gemmatalis). *Toxins* **2018**, *10*, 95. [CrossRef]

46. Jiang, K.; Zhang, Y.; Chen, Z.; Wu, D.; Cai, J.; Gao, X. Structural and Functional Insights into the C-terminal Fragment of Insecticidal Vip3A Toxin of Bacillus thuringiensis. *Toxins* **2020**, *12*, 438. [CrossRef]

47. Núñez-Ramírez, R.; Huesa, J.; Bel, Y.; Ferré, J.; Casino, P.; Arias-Palomo, E. Molecular architecture and activation of the insecticidal protein Vip3Aa from Bacillus thuringiensis. *Nat. Commun.* **2020**, *11*, 3974. [CrossRef] [PubMed]

48. Crickmore, N.; Berry, C.; Panneerselvam, S.; Mishra, R.; Connor, T.R.; Bonning, B.C. A structure-based nomenclature for Bacillus thuringiensis and other bacteria-derived pesticidal proteins. *J. Invertebr. Pathol.* **2020**, 107438. [CrossRef] [PubMed]

49. Leuber, M.; Orlik, F.; Schiffler, B.; Sickmann, A.; Benz, R. Vegetative insecticidal protein (Vip1Ac) of Bacillus thuringiensis HD201: Evidence for oligomer and channel formation. *Biochemistry* **2006**, *45*, 283–288. [CrossRef] [PubMed]

50. Geng, J.; Jiang, J.; Shu, C.; Wang, Z.; Song, F.; Geng, L.; Duan, J.; Zhang, J. Bacillus thuringiensis Vip1 Functions as a Receptor of Vip2 Toxin for Binary Insecticidal Activity against Holotrichia parallela. *Toxins* **2019**, *11*, 440. [CrossRef]

51. Abd El-Ghany, N.; Saker, M.; Salama, H.; Ragaie, M. Histopathology of the larval midgut of Helicoverpa armigera (Hübner) fed on Bacillus thuringiensis crystals and Bt-tomato plants. *J. Genet. Eng. Biotechnol.* **2015**, *13*, 221–225. [CrossRef]

52. Chen, W.; Liu, C.; Lu, G.; Cheng, H.; Shen, Z.; Wu, K. Effects of Vip3AcAa+ Cry1Ac cotton on midgut tissue in Helicoverpa armigera (Lepidoptera: Noctuidae). *J. Insect Sci.* **2018**, *18*, 13. [CrossRef]

53. Banyuls, N.; Hernández-Martínez, P.; Quan, Y.; Ferré, J. Artefactual band patterns by SDS-PAGE of the Vip3Af protein in the presence of proteases mask the extremely high stability of this protein. *Int. J. Biol. Macromol.* **2018**, *120*, 59–65. [CrossRef]

54. Hernández-Martínez, P.; Hernández-Rodríguez, C.S.; Van Rie, J.; Escriche, B.; Ferré, J. Insecticidal activity of Vip3Aa, Vip3Ad, Vip3Ae, and Vip3Af from Bacillus thuringiensis against lepidopteran corn pests. *J. Invertebr. Pathol.* **2013**, *113*, 78. [CrossRef]

55. Bel, Y.; Banyuls, N.; Chakroun, M.; Escriche, B.; Ferré, J. Insights into the structure of the Vip3Aa insecticidal protein by protease digestion analysis. *Toxins* **2017**, *9*, 131. [CrossRef]

56. Gomis-Cebolla, J.; de Escudero, I.R.; Vera-Velasco, N.M.; Hernández-Martínez, P.; Hernández-Rodríguez, C.S.; Ceballos, T.; Palma, L.; Escriche, B.; Caballero, P.; Ferré, J. Insecticidal spectrum and mode of action of the Bacillus thuringiensis Vip3Ca insecticidal protein. *J. Invertebr. Pathol.* **2017**, *142*, 60–67. [CrossRef] [PubMed]

57. Kunthic, T.; Watanabe, H.; Kawano, R.; Tanaka, Y.; Promdonkoy, B.; Yao, M.; Boonserm, P. pH regulates pore formation of a protease activated Vip3Aa from Bacillus thuringiensis. *Biochim. Biophys. Acta Biomembr.* **2017**, *1859*, 2234–2241. [CrossRef] [PubMed]

58. Shao, E.; Zhang, A.; Yan, Y.; Wang, Y.; Jia, X.; Sha, L.; Guan, X.; Wang, P.; Huang, Z. Oligomer Formation and Insecticidal Activity of Bacillus thuringiensis Vip3Aa Toxin. *Toxins* **2020**, *12*, 274. [CrossRef] [PubMed]

59. Singh, G.; Sachdev, B.; Sharma, N.; Seth, R.; Bhatnagar, R.K. Interaction of Bacillus thuringiensis vegetative insecticidal protein with ribosomal S2 protein triggers larvicidal activity in Spodoptera frugiperda. *Appl. Environ. Microbiol.* **2010**, *76*, 7202–7209. [CrossRef]

60. Osman, G.H.; Soltane, R.; Saleh, I.; Abulreesh, H.H.; Gazi, K.S.; Arif, I.A.; Ramadan, A.M.; Alameldin, H.F.; Osman, Y.A.; Idriss, M. Isolation, characterization, cloning and bioinformatics analysis of a novel receptor from black cut worm (Agrotis ipsilon) of Bacillus thuringiensis vip 3Aa toxins. *Saudi J. Biol. Sci.* **2019**, *26*, 1078–1083. [CrossRef]

61. Jiang, K.; Hou, X.; Han, L.; Tan, T.; Cao, Z.; Cai, J. Fibroblast growth factor receptor, a novel receptor for vegetative insecticidal protein Vip3Aa. *Toxins* **2018**, *10*, 546. [CrossRef]

62. Jiang, K.; Hou, X.-Y.; Tan, T.-T.; Cao, Z.-L.; Mei, S.-Q.; Yan, B.; Chang, J.; Han, L.; Zhao, D.; Cai, J. Scavenger receptor-C acts as a receptor for Bacillus thuringiensis vegetative insecticidal protein Vip3Aa and mediates the internalization of Vip3Aa via endocytosis. *PLoS Pathog.* **2018**, *14*, e1007347. [CrossRef]

63. Boukedi, H.; Khedher, S.B.; Triki, N.; Kamoun, F.; Saadaoui, I.; Chakroun, M.; Tounsi, S.; Abdelkefi-Mesrati, L. Overproduction of the Bacillus thuringiensis Vip3Aa16 toxin and study of its insecticidal activity against the carob moth Ectomyelois ceratoniae. *J. Invertebr. Pathol.* **2015**, *127*, 127–129. [CrossRef]

64. Nimsanor, S.; Srisaisup, M.; Jammor, P.; Promdonkoy, B.; Boonserm, P. Intracellular localization and cytotoxicity of Bacillus thuringiensis Vip3Aa against Spodoptera frugiperda (Sf9) cells. *J. Invertebr. Pathol.* **2020**, *171*, 107340. [CrossRef]

65. Liu, J.-G.; Yang, A.-Z.; Shen, X.-H.; Hua, B.-G.; Shi, G.-L. Specific binding of activated Vip3Aa10 to Helicoverpa armigera brush border membrane vesicles results in pore formation. *J. Invertebr. Pathol.* **2011**, *108*, 92–97. [CrossRef]

66. Hou, X.; Han, L.; An, B.; Zhang, Y.; Cao, Z.; Zhan, Y.; Cai, X.; Yan, B.; Cai, J. Mitochondria and Lysosomes Participate in Vip3Aa-Induced Spodoptera frugiperda Sf9 Cell Apoptosis. *Toxins* **2020**, *12*, 116. [CrossRef] [PubMed]

67. Bel, Y.; Jakubowska, A.K.; Costa, J.; Herrero, S.; Escriche, B. Comprehensive analysis of gene expression profiles of the beet armyworm Spodoptera exigua larvae challenged with Bacillus thuringiensis Vip3Aa toxin. *PLoS ONE* **2013**, *8*, e81927. [CrossRef] [PubMed]

68. Crava, C.M.; Jakubowska, A.K.; Escriche, B.; Herrero, S.; Bel, Y. Dissimilar regulation of antimicrobial proteins in the midgut of Spodoptera exigua larvae challenged with Bacillus thuringiensis toxins or baculovirus. *PLoS ONE* **2015**, *10*, e0125991. [CrossRef]

69. Hernández-Martínez, P.; Gomis-Cebolla, J.; Ferré, J.; Escriche, B. Changes in gene expression and apoptotic response in Spodoptera exigua larvae exposed to sublethal concentrations of Vip3 insecticidal proteins. *Sci. Rep.* **2017**, *7*, 1–12. [CrossRef]

70. Song, F.; Chen, C.; Wu, S.; Shao, E.; Li, M.; Guan, X.; Huang, Z. Transcriptional profiling analysis of Spodoptera litura larvae challenged with Vip3Aa toxin and possible involvement of trypsin in the toxin activation. *Sci. Rep.* **2016**, *6*, 23861. [CrossRef]

71. Palma, L.; Hernández-Rodríguez, C.S.; Maeztu, M.; Hernández-Martínez, P.; de Escudero, I.R.; Escriche, B.; Muñoz, D.; Van Rie, J.; Ferré, J.; Caballero, P. Vip3C, a novel class of vegetative insecticidal proteins from Bacillus thuringiensis. *Appl. Environ. Microbiol.* **2012**, *78*, 7163–7165. [CrossRef]

72. Yang, J.; Quan, Y.; Sivaprasath, P.; Shabbir, M.Z.; Wang, Z.; Ferré, J.; He, K. Insecticidal activity and synergistic combinations of ten different Bt toxins against Mythimna separata (Walker). *Toxins* **2018**, *10*, 454. [CrossRef]

73. Gomis-Cebolla, J.; Wang, Y.; Quan, Y.; He, K.; Walsh, T.; James, B.; Downes, S.; Kain, W.; Wang, P.; Leonard, K. Analysis of cross-resistance to Vip3 proteins in eight insect colonies, from four insect species, selected for resistance to Bacillus thuringiensis insecticidal proteins. *J. Invertebr. Pathol.* **2018**, *155*, 64–70. [CrossRef]

74. Barkhade, U.P.; Thakare, A. Protease mediated resistance mechanism to Cry1C and Vip3A in Spodoptera litura. *Egypt. Acad. J. Biol. Sci. A Entomol.* **2010**, *3*, 43–50. [CrossRef]

75. Pickett, B.R.; Gulzar, A.; Ferré, J.; Wright, D.J. Bacillus thuringiensis Vip3Aa Toxin Resistance in Heliothis virescens (Lepidoptera: Noctuidae). *Appl. Environ. Microbiol.* **2017**, *83*. [CrossRef]

76. Bernardi, O.; Bernardi, D.; Horikoshi, R.J.; Okuma, D.M.; Miraldo, L.L.; Fatoretto, J.; Medeiros, F.C.; Burd, T.; Omoto, C. Selection and characterization of resistance to the Vip3Aa20 protein from Bacillus thuringiensis in Spodoptera frugiperda. *Pest Manag. Sci.* **2016**, *72*, 1794–1802. [CrossRef] [PubMed]

77. Yang, F.; Morsello, S.; Head, G.P.; Sansone, C.; Huang, F.; Gilreath, R.T.; Kerns, D.L. F2 screen, inheritance and cross-resistance of field-derived Vip3A resistance in Spodoptera frugiperda (Lepidoptera: Noctuidae) collected from Louisiana, USA. *Pest Manag. Sci.* **2018**, *74*, 1769–1778.

78. Mahon, R.J.; Downes, S.J.; James, B. Vip3A resistance alleles exist at high levels in Australian targets before release of cotton expressing this toxin. *PLoS ONE* **2012**, *7*, e39192. [CrossRef] [PubMed]

79. Chakroun, M.; Banyuls, N.; Walsh, T.; Downes, S.; James, B.; Ferré, J. Characterization of the resistance to Vip3Aa in Helicoverpa armigera from Australia and the role of midgut processing and receptor binding. *Sci. Rep.* **2016**, *6*, 1–11.

80. Pinos, D.; Chakroun, M.; Millán-Leiva, A.; Jurat-Fuentes, J.L.; Wright, D.J.; Hernández-Martínez, P.; Ferré, J. Reduced Membrane-Bound Alkaline Phosphatase Does Not Affect Binding of Vip3Aa in a Heliothis virescens Resistant Colony. *Toxins* **2020**, *12*, 409. [CrossRef] [PubMed]

81. Hernández-Rodríguez, C.; Boets, A.; Van Rie, J.; Ferré, J. Screening and identification of vip genes in Bacillus thuringiensis strains. *J. Appl. Microbiol.* **2009**, *107*, 219–225. [CrossRef] [PubMed]

82. Estruch, J.J.; Carozzi, N.B.; Desai, N.; Duck, N.B.; Warren, G.W.; Koziel, M.G. Transgenic plants: An emerging approach to pest control. *Nat. Biotechnol.* **1997**, *15*, 137–141. [CrossRef]

83. Asokan, R. Isolation and characterization of Bacillus thuringiensis Berliner from soil, leaf, seed dust and insect cadaver. *J. Biol. Control* **2007**, *21*, 83–90.

84. Bourque, S.; Valero, J.; Mercier, J.; Lavoie, M.; Levesque, R. Multiplex polymerase chain reaction for detection and differentiation of the microbial insecticide Bacillus thuringiensis. *Appl. Environ. Microbiol.* **1993**, *59*, 523–527. [CrossRef]

85. Neethu, K.; Priji, P.; Unni, K.; Sajith, S.; Sreedevi, S.; Ramani, N.; Anitha, K.; Rosana, B.; Girish, M.; Benjamin, S. New Bacillus thuringiensis strain isolated from the gut of Malabari goat is effective against Tetranychus macfarlanei. *J. Appl. Entomol.* **2016**, *140*, 187–198. [CrossRef]

86. Şahin, B.; Gomis-Cebolla, J.; Güneş, H.; Ferre, J. Characterization of Bacillus thuringiensis isolates by their insecticidal activity and their production of Cry and Vip3 proteins. *PLoS ONE* **2018**, *13*, e0206813. [CrossRef]

87. Güney, E.; Adıgüzel, A.; Demirbağ, Z.; Sezen, K. Bacillus thuringiensis kurstaki strains produce vegetative insecticidal proteins (Vip 3) with high potential. *Egypt. J. Biol. Pest Control* **2019**, *29*, 81. [CrossRef]

88. Lone, S.A.; Malik, A.; Padaria, J.C. Molecular cloning and characterization of a novel vip3-type gene from Bacillus thuringiensis and evaluation of its toxicity against Helicoverpa armigera. *Microb. Pathog.* **2018**, *114*, 464–469. [CrossRef] [PubMed]

89. Liu, Y.; Lai, Q.; Göker, M.; Meier-Kolthoff, J.P.; Wang, M.; Sun, Y.; Wang, L.; Shao, Z. Genomic insights into the taxonomic status of the Bacillus cereus group. *Sci. Rep.* **2015**, *5*, 1–11. [CrossRef] [PubMed]

90. Rabha, M.; Acharjee, S.; Sarmah, B.K. Multilocus sequence typing for phylogenetic view and vip gene diversity of Bacillus thuringiensis strains of the Assam soil of North East India. *World J. Microbiol. Biotechnol.* **2018**, *34*, 103. [CrossRef] [PubMed]

91. Yu, X.; Zheng, A.; Zhu, J.; Wang, S.; Wang, L.; Deng, Q.; Li, S.; Liu, H.; Li, P. Characterization of vegetative insecticidal protein vip genes of Bacillus thuringiensis from Sichuan Basin in China. *Curr. Microbiol.* **2011**, *62*, 752–757. [CrossRef]

92. Kaviyapriya, M.; Lone, R.; Balakrishnan, N.; Rajesh, S.; Ramalakshmi, A. Cloning and characterization of insecticidal cry/vip genes from an indigenous Bacillus thuringiensis isolate T29 and evaluation of its toxicity to maize fall armyworm Spodoptera frugiperda. *J. Entomol. Zool. Stud.* **2019**, *7*, 1314–1321.

93. Chakroun, M.; Banyuls, N.; Bel, Y.; Escriche, B.; Ferré, J. Bacterial vegetative insecticidal proteins (Vip) from entomopathogenic bacteria. *Microbiol. Mol. Biol. Rev.* **2016**, *80*, 329–350. [CrossRef] [PubMed]

94. Riaz, S.; Nasir, I.A.; Bhatti, M.U.; Adeyinka, O.S.; Toufiq, N.; Yousaf, I.; Tabassum, B. Resistance to Chilo infuscatellus (Lepidoptera: Pyraloidea) in transgenic lines of sugarcane expressing Bacillus thuringiensis derived Vip3A protein. *Mol. Biol. Rep.* **2020**, *47*, 2649–2658. [CrossRef]

95. Bett, B.; Gollasch, S.; Moore, A.; James, W.; Armstrong, J.; Walsh, T.; Harding, R.; Higgins, T.J. Transgenic cowpeas (Vigna unguiculata L. Walp) expressing Bacillus thuringiensis Vip3Ba protein are protected against the Maruca pod borer (Maruca vitrata). *Plant Cell Tissue Organ Cult.* **2017**, *131*, 335–345. [CrossRef]

96. Yan, X.; Lu, J.; Ren, M.; He, Y.; Wang, Y.; Wang, Z.; He, K. Insecticidal Activity of 11 Bt toxins and 3 Transgenic Maize Events Expressing Vip3Aa19 to Black Cutworm, Agrotis ipsilon (Hufnagel). *Insects* **2020**, *11*, 208. [CrossRef]

97. Gayen, S.; Samanta, M.K.; Hossain, M.A.; Mandal, C.C.; Sen, S.K. A deletion mutant ndv200 of the Bacillus thuringiensisvip3BR insecticidal toxin gene is a prospective candidate for the next generation of genetically modified crop plants resistant to lepidopteran insect damage. *Planta* **2015**, *242*, 269–281. [CrossRef]

98. Naimov, S.; Weemen-Hendriks, M.; Dukiandjiev, S.; de Maagd, R.A. Bacillus thuringiensis delta-endotoxin Cry1 hybrid proteins with increased activity against the Colorado potato beetle. *Appl. Environ. Microbiol.* **2001**, *67*, 5328–5330. [CrossRef] [PubMed]

99. Fang, J.; Xu, X.; Wang, P.; Zhao, J.-Z.; Shelton, A.M.; Cheng, J.; Feng, M.-G.; Shen, Z. Characterization of chimeric Bacillus thuringiensis Vip3 toxins. *Appl. Environ. Microbiol.* **2007**, *73*, 956–961. [CrossRef] [PubMed]

100. Gomis-Cebolla, J.; Ferreira dos Santos, R.; Wang, Y.; Caballero, J.; Caballero, P.; He, K.; Jurat-Fuentes, J.L.; Ferré, J. Domain Shuffling between Vip3Aa and Vip3Ca: Chimera Stability and Insecticidal Activity against European, American, African, and Asian Pests. *Toxins* **2020**, *12*, 99. [CrossRef]

101. Saraswathy, N.; Nain, V.; Sushmita, K.; Kumar, P.A. A fusion gene encoding two different insecticidal proteins of Bacillus thuringiensis. *Int. J. Bus. Technopreneurship* **2008**, *2*, 204–209.

102. Song, R.; Peng, D.; Yu, Z.; Sun, M. Carboxy-terminal half of Cry1C can help vegetative insecticidal protein to form inclusion bodies in the mother cell of Bacillus thuringiensis. *Appl. Microbiol. Biotechnol.* **2008**, *80*, 647–654. [CrossRef]

103. Ahmad, A.; Javed, M.R.; Rao, A.Q.; Khan, M.A.; Ahad, A.; Shahid, A.A.; Husnain, T. In-silico determination of insecticidal potential of Vip3Aa-Cry1Ac fusion protein against Lepidopteran targets using molecular docking. *Front. Plant Sci.* **2015**, *6*, 1081. [CrossRef]

104. Liu, M.; Liu, R.; Luo, G.; Li, H.; Gao, J. Effects of site-mutations within the 22 kDa no-core fragment of the Vip3Aa11 insecticidal toxin of bacillus thuringiensis. *Curr. Microbiol.* **2017**, *74*, 655–659. [CrossRef]

105. Kahn, T.W.; Chakroun, M.; Williams, J.; Walsh, T.; James, B.; Monserrate, J.; Ferré, J. Efficacy and resistance management potential of a modified Vip3C protein for control of Spodoptera frugiperda in maize. *Sci. Rep.* **2018**, *8*, 1–11. [CrossRef]

106. Yu, X.; Liu, T.; Sun, Z.; Guan, P.; Zhu, J.; Wang, S.; Li, S.; Deng, Q.; Wang, L.; Zheng, A. Co-expression and synergism analysis of Vip3Aa29 and Cyt2Aa3 insecticidal proteins from Bacillus thuringiensis. *Curr. Microbiol.* **2012**, *64*, 326–331. [CrossRef]

107. Bergamasco, V.; Mendes, D.; Fernandes, O.; Desidério, J.; Lemos, M. Bacillus thuringiensis Cry1Ia10 and Vip3Aa protein interactions and their toxicity in Spodoptera spp.(Lepidoptera). *J. Invertebr. Pathol.* **2013**, *112*, 152–158. [CrossRef]

108. Wang, Z.; Fang, L.; Zhou, Z.; Pacheco, S.; Gómez, I.; Song, F.; Soberón, M.; Zhang, J.; Bravo, A. Specific binding between Bacillus thuringiensis Cry9Aa and Vip3Aa toxins synergizes their toxicity against Asiatic rice borer (Chilo suppressalis). *J. Biol. Chem.* **2018**, *293*, 11447–11458. [CrossRef] [PubMed]

109. Figueiredo, C.S.; Lemes, A.R.N.; Sebastião, I.; Desidério, J.A. Synergism of the Bacillus thuringiensis Cry1, Cry2, and Vip3 proteins in Spodoptera frugiperda control. *Appl. Biochem. Biotechnol.* **2019**, *188*, 798–809. [CrossRef] [PubMed]

110. Guo, S.; Ye, S.; Liu, Y.; Wei, L.; Xue, J.; Wu, H.; Song, F.; Zhang, J.; Wu, X.; Huang, D. Crystal structure of Bacillus thuringiensis Cry8Ea1: An insecticidal toxin toxic to underground pests, the larvae of Holotrichia parallela. *J. Struct. Biol.* **2009**, *168*, 259–266. [CrossRef] [PubMed]

111. Lemes, A.R.N.; Davolos, C.C.; Legori, P.C.B.C.; Fernandes, O.A.; Ferre, J.; Lemos, M.V.F.; Desiderio, J.A. Synergism and antagonism between Bacillus thuringiensis Vip3A and Cry1 proteins in Heliothis virescens, Diatraea saccharalis and Spodoptera frugiperda. *PLoS ONE* **2014**, *9*, e107196. [CrossRef] [PubMed]

112. Carrière, Y.; Crickmore, N.; Tabashnik, B.E. Optimizing pyramided transgenic Bt crops for sustainable pest management. *Nat. Biotechnol.* **2015**, *33*, 161–168. [CrossRef]

113. Xu, C.; Cheng, J.; Lin, H.; Lin, C.; Gao, J.; Shen, Z. Characterization of transgenic rice expressing fusion protein Cry1Ab/Vip3A for insect resistance. *Sci. Rep.* **2018**, *8*, 1–8. [CrossRef]

114. Lemes, A.R.N.; Figueiredo, C.S.; Sebastião, I.; da Silva, L.M.; da Costa Alves, R.; de Siqueira, H.Á.A.; Lemos, M.V.F.; Fernandes, O.A.; Desidério, J.A. Cry1Ac and Vip3Aa proteins from Bacillus thuringiensis targeting Cry toxin resistance in Diatraea flavipennella and Elasmopalpus lignosellus from sugarcane. *PeerJ* **2017**, *5*, e2866. [CrossRef]

115. Mamta, B.; Rajam, M. RNAi technology: A new platform for crop pest control. *Physiol. Mol. Biol. Plants* **2017**, *23*, 487–501. [CrossRef]

116. Baum, J.A.; Bogaert, T.; Clinton, W.; Heck, G.R.; Feldmann, P.; Ilagan, O.; Johnson, S.; Plaetinck, G.; Munyikwa, T.; Pleau, M. Control of coleopteran insect pests through RNA interference. *Nat. Biotechnol.* **2007**, *25*, 1322–1326. [CrossRef]

117. Head, G.P.; Carroll, M.W.; Evans, S.P.; Rule, D.M.; Willse, A.R.; Clark, T.L.; Storer, N.P.; Flannagan, R.D.; Samuel, L.W.; Meinke, L.J. Evaluation of SmartStax and SmartStax PRO maize against western corn rootworm and northern corn rootworm: Efficacy and resistance management. *Pest Manag. Sci.* **2017**, *73*, 1883–1899. [CrossRef]

118. Ni, M.; Ma, W.; Wang, X.; Gao, M.; Dai, Y.; Wei, X.; Zhang, L.; Peng, Y.; Chen, S.; Ding, L. Next-generation transgenic cotton: Pyramiding RNAi and Bt counters insect resistance. *Plant Biotechnol. J.* **2017**, *15*, 1204–1213. [CrossRef] [PubMed]

119. Zhang, J.; Khan, S.A.; Hasse, C.; Ruf, S.; Heckel, D.G.; Bock, R. Full crop protection from an insect pest by expression of long double-stranded RNAs in plastids. *Science* **2015**, *347*, 991–994. [CrossRef] [PubMed]

120. International Service for the Acquisition of Agri-biotech Applications (ISAAA), GM Approval Database. Available online: http://www.isaaa.org/gmapprovaldatabase (accessed on 27 June 2020).

121. Current and Previously Registered Section 3 Plant-Incorporated Protectant (PIP) Registrations. Available online: https://www.epa.gov/ingredients-used-pesticide-products/current-and-previously-registered-section-3-plant-incorporated (accessed on 27 June 2020).

Correction

Correction: Syed, T., et al. Current Insights on Vegetative Insecticidal Proteins (Vip) as Next Generation Pest Killers. *Toxins* 2020, *12*, 522

Tahira Syed [†], Muhammad Askari [†], Zhigang Meng, Yanyan Li, Muhammad Ali Abid, Yunxiao Wei, Sandui Guo, Chengzhen Liang *, and Rui Zhang *

Biotechnology Research Institute, Chinese Academy of Agricultural Sciences, Beijing 100081, China; syedtahira98@gmail.com (T.S.); 2017Y90100082@caas.cn (M.A.); mengzhigang@caas.cn (Z.M.); liyanyan01@caas.cn (Y.L.); abid@caas.cn (M.A.A.); weiyunxiao@caas.cn (Y.W.); guosandui@caas.cn (S.G.)
* Correspondence: liangchengzhen@caas.cn (C.L.); zhangrui@caas.cn (R.Z.); Tel.: +86-10-82106127 (R.Z.)
† These authors contributed equally.

Citation: Syed, T.; Askari, M.; Meng, Z.; Li, Y.; Abid, M.A.; Wei, Y.; Guo, S.; Liang, C.; Zhang, R. Correction: Syed, T., et al. Current Insights on Vegetative Insecticidal Proteins (Vip) as Next Generation Pest Killers. *Toxins* 2020, *12*, 522. *Toxins* **2021**, *13*, 200. https://doi.org/10.3390/toxins13030200

Received: 23 February 2021
Accepted: 2 March 2021
Published: 11 March 2021

The authors wish to make the following corrections to their paper [1]:

In the abstract and introduction section, the words "gram negative" should be replaced with "Gram positive" in first sentence.

In Section 6, the first line of the second paragraph, the second word should be replaced with "laboratory evolved resistance". In the third sentence of same paragraph, reference 72 in the original paper should be replaced with reference [2] in this correction.

In the third paragraph of this section, sentence five should be replaced with "Instead, a low proteolytic activity was found in resistant insects". The last reference in this paragraph (reference 78 in the original paper) should be replaced by reference [3] in this correction.

In the fourth paragraph of this section, the first line should be replaced with "no cross resistance to Vip3C has been observed in insects of different species, previously found to be resistant to Cry1A, Cry2Ab, Dipel (Mixture of Cry1 and Cry2)". The third sentence of this paragraph should be replaced with "The biochemical basis of resistance could not be established by the down regulation of membrane bound alkaline phosphatase (mALP) isoform HvmALP1, observed in Vip3 resistant insects, and results does not support it to be the functional receptor of Vip3".

The reference in the second paragraph of Section 9.1 and the seventh row of Table 2 (reference 98 in the original paper) should be replaced by reference [4] in this correction.

We apologize for any inconvenience caused to readers of *Toxins* by this change. The manuscript will be updated and the original will remain online on the article webpage. We have also rearranged all references and citations according to the correct order. We apologize for any inconvenience caused to our readers.

References

1. Syed, T.; Askari, M.; Meng, Z.; Li, Y.; Abid, M.A.; Wei, Y.; Guo, S.; Liang, C.; Zhang, R. Current Insights on Vegetative Insecticidal Proteins (Vip) as Next Generation Pest Killers. *Toxins* **2020**, *12*, 522. [CrossRef] [PubMed]
2. Yang, F.; Morsello, S.; Head, G.P.; Sansone, C.; Huang, F.; Gilreath, R.T.; Kerns, D.L. F2 screen, inheritance and cross-resistance of field-derived Vip3A resistance in *Spodoptera frugiperda* (Lepidoptera: Noctuidae) collected from Louisiana, USA. *Pest Manag. Sci.* **2018**, *74*, 1769–1778. [CrossRef] [PubMed]
3. Chakroun, M.; Banyuls, N.; Walsh, T.; Downes, S.; James, B.; Ferré, J. Characterization of the resistance to Vip3Aa in *Helicoverpa armigera* from Australia and the role of midgut processing and receptor binding. *Sci. Rep.* **2016**, *6*, 1–11. [CrossRef] [PubMed]
4. Gomis-Cebolla, J.; Ferreira dos Santos, R.; Wang, Y.; Caballero, J.; Caballero, P.; He, K.; Jurat-Fuentes, J.L.; Ferré, J. Domain Shuffling between Vip3Aa and Vip3Ca: Chimera Stability and Insecticidal Activity against European, American, African, and Asian Pests. *Toxins* **2020**, *12*, 99. [CrossRef] [PubMed]

Review

Making 3D-Cry Toxin Mutants: Much More Than a Tool of Understanding Toxins Mechanism of Action

Susana Vílchez

Institute of Biotechnology, Department of Biochemistry and Molecular Biology I, Faculty of Science, University of Granada, 18071 Granada, Spain; svt@ugr.es; Tel.: +34-958-240071

Received: 15 July 2020; Accepted: 20 August 2020; Published: 16 September 2020

Abstract: 3D-Cry toxins, produced by the entomopathogenic bacterium *Bacillus thuringiensis*, have been extensively mutated in order to elucidate their elegant and complex mechanism of action necessary to kill susceptible insects. Together with the study of the resistant insects, 3D-Cry toxin mutants represent one of the pillars to understanding how these toxins exert their activity on their host. The principle is simple, if an amino acid is involved and essential in the mechanism of action, when substituted, the activity of the toxin will be diminished. However, some of the constructed 3D-Cry toxin mutants have shown an enhanced activity against their target insects compared to the parental toxins, suggesting that it is possible to produce novel versions of the natural toxins with an improved performance in the laboratory. In this report, all mutants with an enhanced activity obtained by accident in mutagenesis studies, together with all the variants obtained by rational design or by directed mutagenesis, were compiled. A description of the improved mutants was made considering their historical context and the parallel development of the protein engineering techniques that have been used to obtain them. This report demonstrates that artificial 3D-Cry toxins made in laboratories are a real alternative to natural toxins.

Keywords: 3D-Cry toxins; in vitro evolution; rational design; *Bacillus thuringiensis*; toxin enhancement

Key Contribution: Compilation of all of 3D-Cry toxin mutants with enhanced activity made with different molecular techniques.

---

1. Introduction

Sporulating *Bacillus thurigiensis* produces four non-phylogenetically related insecticidal protein families, the three domain Cry toxins or 3D-Cry toxins, the mosquitocidal Mtx, the binary-like (Bin), and Cyt toxins. All of these proteins form crystals (with the exception of Cry1Ia toxin [1]) concomitantly with the sporulation process. Since the discovery of these crystals in 1953 by Hannay [2], and the demonstration one year later [3] that they were responsible for the already-described entomopathogenic activity of *B. thuringiensis* [4], the study of these toxins has not stopped.

All known Cry toxins (3D-Cry, Mtx like, Bin and Cyt toxins) have been compiled in a brand new database [5], maintained by a commission of experts in charge of assigning a name when a novel Cry protein is described, using the recently proposed structure-based nomenclature rules [6] but with the same basic principles of the rules that were established in 1998 [7]. Most of the 3D-Cry toxins are active against insects from different orders, mainly Lepidoptera, Diptera, Coleoptera, Hemiptera (low toxicity for some aphids), and Hymenoptera, but some of them have other targets such as nematodes, snails, and even cancer cells [8]. Recently, a toxin active against Orthopteran insects has been described [9].

3D-Cry toxins are the best-characterized among the Cry proteins. Among them, Lepidoptera-active toxins are the best known from a mechanistical point of view. 3D-Cry toxins, synthesized as inactive protoxins by *B. thuringiensis*, have to undergo a proteolytic activation process in the guts of the susceptible insects to become active. Since the elucidation of the first three-dimensional structure of the active part of the Cry3A toxin by Li et al. in 1991 [10], the structure of nine other members of the family have been elucidated (Cry1Aa [11], Cry3Bb1 [12], Cry1Ac [13], Cry2Aa [14], Cry4Ba [15], Cry4Aa [16], Cry8Ea1 [17], Cry5B [18], and Cry7Ca1 [9]). Recently, the 3D structure of the 120 KDa Cry1Ac1 prototoxin has also been described [19], showing seven different domains (DI–DVII). Although very different at the amino acid sequence level, the structural disposition of 3D-toxins is very conserved. Active toxins present three very distinct structural domains (hence their name of 3D-toxins), each of them with a specific function. Domain I, at the N terminus end of the protein, is comprised of a bundle of seven α-helices and is responsible for the formation of a pore in the midgut cells of susceptible insects. Domain II, the middle domain, formed by three antiparallel β-sheets, plays an important role in receptor recognition. Domain III, a two antiparallel β-sheet sandwich, is thought to be involved in receptor binding and pore formation [20]. Protoxin Domains IV and VI are α-bundles similar to domains present in other proteins such as spectrin or the bacterial fibrinogen-binding complement inhibitor. On the other hand, Domains V and VII are β-rolls resembling carbohydrate-binding proteins such as sugar hydrolases [19]. Although the functions of DIV–DVII are not known, they have been related with crystal formation, toxin stability, and selective solubilization in the insect gut. In addition, it has recently been suggested that Domains V and VII could also be interacting domains with proteins present in gut membranes, and hence be involved in the recognition of receptors of the full toxin [21].

3D-Cry toxins have been exploited in insect pest management since the late 1930s [22] in agriculture and against health-related insect populations. 3D-Cry toxins have been used in agriculture, not only as spray formulations, but also in plant transgenesis to protect plants such as maize, cotton, soybean, potato, and tomato from insects [23]. 3D-Cry toxins are extremely efficient and their main characteristic is the narrow spectrum of action that each toxin shows. Their specificity is due to a complex mechanism of action, and although it is not completely understood and many questions remain unanswered, it is known to involve several steps. Currently, two models to explain the mechanism of 3D-Cry toxins have been proposed: the sequential binding model and the signaling pathway model. The sequential binding model involves crystal solubilization at a specific pH, proteolytic activation by gut digestive enzymes, receptor recognition at the membrane cells of susceptible insects, helix 1 proteolysis, conformational changes of the molecule, polymerization, and finally, membrane insertion. The signaling pathway model shares the steps of crystal solubilization, proteolytic activation, and receptor binding, but death of the cell is not explained by toxin insertion in the membrane but by the activation of cell apoptosis. Both models [22] have in common the need of 3D-Cry toxins to bind to a specific receptor at the enterocytes and possibly many other interacting receptors are required for the toxin mode of action. Among them, aminopeptidase N (APN), cadherin-like proteins (BT-R1, BtR175, HevCaLP), alkaline phosphatase (ALP), and ABC transporters have been described, with the first two receptors being the best characterized.

Everything that we know today about the mechanism of action of 3D-toxins has been mainly obtained from two sources of information: the construction of 3D-Cry mutants and the study of the resistance phenomena that insects have shown to the action of 3D-Cry toxins [24,25]. The study of mutant proteins is one of the pillars for the elucidation of any mechanism of action of any protein. The principle is simple: if one amino acid of the protein is essential for its mechanism (or the structure), when changed for another amino acid, the functionality of the protein is modified. This is a consequence of the biological principle that the structure and function of any protein are always linked. In the case of 3D-Cry toxins, thousands of mutants have been constructed to elucidate which amino acids are important for maintaining the three-dimensional structure of 3D-toxins, which are responsible for the binding to their receptors, and which are relevant for the solubilization or activation of the toxins. The information provided by the behavior of these mutants has allowed researchers to propose models that explain the mechanism of action of these natural machines, specialized in killing insects.

However, constructing mutant proteins also presents the possibility of obtaining functional variant proteins with different behaviors, even with improved activity. By manipulating the DNA sequence that codifies the 3D-toxins, versions of toxins completely novel in nature with an enhanced activity toward a particular insect, with a broader insect target, or with a novel activity against a non-susceptible insect can be obtained. The more we understand 3D-Cry toxins, the more creative we can be in the generation of artificial toxins. The fact that a high number of molecular techniques for DNA manipulation is available ensures that in the generation of 3D-Cry toxins variants, only our imagination is the limit.

The techniques available for DNA manipulation and generation of novel mutants or variants can be separated in two main groups: (i) those where mutations are generated randomly and afterward, there is a screening process and the desired mutant is selected, and (ii) those where there is a rational design behind the mutant construction. A deep understanding of protein structure and function is needed in order to use rational design, as we must decide which amino acid (or amino acids) is (or are) going to be changed and which amino acid is going to be substituted for. In contrast, random mutation can be generated anywhere in the protein sequence without previous knowledge, and selection of the suitable variant is carried out later on. The latest group of techniques is known as directed evolution or in vitro evolution of proteins in protein engineering. It is a process that simulates natural evolution, introducing a mutation and selecting it if it represents an advantage, but carried out in a laboratory context and with human intervention. For this review, I would like to use the term "in vitro evolution" in a much broader sense. I would like to include those mutants obtained by rational design and those obtained when investigating the function of the toxin, but instead of getting an impaired mutant with no function, a better 3D-Cry toxin is generated. If the reader grants me this license, then the purpose of this review makes much more sense as the intention is to compile all of the mutants generated in 3D-Cry toxins, independently of the objective of the work and the technique used. These man-made (and woman-made) mutants presented here collectively represent the "in vitro evolution" that 3D-Cry toxins have experienced thanks to the work of hundreds of researchers worldwide in a relatively short period of time. Although it must sound highly pretentious comparing human work that of Mother Nature, the fact is that some of the mutants obtained in a laboratory sometimes show better characteristics than natural toxins, at least from a practical point of view.

Previous reports have reviewed the enhancement of 3D-Cry toxin activity [26,27], but this time I would like to give a historical perspective of what the methodological context of protein engineering was like when the enhanced toxins were obtained. Therefore, apart from the objective of updating the information to the present time, this review has the purpose of describing the parallel development of molecular techniques that were used for constructing the improved versions of 3D-Cry toxins. The mutants reviewed here represent all successful variations of the 3D-Cry toxins that have been experienced in a laboratory, collectively, through many different techniques, even if the intention of the researchers was not to obtain an improved mutant. In this report are compiled all the enhanced mutants constructed through the history of 3D-Cry toxin research (Table 1; shown at the end of section two), together with all the relevant positions in the toxin molecules. The sequences of all these mutants are detailed in Table S1. I believe that this is a valuable source of information that I hope will contribute to the production of even better molecules in the future.

2. "In Vitro Evolution" of 3D-Cry Toxins: An Historical Perspective

Although Cry toxins are very efficient molecules and only very minute quantities are required for toxicity, the obsession to improve their efficiency has led to the development of diverse strategies [28]. The developed strategies include (i) the combination of several Cry toxins to increase efficiency toward a specific target [29]; (ii) co-expression with other *B. thuringiensis* proteins such as the P20 protein to provide protection in the larval gut environment [29,30], or chitinases for peritrophic membrane degradation [31]; (iii) combination with chemical compounds such as calcofluor for peritrophic membrane digestion [32], or coating with $Mg(OH)_2$ to increase their resistance to UV light [33]; (iv) combination with other insecticidal toxins such as Cyt toxins [34–37], VIP toxins [38], Bin toxins [39], Metalloproteinase Bmp1 [40], or insect-specific scorpion toxins [41] that synergize their effect; (v) combination with insect chaperones such as Hsp90 chaperone [42]; or (vi) expression of Cry toxins in other backgrounds such as baculovirus [43]. Other strategies developed to increase potency include the fusion of 3D-Cry to other toxins such as neurotoxins (Huwentoxin-I [44], ω-ACTX-Hv1a [45], or huwentoxin XI [46]) present in spiders venom, VIP proteins [47], the N-terminal region of PirB toxin from *Photorhabdus luminescens* [48], or the fusion to other proteins that provide interesting domains such as garlic lectin [49,50], cellulase-binding peptides [51], or *Escherichia coli* maltose binding protein (MBP) [52], thus rendering chimeras with improved activity. Lately, the combination or co-expression of Cry toxins with peptides with sequences similar to natural receptors, or other proteins present in the gut cells, is attracting the attention of researchers. Among them, peptides such as HcAPN3E, derived from an APN receptor in *Hyphantria cunea* [53], or cadhering fragments [54–59] can be mentioned.

Some of the mechanisms for enhancing 3D-Cry toxin toxicity above-mentioned are common strategies in the field of protein engineering (like the fusion of 3D-Cry toxin with other proteins), but are out of the scope of this review as the intention here is to describe specific changes carried out in the amino acid sequence of 3D-Cry toxins that are responsible for the improvement of toxicity.

The starting point in the history of the 3D-Cry toxin "in vitro evolution" could be set at the beginning of the 1980s. At that time, researchers started to understand that the differences in activity of the different *B. thuringiensis* isolated strains were due to the expression of different Cry toxin variants. In 1981, the first *cry* gene was cloned and expressed in a heterologous system [60], a milestone that represented the beginning of molecular biology for Cry toxins. A few years later, the sequence of the first *cry* gene and its deduced amino acid sequence were determined [61]. When the sequence of several other 3D-Cry toxins was available, researchers realized that areas with conserved and variable sequences were present in all of the Cry toxins, so conserved and variable blocks were established. Pretty soon, "variable regions" were related to the specificity observed in 3D-Cry toxins and represented a good starting point for the manipulation of the molecules in order to increase the activity toward insects or expand their target insects. In this sense, the patent number EP0228838 was filed in 1986 by the Mycogen Corporation at the European Patent Office [62] to commercially protect the idea that activity of 3D-Cry toxins could be modified and improved by exchanging specific "variable" regions at their sequence, and a novel method with which to do it. Since then, the history of Cry toxin "in vitro evolution" has not stopped and continues until the present day. Cry toxin history is long and exciting, and has been possible thanks to the development of the molecular tools for DNA manipulation. A description of all the molecular techniques used for the improvement of 3D-Cry toxins will also be made.

2.1. Evolution by Chemical Mutagenesis and Homologue Scanning Mutagenesis, the First Molecular Techniques Used for Cry-Toxins "In Vitro Evolution"

In the 80s, tools for molecular biology were extremely limited. To provide the reader with some background context, restriction enzymes, essential for DNA manipulation nowadays, had only just been recently discovered [63] and the number available was very low. The Sanger method for determining DNA nucleotide sequences had just been developed [64,65] and chemical mutagenesis was pretty much the only tool available for the "in vitro evolution" of Cry toxins. The technique consisted in subjecting *cry* genes to the action of mutagenic substances such as bisulfite or formic acid to obtain random mutations. Bisulfite, a single DNA strand mutagen, converts cytosines into uracils by deamination [66], rendering a transition from cytosine to thiamine or guanine to adenine, depending on the mutagenized strand (sense or antisense strand). Formic acid depurinates DNA by hydrolyzing the N-glycosyl bond between the ribose and purines [67], and when polymerization takes place using this mutated DNA as a template, a transversion is produced.

Chemical mutagenesis has been used to improve the activity of 3D-Cry toxins [68], specifically on CryIA(b) (the original names of the Cry toxins are maintained in this review). After cloning the *cry1A(b)* gene in the M13 phage, in order to obtain single stranded DNA, and setting mutagenesis conditions to obtain 2–3 mutations per gene (by limiting the exposure time of the DNA to the mutagen), the authors obtained eleven mutant toxins that were 3–5 times more toxic toward *Heliothis virescens* than the parental toxin. When the DNA sequence of these clones was determined, a wide range of substitutions was observed, all of them at Domain I of the toxin, although the distribution in the molecule was not known at that time, as none of the 3D-Cry toxin three-dimensional structure had been elucidated yet.

Another option that researchers had available at that time was the technique called homologue-scanning mutagenesis. The technique consisted in exchanging equivalent regions of two 3D-Cry toxins in order to create hybrid molecules. The fragments exchanged were limited by the restriction enzymes that these *cry* genes had in common. Without a doubt, these kinds of exchanges would have been done by PCR nowadays, but this technique was not developed until 1988 [69], and it took time until the research community understood its potential in protein engineering. The main objective of using homologue-scanning mutagenesis was to understand the remarkable specificity that Cry toxins had against the same insects (at the beginning most of them Lepidoptera), but soon researchers understood the potential that this knowledge also had from a practical point of view; identifying regions responsible for the specificity was of paramount importance as they could be manipulated and modified to extend or change the specificity toward other insects.

One of the firsts reports using homolog-scanning mutagenesis in a Cry toxin was published in 1989 [70]. Here, the authors used two *cry* genes, called at the time *icpA1* and *icpC73*, as the first systematic nomenclature for Cry toxins had not been implemented [71] (in fact, the first proposed nomenclature was published the same month as this report). The authors exchanged several fragments between ICPA1, which is highly toxic toward *Bombyx mori*, and the non-toxic ICPC73 by making use of the restriction sites that these two *cry* genes have in common. After the resulting novel toxins were bioassayed, the authors observed that when conserved blocks were exchanged, the activity of the hybrid toxins did not change. However, when variable regions were exchanged, the activity of the toxins dramatically changed and could be redirected toward other insects. In other words, ICPC73 became toxic toward *B. mori* when certain regions from ICPA1 were substituted in its sequence. The region responsible for *B. mori* specificity in the ICPA1 toxin was even narrowed down to the region between residues 332 and 450, a region that we know today includes loop 1, loop 2, and loop 3 in most of the 3D-Cry toxins and has been proven to be involved in receptor recognition and specificity. These authors suggested that if these regions were indeed responsible for specificity in other "IPC" toxins, that they would be an excellent area for mutation in order to redirect activity toward other insects.

The same strategy and the same toxins (now called CryIA(a) and CryIA(c)) were tested in other insects by Schnepf et al. [72]. Both toxins showed a similar activity against *Manduca sexta*, but very different activities against *H. virescens* (as CryIA(c) is 50 times more potent than CryIA(a) for this insect). With these models, the specificity determinant regions for Lepidoptera were determined and it was demonstrated that Cry toxins could be "in vitro evolved" and their specificity could be changed completely. The same year, the specificity-determining region for a Dipteran toxin was determined [73] using homologue-scanning mutagenesis with the CryIIA toxin (active against Diptera and Lepidoptera) and CryIIB (active against Lepidoptera only). They determined that when a 241 nt segment from *cryIIA* was inserted on the *cryIIB* gene, the lepidopteran toxin showed a broader insect spectrum, also becoming active against Diptera. They even narrowed down the region responsible for the specificity of CryIIA protein toward mosquitoes to 76 amino acids.

A following work [74], done by the pioneers in the use of homologue-scanning mutagenesis in a 3D-Cry toxin, demonstrated the same effect in the CryIA(c) toxin and two other economically important pests (*H. virescens* and *Trichoplusia ni*). They defined the minimal region responsible for the toxicity of CryIA(c) as the region between amino acid 332 and 450, an equivalent region described in CryIA(a). Surprisingly, one of the hybrids obtained (hybrid 4109) showed an enhanced activity compared to the parental toxins, being 30 times more toxic than the most potent natural toxin known for *H. virescens*. This represented the first proof that by changing specific areas in the sequence, not only was it a means of modifying the specificity of the 3D-Cry toxins, but also a way of increasing their activity.

At this point in 3D-Cry toxin history, researchers started to have the overall view that 3D-Cry toxins were modular structures and that their function could easily be manipulated by exchanging parts of the molecules.

2.2. Evolution by Domain Swapping

With the elucidation of the first three-dimensional structure of a Cry toxin (Cry3A; PDB: 1DLC) by Li et al. [10], researchers had the opportunity to "see" the spatial distribution of amino acids in a 3D-Cry molecule and to verify that active 3D-Cry toxins showed three very well defined domains (hence their name). Through comparison with other proteins, hypothetical functions for some domains were described. For example, Domain I, with seven long α-helices, long enough to span the lipid bilayer of the cell the membrane, was associated with the pore formation function. Domain II, with three loops at the apical part showing highly variable amino acid sequences, was suggested to be responsible for specificity.

Once the structure of Cry toxins was elucidated and the concept that each domain had a function was established, the idea of improving and redirecting the toxicity of a Cry toxin by exchanging complete domains was reinforced. Many reports of what was called domain swapping were produced using molecular strategies such as in vivo intramolecular recombination, cloning, or overlapping PCR.

The technique known as in vivo intramolecular recombination [75] is based on the construction of a "tandem plasmid" where two truncated proteins, in direct repeated orientations, are cloned (Figure 1). The truncated genes (one lacking the 3′ end of the gene and the other the 5′ end) only overlap in an area where the recombination is intended. Tandem plasmid contains an enzyme restriction site to further discriminate between recombinant plasmids (where the restriction site is lost) and the parental one. Once the tandem plasmid is introduced in a recombinase positive *E. coli* strain (Rec$^+$), random recombination takes place at the homologue regions and novel hybrids or chimeras are produced.

One of the first works using in vivo intramolecular recombination to obtain 3D-Cry toxin hybrids was reported by Caramori et al. [76]. They cloned two truncated toxin genes (for CryIA(a) and CryIA(c)) with overlapping variable regions (63% identity). After the hybrid toxins were bioassayed against several Lepidoptera, it was determined that hybrid 32 (pHy32) and hybrid 45 (pHy45) were more active toward *T. ni* and *Heliotis* sp., respectively, than any of the parental toxins. Two of the hybrid toxins (pHy104 and pHy122) with the same amino acidic sequence even gained a novel activity against *Spodoptera littoralis* as none of the parental toxins were active against this insect.

In vivo intramolecular recombination has been used for the combination of other 3D-Cry toxins and some of them rendered enhanced toxins. This is the case for the combination of CryIC and CryIE toxins [77], both active against Lepidoptera, but with different specificities. CryIC is particularly active against *Spodoptera exigua* (LC$_{50}$ 26 ng/cm^2) and *Mamestra brassicae* (LC$_{50}$ 8 ng/cm^2) while CryIE is not (LC$_{50}$ > 1000 ng/cm^2). The authors constructed two tandem plasmids with truncated genes overlapping at Domain II and III of the toxins to construct CryIC-CryIE and CryIE-CryIC hybrid proteins. One of the resulting hybrids, named G27 and containing Domain I and II from CryIE and Domain III from CryIC, was toxic to *S. exigua* (LC$_{50}$ 2 ng/cm^2). The reverse hybrid (DI and DII from CryIC and DIII from CryIE) was not toxic at all, meaning that the CryIC Domain III was involved in the specificity against this insect.

Using in vivo intramolecular recombination, de Maagd et al. [78] demonstrated that the moderate toxicity of CryIA(b) toward *S. exigua* could be enhanced by constructing a hybrid between DI and DII from CryIA(b) and DIII from a highly active toxin (CryIC). Hybrid H04 showed an increase in toxicity of more than 66-fold (CL$_{50}$ from > 100 μg of toxin/g of diet to 1.66 μg/g) compared to Cry1Ab. The activity of the novel hybrid was even better than the parental CryIC toxin (LC$_{50}$ 11 μg/g), showing a 6.6-fold increase in toxicity. The authors also determined that binding sites of CryIA(b) and CryIC to

S. exigua Brush Border Membrane Vesicles (BBMVs) were different for both toxins and argued that in the case of insect resistance, hybrid H04 could be of great use.

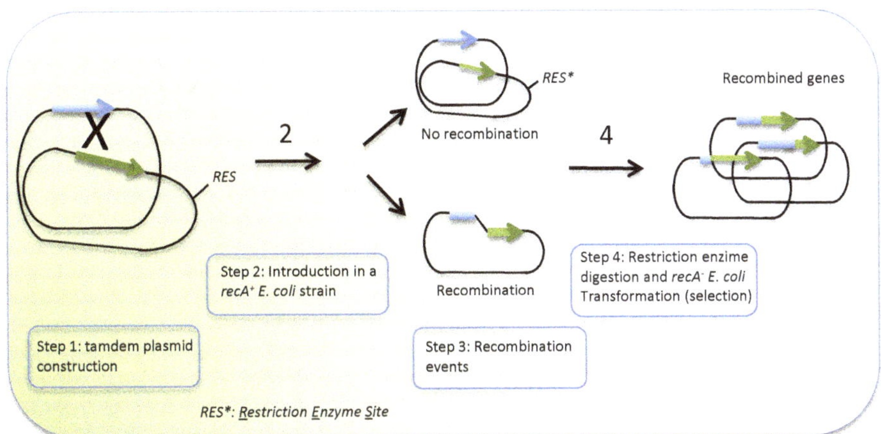

Figure 1. In vivo intramolecular recombination. A tandem plasmid is constructed by the cloning of two truncated genes, one truncated at the 3′ end (blue gene) and the other one at the 5′ end (green gene), leaving an overlapping region were recombination is desired. The plasmid contains a restriction site to further discriminate if recombination took place. Plasmid is then introduced in a *recA*⁺ *E. coli* strain and DNA recombination is allowed. After plasmid extraction from the *E. coli recA*⁺ strain, the pool of plasmids are digested with a restriction enzyme and selected in a *recA*⁻ *E. coli* strain. Each clone represents a recombination event where the two toxins genes have been fused [77].

Other hybrid toxins obtained by in vivo recombination from the less studied Cry1Ba, Cry1Da, and Cry1Fa toxins (Cry toxins nomenclature updated following [7] rules) were reported to have an improved activity [79]. This was the case of hybrid BBC13 (DI-DII from Cry1Ba and DIII from Cry1Ca) that showed an 11.8-fold increase in toxicity toward *M. sexta*, or hybrid BBC15 (DI-DII of Cry1Ba and DIII of Cry1Ca), which showed 8.3-fold and 7.8-fold increases in toxicity against *S. exigua* and *M. sexta*, respectively, or hybrid FFC1 (DI-DII from Cry1Fa and DIII from Cry1Ca) that showed a 5.5-fold increase toward *S. exigua*.

The latest example of hybrids made by in vivo recombination were made from Cry1Ca and Cry1Fb toxins, and DIII from Cry1Ac [80]. Hybrids RK15 (Cry1Ca/Cry1Ca/Cry1Ac) and RK12 (Cry1Fb/Cry1Fb/Cry1Ac) showed an increase in activity, compared to wild type toxins Cry1Ca and Cry1Fb, of more than 172 and 69 times toward *H. virescens*, respectively.

Another way of performing domain swapping and constructing hybrid toxins has been through standard cloning using restriction enzymes (already present or expressly created). This is the case of the hybrid Cry1C/Ab [81], constructed with the first 2194 nucleotides from *cry1C* (731 aa) and the last 1295 nucleotides (432 aa) of the 3′ end of the *cry1Ab* gene. The fusion was possible thanks to the presence of a unique *Kpn*I site in a conserved region of the two genes. The Cry1C/Ab hybrid was 3, 4, and 35 times more active against *S. littoralis*, *Ostrinia nubilalis*, and *Plutella xylostella*, respectively, than the parental Cry1C toxin.

Domain swapping of Coleoptera active toxins has also been successfully accomplished by standard cloning [82], although in this case, it was necessary to introduce restriction sites (*Rsr*II) to obtain the hybrids. Hybrid 1Ia/1Ia/1Ba (DI-DII from Cry1Ia and DIII from Cry1Ba; LC$_{50}$ 22.4 µg/mL) was 2.5 and 7.5 times more toxic than their parental toxins, respectively. Hybrid 1Ba/1Ia/1Ba (DI from Cry1Ba, DII from Cry1Ia, and DIII from Cry1Ba, LC$_{50}$ 7.94 µg/mL) increased its activity even further, showing 17.9 times more potency than the parental Cry1Ba toxin. The latest hybrid toxin was almost as toxic as the Cry3Aa toxin, the most active natural protein for the Colorado potato beetle.

More recently, and thanks to the determination of the Cry1Ac1 full-length toxin 3D structure ([19]; PDB 4W8J), Zghal et al. [21] constructed a 116 KDa chimeric toxin called Cry(4Ba-1Ac) by fusing DI–DIV from Cry4Ba to DV–DVII from Cry1Ac1, using PCR amplification and cloning techniques. This represents a unique case in which other domains, apart from Cry toxin toxic domains, have been swapped. The chimeric toxin showed low toxicity toward *Culex pipiens* when expressed in an acrystalipherous *B. thuringiensis* strain (HD1 CryBpHcry(4Ba-1Ac)), but when co-expressed in a Cry2Aa producing strain (BNS3pHTcry4BLB), the activity increased from 10% mortality to 100% mortality at 200 µg/L. The LC$_{50}$ of the strain bearing only Cry2Aa switched from >>200 µg/L to 0.84 µg/L when co-expressed with the chimeric toxin, meaning that an increase in toxicity of more than 238-fold was produced. This synergy was also observed when Cry2Aa and the chimeric toxin Cry(4Ba-1Ac) were bioassayed in combination. The authors suggested that the increased toxicity could be explained by a better solubilization of the crystals and also proved the importance of the protoxin domains (DIV–DVII) in the stability and the activity of the Cry toxins, a fact that will possibly be exploited in the future.

3D-Cry toxin domain swapping has also been obtained by overlapping PCR [83]. In this case, a hybrid using toxins from different classes, one coleopteran (mCry3A) and one lepidopteran specific (Cry1Ab), was obtained. DNA regions codifying for DI and DII from mCry3A [84] and DIII from Cry1Ab were amplified, containing an overlapping region at the 3′ and 5′ ends, respectively. Amplicons were used as a template in an overlapping PCR with flanking primers. The resulting chimera (called eCry3.1Ab; GenBank GU327680) was highly active (93% mortality) against *Diabrotica virgifera virgifera* when bioassayed with a toxin concentration ranging from 5 to 10 µg of toxin per mL of diet. Although the authors did not determine the LC$_{50}$ of eCry3.1Ab, the toxicity observed was much higher than the previously reported parental mCry3A toxin (LC$_{50}$ 65 µg/mL; [84]).

A more recent example of constructing improved mutants by domain swapping using overlapping PCR [85] was the hybrid toxin Cry1Ac-Cry9Aa containing DI from Cry1Ac and DII and DIII from Cry9Aa. The hybrid showed 4.9 times more activity against *Helicoverpa armigera* than the wild type Cry9Aa. In addition, a Cry1Ac-Cry9AaMod toxin, where helix 1 was proteolytically removed, showed a 5.1-fold increase in toxicity.

2.3. Evolution by Site-Directed Mutagenesis

Site directed mutagenesis is a molecular strategy used to create specific changes in the amino acid sequence of a protein in order to evaluate its role in the molecule. If an amino acid is involved in the mechanism of action of a protein, when changed, the function of the protein is modified and normally abolished. However, the technique could also be used as a means of obtaining novel proteins with improved characteristics.

Site-directed mutagenesis consists of the in vitro synthesis of the codifying DNA of a protein in which one or several nucleotides are changed in a specific site in order to produce a mutant protein. The changed nucleotide is normally introduced using a mutant primer that is in vitro extended thanks to a DNA polymerase (the *E. coli* DNA polymerase Klenow fragment in the early days, and *Taq* polymerase lately when PCR [69] and site-directed mutagenesis by PCR [86] were developed). The in vitro synthesized DNA (mutant DNA) is counter selected from the wild-type DNA, and the mutant protein is expressed and tested for activity.

The elucidation of the three-dimensional structure of a 3D-Cry toxin facilitated the design of mutants for site-directed mutagenesis as the position of the amino acid to be changed was localized in the space together with its interactions with other amino acids in the molecule. Hundreds of site-directed mutagenesis studies in many Cry toxins have been performed, allowing the elucidation of the function of each domain of the protein. Although most of the mutants obtained by site-directed mutagenesis showed an impaired or diminished toxicity, in some cases, the activity of the resulting mutants was higher than the parental toxins. These mutants were not as useful as the impaired mutant to elucidate the mechanism of action of 3D-Cry toxins, but served to settle the concept that activity improvement was possible with only the change of a single amino acid.

This was the case in the work reported by Wu et al. [87] in which the authors constructed 31 mutants at two conserved regions at CryIA(c) Domain I (residues from 84 to 93 and from 160 to 177). Although most of the mutant toxins showed no toxicity, or no change in toxicity at all, one of them, the mutant H168R, localized at the hydrophobic face of the amphypatic α-helix 5, showed a 3–5-fold increase in toxicity toward *M. sexta* compared to the wild type toxin. Another example of Domain I improved mutants was obtained by investigating the role of nine tryptophan residues in the toxicity of Cry1Ab toward *M. sexta* [88]. These authors found that a conservative change to phenylalanine (W73F, W210F, W219F) produced mutant toxins 3.3, 1.5, and 2.3 times more toxic than the parental toxin. In a similar study [89], two α-helix 5 Cry1Ab mutants (V171C and L157C) with a 25-fold and 4-fold increase in toxicity toward *Lymantria dispar*, respectively, were reported.

Site directed mutagenesis studies carried out in 3D-Cry toxins showed that Domain II was particularly sensitive to amino acid changes [90,91]. Some of the mutations performed in this domain were very successful in improving the toxicity against certain insects, and represented an excellent place for redesigning the activity of 3D-Cry toxins. This was the case reported by Rajamohan in 1996 [92] where two single mutants N372A, N372G, and a triple mutant DF-1 (N372A, A282G, L283S) of the Cry1Ab toxin were constructed by site-directed mutagenesis. These residues were localized between the Cry1Ab α-helix 8a and α-helix 8b at Domain II. When bioassayed, N372A and N372G were, respectively, 8.5 and 8.3 times more potent for neonates of *L. dispar* than the parental Cry1Ab toxin, and 9.61 and 9.51 times more potent for fourth instar larvae, respectively. The DF-1 mutant showed an increase in activity of 36- and 17-fold toward neonates, and fourth instar larvae, respectively. The triple mutant was even more toxic (4-fold) than Cry1Aa, the most potent natural toxin active against *L. dispar*.

Loop 1 from Domain II has also been a successful place for mutagenesis, rendering improved mutants in the coleopteran active Cry3A toxin [93]. Site-directed mutagenesis at positions R345, Y350, and Y351 rendered eight single mutants, four double mutants, and three triple mutants, two of them showing an enhanced activity against *Tenebrio molitor*. Mutant A1 (R345A, Y350F, Y351F) and mutant A2 (R345A, ΔY350, ΔY351) were 11.4 and 2.7 times more active than wild type Cry3A, respectively. In addition, an enhanced activity of these two mutants against two other Coleoptera species, *Leptinotarsa decemlineata* and *Chrysomela scripta*, was also observed. Although changes introduced in these mutants were not very drastic (as Y and F differ only by an OH group), differences in toxicity were remarkable. In addition, loop 3 of the Cry3A toxin also proved to be relevant for toxicity enhancement [91]. The triple mutant S484A, R485A, G486A, showed a 2.4-fold increase in toxicity toward *T. molitor* compared to the wild type toxin.

Another place in Domain II where the substitutions introduced by site-directed mutagenesis rendered an enhanced toxin was the area known as D block or Dipteran specific block [73] in the dual toxin Cry2A (dual as it is also active against Lepidoptera). Cry2Ab is not a very potent toxin toward *Anopheles gambiae* (LC$_{50}$ 540 ng/mL), but when residues from 307–337 were mutated, three mutants (N309S, F311I, and A334S) with enhanced toxicity (1.17, 3.17, and 6.75-fold, respectively) were obtained [94]. One of the mutants, A334S, was even more toxic than the highly active natural toxin Cry2Aa [95].

Recently, 3D-Cry toxin Domain III has also been targeted for mutation by site-directed mutagenesis, and mutants with improved characteristics have been found. This is the case of the work reported by Lv et al. [96] in Cry1Ac5. Although the structure of this 3D-Cry toxin is not elucidated, 3D structure modeling using known structures (Cry1Aa (PDB:1CIY), Cry2Aa (PDB:1I5P), Cry3Aa (PDB:1DLC), Cry3Bb1 (PDB:1JI6), and Cry4Ba (PDB:1W99) was useful in localizing the loop sequence between β-20 and β-21 (^{576}NFTSSLGNIV^{586}). Two mutants obtained by site directed mutagenesis, S581A and I585A, showed 1.72- and 1.89-fold increases in toxicity toward *H. armigera*.

Another area that has recently been demonstrated to be susceptible to improvement by site directed mutagenesis is the β sheet 16 in Domain III. A study of alanine substitution in Cry1Ab [97] performed on this area rendered mutants S509A, V513A, and N514A, which showed an increase in toxicity toward *Spodoptera frugiperda* of more than 9.5-, 12.7-, and 51-fold, respectively. As N514 was the most relevant position experiencing toxicity enhancement in β-16, saturation mutagenesis was performed at this position (N was changed for any of the other 19 amino acids). Some of the obtained mutants, N514F, N514H, N514K, N514L, N514Q, N514S, and N514V showed a 44-, 16-, 7-, 9-, 26-, 23-, and 9-fold increase in toxicity, respectively, when compared to the wild type. An equivalent mutation in Cry1Fa (N504A), a more potent toxin than Cry1Ab for *S. frugiperda*, rendered a mutant 6–11 times more toxic than the wild type Cry1Fa in different populations of *S. frugiperda* from different countries. The authors suggested that the increase in toxicity correlated with an increase in the stability of the mutants toward gut proteases and an increase in BBMV binding.

The fact that two other mutants in β-16, this time in Cry1C [98], showed a slight increase in toxicity compared to the wild type (mutant V509A was 1.6 times more toxic than Cry1C for *M. sexta* and mutant N510A was 1.5 times more toxic toward *S. frugiperda*) suggest that β-16 is a recently discovered site for toxicity improvement.

The loop sequence between β-18 and β-19 in Domain III also seems to be a relevant region for toxicity enhancement. Mutant N546A [99] showed a slight increase in toxicity (1.8-fold) in Cry1Ac toxin, which correlated with a binding increase toward *H. armigera* BBMVs [100].

Single mutations in Domain III have also been described to be useful for toxicity enhancement. This was demonstrated in the nematocidal toxin Cry5Ba [101]. Investigating the role of the 3 asparagines present in block 3 of the toxin found that alanine substitution, by site directed mutagenesis, rendered a mutant (N586A) with a 9-fold increase in toxicity (GIC_{50} from 42.11 to 4.75 ng/mL) toward *Caenorhabditis elegans*. Mutant N586A was surprisingly soluble in a wide range of pHs (from pH 5 to pH 12), which correlated with the observed increase in toxicity.

2.4. Evolution by Rational Design

Rational design is a particular case of site-directed mutagenesis performed only when the knowledge of the structure and the function of the protein under study are very deep. In rational design, a hypothesis is formulated and proven by the construction of mutants with single amino acids changes, deletions, or insertions, normally performed by site-directed mutagenesis. Several works have described successful evolution of 3D-Cry toxins by rational design, although some of these were found by accident trying to prove a different hypothesis. This was the case in the work reported by Angsuthanasombat et al. in 1993 [102] where the authors were interested in demonstrating that R203 in Cry4B toxin was essential for proteolytic processing and the α-helices mobility of Domain I. For that, they replaced the R (proteolytic site) with an A, expecting that toxicity would be completely lost as a consequence of the impossibility of the helices to move properly. The effect was completely the opposite as the R203A mutant was 2.8 times more toxic to *Aedes aegypti* than the wild type.

An example of rational design where the researcher did prove their hypothesis was that reported for the evolution of Cry4Ba toxicity [103]. The Cry4Aa toxin is highly active against four species of mosquito (*Ae. aegypti*, *Anopheles quadrimaculatus*, *Culex quinquefasciatus*, and *C. pipiens*), however, the closely related Cry4Ba toxin (the one to be evolved) showed toxicity toward *Ae. aegypti* and *An. quadrimaculatus*, but not to *C. quinquefasciatus* and *C. pipiens*. Through site-directed mutagenesis, the authors delimited the putative loop 3 in Cry4Ba (VIDYNS) and in Cry4Aa (IPATYK), as the Cry4Ba and Cry4Aa 3D structures were not determined until 2005 and 2006, respectively [15,16]. The authors mutated the Cry4Ba loop 3 by replacing D454 with a P and inserting AT after position 454 to yield a novel toxin (named 4BL3PAT) with a novel loop 3 sequence (VIPATYNS). This small change increased Cry4Ba activity 700-fold toward *C. quinquefasciatus*, and by 285-fold to *C. pipiens*. Other versions of the novel toxin were also constructed in loop 3 by substitution of the PAT motive to other motives (4BL3AAT, 4BL3GAT, 4BL3GAV, 4BL3PAA, and 4BL3AAA), showing an activity gain toward *C. quinquefasciatus* and *C. pipiens* in almost all variants (with the exception of 4BL3AAA toward *C. quinquefasctiatus*). The study of the mechanism of the 4BL3PAT mutant, compared to the wild type toxin, showed that both toxins had little difference in the ability of reversible binding to *Culex* BBMVs, a similar capability of irreversible binding, but 4BL3PAT showed a higher pore-forming ability than the Cry4Ba parental toxin. The authors identified two novel proteins in the BBMVs, which are 35 and 36 KDa in size, that the novel mutant bonds to instead of the parental toxin, proposing that these two proteins could be functional receptors and explain why the 4BL3PAT variant is toxic to the *Culex* species (although it was not proven). One year later [104], the same authors reported the evolution of the mosquitocidal Cry19Aa toxin by rational design. The Cry19Aa toxin, active against mosquito species such as *An. quadrimaculatus* (LC_{50} 3 ng/mL) and *C. pipiens* (LC_{50} 6 ng/mL), showed low activity against *Ae. aegypti* (LC_{50} 1.4×10^5 ng/mL). After in silico modeling of the Cry4Ba structure and Cry19Aa using Cry3Aa and Cry4Aa as templates, the following changes in Cry19Aa toxin were introduced: (i) Cry19Aa Loop 1 (^{355}SYWT^{358}) was substituted by the Cry4Ba loop 1 (^{332}YQDLR^{336}), and (ii) Cry19Aa loop 2 (^{414}YPWGD^{418}) was completely deleted to mimic the length present in the Cry4Ba toxin. The resulting mutant 19AL1L2 was >42,000 times more toxic to *Ae. aegypti* (LC_{50} 3.3 ng/mL) than the parental toxin Cry19Aa, being one of the highest activity enhancements in the history of 3D-Cry toxin evolution. The rationale behind these substitutions was to test the hypothesis that changing loop sequences involved in receptor binding could be useful to enhance toxicity by increasing the affinity of the toxin to its receptor. Unfortunately, no differences between Cry19Aa and 19AL1L2 were detected in either reversible or irreversible binding to BBMVs, so the hypothesis was not correct, but it was useful to demonstrate that "in vitro evolution" of Cry-toxins could be efficiently performed by rational design.

However, the greatest achievement in 3D-Cry toxin evolution by rational design was that reported by Liu and Dean in 2006 [105]. These authors were able to redirect the toxicity of a lepidopteran active toxin toward an insect from a different order, the dipteran *C. pipiens*. The mutant was constructed using the lepidopteran active Cry1Aa toxin as a scaffold, and changing loops 1 and 2 of the molecule. Loop 1 from Cry1Aa (^{311}RG^{312}) was enlarged and substituted by the Loop 1 (^{332}YQDL^{335}) from the mosquito active toxin Cry4Ba (although not active against *C. pipiens*). In addition, part of loop 2 in Cry1Aa (LY^{367}RRIILGSGPNNQ^{378}) was deleted (LY^{365}RRIIL), and the ^{376}NNQ^{378} sequence was replaced by a single G, resulting in the mutant 1AaMosq (L1:^{311}YQDL^{314}; L2:^{367}GSGPG^{371}), which gained activity against *C. pipiens* while the activity against Lepidoptera *M. sexta* was abolished. The novel toxin 1AaMosq showed an LC_{50} of 45.73 μg/mL when bioassayed against *C. pipiens*, while the parental toxin Cry1Aa was not toxic to this mosquito at toxin concentrations of 100 μg/mL.

A similar rational design performed in the loops of the Domain II has recently been reported in another 3D-Cry toxin [106]. Cry1Ah and Cry1Ai show high sequence similarity (77% identity at amino acid level) but very different specificity. Cry1Ah is toxic to *H. armigera* but non-toxic to *B. mori* and conversely, Cry1Ai is highly toxic to *B. mori* but has no activity against *H. armigera*. As loops in Domain II of the 3D-Cry toxins are involved in specificity, the authors exchanged loops in the two toxins by reverse PCR in order to evolve the activity of Cry1Ai toward *H. armigera*. One of the obtained mutants, Cry1Ai-h-loop2 (Cry1Ai toxin with loop 2 from Cry1Ah toxin), showed a change of specificity. Toxicity against *H. armigera* increased more than 7.8-fold (from LC_{50} > 500 μg/g to 64.23 μg/g). When the exchange was done in loop 2 and loop 3, the resulting Cry1Ai-h-loop2&3 mutant showed an increase in toxicity even higher, around 58 times (from LC_{50} > 500 μg/g to 8.61 μg/g).

Domain I has also been the subject of evolution by rational design [107]. Through bioinformatic analysis, it was found that the first 42 amino acids of Cry2A toxins interacted with a predicted transmembrane (TM) domain (amino acids 51–62) in helix 2 of the toxin. In addition, it was observed that the predicted TM in Cry2A was shorter than the equivalent TMs domains found in other Cry toxins, as a consequence of the presence of two lysines at positions 63 and 64. They predicted that this interaction could be the reason for the Cry2A toxin being less active against lepidopteran pests compared to Cry1A type toxins, so they made the hypothesis that by removing these 42 amino acids from the molecule, a Cry2A variant with increased activity could be obtained. The deleted mutant,

D42, showed an increase in activity ranging from 2–3-fold toward Lepidoptera *S. littoralis*, *H. armigera*, and *Agrotis ipsilon*. In addition, when lysines 63 and 64 were substituted by a conserved amino acid present in other toxins (F and P), making the TM domain as long as the one present in other Cry toxins, the activity increased even further. Lysine 63 and 64 in the deleted mutant D42 were replaced by phenylalanine/proline by site-directed mutagenesis, and the mutant toxins D42/K63F, D42/K64F, D42/K63F/K64F, and D42/K63F/K64P were obtained. Single mutant toxins showed the same toxicity as D42, but double mutants increased their toxicity toward the tested Lepidoptera between 1.3 and 2.3 times compared to D42 toxicity, and between 4 and 6.5 times compared to the Cry2A wild type, as predicted.

Another successful example of rational design was performed on the Cry3A toxin [84]. In this work, a chymotrypsin/cathepsin G site (AAPF) was introduced into the loop region between α-helix 3 and α-helix 4 in Cry3A Domain I in order to increase the proteolytic efficiency and hence toxicity. The resulting mutant, mCry3, with a loop sequence ^{153}NPAAPFRN^{160}, was active against *D. virgifera virgifera* larvae (LC$_{50}$ 65 μg/mL) compared to the residual activity of the parental toxin Cry3A (LC$_{50}$ >> 100 μg/mL). The authors determined that the increase in activity was due to several factors such as a higher solubility at neutral pH, an increase in the efficiency of the proteolytic process, and an increase of specific membrane binding. The introduced mutation did not alter the activity against the Colorado potato beetle larvae, a susceptible insect for Cry3A, but extended activity toward other coleopteran (*D. virgifera virgifera*).

2.5. Evolution by Random Mutagenesis

Random mutagenesis is one of the strategies that molecular biologists have available to obtain protein variants. In opposition to site-directed mutagenesis, the position of the mutation in random mutagenesis is not controlled. It is frequently used to discover relevant amino acids in a protein when not enough information on the function of the protein is available [90,108], or to create novel variants of a protein with novel functions. There are several techniques that have been used to perform random mutagenesis in 3D-Cry toxins such as in vitro DNA amplification with degenerated primers and error-prone PCR.

The development of the synthesis of degenerated oligonucleotides in 1988 [109] allowed researchers to introduce random mutations in a specific region of the DNA. In 1999, Kumar et al. [110] used a mixture of degenerated primers for the random mutation of a 3D-Cry toxin using a M13mp19 system that provided ssDNA. The researchers' objective was to introduce variability at α-helix 4, and at the loop between α-helix 4 and α-helix 5 of the Cry1Ac toxin. For that, a mix of primers with the wild-type sequence (97% of the primers) and degenerated oligonucleotides (3% of the primers) was used to ensure only one mutation per cycle of amplification. Using this technique, the authors obtained the mutant F134L with a 3-fold enhanced toxicity toward *M. sexta* and *H. virescens*.

Error-prone PCR is an in vitro evolution technique that generates variants using the property of the *Taq* polymerase of introducing substitutions in the DNA amplification process when subjected to certain conditions. Error-prone PCR is the most common method for creating combinatorial libraries based on a single gene. Since its development in 1989 [111], it has been used in many applications, not only to mutate DNA codifying for proteins, but also non-coding DNA regions such as promoter regions [112]. The technique is simple and only requires a PCR mixture slightly different from a standard PCR. The gene subjected to mutagenesis is amplified by upstream and downstream primers in a reaction mix containing Mn^{2+} ions, an unbalanced ratio of dNTPs, and/or a higher concentration of Mg^{2+}. Depending on the conditions, the overall mutagenesis rate can be controlled. Error-prone PCR mutagenesis has not been extensively used in obtaining Cry toxin variants. Only a few examples of using this technique are found in the bibliography, even though the technique has rendered improved versions of a 3D-Cry toxin. This was the case of the work reported by Shu et al. [113] made in Cry8Ca2 toxin (accession number AY518201), which was active against *Anomala corpulenta*. Two mutants (M100 and M102) showed 5- and 4.4-fold increases in toxicity, respectively, against the larva of this Coleoptera

(LC$_{50}$ 0.23 × 10^8 CFU/g and 0.26 × 10^8 CFU/g, respectively, compared to LC$_{50}$ 1.15 × 10^8 CFU/g of the parental toxin). The sequence analysis of the novel variants showed that only a single mutation in each mutant (E642G in M100 and Q439P in M102) was enough to enhance toxicity.

A more recent example of the use of error-prone PCR, although combined with other techniques, is the random evolution of Cry1Ac5, rendering the T525N mutant with a slight increase in toxicity (1.5-fold) toward *S. exigua* [114].

2.6. Evolution by Mixing Cry Genes: DNA Shuffling, In Vitro Recombination, and StEP (Staggered Extension Process)

DNA shuffling is a powerful in vitro evolution tool for generating artificially and highly diversified sequences by homologous gene recombination (Figure 2a). Although this technique is normally used in proteins of unknown three-dimensional structure, it can be used for any protein. The technique involves random DNA fragmentation of two or more homologous genes with DNAse I, and fragment reassembly in a primer-less PCR. After the generation of a variant library, a screening and selection process of functional variants is conducted.

Since the development of this technique [115,116], DNA shuffling has been used to evolve thousands of proteins, mainly enzymes, to modify their function or activity. The convenience of this powerful technique has also been applied for the in vitro evolution of several 3D-Cry toxins.

Although in vitro evolution by DNA shuffling has not been always successful [117], and sometimes no improved 3D-Cry toxins have been obtained, in several other cases, it has been a suitable technique for obtaining variants with higher activity. For example, DNA shuffling of *cry11Aa*, *cry11Ba*, and *cry11Bb* genes, codifying for toxins active against *Ae. aegypti* and *C. quinquefasciatus*, rendered a mutant (Variant 8) 3.8 times more toxic toward *Ae. aegypti* than the parental toxin Cry11Bb (LC$_{50}$ 22.9 ng/mL), and 6.09 times more toxic than the Cry11Aa toxin (LC$_{50}$ 36.9 ng/mL) [118]. DNA sequence analysis of Variant 8 showed that the mutant contained a deletion of 219 nucleotides (73 aa) at the N-terminal end of the molecule (Domain I), and 6 and 13 nucleotide substitutions in Domain II and III, respectively. The comparative analysis at the protein level between Variant 8 and its parental toxins (Cry11Aa and Cry11Bb) showed 13 amino acid substitutions (GenBank access number MH068787).

Very recently, DNA shuffling has been used to increase the toxicity of an already improved mutant derived from Cry3Aa toxin [52]. The mutant IP3-1, engineered by rational design, contained 15 mutations over the three domains of the toxin (W106L, M117I, V140F, I186V, F206L, K230H, S258T, P292S, E294G, F346L, G468A, L491F, M503T, R531G, and I593M) and showed a higher activity toward *D. virgifera* (LC$_{50}$ 214 ppm) than the parental toxin. To further increase its activity, in vitro evolution by DNA shuffling was carried out and six novel mutant toxins (IP3-2, IP3-3, IP3-4, IP3-5, IP3-6, IP3-7) showed more activity than the parental toxin IP3-1 (LC$_{50}$ 19, 14.7, 13.7, 11.3, 11.6, 7.3 ppm, respectively). An analysis of their sequences showed that mutant toxins contained between six and nine additional mutations. From the six mutants obtained, the IP3-7 variant showed the best increase in toxicity of all (LC$_{50}$ from 214 to 7.33 ppm). Most of the mutations obtained after DNA shuffling resulted in a reduction of positively charged residues such as lysine and arginine, making novel toxins more acidic and more soluble at neutral pH (*D. virgifera* gut juice is weakly acidic) and hence more active. In addition, all the selected variants showed a similar mutation at two different positions (K152E and R158E), located in the α-helices 3 and 4 loop. According to the authors, these mutations made the loops more resistant to *D. virgifera* gut proteases, contributing to the increase in toxicity compared to the parental toxin IP3-1.

A more sophisticated way of obtaining chimeras is by in vitro recombination using the technique called in vitro template-change PCR. With this strategy, a library of recombinant toxins made from the lepidopteran active toxin Cry2Aa (active toward *Ostrinia furnacalis*, *P. xylostella*, *Chilo suppressalis*, and *H. armigera*) and the low toxicity Cry2Ac [119] was made. The strategy involved four steps: (i) ssDNA amplification of the complementary strand of both genes by asymmetric PCR (amplification using a single reverse primer); (ii) synthesis of the coding strand using the ssDNA from gene 1 as a template in the presence of ddATP, which avoids further extension once it is incorporated in the polymerized strand; (iii) DNA extension of the randomly truncated library using gene 2 as a template, which is achieved thanks to the use of the KOD DNA polymerase, shows 3′–5′ exonuclease proofreading activity, and is able to remove the ddATP from the truncated molecule and carry on with the extension of the DNA strand; and (iv) amplification of the full toxin fragment with flanking primers, cloning, and expression in a heterologous system. With this strategy, the authors obtained 37 chimeras (named R1–R37) showing recombination events at 37 different regions of the toxin. When recombination occurred at Domain I or Domain III, no change in specificity was observed. However, when recombination took place at Domain II, toxin specificity drastically changed. The Cry2Ad toxin gained toxicity toward *O. furnacalis* when recombination was in the ^{416}NY^{417} region (recombinant R24), toward *P. xylostella* when it occurred at ^{440}RPL^{442} (recombinant R26), and toward *C. suppressalis* and *H. armigera* when it took place at ^{455}GTPGGA^{460} (recombinant R27).

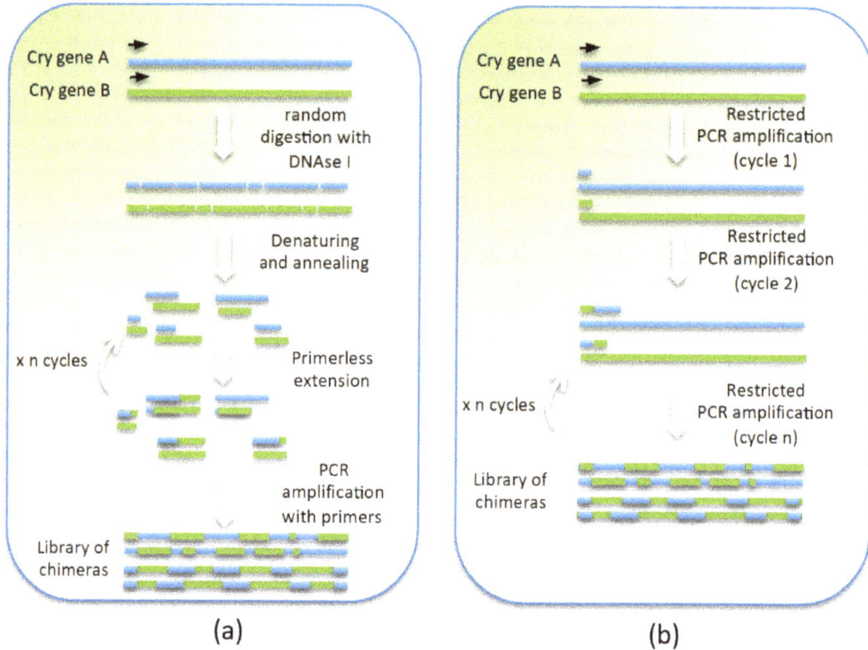

Figure 2. In vitro recombination techniques. (**a**) In DNA shuffling [116], two or more homologous genes are randomly digested with DNAse I (only one strand is represented for simplification purposes). The resulting fragments are extended in a primer-less PCR using homologue fragments as templates. Finally, a PCR using flanking primers is performed in order to obtain a full size library of chimeras, after few rounds of primer-less extension. (**b**) In the staggered extension process or StEP [120], two homologous genes are PCR amplified under restricted conditions (short extension times, and low extension temperature). In cycle 1, a short fragment is extended from a primer in both genes (only one strand is represented). After denaturing, the generated fragments can anneal in the opposite homologous gene, and be extended in cycle 2. After cycle n, a library of recombined chimeras is generated.

The technique, called the staggered extension process or StEP [120], has the same objective as DNA shuffling of producing an in vitro recombination of two or more genes, but with a slightly different methodology. In this technique (Figure 2b), two (or more) homologous genes are denatured and extended from the same primer, using a thermo-cycling program in which the extension step is highly limited in time (few seconds (5 s)) and temperature (extension is carried out at 55 °C, temperature in which *Taq* polymerase has a low extension rate). After 70–80 cycles of denaturing and priming-extension, a library with full-length recombined genes that is cloned and screened can be obtained. This technique has been used for in vitro evolution of the active part (DI, DII, and DIII) of the Cry1Ac5 toxin (GenBank acc. Number M73248) in combination with error-prone PCR. The variants were cloned into a plasmid containing the pro-toxin C-terminal end by Red/ET homologous recombination [121,122]. From the 57 variants obtained, only one was expressed as a full-length 130 kDa toxin containing the mutation T524N in the β-16 and β-17 loops in Domain III. The variant produced more crystals than the wild type, but slightly smaller. When bioassayed toward *S. exigua*, a toxicity 1.5 times higher (LC$_{50}$ 9.6 µg/mL compared to 14.1 µg/mL of the wild type) was observed.

2.7. Evolution by Phage Display

As has been reviewed thus far, generation of 3D-Cry toxin variants does not represent a big challenge nowadays as many molecular techniques for constructing libraries with a high number of mutants are available. The real challenge is to find which of the variants in the library is useful and has the desired properties. Therefore, the key question of in vitro evolution of a protein is the library screening. In the particular case of 3D-Cry toxins, this screening is labor-intensive as every single mutant has to be expressed, solubilized, and bioassayed, representing a time consuming task. As a consequence, only a reduced number of variants from the library are normally tested, and on many occasions, no improved mutants are found. To overcome this problem, approaches such as phage display have been explored to provide a means for the selection of potentially useful 3D-Cry mutants. Phage display is a molecular tool for the screening of library variants with a specific binding characteristic. A phage displayed protein (or a mutant library) consists of its expression on the surface of a bacteriophage in such a way that the protein is available to interact with other proteins, while it is bound to the virus. Display is achieved thanks to the fusion of the protein of interest (or library) to one of the proteins on the surface of the phage. When the phage replicates and viral particles assemble, the protein of interest is also assembled on the surface, being available to interact with other proteins. Displayed libraries can be screened by a process called biopanning, a methodology that allows for the selection of those phages with the desired binding properties (Figure 3). As the phenotype of the selected variant and the genotype in a phage display system are linked, once a phage is selected, the coding DNA for the protein variant can be obtained from the phage genome. Given that one of the premises for 3D-Cry toxin toxicity is to bind to a receptor, this makes phage display a suitable molecular tool for the screening of variants with novel binding characteristics.

Since the invention of the phage display technique [123] and its use for the selection of a peptide from a library with an antibody, many other applications have been developed [124–127]. 3D-Cry toxins have also benefited from the advantages of phage display technology, although several technical limitations had to be overcome before the methodology could be used for big proteins such as 3D-Cry toxins. The first attempt at displaying a 3D-Cry toxin on the surface of a phage was reported by Marzari et al. [128] using the Cry1Aa toxin and M13 phage. Unfortunately, the toxin was not properly displayed and deletions on the fusion protein were observed. One year later, Kasman et al. [129] reported the successful display of the Cry1Ac toxin on the surface of the M13 phage, although the toxin was unable to bind to the APN receptor in in vitro experiments. Later on, other display systems based on the λ and T7 phage, which are assembled in the cytoplasm of *E. coli* and released after lysis instead of being secreted through the bacterial membrane as in M13, were proven to be more appropriate for 3D-Cry toxins. The first successful phage display system, in which the 3D-Cry toxin was able to bind to natural receptors, was described by Vilchez et al. [130]. In this study, the Cry1Ac1 toxin was fused to the gpD protein, an auxiliary protein that represents one of the major components of the λ phage capsid. The displayed toxin was able to selectively recognize and bind proteins present in *M. sexta* BBMVs. Later on, other display systems using the T7 phage [131] and M13 [132] were described. Once the problem of displaying a 3D-Cry toxin was overcome, library variants were developed.

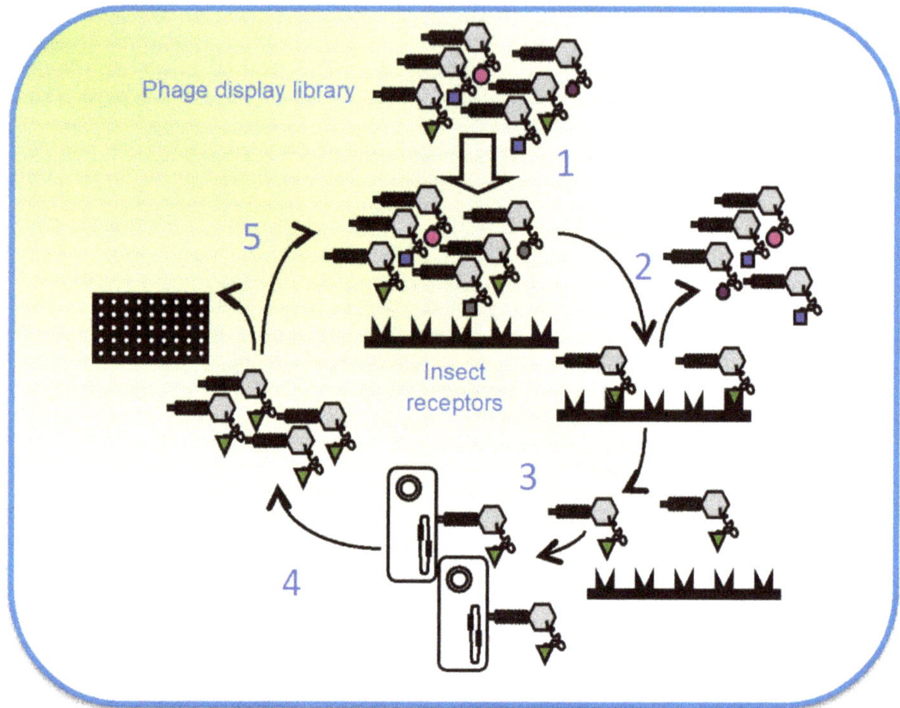

Figure 3. Biopanning of a phage display toxin library. The phage displayed toxin library is biopanned against a specific insect receptor (1). Those phage displaying toxins with affinity to the insect receptor will be retained and those without affinity will be washed out (2). Bond phage will be recovered (3) and amplified (4) by a susceptible *E. coli* strain, making possible to repeat the process (5) in order to obtain higher affinity toxins from the selected pool.

Mutant libraries have been constructed in specific areas of the 3D-Cry toxin using several molecular approaches such as degenerated primers [133–135], DNA shuffling [136,137], or using a previously constructed antibodies library [138]. All these libraries were screened for variants showing high binding affinity toward two of the most well-known receptors (cadherin like receptor and APN), and although an increase of binding affinity is not a guarantee for increased toxicity [134,135], some authors have managed to obtain enhanced 3D-Cry toxin variants compared to the parental toxin.

The first successful report describing the use of a phage display library for the evolution of a 3D-Cry toxin was made by Ishikawa et al. [133] using the T7 phage. They constructed a library of Cry1Aa1 variants at the loop 2 of Domain II, one of the main determinants for specificity in 3D-Cry toxins. Loop 2 variants were constructed by PCR with the degenerated primer Aa369(IILGSGP)375-degenerate-sense (5'TTATATAGAAGANNNNNNNNNNNNNNNNNNNNNNNNNAATAATCAGGAACTGTTTG3'), which could theoretically introduce 1.28×10^9 possible combinations on the seven amino acid residues of the loop. However, in practice, the library contained only 5.0×10^5 variants, less than 0.04% of the possible mutations. Despite the reduced number of variants, the authors managed to select a toxin mutant (R5–51) with strong binding affinity to the bead-immobilized cadherin-like protein BtR175. The selected variant, R5–51, was four times more toxic (LC_{50} 1.6 μg/g diet) than the Cry1Aa1 wild type toxin (LC_{50} 6.3 μg/g diet) toward *B. mori*.

Another case of success in the quest of improving the toxicity of a natural 3D-Cry toxin by phage display technology was the in vitro evolution performed on the moderately active Cry8Ka1 toxin toward the Coleoptera *Anthonomus grandis* [137]. This time, variability was obtained by *cry8Ka1* gene shuffling, which was cloned in the pComb3X phagemid [139] fused to the pIII protein in a M13 system. The resulting library, pCOMBcry8Ka1var, containing 1.0×10^5 cfu/mL variants, was screened toward *A. grandis* BBMVs. Biopanning rendered one variant (Cry8Ka5 mutant) that showed a 3-fold increase in toxicity. Sequence analysis of the Cry8Ka5 variant demonstrated that mutations were randomly introduced at different positions in Domain I (R82Q), Domain II (Y260C, P321A), and Domain III (R508G, K538E, E594N) of the toxin. In addition, a deletion of 16 residues at the N-terminal end was observed.

Non-natural 3D-Cry toxins have also been evolved by DNA shuffling and phage display technology [136]. This is the case of the Cry1Ia12synth toxin (NCBI gene bank accession number FJ938022), a synthetically derived toxin from Cry1Ia12 with a modified codon usage for plant expression optimization. Cyr1Ia12synth is toxic for *S. frugiperda*, but not for the sugarcane giant borer *Telchin licus licus*. From the 30 variants selected by phage display using *T. l. licus* BBMVs, four of them showed higher activity toward *T. l. licus* compared to the wild type toxin. Variant 1 (D233N, E639G), variant 2 (D233N), variant 3 (I116T, L266F, K580R), and variant 4 (M45V, D233N) showed 61%, 75% 56%, and 58% mortality, respectively, higher than the wild type and the negative control (25% mortality). This represents an example of the in vitro evolution of a 3D-Cry toxin in order to be active toward another Lepidoptera specie.

The latest report of a 3D-Cry toxin in vitro evolution using phage display technology was by Dominguez-Flores et al. [140]. In this case, a library of "Crybodies" was displayed on a λ phage system similar to the one reported by Vilchez et al. [130]. Crybodies are molecules derived from the lepidopteran active Cry1Aa13 toxin where loop 2 of Domain II has been replaced by the hypervariable region contained at the complementary determinant region 3 (CDR-H3) of a human antibody library [138]. The Crybody library was biopanned to *Ae. aegypti* larvae guts homogenates, and the selected phage, with high affinity toward gut proteins, was used to obtain the novel Crybodies. Crybodies Cry1Aa13-A8 (L2:^{367}GAREGSSSAYDYW^{379}) and Cry1Aa13-A12, (L2:^{367}GARGDPDFDHSTSYYLDYC^{385}) showed significant mortality (around 90%) after 120 h at 20 µg/mL, while no toxicity was observed in the parental Cry1Aa13 toxin. Concomitantly, both variants showed a 50% decrease in toxicity toward their natural lepidopteran target (*B. mori*). In this case, phage display was proven to be useful not only for improving toxicity against an insect or related species, but also to select variants active against insects from a different order.

Another example that used phage display technology in the field of 3D-Cry toxin evolution is the work reported by Shao et al. [141]. This work describes the construction of six 3D-toxin mutants, obtained by replacing loop 1, loop 2, and loop 3 in Domain II of the Cry1Ab toxin with what they called "gut-binding peptides" or GBPs. These peptides were obtained from a random peptide library displayed on a phage that was biopanned against BBMVs obtained from the hemipteran *Nilaparvata lugens* [141]. P2S (CLMSSQAAC) and P1Z (CHLMSSQAAC) were introduced by overlapping PCR in substitution of loop 1 (^{278}RG^{279}), loop 2 (^{335}RRPFNIGINNQ^{345}), and loop 3 (^{401}SMFRSGFSNSSVS^{413}) in the Cry1Ab toxin. *N. lugens* nymph bioassays showed increased toxicity in five of the six variants selected. Only mutant L3-P1Z was less toxic than the wild type (with an LC_{50} of 189.83 µg/mL). The rest of the mutant toxins (L1-P2S, L2-P2S, L3-P2S, L1-P1Z, and L2-P17) were 5, 9, 5, 1.4, and 2.5 times more toxic, respectively, than the parental toxin. Substitution of loops 1, 2, and 3 was concomitant with a loss in toxicity of Cry1Ab toward *P. xylostella*. This work demonstrated that the in vitro evolution of Cry toxins is not only restricted to the selection of variants with an improved binding to natural receptors, but also evolution can be directed to bind other molecules in the insect guts.

2.8. Evolution by PACE (Phage-Assisted Continuous Evolution)

Phage-assisted continuous evolution is a technique developed at Harvard University by David L. Liu´s research group [142]. It is one of the latest techniques developed in the field of the in vitro evolution of proteins. It is a complex technique, that, as its name implies, is performed with the assistance of a phage. Strictly speaking, it is not an in vitro technique, as evolution is performed inside of a highly engineered *E. coli* strain, but as it is performed in a laboratory, it is considered to be an in vitro evolution technique. The evolution is carried out in what is called "the lagoon" (Figure 4a), an *E. coli* culture with a constant inflow and outflow of growing media. The flow is set up at the appropriate speed to serve as the selection process for the mutants generated in the lagoon, as only the suitable and fast growing mutants stay in the lagoon, while the non-useful mutants are washed out from the growing flask. The average residence time of the cells is less than the *E. coli* replication time. The *E. coli* strain where the evolution takes place contains three plasmids (Figure 4b): (i) an arabinose-inducible mutagenic plasmid (MP), that contains proteins that disrupt the proofreading activity of DNA polymerases, so increasing the error rate in replication; (ii) a selection plasmid (SP), that contains the protein to be evolved and all the genes necessary for M13 phage replication except for *gen III*, essential for host infection; and (iii) an accessory plasmid (AP) where the essential gen III for M13 phage replication is expressed, but only if the right mutant is generated. With this system, the mutation process and the selection process are coupled as only the desired mutants allowing the expression of protein III are replicated. Non-useful mutations produce non-infective phage, so they will be unable to reproduce and will be washed out from the lagoon. The system solves the cumbersome need to screen the entire library in each round of evolution and, given the life cycle of M13 is just 10 min, a high number of rounds of protein evolution could be conducted in a single week. PACE has been used to evolve proteins such as polymerases [143], proteases [144], and genome-editing proteins [145], obtaining variants with completely novel activities and specificities.

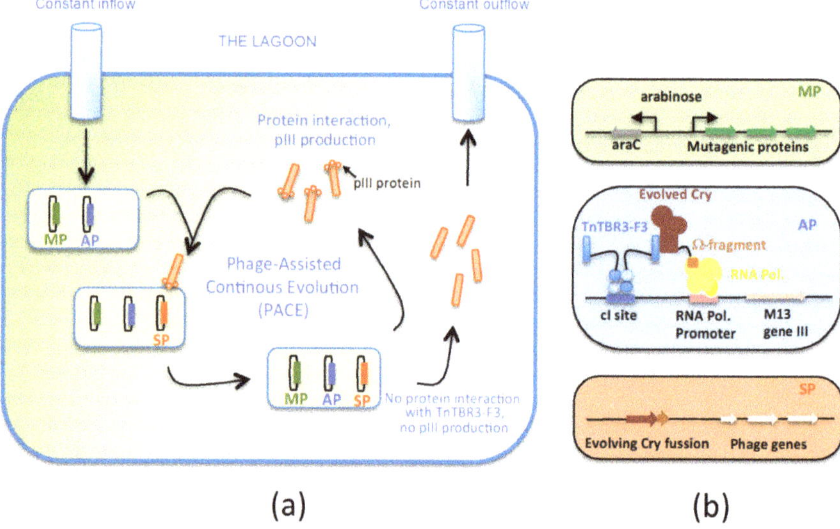

Figure 4. Phage assisted continuous evolution (PACE). In the PACE technique, the evolution occurs in a continuous culture of a highly engineered *E. coli* strain called the lagoon (**a**). The *E. coli* strain contains three plasmids (**b**), the mutagenic plasmid (MP), containing mutagenic proteins induced by arabinose, the selection plasmid (SP), which contains all the M13 genes for phage replication, except for *gene III*, and a transcriptional fusion of the evolving *cry* gene toxin and the *rpoZ* gene, codifying for the omega sub-unit of the RNA polymerase, and the accessory plasmid (AP). AP plasmid contains the M13 *gene III* downstream of a cI site and a promoter. The fusion protein between cI and a fragment from *T. ni* cadherin like receptor (TnTBR3-F3) binds to the cI site. Only Cry toxin variants interacting with TnTBR3-F3 will bind to the proximity of the promoter site so the *gen III* expression will be possible, rendering viable infecting M13 particles. However, if Cry toxins with no affinity toward the TnTBR3-F3 fragment are produced, *gene III* will not be expressed and no infecting M13 particles will be generated. Reproduced from [146] Copyright 2016, Springer Nature.

Table 1. Compilation of all the improved toxin mutants obtained throughout the molecular evolution of 3D-Cry toxin history. Parental toxin, mutants, and results obtained, together with the methodology used for protein evolution, are detailed in chronological order.

Molecular Technique Used	Parental Toxin Evolved	Mutant Name	Evolution Result (Insect)/Evolution Level [1]	Activity Enhancement	Domain Evolved [2]	Reference
Random mutagenesis with mutagens	CryIA(b)	P26-3 P48a14 P48c5 P36a65 P95a76 P95a86 P98c1 P99c62 P107c22 107c25 114a30	Enhanced toxicity (*H. virescens*)/SS	3–5-fold	DI *	[68]
Homolog-scanning mutagenesis	ICPC73	OSU 4205	Novel activity (*B. mori*)/SL		DII *	[70]
Homolog-scanning mutagenesis	CryIIB	Hybrid 513	Novel activity (from Lepidoptera to a dual Lepidoptera and Diptera)/OL		DII *	[73]
Domain swapping (in vivo recombination)	CryIA(a) and CryIA(c)	pHy32 pHy45 pHy104	Enhanced toxicity (*T. ni* and *Heliothis* sp)/SS Novel activity (*S. littoralis*)/SL	2–37-fold 3–7-fold	DII *	[76]
Homolog-scanning mutagenesis	CryIA(c)	Hybrid 4109	Enhanced toxicity (*H. virescens*)/SS	30-fold	DII *	[74]
Site directed mutagenesis	CryIA(c)	H168R	Enhanced toxicity (*M. sexta*)/SS	3–5-fold	DI	[87]
Rational design (site directed mutagenesis)	Cry4B	R203A	Enhanced toxicity (*Ae. aepypti*)/SS	2.8-fold	DI	[102]
Domain swapping (in vivo recombination)	CryIE	G27	Enhanced toxicity (*S. exigua*)/SS	>50-fold	DIII	[77]

Table 1. *Cont.*

Molecular Technique Used	Parental Toxin Evolved	Mutant Name	Evolution Result (Insect)/Evolution Level [1]	Activity Enhancement	Domain Evolved [2]	Reference
Domain swapping (cloning)	CryIA(b)	H04	Enhanced toxicity (*S. exigua*)/SS	More than 60-fold	DIII	[78]
Site directed mutagenesis (alanine scanning mutagenesis)	CryIIIA	Triple mutant: S484A, R485A, G486A	Enhanced toxicity (*T. molitor*)/SS	2.4-fold	DII (Loop 3)	[91]
Site directed mutagenesis	Cry1Ab	N372A N372G	Enhanced toxicity (*L. dispar*)/SS	8.53-fold 9.61-fold	DII	[92]
Site directed mutagenesis	Cry1Ab	DF-1: Triple mutant N372A, A282G, L283S	Enhanced toxicity (*L. dispar*)/SS	36-fold	DII	[92]
Domain Swapping (cloning)	Cry1C	Cry1C/Ab hybrid	Enhanced toxicity (*S. littoralis*, *O. nubilalis*, and *P. xylostella*)/SS	3-, 4- and 35-fold respectively	DI-DVII	[81]
Random mutagenesis	Cry1Ac1	F134L	Enhanced toxicity (*M. sexta* and *H. virescens*)/SS	3-fold	DI α-Helix 4)	[110]
Domain swapping (in vivo recombination)	Cry1Ba	BBC13 BBC15	Enhanced toxicity (*M. sexta*)/SS Enhanced toxicity (*S. exigua*) /SS Enhanced toxicity (*M. sexta*) /SS	11.8-fold 8.3-fold 7.8-fold	DIII	[79]
Domain swapping (in vivo recombination)	Cry1Fa	FFC1	Enhanced toxicity (*S. exigua*) /SS	5.5-fold	DIII	[79]
Rational design (site directed mutagenesis)	Cry3A loop 1	A1	Enhanced toxicity (*T. molitor*) /SS	11.4-fold	DII	[93]
Rational design (site directed mutagenesis)	Cry3A loop 1	A2	Enhanced toxicity (*T. molitor*)/SS	2.7-fold	DII	[93]
Domain swapping (cloning)	Cry1Ia	1Ia/1Ia/1Ba hybrid	Enhanced toxicity (*L. decemlineata*)/SS	2.5-fold respect Cry1Ia and 7.5-fold respect Cry1Ba	DI, DII, DIII	[82]

Table 1. *Cont.*

Molecular Technique Used	Parental Toxin Evolved	Mutant Name	Evolution Result (Insect)/Evolution Level [1]	Activity Enhancement	Domain Evolved [2]	Reference
	Cry1Ba	1Ba/1Ia/1Ba hybrid	Enhanced toxicity (*L. decemlineata*)/SS	17.9-fold	DI, DII, DIII	[82]
Rational design (Site directed mutagenesis)	Cry4Ba using Loop3 from Cry4Aa	4BL3PAT	Evolution from *Anopheles* and *Aedes* to *Culex*/SL	700- and 285-fold increase	DII	[103]
Rational design (Site directed mutagenesis)	Cry19Aa using loop from Cry4Ba	19AL1L2	Evolution from *Anopheles* and *Culex* to *Aedes*/SL	42,000-fold increase	DII	[104]
Domain swapping (in vivo recombination)	Cry1Ca and Cry1Fb using DIII of Cry1Ac	RK15 RK12	Enhanced toxicity (*H. virescens*)/SS Enhanced toxicity (*H. virescens*)/SS	172-fold 69.6-fold		[80]
Rational design (Site directed mutagenesis)	Cry1Aa using loop 1 from Cry4Ba	1AaMosq	Evolution from Lepidoptera to Diptera (mosquito)/OL	From no activity at 100 ug/mL to an LC_{50} 45.73 of ug/mL	DII	[105]
Site directed mutagenesis	Cry1Ab	W73F W210F W219F W455F	Enhanced toxicity (*M. sexta*)/SS Enhanced toxicity (*M. sexta*)/SS Enhanced toxicity (*M. sexta*)/SS Enhanced toxicity (*M. sexta*)/SS	3.3-fold 1.5-fold 2.3-fold 1.4-fold	DI and DII	[88]
Error prone PCR	Cry8Ca2	M100 M102	Enhanced toxicity (*A. corpulenta*)/SS	5-fold 4.4-fold	DIII DII	[113]
Rational design (Site-directed mutagenesis)	Cry2A	D42	Enhanced toxicity (*S. littoralis*)/SS Enhanced toxicity (*H. armigera*)/SS Enhanced toxicity (*A. ipsilon*)/SS	2.85-fold 1.99-fold 2.87-fold	DI	[107]

Table 1. *Cont.*

Molecular Technique Used	Parental Toxin Evolved	Mutant Name	Evolution Result (Insect)/Evolution Level [1]	Activity Enhancement	Domain Evolved [2]	Reference
Rational design (Site-directed mutagenesis)	Cry2A	D42/K63F/K64F	Enhanced toxicity (*S. littoralis*)/SS Enhanced toxicity (*H. armigera*)/SS Enhanced toxicity (*A. ipsilon*)/SS	4.5-fold 2.9-fold 3.7-fold	DI	[107]
Rational design (Site-directed mutagenesis)	Cry2A	D42/K63F/K64P	. Enhanced toxicity (*S. littoralis*)/SS Enhanced toxicity (*H. armigera*)/SS Enhanced toxicity (*A. ipsilon*)/SS	6.6-fold 4.1-fold 4.9-fold	DI	[107]
Phage display	Cry1Aa1	R5-51	Enhanced toxicity (*B. mori*)/SS	4-fold	DII (loop 2)	[133]
Site directed mutagenesis	Cry3A	mCry3A	Novel activity (*D. virgifera virgifera*)/SL	From LC_{50} >> 100 µg/mL to 65 µg/ml	DI (Loop α-helix 3 and 4)	[84]
Site directed mutagenesis	Cry1Ac	N546A	Enhanced toxicity (*H. armigera*)/SS	1.8-fold	DIII	[99]
Site directed mutagenesis	Cry1Ab	V171C L157C	Enhanced toxicity (*L. dispar*)/SS	25-fold 4-fold	DI	[89]
DNA Shuffling and Phage display	Cry1Ia12synth	Variant 1 Variant 2 Variant 3 Variant 4	Novel toxicity (*T. licus licus*)/SL	LC_{50} not determined	DI, DII, DIII	[136]
Domain swapping (Overlapping PCR)	mCry3A	eCry3.1Ab	Enhanced toxicity (*D. virgifera virgifera*)/SS	From low toxicity to 93% mortality at 7.5 µg/mL	DIII	[83]

Table 1. *Cont.*

Molecular Technique Used	Parental Toxin Evolved	Mutant Name	Evolution Result (Insect)/Evolution Level [1]	Activity Enhancement	Domain Evolved [2]	Reference
Error prone PCR and StEP shuffling	Cry1Ac5	T524N	Enhanced toxicity (*S. exigua*)/SS	1.5-fold	DIII	[114]
DNA shuffling and phage display	Cry8Ka1	Cry8Ka5	Enhanced toxicity (*A. grandis*)/SS	3-fold	DI, DII, DIII	[137]
Site directed mutagenesis	Cry1Ac5	S581A I585A	Enhanced toxicity (*H. armigera*)/SS Enhanced toxicity (*H. armigera*)/SS	1.72-fold 1.89-fold	DIII	[45]
Rational design (Site directed mutagenesis)	Cry2Ab	N309S F311I A334S	Enhanced toxicity (*An. gambiae*)/SS Enhanced toxicity (*An. gambiae*)/SS Enhanced toxicity (*An. gambiae*)/SS	1.17-fold 3.17-fold 6.75-fold	DII	[94]
Site directed mutagenesis	Cry5Ba	N586A	Enhanced toxicity (*C. elegans*)/SS	9-fold	DIII	[101]
In vitro template-change PCR (TC-PCR)	Cry2Ad	R24 R26 R27 R27	Novel activity (*O. furnacalis*)/SL Novel activity (*P. xylostella*)/SL Novel activity (*C. suppressalis*)/SL Novel activity (*H. armigera*)/SL	From 0% to 26.7% mortality From 4.6 to 75.6% From 6.7 to 76.7% From 2.2 to 84.1%	DII	[119]

Table 1. *Cont.*

Molecular Technique Used	Parental Toxin Evolved	Mutant Name	Evolution Result (Insect)/Evolution Level [1]	Activity Enhancement	Domain Evolved [2]	Reference
Phage display	Cry1Ab	L1-P2S L2-P2S L3-P2S L1-P1Z L2-P1Z	Enhanced toxicity (*N. lugens*)/SS Enhanced toxicity (*N. lugens*)/SS Enhanced toxicity (*N. lugens*)/SS Enhanced toxicity (*N. lugens*)/SS Enhanced toxicity (*N. lugens*)/SS	5-fold 8.9-fold 5-fold 1.4-fold 2.5-fold	DII	[141]
PACE	Cry1Ac	A01s C04s C05s	Enhanced toxicity (*T. ni* /Cry1Ac resistant *T. ni*)/SS Enhanced toxicity (*T. ni* /Cry1Ac resistant *T. ni*)/SS Enhanced toxicity (*T. ni* /Cry1Ac resistant *T. ni*)/SS	2.2/334-fold 1.1/27.8-fold 1.8/26.4-fold	Not available	[146]
Rational design (reverse PCR)	Cry1Ai	Cry1Ai-h-loop2 Cry1Ai-h-loop2&3	Activity redirected from *B. mori* to *H. armigera*/SL	>7.8-fold >58-fold	DII	[106]
Domain swapping	Cry9Aa	Cry1Ac-Cry9Aa Cry1Ac-Cry9AaMod	Enhanced toxicity (*H. armigera*)/SS Enhanced toxicity (*H. armigera*)/SS	4.9-fold 5.1-fold	DI, DII, DIII	[85]
Gene fusion	Chimeric protein Cry4Ba and Cry1Ac	Cry(4Ba-1Ac)	Enhanced toxicity (*Culex pipiens*)/SS	>238-fold	DI-DVII	[21]
Phage display	Cry1Aa13	Cry1Aa13-A8 Cry1Aa13-A12	Activity redirected from *B. mori* to *Ae. aegypti*/OL	From 0% activity to 90% activity at 20 µg/mL	DII	[140]

Table 1. *Cont.*

Molecular Technique Used	Parental Toxin Evolved	Mutant Name	Evolution Result (Insect)/Evolution Level [1]	Activity Enhancement	Domain Evolved [2]	Reference
Site directed mutagenesis (Alanine scanning)	Cry1Ab	S509A V513A N514A	Enhanced toxicity (*S. frugiperda*)/SS	9.5-fold 12.7-fold 51-fold	DIII (β-16)	[97]
Site directed mutagenesis (saturation mutagenesis)	Cry1Ab	N514F N514H N514K N514L N514Q N514S N514V	Enhanced toxicity (*S. frugiperda*)/SS	44-fold 16-fold 7-fold 9-fold 26-fold 23-fold 9-fold	DIII (β-16)	[97]
Site directed mutagenesis (Alanine scanning)	Cry1Fa	N504A	Enhanced toxicity (*S. frugiperda*)/SS	11-fold	DIII (β-16)	[97]
Site directed mutagenesis	Cry1Ca	V509A N510A	Enhanced toxicity (*S. frugiperda*)/SS Enhanced toxicity (*S. frugiperda*)/SS	1.6-fold 1.5-fold	DIII (β-16) DIII (β-16)	[98]
DNA shuffling	Cry11Aa, Cry11Ba, and Cry11Bb	Variant 8	Enhanced toxicity (*Ae. aegypti*)/SS	3.8-fold increase compared to Cry11Bb and 6.09-fold increase compared to Cry11Aa	DI, DII, DIII	[118]
Rational design and DNA Shuffling	IP3-1: an artificial mutant derived from Cry3Aa1	IP3-2 IP3-3 IP3-4 IP3-5 IP3-6 IP3-7	Enhanced toxicity (*D. virgifera virgifera*)/SS	11-fold 14.6-fold 15.6-fold 19-fold 18.4-fold 29.3-fold	DI, DII, DIII	[52]

[1] Evolution level: SS: Same Specie (evolution toward the same specie; activity enhancement); SL: Specie Level (toxicity evolved to other specie insect from the same order); OL: Order Level (toxicity evolved toward other insect from a different order). [2] DI: Domain 1; DII: Domain 2; DIII: Domain 3. Di *: Indicated the domain evolved although it was not known at the time.

PACE has recently been adapted for the evolution of a 3D-Cry toxin, specifically the lepidopteran Cry1Ac toxin [146]. The cabbage looper *T. ni*, naturally susceptible to Cry1Ac, has developed resistance to the toxin, a fact that has been associated with the mutation of the *ABCC2* transporter gene [147] and the downregulation of the expression of *APN1* [148]. There is no evidence that *T. ni* uses any cadherin-like receptors for its function, so the objective of this work was to evolve a Cry1Ac toxin to specifically bind to a cadherin-like protein (TnCAD), present in the insect cell membrane of *T. ni*, in order to be used as a toxin receptor. For that, the SP plasmid contained, apart from all phage genes (except for *gen III*), a transcriptional fusion of *cry1Ac* with the *rpoZ* gene codifying for the omega sub-unit of the RNA polymerase (Figure 4b). The omega sub-unit is essential for the activity of the RNA polymerase and unless it is present in the RNA polymerase enzymatic complex, the transcription is not possible. The AP plasmid contains M13 *gen III* with an upstream promoter region for the RNA polymerase, and the binding site for the cI protein, a phage repressor protein. In addition, a transcriptional fusion between the cI protein and a fragment of the TnCAD cadherin-like protein, called TnTBR3-F3, was included in the AP plasmid. This fusion protein is able to recognize the cI binding site, allowing the TnTBR3-F3 receptor to interact and bind to other proteins.

In this system, if a Cry toxin variant has the ability of interacting with the TnTBR3-F3 receptor as a consequence of the generated mutations, then it will bind at the promoter region of the M13 *gen III* through the omega sub-unit of the RNA polymerase, allowing the expression of the essential *gen III* for phage replication. If the introduced mutation is not suitable for TnTBR3-F3 binding, then the mutant will not be present in the promoter region, the protein III will not be produced, and the resulting phage will not be infective and it will be lost. The PACE system adapted to the evolution of a Cry1Ac toxin, rendered A01s, C03s, and C05s variants with high binding affinities to the membrane protein TnCAD, in opposition to the wild type toxin Cry1Ac. In addition, when these toxins were bioassayed against *T. ni*, they were 2.2, 1.1, and 1.8 times more active compared to the wild type Cry1Ac, respectively, indicating that toxin-receptor evolution had been taking place. Furthermore, when A01s, C03s and C05s were bioassayed against Cry1Ac resistant *T. ni*, toxin variants showed an increase in toxicity of 334-, 27.8-, and 26.4-fold compared to the wild type Cry1Ac toxin. This work represents a proof of concept that evolution of 3D-Cry toxins to bind novel receptors is possible through PACE and that the technique could be useful in cases where insects have developed resistance to natural toxins.

3. Concluding Remarks

The use of several molecular techniques has allowed researchers to obtain 3D-Cry toxin mutants with improved activities compared to natural toxins. Although in many cases the reason behind this enhancement is not known, the reality is that molecular techniques have been proven to be useful to develop artificial variants. From a practical point of view, these variants represent a real alternative to (i) the intrinsic limitation that 3D-Cry toxins show, as they are only active against a narrow range of insects, and (ii) the resistance phenomena that insects have experienced as a consequence of the extensive use of natural 3D-Cry toxins. This report is proof that minimal changes in the amino acid sequence of a 3D-Cry toxin can lead to a great improvement in toxicity, and that protein engineering, rational design, and in vitro evolution are powerful tools to develop artificial 3D-Cry toxins with surprising and novel activities. The compilation of all of these successful examples and the description of all the sensitive positions that have been used to obtain 3D-Cry toxin variants represents a valuable source of information for the further manipulation of natural toxins.

Supplementary Materials: The following are available online at http://www.mdpi.com/2072-6651/12/9/600/s1, Table S1: Sequence of some mutant toxins mentioned in the review.

Funding: This research was funded by the Andalusian Operative Program, Grant No. B-BIO-081-UGR18.

Acknowledgments: I would like to thank Guest Editors Juan Ferré, Yolanda Bel, and Patricia Hernandez-Martínez from the University of Valencia, Spain for the invitation to write this review for the Special Issue "*Bacillus thuringiensis* Toxins: Functional Characterization and Mechanism of Action". I wish to thank my mentor Prof. David Ellar from the University of Cambridge, UK, who introduced me to the exciting field of Cry toxin research.

Conflicts of Interest: The author declares no conflict of interest.

References

1. Dammak, M.; Jaoua, S.; Tounsi, S. Construction of a *Bacillus thuringiensis* genetically-engineered strain harbouring the secreted Cry1Ia delta-endotoxin in its crystal. *Biotechnol. Lett.* **2011**, *33*, 2367–2372. [CrossRef]
2. Hannay, C. Crystalline inclusions in aerobic sporeforming bacteria. *Nature* **1953**, *172*, 1004. [CrossRef]
3. Angus, A. Bacterial toxin paralysing silkworm larvae. *Nature* **1954**, *4403*, 545. [CrossRef]
4. Ishiwata, S. On a type of severe flacherie (sotto disease). *Dainihon Sanshi Kaiho* **1901**, *114*, 1–5.
5. Bacterial Pesticidal Protein Resource Center. Available online: https://www.bpprc.org (accessed on 15 July 2020).
6. Crickmore, N.; Berry, C.; Panneerselvam, S.; Mishra, R.; Connor, T.R.; Bonning, B.C. A structure-based nomenclature for *Bacillus thuringiensis* and other bacteria-derived pesticidal proteins. *J. Invertebr. Pathol.* **2020**, 107438. [CrossRef]
7. Crickmore, N.; Zeigler, D.R.; Feitelson, J.; Schnepf, E.; Van Rie, J.; Lereclus, D.; Baum, J.; Dean, D.H. Revision of the nomenclature for the *Bacillus thuringiensis* pesticidal crystal proteins. *Microbiol. Mol. Biol. Rev.* **1998**, *62*, 807–813. [CrossRef]
8. Palma, L.; Muñoz, D.; Berry, C.; Murillo, J.; Caballero, P. *Bacillus thuringiensis* toxins: An overview of their biocidal activity. *Toxins* **2014**, *6*, 3296–3325. [CrossRef]
9. Jing, X.; Yuan, Y.; Wu, Y.; Wu, D.; Gong, P.; Gao, M. Crystal structure of *Bacillus thuringiensis* Cry7Ca1 toxin active against *Locusta migratoria* manilensis. *Protein Sci.* **2019**, *28*, 609–619. [CrossRef]
10. Li, J.; Carroll, J.; Ellar, D. Crystal structure of insecticidal delta-endotoxin from *Bacillus thuringiensis* at 2.5 Å resolution. *Nature* **1991**, *353*, 815–821. [CrossRef]
11. Grochulski, P.; Masson, L.; Borisova, S.; Pusztai-Carey, M.; Schwartz, J.-L.; Brousseau, R.; Cygler, M. *Bacillus thuringiensis* CryIA(a) Insecticidal toxin: Crystal structure and channel formation. *J. Mol. Biol.* **1995**, *254*, 447–464. [CrossRef]
12. Galitsky, N.; Cody, V.; Wojtczak, A.; Ghosh, D.; Luft, J.R.; Pangborn, W.; English, L. Biological crystallography structure of the insecticidal bacterial δ-endotoxin Cry3Bb1 of *Bacillus thuringiensis*. *Acta Cryst.* **2001**, *57*, 1101–1109. [CrossRef]
13. Derbyshire, D.; Ellar, D.; Li, J. Crystallization of the *Bacillus thuringiensis* toxin Cry1Ac and its complex with the receptor ligand N-acetyl-D-galactosamine. *Acta Crystallogr. D Biol. Crystallogr.* **2001**, *57*, 1938–1944. [CrossRef]
14. Morse, R.J.; Yamamoto, T.; Stroud, R.M. Structure of Cry2Aa suggests an unexpected receptor binding epitope. *Structure* **2001**, *9*, 409–417. [CrossRef]
15. Boonserm, P.; Davis, P.; Ellar, D.J.; Li, J. Crystal Structure of the mosquito-larvicidal toxin Cry4Ba and its biological implications. *J. Mol. Biol.* **2005**, *348*, 363–382. [CrossRef]
16. Boonserm, P.; Mo, M.; Angsuthanasombat, C.; Lescar, J. Structure of the functional form of the mosquito larvicidal Cry4Aa toxin from *Bacillus thuringiensis* at a 2.8-angstrom resolution. *J. Bacteriol.* **2006**, *188*, 3391–3401. [CrossRef]
17. Guo, S.; Ye, S.; Liu, Y.; Wei, L.; Xue, J.; Wu, H.; Song, F.; Zhang, J.; Wu, X.; Huang, D.; et al. Crystal structure of *Bacillus thuringiensis* Cry8Ea1: An insecticidal toxin toxic to underground pests, the larvae of *Holotrichia parallela*. *J. Struct. Biol.* **2009**, *168*, 259–266. [CrossRef]
18. Hui, F.; Scheib, U.; Hu, Y.; Sommer, R.J.; Aroian, R.V.; Ghosh, P. Structure and glycolipid binding properties of the nematicidal protein Cry5B. *Biochemistry* **2012**, *51*, 9911–9921. [CrossRef]
19. Evdokimov, A.G.; Moshiri, F.; Sturman, E.J.; Rydel, T.J.; Zheng, M.; Seale, J.W.; Franklin, S. Structure of the full-length insecticidal protein Cry1Ac reveals intriguing details of toxin packaging into in vivo formed crystals. *Protein Sci.* **2014**, *23*, 1491–1497. [CrossRef]
20. Pardo-López, L.; Soberón, M.; Bravo, A. *Bacillus thuringiensis* insecticidal three-domain Cry toxins: Mode of action, insect resistance and consequences for crop protection. *FEMS Microbiol. Rev.* **2013**, *37*, 3–22. [CrossRef]
21. Zghal, R.Z.; Elleuch, J.; Ben Ali, M.; Darriet, F.; Rebaï, A.; Chandre, F.; Jaoua, S.; Tounsi, S. Towards novel Cry toxins with enhanced toxicity/broader: A new chimeric Cry4Ba/Cry1Ac toxin. *Appl. Microbiol. Biotechnol.* **2017**, *101*, 113–122. [CrossRef]

22. Melo, A.L.D.A.; Soccol, V.T.; Soccol, C.R. *Bacillus thuringiensis*: Mechanism of action, resistance, and new applications: A review. *Crit. Rev. Biotechnol.* **2016**, *36*, 317–326. [CrossRef] [PubMed]

23. Lucena, W.A.; Pelegrini, P.B.; Martins-de-Sa, D.; Fonseca, F.C.A.; Gomes, J.E.; de Macedo, L.L.P.; da Silva, M.C.M.; Sampaio, R.; Grossi-de-Sa, M.F. Molecular approaches to improve the insecticidal activity of *Bacillus thuringiensis* Cry toxins. *Toxins* **2014**, *6*, 2393–2423. [CrossRef] [PubMed]

24. Coates, B.S. *Bacillus thuringiensis* toxin resistance mechanisms among Lepidoptera: Progress on genomic approaches to uncover causal mutations in the European corn borer, *Ostrinia nubilalis*. *Curr. Opin. Insect Sci.* **2016**, *15*, 70–77. [CrossRef] [PubMed]

25. Peterson, B.; Bezuidenhout, C.; Van den Berg, J. An overview of mechanisms of Cry toxin resistance in Lepidopteran insects. *J. Econ. Entomol.* **2017**, *110*, 362–377. [CrossRef] [PubMed]

26. Bravo, A.; Gómez, I.; Porta, H.; García-Gómez, B.I.; Rodriguez-Almazan, C.; Pardo, L.; Soberón, M. Evolution of *Bacillus thuringiensis* Cry toxins insecticidal activity. *Microb. Biotechnol.* **2013**, *6*, 17–26. [CrossRef]

27. Deist, B.R.; Rausch, M.A.; Fernandez-Luna, M.T.; Adang, M.J.; Bonning, B.C. Bt toxin modification for enhanced efficacy. *Toxins* **2014**, *6*, 3005–3027. [CrossRef]

28. Pardo-López, L.; Muñoz-Garay, C.; Porta, H.; Rodríguez-Almazán, C.; Soberón, M.; Bravo, A. Strategies to improve the insecticidal activity of Cry toxins from *Bacillus thuringiensis*. *Peptides* **2009**, *30*, 589–595. [CrossRef]

29. Khasdan, V.; Sapojnik, M.; Zaritsky, A.; Horowitz, A.R.; Boussiba, S.; Rippa, M.; Manasherob, R.; Ben-Dov, E. Larvicidal activities against agricultural pests of transgenic *Escherichia coli* expressing combinations of four genes from *Bacillus thuringiensis*. *Arch. Microbiol.* **2007**, *188*, 643–653. [CrossRef]

30. Elleuch, J.; Jaoua, S.; Ginibre, C.; Chandre, F.; Tounsi, S.; Zghal, R.Z. Toxin stability improvement and toxicity increase against dipteran and lepidopteran larvae of *Bacillus thuringiensis* crystal protein Cry2Aa. *Pest Manag. Sci.* **2016**, *72*, 2240–2246. [CrossRef]

31. Hu, S.B.; Liu, P.; Ding, X.Z.; Yan, L.; Sun, Y.J.; Zhang, Y.M.; Li, W.P.; Xia, L.Q. Efficient constitutive expression of chitinase in the mother cell of *Bacillus thuringiensis* and its potential to enhance the toxicity of Cry1Ac protoxin. *Appl. Microbiol. Biotechnol.* **2009**, *82*, 1157–1167. [CrossRef]

32. Leetachewa, S.; Khomkhum, N.; Sakdee, S.; Wang, P.; Moonsom, S. Enhancement of insect susceptibility and larvicidal efficacy of Cry4Ba toxin by calcofluor. *Parasites Vectors* **2018**, *11*, 1–9. [CrossRef] [PubMed]

33. Pan, X.; Xu, Z.; Li, L.; Shao, E.; Chen, S.; Huang, T.; Chen, Z.; Rao, W.; Huang, T.; Zhang, L.; et al. Adsorption of Insecticidal Crystal Protein Cry11Aa onto Nano-Mg(OH)2: Effects on bioactivity and anti-ultraviolet ability. *J. Agric. Food Chem.* **2017**, *65*, 9428–9434. [CrossRef] [PubMed]

34. Pérez, C.; Fernández, L.E.; Sun, J.; Folch, J.L.; Gill, S.S.; Soberón, M.; Bravo, A. *Bacillus thuringiensis* subsp. israelensis Cyt1Aa synergizes Cry11Aa toxin by functioning as a membrane-bound receptor. *Proc. Natl. Acad. Sci. USA* **2005**, *102*, 18303–18308. [CrossRef] [PubMed]

35. Cantón, P.E.; Reyes, E.Z.; De Escudero, I.R.; Bravo, A.; Soberón, M. Binding of *Bacillus thuringiensis* subsp. israelensis Cry4Ba to Cyt1Aa has an important role in synergism. *Peptides* **2011**, *32*, 595–600. [CrossRef]

36. Park, H.W.; Pino, B.C.; Kozervanich-Chong, S.; Hafkenscheid, E.A.; Oliverio, R.M.; Federici, B.A.; Bideshi, D.K. Cyt1Aa from *Bacillus thuringiensis* subsp. israelensis enhances mosquitocidal activity of B. thuringiensis subsp. kurstaki HD-1 against Aedes aegypti but not Culex quinquefasciatus. *J. Microbiol. Biotechnol.* **2013**, *23*, 88–91. [CrossRef]

37. Elleuch, J.; Zghal, R.Z.; Jemaà, M.; Azzouz, H.; Tounsi, S.; Jaoua, S. New *Bacillus thuringiensis* toxin combinations for biological control of lepidopteran larvae. *Int. J. Biol. Macromol.* **2014**, *65*, 148–154. [CrossRef]

38. Yang, J.; Quan, Y.; Sivaprasath, P.; Shabbir, M.Z.; Wang, Z.; Ferré, J.; He, K. Insecticidal activity and synergistic combinations of ten different Bt toxins against *Mythimna separata* (Walker). *Toxins* **2018**, *10*, 454. [CrossRef]

39. Wirth, M.C.; Jiannino, J.A.; Federici, B.A.; Walton, W.E. Synergy between toxins of *Bacillus thuringiensis* subsp. israelensis and Bacillus sphaericus. *J. Med. Entomol.* **2004**, *41*, 935–941. [CrossRef]

40. Luo, X.; Chen, L.; Huang, Q.; Zheng, J.; Zhou, W.; Peng, D.; Ruan, L.; Sun, M. *Bacillus thuringiensis* metalloproteinase Bmp1 functions as a nematicidal virulence factor. *Appl. Environ. Microbiol.* **2013**, *79*, 460–468. [CrossRef]

41. Matsumoto, R.; Shimizu, Y.; Howlader, M.T.H.; Namba, M.; Iwamoto, A.; Sakai, H.; Hayakawa, T. Potency of insect-specific scorpion toxins on mosquito control using *Bacillus thuringiensis* Cry4Aa. *J. Biosci. Bioeng.* **2014**, *117*, 680–683. [CrossRef]

42. García-Gómez, B.I.; Cano, S.N.; Zagal, E.E.; Dantán-Gonzalez, E.; Bravo, A.; Soberón, M. Insect Hsp90 chaperone assists *Bacillus thuringiensis* Cry toxicity by enhancing protoxin binding to the receptor and by protecting protoxin from gut protease degradation. *mBio* **2019**, *10*, 1–12. [CrossRef] [PubMed]

43. El-Menofy, W.; Osman, G.; Assaeedi, A.; Salama, M. A novel recombinant baculovirus overexpressing a *Bacillus thuringiensis* Cry1Ab toxin enhances insecticidal activity. *Biol. Proced. Online* **2014**, *16*, 1–7. [CrossRef]

44. Xia, L.; Long, X.; Ding, X.; Zhang, Y. Increase in insecticidal toxicity by fusion of the *cry1Ac* gene from *Bacillus thuringiensis* with the neurotoxin gene *hwtx-I*. *Curr. Microbiol.* **2009**, *58*, 52–57. [CrossRef] [PubMed]

45. Li, W.P.; Xia, L.Q.; Ding, X.Z.; Lv, Y.; Luo, Y.S.; Hu, S.B.; Yin, J.; Yan, F. Expression and characterization of a recombinant Cry1Ac crystal protein fused with an insect-specific neurotoxin ω-ACTX-Hv1a in *Bacillus thuringiensis*. *Gene* **2012**, *498*, 323–327. [CrossRef]

46. Sun, Y.; Fu, Z.; He, X.; Yuan, C.; Ding, X.; Xia, L. Enhancement of *Bacillus thuringiensis* insecticidal activity by combining Cry1Ac and bi-functional toxin HWTX-XI from spider. *J. Invertebr. Pathol.* **2016**, *135*, 60–62. [CrossRef] [PubMed]

47. Sellami, S.; Jemli, S.; Abdelmalek, N.; Cherif, M.; Abdelkefi-Mesrati, L.; Tounsi, S.; Jamoussi, K. A novel Vip3Aa16-Cry1Ac chimera toxin: Enhancement of toxicity against *Ephestia kuehniella*, structural study and molecular docking. *Int. J. Biol. Macromol.* **2018**, *117*, 752–761. [CrossRef]

48. Hu, X.; Liu, Z.; Li, Y.; Ding, X.; Xia, L.; Hu, S. PirB-Cry2Aa hybrid protein exhibits enhanced insecticidal activity against *Spodoptera exigua* larvae. *J. Invertebr. Pathol.* **2014**, *120*, 40–42. [CrossRef]

49. Tajne, S.; Sanam, R.; Gundla, R.; Gandhi, N.S.; Mancera, R.L.; Boddupally, D.; Vudem, D.R.; Khareedu, V.R. Molecular modeling of Bt Cry1Ac (DI-DII)-ASAL (*Allium sativum* lectin)-fusion protein and its interaction with aminopeptidase N (APN) receptor of *Manduca sexta*. *J. Mol. Graph. Model.* **2012**, *33*, 61–76. [CrossRef]

50. Tajne, S.; Boddupally, D.; Sadumpati, V.; Vudem, D.R.; Khareedu, V.R. Synthetic fusion-protein containing domains of Bt Cry1Ac and *Allium sativum* lectin (ASAL) conferred enhanced insecticidal activity against major lepidopteran pests. *J. Biotechnol.* **2013**, *171*, 71–75. [CrossRef] [PubMed]

51. Guo, C.H.; Zhao, S.T.; Ma, Y.; Hu, J.J.; Han, X.J.; Chen, J.; Lu, M.Z. *Bacillus thuringiensis* Cry3Aa fused to a cellulase-binding peptide shows increased toxicity against the longhorned beetle. *Appl. Microbiol. Biotechnol.* **2012**, *93*, 1249–1256. [CrossRef]

52. Hou, J.; Cong, R.; Izumi-Willcoxon, M.; Ali, H.; Zheng, Y.; Bermudez, E.; McDonald, M.; Nelson, M.; Yamamoto, T. Engineering of *Bacillus thuringiensis* Cry proteins to enhance the activity against western corn rootworm. *Toxins* **2019**, *11*, 162. [CrossRef] [PubMed]

53. Zhang, Y.; Zhao, D.; Yan, X.; Guo, W.; Bao, Y.; Wang, W.; Wang, X. Identification and characterization of *Hyphantria cunea* aminopeptidase N as a binding protein of *Bacillus thuringiensis* Cry1Ab35 toxin. *Int. J. Mol. Sci.* **2017**, *18*, 2575. [CrossRef]

54. Park, Y.; Abdullah, M.A.F.; Taylor, M.D.; Rahman, K.; Adang, M.J. Enhancement of *Bacillus thuringiensis* Cry3Aa and Cry3Bb toxicities to coleopteran larvae by a toxin-binding fragment of an insect cadherin. *Appl. Environ. Microbiol.* **2009**, *75*, 3086–3092. [CrossRef]

55. Peng, D.; Xu, X.; Ruan, L.; Yu, Z.; Sun, M. Enhancing Cry1Ac toxicity by expression of the *Helicoverpa armigera* cadherin fragment in *Bacillus thuringiensis*. *Res. Microbiol.* **2010**, *161*, 383–389. [CrossRef]

56. Gao, Y.; Jurat-Fuentes, J.L.; Oppert, B.; Fabrick, J.A.; Liu, C.; Gao, J.; Lei, Z. Increased toxicity of *Bacillus thuringiensis* Cry3Aa against *Crioceris quatuordecimpunctata*, *Phaedon brassicae* and *Colaphellus bowringi* by a *Tenebrio molitor* cadherin fragment. *Pest Manag. Sci.* **2011**, *67*, 1076–1081. [CrossRef] [PubMed]

57. Rahman, K.; Abdullah, M.A.F.; Ambati, S.; Taylor, M.D.; Adang, M.J. Differential protection of Cry1Fa toxin against *Spodoptera frugiperda* larval gut proteases by cadherin orthologs correlates with increased synergism. *Appl. Environ. Microbiol.* **2012**, *78*, 354–362. [CrossRef]

58. Park, Y.; Hua, G.; Taylor, M.D.; Adang, M.J. A coleopteran cadherin fragment synergizes toxicity of *Bacillus thuringiensis* toxins Cry3Aa, Cry3Bb, and Cry8Ca against lesser mealworm, *Alphitobius diaperinus* (Coleoptera: Tenebrionidae). *J. Invertebr. Pathol.* **2014**, *123*, 1–5. [CrossRef]

59. Park, Y.; Hua, G.; Ambati, S.; Taylor, M.; Adang, M.J. Binding and synergizing motif within coleopteran cadherin enhances Cry3Bb toxicity on the Colorado Potato Beetle and the lesser mealworm. *Toxins* **2019**, *11*, 386. [CrossRef]

60. Schnepf, H.E.; Whiteley, H.R. Cloning and expression of the *Bacillus thuringiensis* crystal protein gene in *Escherichia coli*. *Proc. Natl. Acad. Sci. USA* **1981**, *78*, 2893–2897. [CrossRef]

61. Schnepf, H.E.; Wongs, H.C.; Whiteley5, H.R. The amino acid sequence of a crystal protein from *Bacillus thuringiensis* deduced from the DNA base sequence. *J. Biol. Chem.* **1985**, *260*, 6264–6272.

62. Wong, S.; Wilcox, E.; Edwards, D.; Herrnstadt, C. Process for Altering the Host Range of *Bacillus thuringiensis* Toxins, and Novel Toxins Produced Thereby 1986. Pattent Application No. CA19860617139 14 October 1986.

63. Smith, H.O.; Wilcox, K.W. A restriction enzyme from *Hemophilus influenzae* I. Purification and general properties. *J. Mol. Biol.* **1970**, *51*, 379–391. [CrossRef]

64. Sanger, F.; Coulson, A. A rapid method for determining sequences in DNA by primed synthesis with DNA polymerase. *J. Mol. Biol.* **1975**, *94*, 441–448. [CrossRef]

65. Sanger, F.; Nicklen, S.; Coulson, A.R. DNA sequencing with chain-terminating inhibitors. *Proc. Natl. Acad. Sci. USA* **1977**, *74*, 5463–5467. [CrossRef] [PubMed]

66. Shortle, D.; Botstein, D. Single-stranded gaps as localized targets for in vitro mutagenesis. *Basic Life Sci.* **1982**, *20*, 147–155. [CrossRef] [PubMed]

67. Myers, R.; Lerman, L.; Maniatis, T. A general method for saturation mutagenesis of cloned DNA fragments. *Science* **1985**, *229*, 242–247. [CrossRef]

68. Jellis, C.; Bass And, D.; Beerman, N.; Dennis, C.; Farrell, K.; Piot, J.C.; Rusche, J.; Carson, H.; Witt, D. Molecular biology of *Bacillus thuringiensis* and potential benefits to agriculture. *Isr. J. Entomol.* **1989**, *23*, 189–199.

69. Saiki, R.; Gelfand, D.; Stoffel, S.; Scharf, S.; Higuchi, R.; Horn, G.; Mullis, K.; Erlich, H. Primer-directed enzymatic amplification of DNA with a thermostable DNA polymerase. *Science* **1988**, *239*, 487–491. [CrossRef]

70. Ge, A.Z.; Shivarova, N.I.; Deant, D.H. Location of the *Bombyx mori* specificity domain on a *Bacillus thuringiensis* 6-endotoxin protein. *Proc. Nat. Acad. Sci. USA* **1989**, *86*, 4037–4041. [CrossRef]

71. Höfte, H.; Whiteley, H. Insecticidal crystal proteins of *Bacillus thuringiensis*. *Microbiol. Rev.* **1989**, *52*, 242–255. [CrossRef]

72. Schnepf, H.E.; Tomczak, K.; Ortega, J.P.; Whiteleys, H.R. Specificity-determining regions of a Lepidopteran-specific insecticidal protein produced by *Bacillus thuringiensis*. *J. Biol. Chem.* **1990**, *265*, 20923–20930.

73. Widner, W.R.; Whiteley, H.R. Location of the Dipteran specificity region in a Lepidopteran-Dipteran crystal protein from *Bacillus thuringiensis*. *J. Bacteriol.* **1990**, *172*, 2826–2832. [CrossRef]

74. Ge, A.Z.; Rivers, D.; Milne, R.; Dean, D.H. Functional domains of *Bacillus thuringiensis* insecticidal crystal proteins. *J. Biol. Chem.* **1991**, *266*, 17954–17958. [PubMed]

75. Weber, H.; Weissmann, C. Formation of genes coding for hybrid proteins by recoinbination between related, cloned genes in *Escherichia coli*. *Nucl. Acids Res.* **1983**, *11*, 5661–5669. [CrossRef] [PubMed]

76. Caramori, T.; Albertini, A.M.; Galizzi, A. In vivo generation of hybrids between two *Bacillus thuringiensis* insect-toxin-encoding genes. *Gene* **1991**, *98*, 37–44. [CrossRef]

77. Bosch, D.; Schipper, B.; Van Der Kleij, H.; De Maagd, R.A.; Stiekema, W.J. Recombinant *Bacillus thuringiensis* crystal proteins with new properties: Possibilities for resistance management. *Biotechnology* **1994**, *12*, 915–918. [CrossRef] [PubMed]

78. De Maagd, R.A.; Kwa, M.S.G.; Van Der Klei, H.; Yamamoto, T.; Schipper, B.; Vlak, J.M.; Stiekema, W.J.; Bosch, D. Domain III substitution in *Bacillus thuringiensis* delta-endotoxin CryIA(b) results in superior toxicity for *Spodoptera exigua* and altered membrane protein recognition. *Appl. Environ. Microbiol.* **1996**, *62*, 1537–1543. [CrossRef] [PubMed]

79. De Maagd, R.A.; Weemen-Hendriks, M.; Stiekema, W.; Bosch, D. *Bacillus thuringiensis* delta-endotoxin Cry1C domain III can function as a specificity determinant for *Spodoptera exigua* in different, but not all, Cry1-Cry1C hybrids. *Appl. Environ. Microbiol.* **2000**, *66*, 1559–1563. [CrossRef] [PubMed]

80. Karlova, R.; Weemen-Hendriks, M.; Naimov, S.; Ceron, J.; Dukiandjiev, S.; De Maagd, R.A. *Bacillus thuringiensis* δ-endotoxin Cry1Ac domain III enhances activity against *Heliothis virescens* in some, but not all Cry1-Cry1Ac hybrids. *J. Invertebr. Pathol.* **2005**, *88*, 169–172. [CrossRef]

81. Sanchis, V.; Gohar, M.; Chaufaux, J.; Arantes, O.; Meier, A.; Agaisse, H.; Cayley, J.; Lereclus, D. Development and field performance of a broad-spectrum nonviable asporogenic recombinant strain of *Bacillus thuringiensis* with greater potency and UV resistance. *Appl. Environ. Microbiol.* **1999**, *65*, 4032–4039. [CrossRef]

82. Naimov, S.; Weemen-Hendriks, M.; Dukiandjiev, S.; De Maagd, R.A. *Bacillus thuringiensis* delta-endotoxin Cry1 hybrid proteins with increased activity against the Colorado potato beetle. *Appl. Environ. Microbiol.* **2001**, *67*, 5328–5330. [CrossRef]

83. Walters, F.S.; Defontes, C.M.; Hart, H.; Warren, G.W.; Chen, J.S. Lepidopteran-active variable-region sequence imparts coleopteran activity in eCry3.1Ab, an engineered *Bacillus thuringiensis* hybrid insecticidal protein. *Appl. Environ. Microbiol.* **2010**, *76*, 3082–3088. [CrossRef] [PubMed]

84. Walters, F.S.; Stacy, C.M.; Mi, K.L.; Palekar, N.; Chen, J.S. An engineered chymotrypsin/cathepsin G site in domain I renders *Bacillus thuringiensis* Cry3A active against western corn rootworm larvae. *Appl. Environ. Microbiol.* **2008**, *74*, 367–374. [CrossRef]

85. Shah, J.V.; Yadav, R.; Ingle, S.S. Engineered Cry1Ac-Cry9Aa hybrid *Bacillus thuringiensis* delta-endotoxin with improved insecticidal activity against *Helicoverpa armigera*. *Arch. Microbiol.* **2017**, *199*, 1069–1075. [CrossRef]

86. Dulau, L.; Cheyrou, A.; Aigle, M. Directed mutagenesis using PCR. *Nucleic Acids Res.* **1989**, *17*, 2873. [CrossRef] [PubMed]

87. Wu, D.; Aronson, A.I. Localized mutagenesis defines regions of the *Bacillus thuringiensis* delta-endotoxin involved in toxicity and specificity. *J. Biol. Chem.* **1992**, *267*, 2311–2317. [PubMed]

88. Padilla, C.; Pardo-López, L.; De La Riva, G.; Gómez, I.; Sánchez, J.; Hernandez, G.; Nuñez, M.E.; Carey, M.P.; Dean, D.H.; Alzate, O.; et al. Role of tryptophan residues in toxicity of Cry1Ab toxin from *Bacillus thuringiensis*. *Appl. Environ. Microbiol.* **2006**, *72*, 901–907. [CrossRef]

89. Alzate, O.; Osorio, C.; Florez, A.M.; Dean, D.H. Participation of valine 171 in α-helix 5 of *Bacillus thuringiensis* Cry1Ab δ-endotoxin in translocation of toxin into *Lymantria dispar* midgut membranes. *Appl. Environ. Microbiol.* **2010**, *76*, 7878–7880. [CrossRef] [PubMed]

90. Smedley, D.P.; Ellar, D.J. Mutagenesis of three surface-exposed loops of a *Bacillus thuringiensis* insecticidal toxin reveals residues important for toxicity, receptor recognition and possibly membrane insertion. *Microbiology* **1996**, *142*, 617–624. [CrossRef]

91. Wu, S.J.; Dean, D.H. Functional significance of loops in the receptor binding domain of *Bacillus thuringiensis* CryIIIA delta-endotoxin. *J. Mol. Biol.* **1996**, *255*, 628–640. [CrossRef]

92. Rajamohan, F.; Alzate, O.; Cotrill, J.A.; Curtiss, A.; Dean, D.H. Protein engineering of *Bacillus thuringiensis* endotoxin: Mutations at domain II of CryIAb enhance receptor affinity and toxicity toward gypsy moth larvae. *Proc. Natl. Acad. Sci. USA* **1996**, *93*, 14338–14343. [CrossRef]

93. Wu, S.-J.; Koller, C.N.; Miller, D.L.; Bauer, L.S.; Dean, D.H. Enhanced toxicity of *Bacillus thuringiensis* Cry3A N-endotoxin in coleopterans by mutagenesis in a receptor binding loop. *FEBS Lett.* **2000**, *473*, 227–232. [CrossRef] [PubMed]

94. Mcneil, B.C.; Dean, D.H. *Bacillus thuringiensis* Cry2Ab is active on *Anopheles* mosquitoes: Single D block exchanges reveal critical residues involved in activity. *FEMS Microbiol. Lett.* **2011**, *325*, 16–21. [CrossRef] [PubMed]

95. Nicholls, C.N.; Ahmad, W.; Ellar, D.J. Evidence for two different types of insecticidal P2 toxins with dual specificity in *Bacillus thuringiensis* subspecies. *J. Bacteriol.* **1989**, *171*, 5141–5147. [CrossRef] [PubMed]

96. Lv, Y.; Tang, Y.; Zhang, Y.; Xia, L.; Wang, F.; Ding, X.; Yi, S.; Li, W.; Yin, J. The role of β20-β21 loop structure in insecticidal activity of Cry1Ac toxin from *Bacillus thuringiensis*. *Curr. Microbiol.* **2011**, *62*, 665–670. [CrossRef]

97. Gómez, I.; Ocelotl, J.; Sánchez, J.; Lima, C.; Martins, E.; Rosales-Juárez, A.; Aguilar-Medel, S.; Abad, A.; Dong, H.; Monnerat, R.; et al. Enhancement of *Bacillus thuringiensis* Cry1Ab and Cry1Fa toxicity to *Spodoptera frugiperda* by domain III mutations indicates there are two limiting steps in toxicity as defined by receptor binding and protein stability. *Appl. Environ. Microbiol.* **2018**, *84*, 1–11. [CrossRef]

98. Gómez, I.; Rodríguez-Chamorro, D.E.; Flores-Ramírez, G.; Grande, R.; Zúñiga, F.; Portugal, F.J.; Sánchez, J.; Pacheco, S.; Bravo, A.; Soberón, M. *Spodoptera frugiperda* (J. E. Smith) aminopeptidase N1 is a functional receptor of the *Bacillus thuringiensis* Cry1Ca toxin. *Appl. Environ. Microbiol.* **2018**, *84*. [CrossRef]

99. Xia, L.; Wang, F.; Ding, X.; Zhao, X.; Fu, Z.; Quan, M.; Yu, Z. The role of β18–β19 lobop structure in insecticidal activity of Cry1Ac toxin from *Bacillus thuringiensis*. *Chin. Sci. Bull.* **2008**, *53*, 3178–3184. [CrossRef]

100. Xiang, W.F.; Qiu, X.L.; Zhi, D.X.; Min, Z.X.; Yuan, L.; Quan, Y.Z. N546 in β18-β19 loop is important for binding and toxicity of the *Bacillus thuringiensis* Cry1Ac toxin. *J. Invertebr. Pathol.* **2009**, *101*, 119–123. [CrossRef]

101. Wang, F.; Liu, Y.; Zhang, F.; Chai, L.; Ruan, L.; Peng, D.; Sun, M. Improvement of crystal solubility and increasing toxicity against *Caenorhabditis elegans* by asparagine substitution in block 3 of *Bacillus thuringiensis* crystal protein Cry5Ba. *Appl. Environ. Microbiol.* **2012**, *78*, 7197–7204. [CrossRef]

102. Angsuthanasombat, C.; Crickmore, N.; Ellar, D.J. Effects on toxicity of eliminating a cleavage site in a predicted interhelical loop in *Bacillus thuringiensis* CryIVB δ-endotoxin. *FEMS Microbiol. Lett.* **1993**, *111*, 378–1097. [CrossRef]

103. Abdullah, M.A.F.; Alzate, O.; Mohammad, M.; McNall, R.J.; Adang, M.J.; Dean, D.H. Introduction of *Culex* toxicity into *Bacillus thuringiensis* Cry4Ba by protein engineering. *Appl. Environ. Microbiol.* **2003**, *69*, 5343–5353. [CrossRef] [PubMed]

104. Abdullah, M.A.F.; Dean, D.H. Enhancement of Cry19Aa mosquitocidal activity against *Aedes aegypti* by mutations in the putative loop regions of domain II. *Appl. Environ. Microbiol.* **2004**, *70*, 3769–3771. [CrossRef] [PubMed]

105. Liu, X.S.; Dean, D.H. Redesigning *Bacillus thuringiensis* Cry1Aa toxin into a mosquito toxin. *Protein Eng. Des. Sel.* **2006**, *19*, 107–111. [CrossRef] [PubMed]

106. Zhou, Z.; Liu, Y.; Liang, G.; Huang, Y.; Bravo, A.; Soberón, M.; Song, F.; Zhou, X.; Zhang, J. Insecticidal specificity of Cry1Ah to *Helicoverpa armigera* is determined by binding of APN1 via domain II loops 2 and 3. *Appl. Environ. Microbiol.* **2017**, *83*. [CrossRef]

107. Mandal, C.C.; Gayen, S.; Basu, A.; Ghosh, K.S.; Dasgupta, S.; Maiti, M.K.; Sen, S.K. Prediction-based protein engineering of domain I of Cry2A entomocidal toxin of *Bacillus thuringiensis* for the enhancement of toxicity against lepidopteran insects. *Protein Eng. Des. Sel.* **2007**, *20*, 599–606. [CrossRef]

108. Smith, G.P.; Ellar, D.J. Mutagenesis of two surface-exposed loops of the *Bacillus thuringiensis* CrylC δ-endotoxin affects insecticidal specificity. *Biochem. J.* **1994**, *302*, 611–616. [CrossRef]

109. Ner, S.S.; Goodin, D.B.; Smith, M. A simple and efficient procedure for generating random point mutations and for codon replacements using mixed oligodeoxynucleotides. *DNA* **1988**, *7*, 127–134. [CrossRef]

110. Kumar, A.S.M.; Aronson, A.I. Analysis of mutations in the pore-forming region essential for insecticidal activity of a *Bacillus thuringiensis* delta-endotoxin. *J. Bacteriol.* **1999**, *181*, 6103–6107. [CrossRef]

111. Leung, D.; Chen, E.; Goeddel, D. A method for random mutagenesis of a defined DNA segment using a modified polymerase chain reaction. *Technique* **1989**, *1*, 11–15.

112. Van Dillewijn, P.; Vilchez, S.; Paz, J.A.; Ramos, J.L. Plant-dependent active biological containment system for recombinant rhizobacteria. *Environ. Microbiol.* **2004**, *6*. [CrossRef]

113. Shu, C.; Liu, R.; Wang, R.; Zhang, J.; Feng, S.; Huang, D.; Song, F. Improving toxicity of *Bacillus thuringiensis* strain contains the *cry8Ca* gene specific to *Anomala corpulenta* larvae. *Curr. Microbiol.* **2007**, *55*, 492–496. [CrossRef] [PubMed]

114. Shan, S.; Zhang, Y.; Ding, X.; Hu, S.; Sun, Y.; Yu, Z.; Liu, S.; Zhu, Z.; Xia, L. A Cry1ac toxin variant generated by directed evolution has enhanced toxicity against lepidopteran insects. *Curr. Microbiol.* **2011**, *62*, 358–365. [CrossRef] [PubMed]

115. Stemmer, W. Rapid evolution of a protein in vitro by DNA shuffling. *Nature* **1994**, *370*, 389–390. [CrossRef] [PubMed]

116. Stemmer, W.P.C. DNA shuffling by random fragmentation and reassembly: In vitro recombination for molecular evolution. *Proc. Natl. Acad. Sci. USA.* **1994**, *91*, 10747–10751. [CrossRef]

117. Knight, J.S.; Broadwell, A.H.; Grant, W.N.; Shoemaker, C.B. A strategy for shuffling numerous *Bacillus thuringiensis* crystal protein domains. *J. Econ. Entomol.* **2004**, *97*, 1805–1813. [CrossRef]

118. Florez, A.M.; Suarez-Barrera, M.O.; Morales, G.M.; Rivera, K.V.; Orduz, S.; Ochoa, R.; Guerra, D.; Muskus, C. Toxic activity, molecular modeling and docking simulations of *Bacillus thuringiensis* Cry11 toxin variants obtained via DNA shuffling. *Front. Microbiol.* **2018**, *9*, 1–14. [CrossRef] [PubMed]

119. Shu, C.; Zhou, J.; Crickmore, N.; Li, X.; Song, F.; Liang, G.; He, K.; Huang, D.; Zhang, J. In vitro template-change PCR to create single crossover libraries: A case study with *Bacillus thuringiensis* Cry2A toxins. *Sci. Rep.* **2016**, *6*. [CrossRef]

120. Zhao, H.; Giver, L.; Shao, Z.; Affholter, J.A.; Arnold, F.H. Molecular evolution by staggered extension process (StEP) in vitro recombination. *Nat. Biotechnol.* **1998**, *16*, 258–261. [CrossRef] [PubMed]

121. Zhang, Y.; Buchholz, F.; Muyrers, J.P.P.; Stewart, A.F. A new logic for DNA engineering using recombination in *Escherichia coli*. *Nat. Genet.* **1998**, *20*, 123–128. [CrossRef]

122. Zhang, Y.; Muyrers, J.; Testa, G.; Stewart, A. DNA cloning by homologous recombination in *Escherichia coli*. *Nat. Biotechnol.* **2000**, *18*, 1314–1317. [CrossRef]

123. Smith, G. Filamentous fusion phage: Novel expression vectors that display cloned antigens on the virion surface. *Science* **1985**, *228*, 1315–1317. [CrossRef] [PubMed]

124. De La Cruz, V.F.; Lal, A.A.; McCutchan, T.F. Immunogenicity and epitope mapping of foreign sequences via genetically engineered filamentous phage. *J. Biol. Chem.* **1988**, *263*, 4318–4322. [PubMed]

125. Scott, J.; Smith, G. Searching for peptide ligands with an epitope library. *Science* **1990**, *249*, 386–390. [CrossRef] [PubMed]

126. Landon, L.; Deutscher, S. Combinatorial discovery of tumor targeting peptides using phage display. *J. Cell. Biochem.* **2003**, *90*, 509–517. [CrossRef]

127. Luzar, J.; Štrukelj, B.; Lunder, M. Phage display peptide libraries in molecular allergology: From epitope mapping to mimotope-based immunotherapy. *Eur. J. Allergy Clin. Immunol.* **2016**, *71*, 1526–1532. [CrossRef]

128. Marzari, R.; Edomi, P.; Bhatnagar, R.K.; Ahmad, S.; Selvapandiyan, A.; Bradbury, A. Phage display of *Bacillus thuringiensis* CryIA(a) insecticidal toxin. *FEBS Lett.* **1997**, *411*, 27–31. [CrossRef]

129. Kasman, L.M.; Lukowiak, A.A.; Garczynski, S.F.; Mcnall, R.J.; Youngman, P.; Adang, M.J. Phage display of a biologically active *Bacillus thuringiensis* toxin. *Appl. Environ. Microbiol.* **1998**, *64*, 2995–3003. [CrossRef]

130. Vilchez, S.; Jacoby, J.; Ellar, D.J. Display of biologically functional insecticidal toxin on the surface of lambda phage. *Appl. Environ. Microbiol.* **2004**, *70*, 6587–6594. [CrossRef]

131. Pacheco, S.; Gómez, I.; Sato, R.; Bravo, A.; Soberón, M. Functional display of *Bacillus thuringiensis* Cry1Ac toxin on T7 phage. *J. Invertebr. Pathol.* **2006**, *92*, 45–49. [CrossRef]

132. Pacheco, S.; Cantón, E.; Zuñiga-Navarrete, F.; Pecorari, F.; Bravo, A.; Soberón, M. Improvement and efficient display of *Bacillus thuringiensis* toxins on M13 phages and ribosomes. *AMB Express* **2015**, *5*, 73. [CrossRef]

133. Ishikawa, H.; Hoshino, Y.; Motoki, Y.; Kawahara, T.; Kitajima, M.; Kitami, M.; Watanabe, A.; Bravo, A.; Soberon, M.; Honda, A.; et al. A system for the directed evolution of the insecticidal protein from *Bacillus thuringiensis*. *Mol. Biotechnol.* **2007**, *36*, 90–101. [CrossRef] [PubMed]

134. Fujii, Y.; Tanaka, S.; Otsuki, M.; Hoshino, Y.; Morimoto, C.; Kotani, T.; Harashima, Y.; Endo, H.; Yoshizawa, Y.; Sato, R. Cry1Aa binding to the cadherin receptor does not require conserved amino acid sequences in the domain II loops. *Biosci. Rep.* **2013**, *33*, 103–112. [CrossRef] [PubMed]

135. Endo, H.; Kobayashi, Y.; Hoshino, Y.; Tanaka, S.; Kikuta, S.; Tabunoki, H.; Sato, R. Affinity maturation of Cry1Aa toxin to the *Bombyx mori* cadherin-like receptor by directed evolution based on phage display and biopanning selections of domain II loop 2 mutant toxins. *Microbiologyopen* **2014**, *3*, 568–577. [CrossRef]

136. Craveiro, K.I.C.; Gomes, J.E.; Silva, M.C.M.; Macedo, L.L.P.; Lucena, W.A.; Silva, M.S.; de Souza, J.D.A.; Oliveira, G.R.; Quezado de Magalhães, M.T.; Santiago, A.D.; et al. Variant Cry1Ia toxins generated by DNA shuffling are active against sugarcane giant borer. *J. Biotechnol.* **2010**, *145*, 215–221. [CrossRef]

137. Oliveira, G.R.; Silva, M.C.; Lucena, W.A.; Nakasu, E.Y.; Firmino, A.A.; Beneventi, M.A.; Souza, D.S.; Gomes, J.E.; Da De Souza, J.; Rigden, D.J.; et al. Improving Cry8Ka toxin activity towards the cotton boll weevil (*Anthonomus grandis*). *BMC Biotechnol.* **2011**, *11*, 1–13. [CrossRef]

138. Pigott, C.R.; King, M.S.; Ellar, D.J. Investigating the properties of *Bacillus thuringiensis* Cry proteins with novel loop replacements created using combinatorial molecular biology. *Appl. Environ. Microbiol.* **2008**, *74*, 3497–3511. [CrossRef]

139. Andris-Widhopf, J.; Rader, C.; Steinberger, P.; Barbas, C., 3rd; Fuller, R. Methods for the generation of chicken monoclonal antibody fragments by phage display. *J. Immunol. Methods* **2000**, *242*, 159–181. [CrossRef]

140. Domínguez-Flores, T.; Romero-Bosquet, M.D.; Gantiva-Díaz, D.M.; Luque-Navas, M.J.; Berry, C.; Osuna, A.; Vílchez, S. Using phage display technology to obtain Crybodies active against non-target insects. *Sci. Rep.* **2017**, *7*. [CrossRef]

141. Shao, E.; Lin, L.; Chen, C.; Chen, H.; Zhuang, H.; Wu, S.; Sha, L.; Guan, X.; Huang, Z. Loop replacements with gut-binding peptides in Cry1Ab domain II enhanced toxicity against the brown planthopper, *Nilaparvata lugens* (Stål). *Sci. Rep.* **2016**, *6*, 20106. [CrossRef]

142. Esvelt, K.M.; Carlson, J.C.; Liu, D.R. A system for the continuous directed evolution of biomolecules. *Nature* **2011**, *472*, 499–503. [CrossRef]

143. Leconte, A.M.; Dickinson, B.C.; Yang, D.D.; Chen, I.A.; Allen, B.; Liu, D.R. A population-based experimental model for protein evolution: Effects of mutation rate and selection stringency on evolutionary outcomes. *Biochemistry* **2013**, *52*, 1490–1499. [CrossRef] [PubMed]

144. Dickinson, B.C.; Packer, M.S.; Badran, A.H.; Liu, D.R. A system for the continuous directed evolution of proteases rapidly reveals drug-resistance mutations. *Nat. Commun.* **2014**, *5*. [CrossRef]

145. Hubbard, B.P.; Badran, A.H.; Zuris, J.A.; Guilinger, J.P.; Davis, K.M.; Chen, L.; Tsai, S.Q.; Sander, J.D.; Joung, J.K.; Liu, D.R. Continuous directed evolution of DNA-binding proteins to improve TALEN specificity. *Nat. Methods* **2015**, *12*, 939–942. [CrossRef]

146. Badran, A.H.; Guzov, V.M.; Huai, Q.; Kemp, M.M.; Vishwanath, P.; Kain, W.; Nance, A.M.; Evdokimov, A.; Moshiri, F.; Turner, K.H.; et al. Continuous evolution of *Bacillus thuringiensis* toxins overcomes insect resistance. *Nature* **2016**, *533*, 58–63. [CrossRef]

147. Baxter, S.W.; Badenes-Pérez, F.R.; Morrison, A.; Vogel, H.; Crickmore, N.; Kain, W.; Wang, P.; Heckel, D.G.; Jiggins, C.D. Parallel evolution of *Bacillus thuringiensis* toxin resistance in Lepidoptera. *Genetics* **2011**, *189*, 675–679. [CrossRef]

148. Tiewsiri, K.; Wang, P. Differential alteration of two aminopeptidases N associated with resistance to *Bacillus thuringiensis* toxin Cry1Ac in cabbage looper. *Proc. Natl. Acad. Sci. USA* **2011**, *108*, 14037–14042. [CrossRef]

Review

The Cytocidal Spectrum of *Bacillus thuringiensis* Toxins: From Insects to Human Cancer Cells

Gretel Mendoza-Almanza [1], Edgar L. Esparza-Ibarra [2], Jorge L. Ayala-Luján [3], Marisa Mercado-Reyes [2], Susana Godina-González [3], Marisa Hernández-Barrales [3] and Jorge Olmos-Soto [4,*]

[1]   National Council of Science and Technology, Autonomous University of Zacatecas, Zacatecas 98000, Zacatecas, Mexico; grmendoza@conacyt.mx
[2]   Academic Unit of Biological Sciences, Autonomous University of Zacatecas, Zacatecas 98068, Zacatecas, Mexico; lesparza@uaz.edu.mx (E.L.E.-I.); marisamercado@uaz.edu.mx (M.M.-R.)
[3]   Academic Unit of Chemical Sciences, Autonomous University of Zacatecas, Zacatecas 98160, Zacatecas, Mexico; jayala69@uaz.edu.mx (J.L.A.-L.); sgodina@uaz.edu.mx (S.G.-G.); marisahb@uaz.edu.mx (M.H.-B.)
[4]   Department of Marine Biotechnology, Center for Scientific Research and Higher Education of Ensenada, Ensenada 22860, Baja California, Mexico
*   Correspondence: jolmos@cicese.mx; Tel.: +52-646-175-0500

Received: 25 March 2020; Accepted: 2 May 2020; Published: 6 May 2020

Abstract: *Bacillus thuringiensis* (Bt) is a ubiquitous bacterium in soils, insect cadavers, phylloplane, water, and stored grain, that produces several proteins, each one toxic to different biological targets such as insects, nematodes, mites, protozoa, and mammalian cells. Most Bt toxins identify their particular target through the recognition of specific cell membrane receptors. Cry proteins are the best-known toxins from Bt and a great amount of research has been published. Cry are cytotoxic to insect larvae that affect important crops recognizing specific cell membrane receptors such as cadherin, aminopeptidase-N, and alkaline phosphatase. Furthermore, some Cry toxins such as Cry4A, Cry4B, and Cry11A act synergistically with Cyt toxins against dipteran larvae vectors of human disease. Research developed with Cry proteins revealed that these toxins also could kill human cancer cells through the interaction with specific receptors. Parasporins are a small group of patented toxins that may or may not have insecticidal activity. These proteins could kill a wide variety of mammalian cancer cells by recognizing specific membrane receptors, just like Cry toxins do. Surface layer proteins (SLP), unlike the other proteins produced by Bt, are also produced by most bacteria and archaebacteria. It was recently demonstrated that SLP produced by Bt could interact with membrane receptors of insect and human cancer cells to kill them. Cyt toxins have a structure that is mostly unrelated to Cry toxins; thereby, other mechanisms of action have been reported to them. These toxins affect mainly mosquitoes that are vectors of human diseases like *Anopheles spp* (malaria), *Aedes spp* (dengue, zika, and chikungunya), and *Culex spp* (Nile fever and Rift Valley fever), respectively. In addition to the Cry, Cyt, and parasporins toxins produced during spore formation as inclusion bodies, Bt strains also produce Vip (Vegetative insecticidal toxins) and Sip (Secreted insecticidal proteins) toxins with insecticidal activity during their vegetative growth phase.

Keywords: *Bacillus thuringiensis*; Cry; Cyt; parasporins; S-layer proteins; Vip; Sip; membrane receptors; insecticidal activity; anticancer activity

Key Contribution: This review focused on describing *Bacillus thuringiensis* proteins and their toxic and cytotoxic activity.

---

1. Introduction

Bacillus thuringiensis is a Gram-positive and sporulated bacterium that is widely distributed in soils, plants, and insects around the world [1,2]. Bt is well known because it produces a great variety of useful proteins for pest control in agriculture (Cry, Vip, Sip) [3,4], minimizes diseases transmitted by mosquitoes (Cyt) [5], inhibits pathogens development in animals [6], and because it induces cytotoxicity in human cancer cells (PS, SLP, and Cry) [7,8].

In 1901, Ishiwata found Bt for the first time in *Bombyx mori* and called it *Bacillus sotto*. In 1915, in Thuringia, Berliner isolated this bacterium from the moth *Ephestia kuehniella* and called it *Bacillus thuringiensis* [4].

During the sporulation process, Bt produces inclusion bodies (IB) in parasporal position with cubic, bipyramidal, spherical, oval, and irregular shapes that can be distinguished by scanning electron microscopy (SEM) (Figure 1) [9–11]. The IB are formed by a conglomeration of delta-endotoxin monomers classified according to the sequence similarities between two significant families; Cry and Cyt toxins [10–12]. In addition, Bt produces other important toxins such as parasporins, S-layer, Vip, and Sip proteins that will be discussed in this review.

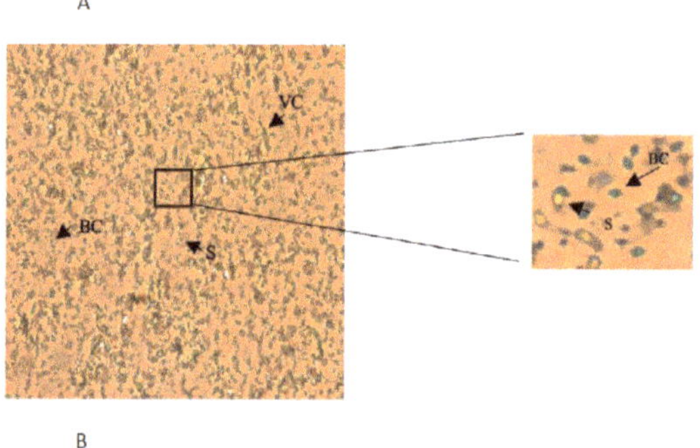

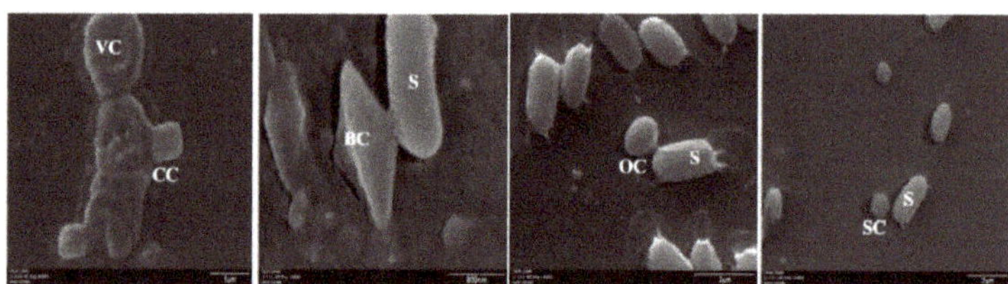

Figure 1. Different morphologies of *Bacillus thuringiensis* crystals; (**A**) image observed at 40× in an optical microscope, the crystal was stained with malachite green. (**B**) Image observed at SEM with different magnifications from the left to the right (1) 20,000×, (2) 15,000×, (3) 15,000×, and (4) 50,000×. VC: vegetative cell; CC: cubic crystal; S: spore; BC: bipyramidal crystal; OC: ovoid crystal; SC: spherical crystal.

Cry and Cyt proteins received their current nomenclature after creation of the *Bacillus thuringiensis* Toxin Nomenclature Committee [13]. This Committee proposed a classification system of four hierarchical ranks based on the place each toxin occupies in the phylogenetic tree. Cry and Cyt delta-endotoxins with less than 45% sequence identity differ in primary level and are classified as Cyt1, Cry1, Cry2, etc. Cry and Cyt delta-endotoxins with 78% sequence identity differ in secondary rank and a capital letter is added to their name, e.g., Cyt1A, Cry1A, Cry2A. Toxins with 95% identity constitute the border for a tertiary rank and small letters differentiate these proteins from each other, e.g., Cyt1Aa, Cry1Aa, Cry1Ab, Cry1Ac [13–15].

Parasporins have less than 25% amino acid sequence homology with Cry toxins [7]. However, their mechanism of action is very similar; both families recognize specific membrane receptors on cancer cells to trigger cell death [6]. Parasporins do not induce hemolytic activity but may or may not have insecticidal activity, nevertheless, they show preferential cytotoxicity against human cancer cells instead of healthy human cells in vitro [16].

Surface layer proteins (SLP) are embedded into cell membranes of many Gram-negative and Gram-positive bacteria; they are commonly associated with polysaccharides and peptidoglycans, respectively [9]. Main functions of SLP proteins are: (1) interaction with extracellular proteins, (2) protection against pathogens, (3) phagocytosis, (4) stabilization of membranes, and (5) adhesion, among others [17]. Unlike *cry* genes, which are expressed during the sporulation process, *s-layer* genes are constitutively expressed throughout the entire cell life cycle. According to previous reports, S-layer proteins from Bt are associated with toxicity against *Epilachna varivestis* [9,18]. Furthermore, there is a report from an S-layer protein with selective cytotoxic activity against MDA-MB-231 breast cancer cells line [8].

Bt synthesizes and secretes to the medium Sip (secreted insecticidal protein) and Vip (vegetative insecticidal protein) proteins during the exponential growth phase. There are reports about the insecticidal activity of these proteins against some coleopterans and lepidopterans [19–21].

Bt toxins have been isolated and classified into at least 78 Cry [13], 3 Cyt [13], 6 parasporins [7], 1 SLP [18], 1 Sip [11], and 4 Vip families [11].

2. Leading Toxic Proteins of *Bacillus thuringiensis* and their Mechanism of Action

2.1. Cry Toxins

Cry proteins are widely known by their toxic activity against insects belonging to orders such as *Hymenoptera, Coleoptera, Homoptera, Orthoptera*, and *Mallophaga*, as well as nematodes, mites, and protozoa [2,11,22,23]. Their toxic activity against insect larvae has allowed these toxins to be used for biological control of pests through spray formulations and transgenic crops that incorporate Cry proteins or some active fragment [22,24–28]. Tobacco was the first Bt crop produced by "Plant Genetic Systems" in Belgium in 1985 [29]. Since then, other crops, such as corn, cotton, potato, rice, brinjal, and soybean, have been genetically modified with Bt toxins to resist insect pests [30,31].

Cry toxins are highly specific to their target organisms. It is unusual to find a Cry toxin that targets more than one insect order, as is the case of Cry1Ba which is toxic to moths, flies, and beetles larvae [32]. *Cry* genes reside on plasmids that are naturally transferred from one Bt strain to another by conjugation or recombination [33,34]. This transfer of information plays an essential role in the biodiversity of Bt strains [34]. The final composition of *cry* genes in a strain determines the specificity and toxicity against biological targets, including human cells [3,15,34]. More than 700 *cry* genes have been classified into groups and subgroups, according to their amino acid sequence similarity [11,13].

X-ray crystallography of Cry proteins has evidenced three structural domains; hence, Cry toxins are also known as 3d-Cry toxins. The N-terminal Domain I is formed by seven α-helices, with a conserved hydrophobic helix α-5 in the center, which is related to oligomerization of the toxin [3,4,35]. Helix α-5 is also responsible for pore-formation in the membrane of susceptible cells, and for toxin insertion into the cell. Given these characteristics, Cry proteins are classified as pore-forming toxins (PFT). In this sense, Domain I is the most conserved among all Cry toxins, sharing some structural similarity with Colicin Ia, another PFT [3,4,36,37].

Domain II is composed of three antiparallel β-sheets that form a hydrophobic core, with highly variable exposed loop regions. This domain is responsible for toxin specificity; therefore, indicates the binding sites into receptors in susceptible larvae [3,38–40].

Domain III is composed of antiparallel β sheets that form a β sandwich structure. This domain is also involved in receptor binding specificity; additionally, it is also associated with pore formation in cell membranes [3,38–40].

Cry toxins belong to the PFT family due to their mechanism of action, in which Domain I is inserted into the membrane of target cells, creating a trans-membrane ion channel and triggering the host's death [2,3,41]. The PFT family comprises different types of toxins, including, among others, Colicin family, produced by Escherichia coli [42]. The ClyA family is produced by *Escherichia coli* and *Salmonella* enteric strains [43]. The Actinoporin family is produced by sea anemones [44]. The Haemolysin family, produced by *Staphylococcus aureus* [45]. The Aerolysin family, produced by *Aeromonas hydrophila* [46]. The CDC family, produced by pathogenic Gram-positive bacteria such as *Clostridium perfringens, Bacillus anthracis*, and *Streptococcus pneumoniae* [47]. The toxins from the PFT family have several characteristics in common: (1) The way they fold, which suggests all share a similar mechanism of action [48–51]; (2) All recognize specific receptors on cell membranes [48–51]; (3) Promote oligomerization at the interaction site after receptor recognition [48–51].

PFT are classified into two main groups according to their secondary structures, which are responsible to the formation of pores: toxins from α-helical group includes Colicin, Exotoxin A, Diphtheria, and Cry toxins. In this group, the α-helix region is responsible for the trans-membrane ion channel formation [48–50]. On the other hand, β-barrel toxins include Aerolysin [50], Hemolysin [45], Perfringolysin O [52], and Cyt toxins [53]. These toxins insert themselves into the cell membrane, forming a barrel composed of β sheets hairpin monomers [48–50].

Mechanism of Action from Cry Toxins

3d-Cry proteins are produced as large protoxins with a molecular weight around 130 kDa, such as Cry1Aa protein [3,4,11,14], or short protoxins between 65 and 70 kDa, such as Cry11Aa protein [4,14]. The large protoxins are processed by insect midgut proteases at C-terminal and N-terminal ends [3,4,11,14], while short protoxins are processed only at the N-terminal end [4,14]. In both cases, protease-resistant core results in an active Cry toxin of 60 and 70 kDa approximately which retains the 3d structure [1,4,11,14]. The resistant fragment is responsible for cytotoxicity against larvae insects, nematodes, protozoans, and human cancer cells [1,4,11,14]. However, incorrect or deficient protoxin activation, and/or rapid degradation of toxins by other proteases, could reduce the toxicity of Cry proteins against their target [4,14].

(a) The Pore-Forming Model

The most accepted mechanism of action of Cry toxins against insect larvae is the pore-forming model [33,37,54,55], which is summarized in Figure 2A. Once larvae ingest toxin crystals, these solubilized at extreme pH (acid or alkaline, depending on the Cry toxin) and proteolyzed by proteases under suitable physicochemical conditions on the midgut. The activated toxins can reach the apical brush border membrane (microvilli) of the insect's midgut by crossing the peritrophic matrix. Cry toxins must recognize receptors in brush-border membranes to form pores; therefore, specificity is crucial for Cry proteins toxicity [3,55,56]. In this sense, Domain I needs to be inserted into the cell membrane through its hydrophobic helical hairpin [55,57]. The amphipathic helices attach to the surface of membranes using hydrophobic helices α-4 and α-5 to enter into the phospholipid bilayer [55,57]. In addition, highly variable and exposed loops from Domain II also participate in binding to receptors, a process that apparently involves two steps. The first step consists of recognition of aminopeptidase N (APN) and alkaline phosphatase (ALP) receptors and the formation of a weak binding with Cry toxins [11,55–59], which produces a reversible reaction [10,55].

The second step consist in the formation of an irreversible binding (Kd 19 nM) through recognition of a 12 amino acid ectodomain region (EC12) from the cadherin receptor (BT-R1) [10,60–62]. Conserved sequence motifs near the N and C ends of EC12 have been reported to be crucial for binding of toxins in insect cells [10,61,62]. Bt-R1 is a highly specific and selective binding receptor to Cry1Ab toxin, it was identified for the first time in the midgut of *Manduca sexta* larvae; this receptor is also responsible for Cry1Ab oligomerization [60–62].

BT-R1 is a protein of 210 kDa composed of four domains: (1) an ectodomain with twelve modules (EC1-EC12) composed of β-barrel cadherin repeats; (2) a membrane-proximal extracellular domain; (3) a transmembrane receptor; (4) a cytoplasmic domain [10,60,63]. The interaction between Cry1A toxin and BT-R1 facilitate the proteolytic cleavage of helix α1 from Domain I, which is located at the N-terminal end, resulting in the formation of a pre-pore oligomer structure that increases the affinity between oligomer and membrane receptors APN and ALP [35,62,64]. The oligomer inserted into the cell membrane creates an ionic pore that leads to osmotic failure, followed by septicemia and insect death [2,12,23,41]. Other intracellular molecules, such as actin, flotillin, prohibitin, and V-ATPase, have been found to participate in the binding to Cry toxins [11,65].

Domain III is a key structure in toxin stability, it binds to N-acetylgalactosamine (GalNAc) in the APN receptor [55,66,67]. APN has been identified as a binding receptor for Cry1A toxins in *M. sexta*, *H. dispar* [55,66,67], and *B. mori* [55]. ATP-binding cassette transporters (ABC proteins) are also involved in Cry toxicity, especially member 2 of subfamily C (ABCC2). These proteins may help Cry1A toxins carry out their primary task: binding to receptors and inserting oligomers into the cell membrane [68,69].

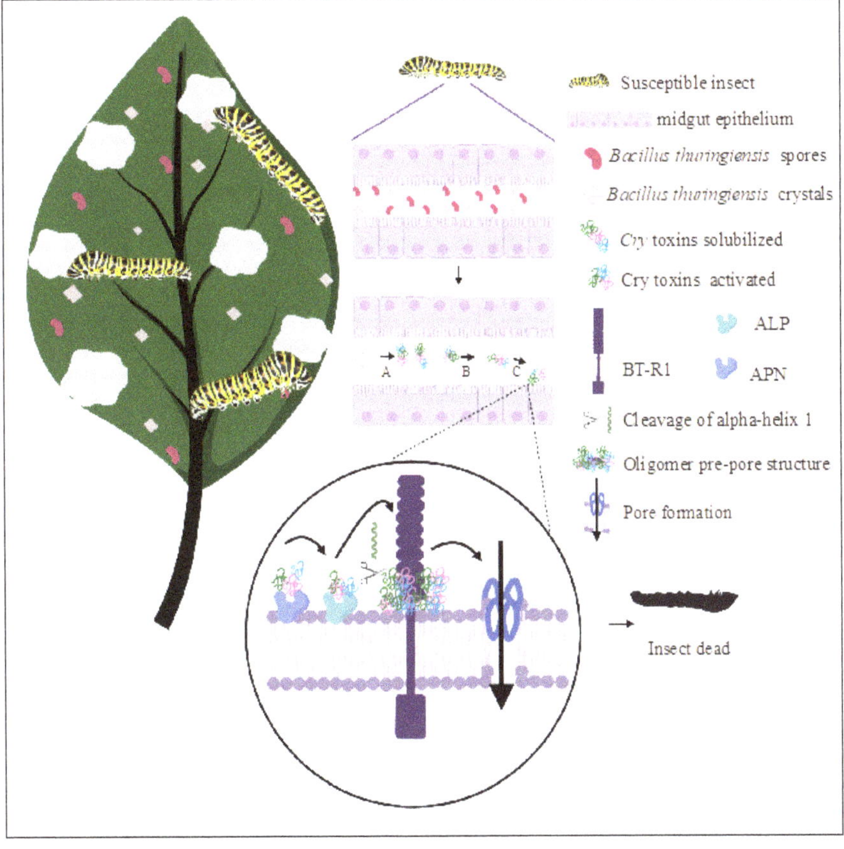

(a)

Figure 2. *Cont.*

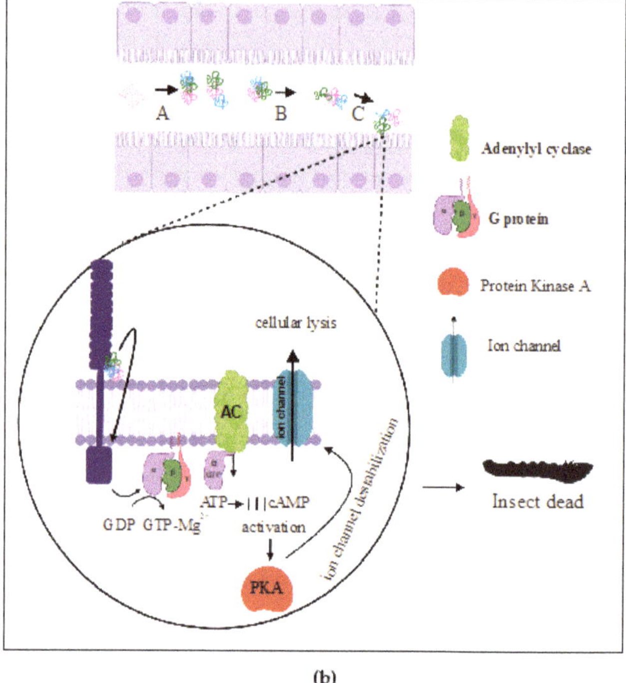

Figure 2. Mechanism of action of Cry proteins. (**a**) Pore-forming model, once larvae ingest crystals, these are solubilized and proteolyzed in larvae midgut. Cry toxins recognize APN, ALP, and EC12/BT-R1 membrane receptors. Cry toxins suffer a proteolytic cleavage on helix α1, resulting in formation of a pre-pore oligomer structure. Posteriorly, oligomer structure is inserted into the cell membrane and creates an ionic pore that leads to osmotic failure, followed by septicemia and insect death. (**b**) Signaling pathway model, once Cry toxins recognize and bind to a cadherin receptor, induces activation of adenylyl cyclase that triggers an increase in cAMP and activates protein kinase A (PKA). This activation will induce a cascade of events that results in an ion channel formation in the membrane, cytoskeleton destabilization, and programmed cell death. A, Solubilization. B, Activation by proteolysis. C, Recognition of membrane receptor. Created with Biorender.com.

However, recently published evidence has shown that toxicity of Cry1AbMod and Cry1AcMod (Cry toxins whose structure has been modified by deleting helix α1) against *M. sexta* that does not involve the expression of the cadherin receptor and still can form toxic oligomeric structures [70–72]. This finding is of great importance for development of strategies to counteract resistance in transgenic crops, and to increase our knowledge of the mechanism of action of Cry toxins.

(b) The Signaling Pathway Model

Cell culture toxicity assays developed to elucidate the action mechanism used by Cry toxins in the signaling pathway model have been carried out in High Five (H5) cell line; which involves expression of a cadherin receptor from *Manduca sexta* [73,74] in a Sf9 cell line from lepidopterans [74,75]. The signaling pathway model postulates that Cry proteins cytotoxicity is mediated by recognizing and binding to a cadherin receptor, which activates an Mg^{2+-} dependent cellular signal cascade pathway that leads to cell death [74]. The binding between Cry toxin and cadherin receptor induces the activation of adenylyl cyclase; which triggers an increase in cAMP and activates protein kinase

A (PKA). These events trigger a cascade that results in an ion channel formation on the membrane, cytoskeleton destabilization, and programmed cell death (Figure 2B) [76].

2.2. Cyt Family

Cyt toxins size ranges from 25 to 28 kDa; their three-dimensional structure shows Cyt proteins have a single α-β domain with low sequence homology to Cry toxins [77]. Cyt toxins affect mainly mosquitoes that are vectors of human diseases; *Anopheles spp* (malaria), *Aedes spp* (dengue, zika, and chikungunya), and *Culex spp* (Nile fever and Rift Valley fever) [78]. Bt subsp. israelensis (Bti) is the most commonly used worldwide to control these vectors [3,78], because it produces Cry4Aa, Cry4Ba, Cry10Aa, Cry11Aa, Cyt1Aa, Cyt2Ba, and Cyt1Ca toxins that together act synergistically to kill mosquito larvae of Culex, Aedes, and Anopheles [79,80] (Table 1).

Table 1. Cyt toxins could synergize with several Cry toxins to act against mosquitoes.

Aedes spp.	*Culex spp.*	*Anopheles spp.*
Cyt1Aa, Cyt2Ba	Cyt1Aa, Cyt2Ba	Cyt1Aa, Cyt2Aa
Cry4Aa	Cry4Aa	Cry4Aa
Cry4Ba	Cry4Ba	Cry10Aa
Cry10Aa	Cry10Aa	Cry11Aa
Cry11Aa	Cry11Aa	

Furthermore, Cyt proteins have other important biologic targets, among others, mammalian and erythrocyte cells [81]; and the pea aphid (*Acyrthosiphon pisum*) and certain weevils (*Diaprepes abbreviates*) [82], both pest insects where Cyt toxins could be used as biocontrol.

Three subfamilies of Cyt toxins have been described so far, all of them sharing a high level of sequence identity: Cyt1 (1Aa, 1Ab, 1Ab, 1Ac, and 1Ad), Cyt2 (2Aa, 2Ba, 2Bb, 2Bc, and 2Ca), and Cyt3Aa1 [10]. Cyt1Aa shows significant similarity with volvatoxin 2 (VVA2), a cardiotoxin isolated from the mushroom *V. volvacea* [83]. VVA1 and VVA2 form the VVA toxin family, a PFT family that has hemolytic and cytotoxic activity in human red blood cells and tumor cells, respectively [84]. Cyt1Ca is different from other members of the Cyt family, since they have an extra domain in the C-terminal end with homology to a carbohydrate-binding domain of ricin. However, there are no reports of larvicidal or hemolytic activity for Cyt1Ca [11,85].

The overall structure of Cyt toxins is formed by a β sheet consisting of six antiparallel strands flanked by a α helix layer and an extra strand β0 at the N-terminal end, which could be involved in dimerization and proteolytic activation [85]. Notoriously, an extra strand at the N-terminal end has not been reported in Cyt2 toxins.

Cyt Proteins Mechanism of Action

Cyt proteins are synthesized as short protoxins [3,86]. The proteolytic cleavage sites in Cyt toxins are found at N-terminal and C-terminal ends [77]. It is widely recognized that loops of the helices are involved in membrane–cell interaction and intermolecular assembly [80]. However, the action mechanism followed by Cyt toxins is still not clear; moreover, it is unknown whether there are specific receptors through which Cyt toxins recognize their target cells. Nevertheless, there are currently two main models of the cytotoxicity mechanism carried out by Cyt toxins.

(a) The Pore-Formation Model

The pore-formation model describes binding of Cyt toxins in their monomer form to specific receptors in the membrane surface, similar to the Cry toxin mechanism. In this case, Cyt toxins interact directly with saturated membrane lipids such as phosphatidylcholine, phosphatidylethanolamine, and sphingomyelin [80,81]. Cyt toxins undergo a conformational change that helps to recruit six monomers of Cyt and assemble them into an open-umbrella structure in which the strands span the lipid bilayer transversely, while alpha helices rest on the membrane surface. The result is a pore-forming model with membrane permeabilization (Figure 3a) [80].

(b) The Detergent-Effect Model

The detergent-effect model suggest that Cyt toxins kill target cells through a solubilization effect on their membrane. In this model, Cyt toxins concentrate on the cell membrane surface and destroy the lipid bilayer in a detergent-like manner (Figure 3b) [80].

Both the pore-formation model and detergent model are not mutually exclusive, as it is thought that, depending on toxin concentration, one or both may act on susceptible cells [80]. Cyt oligomerization and pore formation could be carried out at low Cyt concentration [87], while the detergent effect could be induced only at high toxin concentration [87]. Therefore, the cell membrane from target cells is unable to assemble oligomers at high toxin concentration; instead, it forms a toxin–lipid complex in which the integrity of the membrane is completely lost [80].

Another mechanism of action of Cyt toxins involves a synergistic activity between different members of the family. Thus, when Cyt1Ab and Cyt2Ba act together, they enhance the insecticidal activity against *Aedes aegypti* larvae and resistant *Culex quinquefasciatus* larvae [88]. It has also been found that two known epitopes of Cyt1Aa (196EIKVSAVKE204 and 220NIQSLKFAQ228) binds to Cry4B and Cry11Aa toxins to enhance their toxic effect against the mosquito *Anopheles albimanus* and *Culex quinquefasciatus* [15,84,89]. Epitopes of Cyt1Aa play a receptors role to Cry4B and Cry11Aa, similar to the cadherin receptor of *Manduca sexta*. When Cyt1Aa binds to membrane cell receptors, Cry11Aa or Cry4B binds to this toxin, increasing oligomerization and pore formation effects [3,89,90]. Another synergistic effect occurs when Cyt toxins act together with Mtx1 (another toxin produced by Bt) against *Culex quinquefasciatus* larvae [91].

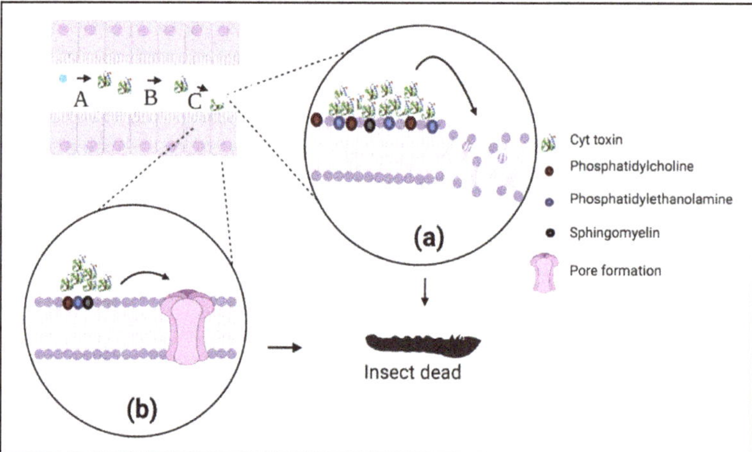

Figure 3. Mechanism of action of Cyt proteins. (**a**) Pore-forming model; once Cyt toxins interact with phosphatidylcholine, phosphatidylethanolamine, and sphingomyelin, they undergo a conformational change that helps recruit six Cyt monomers and assemble them into an open-umbrella structure, which results in a pore-formation and subsequent membrane permeabilization and larva death. (**b**) Detergent effect model; high Cyt toxin concentration binds to the lipid bilayer on cell membrane surface and destroys it through a detergent-effect. Both, the pore-formation model and detergent model are not mutually exclusive, because, the detergent effect acts at high Cyt toxins concentrations, while the pore-forming model acts at low Cyt toxins concentrations. A, Solubilization. B, Activation by proteolysis. C, Recognition of membrane receptor. Created with Biorender.com.

In recent years, several studies have been published related to Cyt proteins modification throughout genetic engineering to produce chimeric toxins [82,92]. These modifications take advantage of small Cyt toxin sizes and low degree complexity of their quaternary structure, as well as their high toxic capacity. The advantages of creating chimeras is to diversify their targets and to increase Cyt toxin potency.

The first successful report of a Cyt chimeric creation describes the insertion of a pea aphid gut-binding peptide GBP3.1, into the amino acid sequence of Cyt2Aa toxin. This peptide prevents the uptake of a plant virus by its vector, the pea aphid. It comprises 12 amino acids (TCSKKYPRSPCM), which bind to alanyl aminopeptidase-N on membrane surface of the aphid gut epithelium. Naturally, Cyt2Aa has low toxicity against the pea aphid, however, Chogule and coworkers succeeded in enhancing the binding of Cyt toxin and increasing its toxicity against aphids, by turning it into a chimeric protein [82].

In another study, Torres-Quintero et al. [92] modified Cyt1Aa by inserting the amino acid sequence of loop3 from Domain II of Cry1Ab (FRSGFSNSSVSI), which induces binding affinity of Cyt1Aa toxin to APN and CAD receptors of *Manduca sexta* [91]. Naturally, Cyt1Aa is not toxic to *M. sexta*, however, chimeric toxin had more significant toxicity to *M. sexta* and *Plutella xylostella* [92]. These results open new possibilities to the application of delta-endotoxins from *Bacillus thuringiensis*, to a new target pest.

2.3. Parasporins

Extensive screening analysis to find new possible targets to Bt strains has shown that non-insecticidal Cry proteins are more widely distributed in nature than insecticidal Cry proteins [16,93,94]. This fact has led researchers to inquire about the possible biological activities or targets of non-insecticidal and non-hemolytic Cry proteins. In this sense, Mizuki et al., in 1999, was the first group to report delta-endotoxins from Bt with selective cytotoxicity against leukemia cells, after a large-scale screening analysis involving protease-digested parasporal proteins from 1744 Bt strains. 1700 strains were isolated in Japan, while 44 were obtained from the Pasteur Institute in Paris [95]. From the isolated strains, 60 presented hemolysis activity and were eliminated by containing *cyt* genes, the rest of strains were tested in vitro for cytocidal activity against MOLT-4 cells and insecticidal activity [95]. At the end of the screening, authors selected only two strains (A1190 and A1462), because they produced toxins that selectively killed leukemic cells instead of normal T-cells [95]. These results inspired an intense and extensive screening of non-insecticidal and non-hemolytic toxins with cytotoxicity against cancer cells throughout the world, which led to the classification of a new type of Cry proteins called parasporins (PS) [96]. At first, this new classification included all non-insecticidal and non-hemolytic Cry toxins with selective cytotoxic activity against cancer cells [7,96]. Time later, it was accepted that non-hemolytic but insecticidal activity could also be part of this classification [96]. Bt strains that produce PS have been isolated from soils of various ecosystems in several countries such as Japan [95,97–120], Vietnam [121,122], India [94,123], Malaysia [124], China [125], Iran [126], and Saudi Arabia [127]. Canada [128] and Caribbean [129] have contributed with new parasporins.

In nature, there are several toxins produced by bacteria capable of killing mammalian cells through pores formation in cell membranes and/or by apoptosis activation [51]. Such is the case of aerolysin from *Aeromonas hydrophila* and alpha-toxin of *Clostridium perfringens*, which are both PFTs that recognize GPI-anchored proteins in the membrane of susceptible cells [51]. Some parasporins share structural homology with both toxins, therefore, it is assumed that PS proteins contain an action mechanism similar to aerolysin and alpha-toxin [130]. So far, 19 parasporins have been identified and organized in six families (Table 2) [7,96].

2.3.1. Parasporin Classification

PS1 Family

PS1 family has cytotoxic effects against certain cancer cell lines such as HeLa (cervix cancer) [99], HL60 (leukemia) [109], Jurkat (leukemia), and HepG2 (liver cancer) [128]. Findings suggest PS1 toxins recognize a common receptor contained between these cell lines, identified as beclin-1 [131]. In healthy cells, beclin-1 exists intracellularly and is involved in autophagia and apoptosis processes, however, in susceptible cell, beclin-1 exists extracellularly and acts as a PS1 receptor [132].

The PS1 family includes, PS1Aa1 (Cry31Aa1), PS1Aa2 (Cry31Aa2), PS1Aa3 (Cry31Aa3), PS1Aa4 (Cry31Aa4), PS1Aa5 (Cry31Aa5), PS1Aa6 (Cry31Aa6), PS1Ab1 (Cry31Ab1), PS1Ab2 (Cry1Ab2), PS1Ac1 (Cry31Ac1), PS1Ac2 (Cry31Ac2) [96].

PS2 Family

PS2Aa1 (Cry46Aa1), PS2Aa2 (Cry46Aa2), and PS2Ab2 (Cry46Ab1) proteins constitute the PS2 family [96]. It is reported that the Bt A1547 strain produce PS2Aa1 and PS2Aa2 [95]. Parasporins from this family are produced as small toxins of around 30 kDa that are cytotoxic to cancer cell lines like HepG2 (liver cancer), Sawano (endometrial cancer), HL60, CaCo-2 (colon cancer), Jurkat (leukemia), and MOLT-4 [133,134].

A peculiarity of PS2 family toxins is that they do not have a typical 3d structure; instead, they are very similar to PFT-aerolysin which is formed mainly by elongated β sheets [133,134]. In 2009, Akiba et al. [134] proposed that PS2 toxins were able to produce oligomers that induce membrane pores and cell death by bind to lipid rafts. In 2017, Abe et al. reported that GPI was an essential co-receptor to PS2 parasporins toxic activity [135]. The action mechanism of PS2 family members is apparent by activating apoptosis, which is associated to an increase in the tumor suppressor gene PAR-4 expression and through inhibition of the PI3K/AKT pathway [136].

PS3 Family

PS3Aa1 (Cry41Aa1) is the single member of this family [96]. This parasporin is the only one with a ricin domain that plays a role in stabilization of the interaction between toxins and carbohydrate residues of the membrane [137]. In 2018, a research group led by Crickmore studied the PS3 protein and observed that it is structurally related to insecticidal toxins, except for the ricin domain. Using site-directed mutagenesis, they concluded that ricin domain is not associated with PS3 selective cytotoxic activity against the HepG2 cancer cell line [138].

In contrast to the mechanism of action used by PS2, it was proposed that PS3 induced cell death by necrosis throughout a pore formation in cancer cell membranes, as was evidenced by an increase in lactate dehydrogenase (LDH) release; mainly in HL60 and HepG2 cancer cell lines [138].

PS4 Family

PS4 is similar in size and structure to PS2 family members, its active form is around 27 kDa and presents a selective cytotoxicity against MOLT-4, CaCo-2, HL60, U937, HepG2, Sawano, DE-4 (leukemia), TS (uterine cancer), and TCS (cervical cancer) cancer cell lines [114,115]. A peculiarity of PS4 toxin is that it can be activated in both alkaline and acid pH [129], while most of parasporins are solubilized in alkaline pH. Actually, acid pH increases their cytotoxic effect against several cancer cell lines [115].

Regarding the mechanism of action of PS4, the evidence found (non-specific binding to the membrane, release of LDH and entrance into the cell of dextrans with different molecular weights) suggests that cell death occurs by necrosis [114,115,130]. The PS4 family includes PS4Aa1 (Cry45Aa1) as the only member [96].

PS5 Family

PS5Aa1 (Cry64Aa1) is the only member of the PS5 family; this protein has been isolated from the *Bt tohokuensis* A1100 strain [96]. The active form of PS5Aa1 has an approximate size of 30 kDa and is selectively toxic to MOLT-4 [120]. Concerning sequence similarity, PS5 shows higher similarity to PFT aerolysin than the other parasporins [120,139]. It has been reported that PS5 is toxic to MOLT-4, Jurkat, HL-60, HepG2, HeLa, Sawano, TCS, CaCo-2, and K562 cancer cell lines, but it also shows potent activity against healthy tissue cells such as UtSMC (normal uterus) and MRC-5 (normal lung) [120]. However, there is still no evidence about its mechanism of action.

PS6 Family

PS6 is closely related to PS1, sharing a conserved sequence of fifty amino acids. PS6 is selectively toxic to HepG2 and HeLa [117], but its mechanism of action is still unknown. The PS6 family includes only PS6Aa1 (Cry63Aa1), which has been isolated from Bt M019 [96,117] and 64-1-94 [129].

2.3.2. Mechanism of Action of Parasporins

The cytocidal activity of PS against cancer cell lines ranges from EC_{50} 0.0017 µg/mL of PS2 against Sawano to 3.0 µg/mL of PS1 against HepG2. Table 2 shows the reported EC_{50} to PS in cancer cell lines.

In management of parasporins, different methods of crystals solubilization and activation should be tried, since effectiveness of their cytocidal activity against cancer cells depends on this. Similar to what happens on insects, a correct activation of protoxins is essential for cell membrane receptors recognition and subsequent triggering of cancer cell death.

As an example, there is a particular case of PS2Aa1 not showing toxic activity when activated using trypsin, but, when activated with proteinase K, it shows activity against human cancer cells [137]. In this sense, it is crucial to know that sites of cleavage to trypsin and proteinase K are different.

The mechanism of action of parasporins against target cancer cells is poorly understood, however, available information has shown that parasporins exhibit several mechanisms of action to kill cancer cells. These proteins act in a similar way to Cry toxins because they are highly specific to a cell type, nevertheless, it is well known that Cry toxins specificity depends on cell membrane receptors (cadherin, aminopeptidase-N, and alkaline phosphatase) recognition. On the other hand, PS interaction with cell membrane receptors is still being investigated, several molecules that act as parasporin receptors have been reported and patented. In this sense, Beclin-1 acts as PSAa1 receptor [131]. Glycosylphosphatidylinositol (GPI)-anchored protein is involved in efficient cytocidal action of PS2Aa1 [140]. GADPH from CEM-SS leukemic cell line acts as receptor from a PS found in Malaysia [124]. One of the most important characteristics of parasporins is their ability to discriminate cancer cells from non-cancer cells, which is directly related to cell membrane receptors recognition.

Table 2. Parasporin characteristics.

PS	Bt Strain	Cry Gene	Protoxin (kDa)	Active Toxin (kDa)	Protease Activation	Main Cellular Target	EC50 [μg/mL]	Mechanism Action	Country	Reference
PS1Aa1	A1190	Cry31Aa1	81	15, 56	Trypsin	HeLa	0.12	Apoptosis	Japan	[99]
PS1Aa2	M15	Cry31Aa2	83	55, 70	Trypsin	Jurkat HepG2	0.02 0.02	ND	Canada	[128]
PS1Aa3	B195	Cry31Aa3	81	56	Trypsin	HeLa	14.7	ND	Japan	[116]
PS1Aa4	Bt79-25	Cry31Aa4	81	NP	Proteinase K	ND	ND	ND	Vietnam	[122]
PS1Aa5	Bt92-10	Cry31Aa5	81	NP	Proteinase K	ND	ND	ND	Vietnam	[122]
PS1Aa6	M019, 64-1-94	Cry31Aa6	70	15, 55	Trypsin	HepG2	0.52	ND	Japan, Caribbean	[117,129]
PS1Ab1	B195	Cry31Ab1	82	56	Trypsin	HeLa	14.7	ND	Japan	[116]
PS1Ab2	Bt31-5	Cry31Ab2	82	NP	Proteinase K	ND	ND	ND	Vietnam	[122]
PS1Ac1	Bt87-29	Cry31Ac1	87	NP	Proteinase K	ND	ND	ND	Vietnam	[122]
PS1Ac2	B0462	Cry31Ac2	81	15, 60	Proteinase K	HeLa	2	Apoptosis	Japan	[118]
PS1Ad1	64-1-94, M15, M019	Cry31Ad1	73	14, 59	Trypsin	HepG2	0.52	ND	Caribbean, Canada, Japan	[117,128,129]
PS2Aa1	A1547	Cry46Aa1	37	30	Proteinase K	HepG2	0.023	Pore-forming	Japan, USA	[105]
PS2Aa2	A1470	Cry31Aa2	30	28	Proteinase K	MOLT-4	0.041	ND	Japan	[119]
PS2Ab1	TK-E6	Cry31Ab1	33	29	Proteinase K	Jurkat	0.545ng/ml	ND	Japan	[113]
PS3Aa1	A1462	Cry41Aa1	88	64	Proteinase K	HL60	1.32	ND	Japan	[100]
PS3Ab1	A1462	Cry41Ab1	88	64	Proteinase K	HL60	1.25	ND	Japan	[100]
PS4Aa1	A1470	Cry45Aa1	31	28	Proteinase K	CaCo2	0.124	Pore-forming	Japan	[111]
PS5Aa1	A1100	Cry64Aa1	33	30	Proteinase K	TCS	0.046	ND	Japan	[120]
PS6Aa1	M019, 64-1-94	Cry63Aa1	85	14, 59	Trypsin	HepG2	2.3	ND	Japan, Caribbean	[117,129]

NP- not published. ND- not determined.

2.4. S-Layer Proteins

Surface layer proteins (SLP) are widely represented, both in Gram-negative and Gram-positive bacteria, including *Bacillus* [17]. Similar to delta-endotoxins, SLP are assembled into parasporal positions with several shapes (oblique, square, or hexagonal) [8]. They have a molecular mass between 40 and 170 kDa [141,142], and are involved mainly in growth, survival, and maintenance of cell integrity [9]. There are also reports of their antiviral and antibacterial activity, as well as of their anti-inflammatory effects [141,142].

SLP toxins activity is still unclear; it has been suggested that SLP have a similar insecticidal activity to Cry proteins but with a different mechanism [141,142]. The SLP obtained from GP1 Bt strain is the only one that has been reported to have pesticidal activity against *Epilachna varivestis* [18].

A recent study reported an SLP from Bt with high selective cytotoxic activity in vitro against the MDA-MB-231 breast cancer cell line. Authors suggest that cadherin-11 receptor present in cancer cells seems to be involved in SLP recognition; however, the mechanism of action is still under study [8].

2.5. Toxins Secreted by Bt

In addition to Cry, Cyt, PS, and SLP toxins produced in parasporal bodies during sporulation, Bt secretes during vegetative growth phase other toxins with insecticidal activity [11]. There are two main families of secreted insecticidal proteins, one is known as vegetative insecticidal proteins (Vip) [20,21] and the other as secreted insecticidal proteins (Sip) [19]. These proteins contain a signal peptide sequence in the N-terminal end that is cleaved after the secretion process is completed [143,144].

2.5.1. Vip Family

Vip toxins are not shaped as parasporal inclusion bodies, instead they are produced and secreted during the vegetative growth phase and their expression ends before the sporulation stage begins [11]. These insecticidal toxins have been characterized as; Vip1, Vip2, Vip3, and Vip4; however, their mechanism of action against insects is not entirely understood yet [20,28].

Vip1 is synthesized as a protoxin of 100 kDa and after secretion; a mature toxin of 80 kDa is produced, additionally, Vip2 releases a trypsin-resistant fragment of 50 kDa [144]. Together, Vip1 and Vip2 produce a Vip binary toxin [145], with synergistic insecticidal activity against some coleopteran pests and the sap-sucking insect pest *Aphis gossypii (Hemiptera)* [146].

Vip2 is similar in structure and behavior to the CdtA toxin from *Clostridium difficile*; this toxin presents an ADP-ribosyltransferase activity and its principal target is the actin protein, therefore, could induce cytoskeletal disruption and cell death when it is activated [147].

In monomer form, Vip1 binds to its receptors and a conformational change is produced because of this interaction; thus, more Vip1 toxins are attracted to form a heptamer that translocates Vip2 into the cytoplasm through acid endosomes [148]. Once inside the cell, Vip2 destroys actin filaments that disrupt the cytoskeleton and eventually induce cell death [148].

Due to its similarity with other A+B binary toxins, it has been concluded that Vip2 is responsible for most of the cytotoxic activity, while Vip1 is responsible for binding to membrane receptors in susceptible insects [149].

Vip3 is a single-chain protein that is toxic to a wide variety of lepidopterans and other insects, such as *Agrotis ipsilon*, *Spodoptera exigua*, and *S. frugiperda*, which are less susceptible to Cry1A toxins [21]. Vip3 are proteins of 88 kDa approximately without homology to any other known insecticidal protein [86,150]. In contrast to Vip1 and Vip2, signal peptide sequences in Vip3 are not processed during secretion and are present in the mature secreted peptide, suggesting they play an important role in protein structure and insecticidal activity. However, cleavage of the N-terminal end activates the protoxin; the 66 kDa active toxin is fragmented from the 22-kDa N-terminal portion [21]. The mechanism of action is still unclear and it has been suggested that Vip3 proteins act in a similar way to PFT, but their membrane receptors are still unknown. In vitro experiments have shown that Vip3 does not compete for binding sites of Cry1A in *Manduca sexta* nor *S. frugiperda* [151].

Vip 4 is the most recently discovered member of the Vip family. It has a molecular mass of ~108 kDa and a 34% identity to Vip1Aa1 protein, specifically to the B component of the binary toxin. For this reason, it has been suggested that Vip4 might interact with unknown A component to produce toxicity; therefore, such information is needed to understand its action mechanism [20].

The Vip proteins that have been reported so far are 15 Vip1 proteins, 20 Vip2 proteins, 101 Vip3 proteins, and 1 Vip4 [13].

2.5.2. Sip Toxins

Secreted insecticidal proteins (Sip) are toxins produced by Bt with an approximate size of 41 kDa. Similarly to Vip proteins, Sip toxins are synthesized containing a signal peptide sequence of 30 amino acids, which is processed by proteases [19] and an active protein is released [11]. It is known that Sip proteins have insecticidal activity against Coleopterans such as *Leptinotarsa decemlineata*, *Diabrotica undecimpunctata howardi*, and *Diabrotica virgifera virgifera* [19]. However, their mechanism of action is still unknown.

3. Conclusions

The horizontal transfer of genetic information through the conjugation of plasmids in *Bacillus thuringiensis* opens up a world of possibilities for the discovery of new toxins, new structures, new targets, and even new classification.

Understanding the structural characteristics of Bt toxins and their mechanism of action will allow us to develop new products for improving pest management and human health. In this sense, new combinations of insecticidal Cry proteins have been recently found, opening new possibilities to pest control without genetic and neither molecular manipulation [152]. Additionally, in 2019, Mendoza and coworkers for first time reported that Cry1A toxins from Bt presented a highly specific anticancer activity in HeLa cells and also against insects. Authors suggested that in both cases, a specific interaction between Cry toxins and cell membrane receptors could be initiating toxicity on insects and in human cancer cells [153].

SLP proteins are other Bt toxins with both pesticide and anticancer activity. Recently studies have shown that these proteins also could be recognizing specific cell membrane receptors in cancer cells line [8], as Cry toxins do [153]. In addition to cytotoxic activity on insects and human cancer cell lines, SLP carry out structural and protection activities in several microorganisms. Furthermore, these proteins could be found in Gram-positive, Gram-negative, and archaebacteria, not only in Bt.

Parasporins also present specific anticancer activity in vitro; these proteins are found in non-insecticidal and non-hemolytic strains, opening the possibility to develop anticancer agents with therapeutic potential and without secondary effects in patients. However, very little is known about their mechanism of action and the receptors recognized to carry out their cytotoxicity. PS1 and PS2 are the most extensively studied parasporin families, therefore, have been used as a model to answers several questions regarding preferential activity of these toxins against cancer cells.

Finally, it is important to mention that more research is needed to understand the mechanisms of action used by Bt toxins. According to reported studies, it seems that most of them recognize specific cell membrane receptors in susceptible cells; however, what is happening inside cells once the interaction has begun still is a mystery. Thus, it is essential to know and understand the signaling pathways involved in toxicity to be capable of developing new anticancer compounds and to improve pest control including resistance developed to Bt toxins.

Author Contributions: G.M.-A., E.L.E.-I., J.L.A.-L., M.M.-R., S.G.-G., M.H.-B., and J.O.-S., were responsible for cited literature and writing the manuscript. G.M.-A. and J.O.S. Writing Original Draft Preparation. G.M.-A. was responsible of the Conceptualization. All authors read and approved the final manuscript.

Funding: This research received no external funding.

Acknowledgments: G.M.-A. thank National Council for Science and Technology, Mexico, for the program Cátedras CONACYT.

Conflicts of Interest: The authors declare no conflict of interest.

References

1. Knowles, B.H.; Dow, J.A.T. The crystal delta-endotoxins of *Bacillus thuringiensis*—Models for their mechanism of action on the insect gut. *Bioessays* **1993**, *15*, 469–476. [CrossRef]
2. Schnepf, E.; Crickmore, N.; Van Rie, J.; Lereclus, D.; Baum, J.F.; Zeigler, D.R.; Dean, D.H. *Bacillus thuringiensis* and its pesticidal crystal proteins. *Microbiol. Mol. Biol. Rev.* **1998**, *62*, 775–806. [CrossRef] [PubMed]
3. Bravo, A.; Gill, S.S.; Soberón, M. Mode of action of *Bacillus thuringiensis* Cry and Cyt toxins and their potential for insect control. *Toxicon* **2007**, *49*, 423–435. [CrossRef] [PubMed]
4. Roh, J.Y.; Choi, J.Y.; Li, M.S.; Jin, B.R.; Je, Y.H. *Bacillus thuringiensis* as a specific, safe, and effective tool for insect pest control. *J. Mol. Biol.* **2007**, *17*, 547–559.
5. Becker, N. Bacterial control of vector-mosquitoes and black flies. In *Entomopathogenic Bacteria: From Laboratory to Field Application*; Charles, J.F., Delécluse, A., Nielsen-LeRoux, C., Eds.; Kluwer Academic Publishers: Berlin, Germany, 2000; pp. 383–398. [CrossRef]
6. Kotze, A.C.; Grady, J.O.; Gough, J.M.; Pearson, R.; Bagnall, N.H.; Kemp, D.H.; Akhurst, R.J. Toxicity of *Bacillus Thuringiensis* to Parasitic and Free-Living Life-Stages of Nematode Parasites of Livestock. *Int. J. Parasitol.* **2005**, *35*, 1013–1022. [CrossRef]
7. Ohba, M.; Mizuki, E.; Uemori, A. Parasporin, a new anticancer protein group from *Bacillus thuringiensis*. *Anticancer Res.* **2009**, *29*, 427–433.
8. Rubio, V.P.; Bravo, A.; Olmos, J. Identification of a *Bacillus thuringiensis* Surface Layer Protein with Cytotoxic Activity against MDA-MB-231 Breast Cancer Cells. *J. Microbiol. Biotechnol.* **2017**, *27*, 36–42. [CrossRef]
9. Gang, G.; Lei, Z.; Zhou, Z.; Qiqi, M.; Jianping, L.; Chenguang, Z.; Lei, Z.; Ziniu, Y.; Ming, S. A new group of parasporal inclusions encoded by the S-layer gene of *Bacillus thuringiensis*. *FEMS. Microbiol. Lett.* **2008**, *282*, 1–7.
10. Liu, L.; Boyd, S.D.; Bulla-Jr, L.A.; Winkler, D.D. The Defined Toxin-binding Region of the Cadherin G-protein Coupled Receptor, BT-R1, for the Active Cry1Ab Toxin of *Bacillus thuringiensis*. *J. Proteomics Bioinform.* **2018**, *11*, 201–210. [CrossRef]
11. Palma, L.; Muñoz, D.; Berry, C.; Murillo, J.; Caballero, P. *Bacillus thuringiensis* toxins: An overview of their biocidal activity. *Toxins* **2014**, *6*, 3296–3325. [CrossRef]
12. Bulla, L.A.; Kramer, K.J.; Cox, D.J.; Jones, B.L.; Davidson, L.I.; Lookhart, G.L. Purification and characterization of the entomocidal protoxin of *Bacillus thuringiensis*. *J. Biol. Chem.* **1981**, *256*, 3000–3004. [PubMed]
13. Crickmore, N.; Zeigler, D.R.; Schnepf, E.; Rie, J.; Lereclus, D.; Baum, J.; Bravo, A.; Dean, D.H. *Bacillus thuringiensis* Toxin Nomenclature. Available online: http://www.lifesci.sussex.ac.uk/home/Neil_Crickmore/Bt/ (accessed on 3 April 2020).
14. Höfte, H.; Whiteley, H.R. Insecticidal crystal proteins of *Bacillus thuringiensis*. *Microbiol. Rev.* **1989**, *5*, 242–255.
15. Van Frankenhuyzen, K. Insecticidal activity of *Bacillus thuringiensis* crystal proteins. *J. Invertebr. Pathol.* **2009**, *101*, 1–16. [CrossRef] [PubMed]
16. Akiba, T.; Okumura, S. Parasporins 1 and 2: Their structure and activity. *J. Invertebr. Pathol.* **2016**, *142*, 44–49. [CrossRef] [PubMed]
17. Sará, M.; Sleytr, U.B. S-Layer Proteins. *J. Bacteriol.* **2000**, *182*, 859–868. [CrossRef]
18. Peña, G.; Miranda-Rios, J.; de la Riva, G.; Pardo-López, L.; Soberón, M.; Bravo, A. A *Bacillus thuringiensis* S-layer protein involved in toxicity against *Epilachna varivestis (Coleoptera: Coccinellidae)*. *Appl. Environ. Microbiol.* **2006**, *72*, 353–360. [CrossRef]
19. Donovan, W.P.; Engleman, J.T.; Donovan, J.C.; Baum, J.A.; Bunkers, G.J.; Chi, D.J.; Clinton, W.P.; English, L.; Heck, G.R.; Ilagan, O.M.; et al. Discovery and characterization of Sip1A: A novel secreted protein from *Bacillus thuringiensis* with activity against coleopteran larvae. *Appl. Microbiol. Biotechnol.* **2006**, *72*, 713–719. [CrossRef]
20. Chakroun, M.; Banyuls, N.; Bel, Y.; Escriche, B.; Ferré, J. Bacterial Vegetative Insecticidal Proteins (Vip) from Entomopathogenic Bacteria. *Microbiol. Mol. Biol. Rev.* **2016**, *80*, 329–350. [CrossRef]

21. Sahin, B.; Gomis-Cebolla, J.; Güneş, H.; Ferré, J. Characterization of *Bacillus thuringiensis* isolates by their insecticidal activity and their production of Cry and Vip3 proteins. *PLoS ONE* **2018**, *13*, e0206813. [CrossRef]

22. Christou, O.; Capell, T.; Kohli, A.; Gatehouse, J.A.; Gatehouse, A.M. Recent developments and future prospects in insect pest control in transgenic crop. *Trend Plant Sci.* **2006**, *11*, 302–308. [CrossRef]

23. Pardo-Lopez, L.; Soberon, M.; Bravo, A. *Bacillus thuringiensis* insecticidal three-domain Cry toxins: Mode of action, insect resistance and consequences for crop protection. *FEMS Microbiol. Rev.* **2013**, *37*, 3–22. [CrossRef]

24. Mendelson, M.; Kough, J.; Vaituzis, Z.; Matthews, K. Are Bt crops safe? *Nat. Biotechnol.* **2003**, *21*, 1003–1009. [CrossRef] [PubMed]

25. Romeis, J.; Meissle, M.; Bigler, F. Transgenic crops expressing *Bacillus thuringiensis* toxins and biological control. *Nat. Biotechnol.* **2006**, *24*, 63–71. [CrossRef] [PubMed]

26. Chen, M.; Zhao, J.Z.; Collins, H.L.; Earle, E.D.; Cao, J.; Shelton, A.M. A critical assessment of the effects of Bt transgenic pants on parasitoids. *PLoS ONE* **2008**, *3*, e2284.

27. Melo, A.L.; Soccol, V.T.; Soccol, C.R. *Bacillus thuringiensis*: Mechanism of action, resistance, and new applications: A review. *Crit. Rev. Biotechnol.* **2016**, *36*, 317–326. [CrossRef] [PubMed]

28. Tabashnik, B.E.; Carrière, Y. Surge in insect resistance to transgenic crops and prospects for sustainability. *Nat. Biotechnol.* **2017**, *35*, 926–935. [CrossRef] [PubMed]

29. Fischhoff, D.A.; Bowdish, K.S.; Perlak, F.J.; Marrone, P.G.; McCormick, S.M.; Niedermeyer, J.G.; Dean, D.A.; Kusano-Kretzmer, K.; Mayer, E.J.; Rochester, D.E.; et al. Insect tolerant transgenic tomato plants. *Nat. Biotechnol.* **1987**, *5*, 807–813. [CrossRef]

30. Vaek, M.; Reynaerts, A.; Hofte, A. Transgenic plants protected from insects. *Nature* **1987**, *325*, 33–37. [CrossRef]

31. Koch, M.S.; Ward, J.M.; Levine, S.L.; Baum, G.A.; Vicini, J.L.; Hammond, B.G. The food and environmental safety of Bt crops. *Front. Plant. Sci.* **2015**, *6*, 283–336. [CrossRef]

32. Zhong, C.; Ellar, D.J.; Bishop, A.; Johnson, C.; Lin, S.; Hart, E.R. Characterization of a *Bacillus thuringiensis* delta endotoxin whic iIs toxic to insects in three orders. *J. Invertebr. Pathol.* **2000**, *76*, b131–b139. [CrossRef]

33. Schnepf, H.E.; Whiteley, H.R. Cloning and expression of the *Bacillus thuringiensis* crystal protein gene in Escherichia coli. *Proc. Natl. Acad. Sci. USA* **1981**, *78*, 2893–2897. [CrossRef]

34. Thomas, D.J.; Morgan, J.A.; Whipps, J.M.; Saunders, J.R. Plasmid Transfer between the *Bacillus thuringiensis* Subspecies *kurstaki* and *tenebrionis* in Laboratory Culture and Soil and in *Lepidopteran* and *Coleopteran* Larvae. *Appl. Environ. Microbiol.* **2000**, *66*, 118–124. [CrossRef] [PubMed]

35. Arenas, I.; Bravo, A.; Soberón, M.; Gómez, I. Role of alkaline phosphatase from *Manduca sexta* in the mechanism of action of *Bacillus thuringiensis* Cry1Ab toxin. *J. Biol. Chem.* **2010**, *285*, 12497–12503. [CrossRef]

36. Parker, M.W.; Feil, S.C. Pore forming protein toxins: From structure to function. *Progress. Biophys. Mol. Biol.* **2005**, *88*, 91–124. [CrossRef] [PubMed]

37. Rodríguez-Almazán, C.; Zavala, L.E.; Muñoz-Garay, C.; Jiménez-Juárez, N.; Pacheco, S.; Masson, L.; Soberón, M.; Bravo, A. Dominant Negative Mutants of *Bacillus thuringiensis* Cry1Ab Toxin Function as Anti-Toxins: Demonstration of the Role of Oligomerization in Toxicity. *PLoS ONE* **2009**, *4*, e5545. [CrossRef] [PubMed]

38. De Maagd, R.A.; Bravo, A.; Crickmore, N. How *Bacillus thuringiensis* has evolved specific toxins to colonize the insect world. *Trends Genet.* **2001**, *17*, 193–199. [CrossRef]

39. Rajamohan, F.; Hussain, S.R.A.; Cotrill, J.A.; Gould, F.; Dean, D.H. Mutations at domain II, loop 3, of *Bacillus thuringiensis* CryIAa and CryIAb δ-endotoxins suggest loop 3 is involved in initial binding to lepidopteran midguts. *J. Biol. Chem.* **1996**, *271*, 25220–25226. [CrossRef]

40. Lee, M.K.; Rajamohan, F.; Jenkins, J.L.; Curtiss, A.S.; Dean, D.H. Role of two arginine residues in domain II, loop 2 of Cry1Ab and Cry1Ac *Bacillus thuringiensis* δ-endotoxin in toxicity and binding to *Manduca sexta* and *Lymantria dispar* aminopeptidase N. *Mol. Microbiol.* **2000**, *38*, 289–298. [CrossRef]

41. De Maagd, R.A.; Bravo, A.; Berry, C.; Crickmore, N.; Schnepf, H.E. Structure, diversity, and evolution of protein toxins from spore-forming entomopathogenic bacteria. *Annu. Rev. Genet.* **2003**, *37*, 409–433. [CrossRef]

42. Lakey, J.H.; van der Goot, F.G.; Pattus, F. All in the family: The toxic activity of pore-forming toxins. *Toxicology* **1994**, *87*, 85–108. [CrossRef]

43. Hunt, S.; Green, J.; Artymiuk, P.J. Hemolysin E (HlyE, ClyA, SheA) and related toxins. *Adv. Exp. Med. Biol.* **2010**, *677*, 116–126.

44. Kristan, K.C.; Viero, G.; Dalla-Serra, M.; Macek, P.; Anderluh, G. Molecular mechanism of pore formation by actinoporins. *Toxicon* **2009**, *54*, 1125–1134. [CrossRef]

45. DuMont, A.L.; Torres, V.J. Cell targeting by the *Staphylococcus aureus* pore-forming toxins: It's not just about lipids. *Trends Microbiol.* **2014**, *22*, 21–27. [CrossRef]

46. Iacovache, I.; Dal Peraro, M.; van der Goot, F.G. *The Comprehensive Sourcebook of Bacterial Protein Toxins*; Elsevier Ltd.: Amsterdam, The Netherlands, 2015.

47. Hotze, E.M.; Le, H.M.; Sieber, J.R.; Bruxvoort, C.; McInerney, M.J.; Tweten, R.K. Identification and characterization of the first cholesterol-dependent cytolysins from Gram-negative bacteria. *Infect. Immun.* **2013**, *81*, 216–225. [CrossRef]

48. Gouaux, E. Channel-forming toxins: Tales of transformation. *Curr. Opin. Struct. Biol.* **1997**, *7*, 566–573. [CrossRef]

49. Lesieur, C.; Vecsey-Semjn, B.; Abrami, L.; Fivaz, M.; van der Goot, F.G. Membrane insertion: The strategy of toxins. *Mol. Membrane Biol.* **1997**, *14*, 45–64. [CrossRef]

50. Iacovache, I.; Bischofberger, M.; van der Goot, F.G. Structure and assembly of pore-forming proteins. *Curr. Opin. Struct. Biol.* **2010**, *20*, 241–246. [CrossRef]

51. Dal Peraro, M.; van der Goot, G. Pore-forming toxins: Ancient, but never really out of fashion. *Nat. Rev. Microbiol.* **2016**, *14*, 77–92. [CrossRef]

52. Shepard, L.A.; Heuck, A.P.; Hamman, B.D.; Rossjohn, J.; Parker, M.W.; Ryan, K.R.; Johnson, A.E.; Tweten, R.K. Identification of a membrane- spanning domain of the thiol-activated pore-forming toxin *Clostridium perfringens* perfringolysin O: An α-helical to β-sheet transition identified by fluorescence spectroscopy. *Biochemistry* **1998**, *37*, 14563–14574. [CrossRef]

53. Xu, C.; Wang, B.C.; Yu, Z.; Sun, M. Structural Insights into *Bacillus thuringiensis* Cry, Cyt and Parasporin Toxins. *Toxins* **2014**, *6*, 2732–2770. [CrossRef]

54. Zhuang, M.; Oltean, D.I.; Gómez, I.; Pullikuth, A.K.; Soberón, M.; Bravo, A.; Gill, S.S. *Heliothis virescens* and *Manduca sexta* lipid rafts are involved in Cry1A toxin binding to the midgut epithelium and subsequent pore formation. *J. Biol. Chem.* **2002**, *277*, 13863–13872. [CrossRef] [PubMed]

55. Bravo, A.; Gómez, I.; Conde, J.; Muñoz-Garay, C.; Sánchez, J.; Miranda, R.; Zhuang, M.; Gill, S.S.; Soberón, M. Oligomerization triggers binding of a Bacillus thuringiensis Cry1Ab pore-forming toxin to aminopeptidase N receptor leading to insertion into membrane microdomains. *Biochim. Biophys. Acta* **2004**, *1667*, 38–46. [CrossRef] [PubMed]

56. Bravo, A.; Soberón, M.; Gill, S.S. Bacillus thuringiensis Mechanisms and Use. *Compr. Mol. Insect Sci.* **2005**, 175–205. [CrossRef]

57. Pardo-López, L.; Gómez, I.; Rausell, C.; Sánchez, J.; Soberón, M.; Bravo, A. Structural changes of the Cry1Ac oligomeric pre-pore from *Bacillus thuringiensis* induced by *N*-acetylgalactosamine facilitates toxin membrane insertion. *Biochemistry* **2006**, *45*, 10329–10336.

58. Pigott, C.R.; Ellar, D.J. Role of receptors in *Bacillus thuringiensis* crystal toxin activity. *Microbiol. Mol. Biol. Rev.* **2007**, *71*, 255–281. [CrossRef]

59. Adang, M.J.; Crickmore, N.; Jurat-Fuentes, J.L. Diversity of *Bacillus thuringiensis* crystal toxins and mechanism of action. *Adv. Insect Physiol.* **2014**, *47*, 39–87.

60. Gómez, I.; Oltean, D.I.; Gill, S.S.; Bravo, A.; Soberón, M. Mapping the epitope in cadherin-like receptors involved in *Bacillus thuringiensis* Cry1A toxin interaction using phage display. *J. Biol. Chem.* **2001**, *276*, 28906–28912. [CrossRef]

61. Gomez, I.; Miranda-Rios, J.; Rudiño-Piñera, E.; Oltean, D.I.; Gill, S.S.; Bravo, A.; Soberón, M. Hydropathic complementarity determines interaction of epitope (869) HITDTNNK (876) in *Manduca sexta* Bt-R(1) receptor with loop 2 of domain II of *Bacillus thuringiensis* Cry1A toxins. *J. Biol. Chem.* **2002**, *277*, 30137–30143. [CrossRef]

62. Gómez, I.; Dean, D.H.; Bravo, A.; Soberón, M. Molecular basis for *Bacillus thuringiensis* Cry1Ab toxin specificity: Two structural determinants in the *Manduca sexta* Bt-R$_1$ receptor interact with loops α8 and 2 in domain II of Cy1Ab toxin. *Biochemistry* **2003**, *42*, 10482–10489.

63. Dorsch, J.A.; Candas, M.; Griko, N.B.; Maaty, W.S.A.; Midboe, E.G.; Vadlamudi, R.K.; Bulla Jr, L.A. Cry1A toxins of Bacillus thuringiensis bind specifically to a region adjacent to the membrane-proximal extracellular domain of BT-R1 in *Manduca sexta*: Involvement of a cadherin in the entomopathogenicity of *Bacillus thuringiensis*. *Insect Biochem. Mol. Biol.* **2002**, *32*, 1025–1036. [CrossRef]

64. Gómez, I.; Sánchez, J.; Miranda, R.; Bravo, A.; Soberón, M. Cadherin-like receptor binding facilitates proteolytic cleavage of helix alpha-1 in domain I and oligomer pre-pore formation of *Bacillus thuringiensis* Cry1Ab toxin. *FEBS Lett.* **2002**, *513*, 242–246. [CrossRef]

65. Kuadkitkan, A.; Smith, D.R.; Berry, C. Investigation of the Cry4B-prohibitin interaction in *Aedes aegypti* cells. *Curr. Microbiol.* **2012**, *65*, 446–454. [CrossRef] [PubMed]

66. Grochulski, P.; Masson, L.; Borisova, S.; Pusztaicarey, M.; Schwartz, J.L.; Brousseau, R.; Cygler, M. *Bacillus thuringiensis* CryIA(a) insecticidal toxin: Crystal structure and channel formation. *J. Mol. Biol.* **1995**, *254*, 447–464. [CrossRef] [PubMed]

67. Pacheco, S.; Gómez, I.; Arenas, I.; Saab-Rincón, G.; Rodríguez-Almazán, C.; Gill, S.S.; Bravo, A.; Soberón, M. Domain II loop 3 of *Bacillus thuringiensis* Cry1Ab toxin is involved in a "ping pong" binding mechanism with *Manduca sexta* aminopeptidase-N and cadherin receptors. *J. Biol. Chem.* **2009**, *284*, 32750–32757. [CrossRef]

68. Ocelotl, J.; Sánchez, J.; Gómez, I.; Tabashnik, B.E.; Bravo, A.; Soberón, M. ABCC2 is associated with *Bacillus thuringiensis* Cry1Ac toxin oligomerization and membrane insertion in diamondback moth. *Sci. Rep.* **2017**, *7*, 2386. [CrossRef]

69. Heckel, D.G. Learning the ABCs of Bt: ABC transporters and insect resistance to *Bacillus thuringiensis* provide clues to a crucial step in toxin mode of action. *Pest. Biochem. Physiol.* **2012**, *104*, 103–110. [CrossRef]

70. Soberón, M.; Pardo-Lopez, L.; Lopez, I.; Gómez, I.; Tabashnik, B.E.; Bravo, A. Engineering modified Bt toxins to counter insect resistance. *Science* **2007**, *318*, 1640–1642.

71. Porta, H.; Jimenez, G.; Cordoba, E.; Leon, P.; Soberón, M.; Bravo, A. Tobacco plants expressing the Cry1AbMod toxin suppress tolerance to Cry1Ab toxin of *Manduca sexta* cadherin-silenced larvae. *Insect Biochem. Mol. Biol.* **2011**, *41*, 513–519. [CrossRef]

72. Tabashnik, B.E.; Huang, F.; Ghimire, M.N.; Leonard, B.R.; Siegfried, B.D.; Rangasamy, M.; Yang, Y.; Wu, Y.; Gahan, L.J.; Heckel, D.G.; et al. Efficacy of genetically modified Bt toxins against insects with different mechanisms of resistance. *Nat. Biotechnol.* **2011**, *29*, 1128–1131. [CrossRef]

73. Wickham, T.J.; Davis, T.; Granados, R.R.; Shuler, M.L.; Wood, H.A. Screening of insect cell lines for the production of recombinant proteins and infectious virus in the baculovirus expression system. *Biotechnol. Prog.* **1992**, *8*, 391–396. [CrossRef]

74. Castella, C.; Pauron, D.; Hilliou, F.; Trang, V.T.; Zucchini-Pascal, N.; Gallet, A.; Barbero, P. Transcriptomic analysis of *Spodoptera frugiperda* Sf9 cells resistant to *Bacillus thuringiensis* Cry1Ca toxin reveals that extracellular Ca2+, Mg2+ and production of cAMP are involved in toxicity. *Biol. Open* **2019**, *18*. [CrossRef]

75. Kwa, M.S.G.; de Maagd, R.A.; Stiekema, W.J.; Vlak, J.M.; Bosh, D. Toxicity and binding properties of the *Bacillus thuringiensis* delta-endotoxin Cry1C to cultured insect cells. *J. Invertebr. Pathol.* **1998**, *71*, 121–127. [CrossRef]

76. Zhang, X.; Candas, M.; Griko, N.B.; Taissing, R.; Bulla, L.A., Jr. A mechanism of cell death involving an adenylyl cyclase/PKA signaling pathway is induced by the Cry1Ab toxin of *Bacillus thuringiensis*. *Proc. Natl. Acad. Sci. USA* **2006**, *103*, 9897–9902. [CrossRef]

77. Li, J.; Koni, P.A.; Ellar, D.J. Structure of the mosquitocidal delta-endotoxin CytB from *Bacillus thuringiensis* ssp. *kyushuensis and implications for membrane pore formation*. *J. Mol. Biol.* **1996**, *257*, 129–1522.

78. Zhang, Q.; Hua, G.; Adang, M.J. Effects and mechanisms of *Bacillus thuringiensis* crystal toxins for mosquito larvae. *Insect Sci.* **2017**, *24*, 714–729. [CrossRef]

79. Berry, C.; O'Neil, S.; Ben-Dov, E.; Jones, A.F.; Murphy, L.; Quail, M.A.; Holden, M.T.; Harris, D.; Zaritsky, A.; Parkhill, J. Complete sequence and organization of pBtoxis, the toxin-coding plasmid of *Bacillus thuringiensis* subsp. *israelensis*. *Appl. Environ. Microbiol.* **2002**, *68*, 5082–5095. [CrossRef]

80. Butko, P. Cytolytic toxin Cyt1A and its mechanism of membrane damage: Data and hypotheses. *Appl. Environ. Microbiol.* **2003**, *69*, 2415–2422. [CrossRef]

81. Thomas, W.E.; Ellar, D.J. *Bacillus thuringiensis var. israeliensis* crystal delta-endotoxin: Effects on insect and mammalian cells in vitro and in vivo. *J. Cell Sci.* **1983**, *60*, 181–197.

82. Chogule, N.P.; Li, H.; Liu, S.; Linz, L.B.; Narva, K.E.; Meade, T.; Bonning, B.C. Retargeting of the *Bacillus thuringiensis* toxin Cyt2Aa against hemipteran insect pests. *Proc. Natl. Acad. Sci. USA* **2013**, *110*, 8465–8470. [CrossRef]

83. Lin, S.C.; Lo, Y.C.; Lin, J.Y.; Liaw, Y.C. Crystal structures and electron micrographs of fungal volvatoxin A2. *J. Mol. Biol.* **2004**, *343*, 477–491. [CrossRef]

84. Lin, J.Y.; Jeng, T.W.; Chen, C.C.; Shi, G.Y.; Tung, T.C. Isolation of a new cardiotoxic protein from the edible mushroom, *Volvariella volvacea*. *Nature* **1973**, *246*, 524–525. [CrossRef] [PubMed]

85. Manasherob, R.; Itsko, M.; Sela-Baranes, N.; Ben-Dov, E.; Berry, C.; Cohen, S.; Zaritsky, A. Cyt1Ca from *Bacillus thuringiensis* subsp. *israelensis*: Production in *Escherichia coli* and comparison of its biological activities with those of other Cyt-like proteins. *Microbiology* **2006**, *152*, 2651–2659. [CrossRef]

86. Bravo, A.; Martínez de Castro, D.; Sánchez, J.; Cantón, P.E.; Mendoza, G.; Gómez, I.; Pacheco, S.; García-Gómez, B.I.; Onofre, J.; Ocelotl, J.; et al. Mechanism of action of Bacillus thuringiensis insecticidal toxins and their use in the control of insect pests. In *Comprehensive Sourcebook of Bacterial Protein Toxins*, 4th ed.; Alouf, J.E., Ed.; Elsevier: Amsterdam, The Netherlands, 2015; pp. 858–873.

87. Butko, P.; Huang, F.; Pusztai-Carey, M.; Surewicz, W.K. Membrane permeabilization induced by cytolytic delta-endotoxin CytA from *Bacillus thuringiensis var. israelensis*. *Biochemistry* **1996**, *35*, 11355–11360. [CrossRef] [PubMed]

88. Wirth, M.C.; Dlécluse, A.; Walton, W.E. Cyt1Ab1 and Cyt2Ba1 from Bacillus thuringiensis subsp. medellin and *B. thuringiensis* subsp. *israelensis* synergize *Bacillus sphaericus* against *Aedes aegypti* and Resistant *Culex quinquefasciatus* (*Diptera: Culicidae*). *Appl. Environ. Microbiol.* **2001**, *67*, 3280–3284. [CrossRef] [PubMed]

89. Soberón, M.; Lopez-Díaz, J.A.; Bravo, A. Cyt toxins produced by *Bacillus thuringiensis*: A protein fold conserved in several pathogenic microorganisms. *Peptides* **2013**, *41*, 87–93.

90. Pérez, C.; Fernández, L.E.; Sun, J.; Folch, J.L.; Gill, S.S.; Soberón, M.; Bravo, A. *Bacillus thuringiensis subsp. israelensis* Cyt1Aa synergizes Cry11Aa toxin by functioning as a membrane-bound receptor. *Proc. Natl. Acad. Sci. USA* **2005**, *102*, 18303–18308.

91. Zhang, B.H.; Liu, M.; Yang, Y.K.; Yuan, Z.M. Cytolytic Toxin Cyt1aa of Bacillus thuringiensis Synergizes the Mosquitocidal Toxin Mtx1 of *Bacillus sphaericus*. *Biosci. Biotech. Bioch.* **2006**, *70*, 2199–2204. [CrossRef]

92. Torres-quintero, M.C.; Gómez, I.; Pacheco, S.; Sánchez, J.; Flores, H.; Osuna, J.; Mendoza, G.; Soberón, M.; Bravo, A. Engineering *Bacillus thuringiensis* Cyt1Aa toxin specificity from dipteran to lepidopteran toxicity. *Sci. Rep.* **2018**, *8*, 4989. [CrossRef]

93. Ohba, M.; Aizawa, K. Insect toxicity of *Bacillus thuringiensis* isolated from soils of Japan. *J. Invertebr. Pathol.* **1986**, *47*, 12–20. [CrossRef]

94. Chubika, T.; Girija, D.; Deepa, K.; Salini, S.; Meera, N.; Raghavamenon, A.; Divya, M.; Babu, T. A parasporin from *Bacillus thuringiensis* native to Peninsular India induces apoptosis in cancer cells through intrinsic pathway. *J. Biosci.* **2018**, *43*, 407–416. [CrossRef]

95. Mizuki, E.; Ohba, M.; Akao, T.; Yamashita, S.; Saitoh, H.; Park, Y.S. Unique activity associated with non-insecticidal *Bacillus thuringiensis* parasporal inclusions: In vitro cell-killing action on human cancer cells. *J. Appl. Microbiol.* **1999**, *86*, 477–486. [CrossRef] [PubMed]

96. Okumura, S.; Ohba, M.; Mizuki, E.; Crickmore, N.; Côté, J.C.; Nagamatsu, Y.; Kitada, S.; Sakai, H.; Harata, K.; Shin, T. Parasporin nomenclature. 2010. Available online: http://parasporin.fitc.pref.fukuoka.jp/list.html (accessed on 3 April 2020).

97. Kim, H.S.; Yamashita, S.; Akao, T.; Saitoh, H.; Higuchi, K.; Park, Y.S.; Mizuki, E.; Ohba, M. In vitro cytotoxicity of non-Cyt inclusion proteins of a *Bacillus thuringiensis* isolate against human cells, including cancer cells. *J. Appl. Microbiol.* **2000**, *89*, 16–23. [CrossRef] [PubMed]

98. Lee, D.W.; Akao, T.; Yamashita, S.; Katayama, H.; Maeda, M.; Saitoh, H.; Mizuki, E.; Ohba, M. Noninsecticidal parasporal proteins of a *Bacillus thuringiensis* serovar *shandongiensis* isolate exhibit a preferential cytotoxicity against human leukemic T cells. *Biochem. Biophys. Res. Commun.* **2000**, *272*, 218–223. [CrossRef] [PubMed]

99. Mizuki, E.; Park, Y.S.; Saitoh, H.; Yamashita, S.; Akao, T.; Higuchi, K.; Ohba, M. Parasporin, a human leukemic cell-recognizing parasporal protein of *Bacillus thuringiensis*. *Clin. Diagn. Lab. Immunol.* **2000**, *7*, 625–634. [CrossRef]

100. Yamashita, S.; Akao, T.; Mizuki, E.; Saitoh, H.; Higuchi, K.; Park, Y.S.; Kim, H.S.; Ohba, M. Characterization of the anti-cancer-cell parasporal proteins of a *Bacillus thuringiensis* isolate. *Can. J. Microbiol.* **2000**, *46*, 913–919. [CrossRef]

101. Lee, D.W.; Katayama, H.; Akao, T.; Maeda, M.; Tanaka, R.; Yamashita, S.; Saitoh, H.; Mizuki, E.; Ohba, M. A 28-kDa protein of the *Bacillus thuringiensis* serovar *shandongiensis* isolate 89-T-34-22 induces a human leukemic cell-specific cytotoxicity. *Biochim. Biophys. Acta* **2001**, *1547*, 57–63. [CrossRef]

102. Yamagiwa, M.; Namba, A.; Akao, T.; Mizuki, E.; Ohba, M.; Sakai, H. Cytotoxicity of *Bacillus thuringiensis* crystal protein against mammalian cells. *Mem. Fac. Eng. Okayama Univ.* **2002**, *36*, 61–66.

103. Namba, A.; Yamagiwa, M.; Amano, H.; Akao, T.; Mizuki, E.; Ohba, M.; Sakai, H. The cytotoxicity of *Bacillus thuringiensis* subsp. *coreanensis* A1519 strain against the human leukemic T cell. *Biochim. Biophys. Acta* **2003**, *1622*, 29–35. [CrossRef]

104. Akiba, T.; Abe, Y.; Kitada, S.; Kusaka, Y.; Ito, A.; Ichimatsu, T.; Katayama, H.; Akao, T.; Higuchi, K.; Mizuki, E.; et al. Crystallization of parasporin-2, a *Bacillus thuringiensis* crystal protein with selective cytocidal activity against human cells. *Acta Cryst.* **2004**, *D60*, 2355–2357.

105. Ito, A.; Sasaguri, Y.; Kitada, S.; Kusaka, Y.; Kuwano, K.; Masutomi, K.; Mizuki, E.; Akao, T.; Ohba, M. Selective cytocidal action of a crystal protein of *Bacillus thuringiensis* on human cancer cells. *J. Biol. Chem.* **2004**, *279*, 21282–21286. [CrossRef]

106. Okumura, S.; Akao, T.; Higuchi, K.; Saitoh, H.; Mizuki, E.; Ohba, M.; Inouye, K. *Bacillus thuringiensis* serovar *shandongiensis* strain 89-T-34-22 produces multiple cytotoxic proteins with similar molecular masses against human cancer cell. *Lett. Appl. Microbiol.* **2004**, *39*, 89–92. [CrossRef] [PubMed]

107. Amano, H.; Yamagiwa, M.; Akao, T.; Mizuki, E.; Ohba, M.; Sakai, H. A novel 29-kDa crystal protein from *Bacillus thuringiensis* induces caspase activation and cell death of Jurkat T cells. *Biosci. Biotechnol. Biochem.* **2005**, *69*, 2063–2072. [CrossRef] [PubMed]

108. Katayama, H.; Yokota, H.; Akao, T.; Nakamura, O.; Ohba, M.; Mekada, E.; Mizuki, E. Parasporin-1, a novel cytotoxic protein to human cells from non-insecticidal parasporal inclusions of *Bacillus thuringiensis*. *J. Biochem.* **2005**, *137*, 17–25. [CrossRef]

109. Okumura, S.; Saitoh, H.; Ishikawa, T.; Wasano, N.; Yamashita, S.; Kusumoto, K.; Akao, T.; Mizuki, E.; Ohba, M.; Inouye, K. Identification of a Novel Cytotoxic Protein, Cry45Aa, from *Bacillus thuringiensis* A1470 and Its Selective Cytotoxic Activity against Various Mammalian Cell Lines. *J. Agric. Food Chem.* **2005**, *53*, 6313–6318. [CrossRef] [PubMed]

110. Yamashita, S.; Katayama, H.; Saitoh, H.; Akao, T.; Park, Y.S.; Mizuki, E.; Ohba, M.; Ito, A. Typical three-domain Cry proteins of *Bacillus thuringiensis* strain A1462 exhibit cytocidal activity on limited human cancer cells. *J. Biochem.* **2005**, *138*, 663–672. [CrossRef] [PubMed]

111. Okumura, S.; Saitoh, H.; Wasano, N.; Katayama, H.; Higuchi, K.; Mizuki, E.; Inouye, K. Efficient Solubilization, activation and purification of recombinant Cry45Aa of *Bacillus thuringiensis* expressed as an *Escherichia coli* inclusion body. *Protein Expres. Purif.* **2006**, *47*, 144–151. [CrossRef]

112. Saitoh, H.; Okumura, S.; Ishikawa, T.; Akao, T.; Mizuki, E.; Ohba, M. Investigation of a novel *Bacillus thuringiensis* gene encoding a parasporal protein, parasporin-4, that preferentially kills leukaemic T cells. *Biosci. Biotechnol. Biochem.* **2006**, *70*, 2935–2941. [CrossRef]

113. Hayakawa, T.; Kanagawa, R.; Kotani, Y.; Kimura, M.; Yamagiwa, M.; Yamane, Y.; Takebe, S.; Sakai, H. Parasporin-2Ab, a newly isolated cytotoxic crystal protein from *Bacillus thuringiensis*. *Curr. Microbiol.* **2007**, *55*, 278–283. [CrossRef]

114. Inouye, K.; Okumura, S.; Mizuki, E. Parasporin-4, A Novel Cancer Cell-killing Protein Produced by *Bacillus thuringiensis*. *Food Sci. Biotechnol.* **2008**, *17*, 219–227.

115. Okumura, S.; Saitoh, H.; Ishikawa, T.; Mizuki, E.; Inouye, K. Identification and characterization of a novel cytotoxic protein, parasporin-4, produced by *Bacillus thuringiensis* A1470 strain. *Biotechnol. Annu. Rev.* **2008**, *14*, 225–252.

116. Uemori, A.; Ohgushi, A.; Yasutake, K.; Maeda, M.; Mizuki, E.; Ohba, M. Parasporin-1Ab, a Novel *Bacillus thuringiensis* Cytotoxin Preferentially Active on Human Cancer Cells In Vitro. *Anticancer Res.* **2008**, *28*, 91–95. [PubMed]

117. Nagamatsu, Y.; Okamura, S.; Saitou, H.; Akao, T.; Mizuki, E. Three Cry toxins in two types from *Bacillus thuringiensis* strain M019 preferentially kill human hepatocyte cancer and uterus cervix cancer cells. *Biosci. Biotechnol. Biochem.* **2010**, *74*, 494–498. [CrossRef] [PubMed]

118. Kuroda, S.; Begum, A.; Saga, M.; Hirao, A.; Mizuki, E.; Sakai, H.; Hayakawa, T. Parasporin 1Ac2, a Novel Cytotoxic Crystal Protein Isolated from *Bacillus thuringiensis* B0462 Strain. *Curr. Microbiol.* **2013**, *66*, 475–480. [CrossRef] [PubMed]

119. Okumura, S.; Ishikawa, T.; Saitoh, H.; Akao, T.; Mizuki, E. Identification of a second cytotoxic protein produced by *Bacillus thuringiensis* A1470. *Biotechnol. Lett.* **2013**, *35*, 1889–1894. [CrossRef] [PubMed]

120. Ekino, K.; Okumura, S.; Ishikawa, T.; Kitada, S.; Saitoh, H.; Akao, T.; Oka, T.; Nomura, Y.; Ohba, M.; Shin, T.; et al. Cloning and Characterization of a Unique Cytotoxic Protein Parasporin-5 Produced by *Bacillus thuringiensis* A1100 Strain. *Toxins* **2014**, *6*, 1882–1895. [CrossRef] [PubMed]

121. Yasutake, K.; Binh, N.D.; Kagoshima, K.; Uemori, A.; Ohgushi, A.; Maeda, M.; Mizuki, E.; Yu, Y.M.; Ohba, M. Occurrence of parasporin-producing *Bacillus thuringiensis* in Vietnam. *Can. J. Microbiol.* **2006**, *52*, 365–372. [CrossRef]

122. Yasutake, K.; Uemori, A.; Binh, N.D.; Mizuki, E.; Ohba, M. Identification of parasporin genes in Vietnamese isolates of *Bacillus thuringiensis. Zeitschrift fur Naturforschung C* **2008**, *63*, 139–143. [CrossRef]

123. Lenina, N.K.; Naveenkuma, A.; Sozhavendan, A.E.; Balakrishnan, N.; Balasubramani, V.; Udayasuriyan, V. Characterization of parasporin gene harboring Indian isolates of *Bacillus thuringiensis. Biotech* **2014**, *4*, 545–551. [CrossRef]

124. Nadarajah, V.D.; Ting, D.; Chan, K.K.; Mohamed, S.M.; Kanakeswary, K.; Lee, H.L. Selective cytotoxic activity against leukemic cell lines from mosquitocidal *Bacillus thuringiensis* parasporal inclusions. *Southeast Asian J. Trop. Med. Public Health* **2008**, *39*, 235–245.

125. Zhu, L.; Li, C.; Wu, J.; Liang, J.; Shi, Y. Apoptosis of HL-60 cells induced by crystal proteins from *Bacillus thuringiensis* Bt9875. *Wei Sheng Wu xue Bao* **2008**, *48*, 690–694.

126. Moazamian, E.; Bahador, N.; Azarpira, N.; Rasouli, M. Anti-cancer Parasporin Toxins of New *Bacillus thuringiensis* Against Human Colon (HCT-116) and Blood (CCRF-CEM) Cancer Cell Lines. *Curr. Microbiol.* **2018**, *75*, 1090–1098. [CrossRef] [PubMed]

127. Aboul-Soud, M.; Al-Amri, M.Z.; Kumar, A.; Al-Sheikh, Y.A.; Ashour, A.E.; El-Kersh, T.A. Specific Cytotoxic Effects of Parasporal Crystal Proteins Isolated from Native Saudi Arabian *Bacillus thuringiensis* Strains against Cervical Cancer Cells. *Molecules* **2019**, *24*, 506. [CrossRef] [PubMed]

128. Jung, Y.C.; Mizuki, E.; Côte, J.C. Isolation and characterization of a novel *Bacillus thuringiensis* strain expressing a novel crystal protein with cytocidal activity against human cancer cells. *J. Appl. Microbiol.* **2007**, *103*, 65–79. [CrossRef] [PubMed]

129. González, E.; Granados, J.C.; Short, J.D.; Ammons, D.R.; Rampersad, J. Parasporins from a Caribbean Island: Evidence for a Globally dispersed *Bacillus thuringiensis* strain. *Curr. Microbiol.* **2011**, *62*, 1643–1648. [CrossRef] [PubMed]

130. Okumura, S.; Saitoh, H.; Ishikawa, T.; Inouye, K.; Mizuki, E. Mode of action of parasporin-4, a cytocidal protein from *Bacillus thuringiensis. Biochim. Biophys. Acta* **2011**, *1808*, 1476–1482. [CrossRef]

131. Katayama, H.; Kusaka, Y.; Mizuk, E. Parasporin-1 Receptor and Use Thereof. U.S. Patent 20110038880 2011.

132. Liang, X.H.; Jackson, S.; Seaman, M.; Brown, K.; Kempkes, B.; Hibshoosh, H.; Levine, B. Induction of autophagy and inhibition of tumorigenesis by beclin 1. *Nature* **1999**, *402*, 672–676. [CrossRef]

133. Kitada, S.; Abe, Y.; Shimada, H.; Kusaka, Y.; Matsuo, Y.; Katayama, H.; Okumura, S.; Akao, T.; Mizuki, E.; Kuge, O.; et al. Cytocidal actions of parasporin-2, an anti-tumor crystal toxin from *Bacillus thuringiensis. J. Biol. Chem.* **2006**, *281*, 26350–26360. [CrossRef]

134. Akiba, T.; Abe, Y.; Kitada, S.; Kusaka, Y.; Ito, A.; Ichimatsu, T.; Katayama, H.; Akao, T.; Higuchi, K.; Mizuki, E.; et al. Crystal structure of the parasporin-2 *Bacillus thuringiensis* toxin that recognizes cancer cells. *J. Mol. Biol.* **2009**, *13*, 121–133. [CrossRef]

135. Abe, Y.; Inoue, H.; Ashida, H.; Maeda, Y.; Kinoshita, T.; Kitada, S. Glycan region of GPI anchored-protein is required for cytocidal oligomerization of an anticancer parasporin-2, Cry46Aa1 protein, from *Bacillus thuringiensis* strain 3. *J. Invertebr. Pathol.* **2017**, *142*, 71–81. [CrossRef]

136. Brasseur, K.; Auger, P.; Asselin, E.; Parent, S.; Côte, J.C.; Sirois, M. Parasporin-2 from a New *Bacillus thuringiensis* 4R2 Strain Induces Caspases Activation and Apoptosis in Human Cancer Cells. *PLoS ONE* **2015**, *10*, e0135106. [CrossRef]

137. Hazes, B. The (QxW)3 domain: A flexible lectin scaffold. *Protein Sci.* **1996**, *5*, 1490–1501. [CrossRef] [PubMed]

138. Krishnan, V.; Domanska, B.; Elhigazi, A.; Afolabi, F.; West, M.J.; Crickmore, N. The human cancer cell active toxin Cry41Aa from *Bacillus thuringiensis* acts like its insecticidal counterparts. *Biochem. J.* **2017**, *474*, 1591–1602. [CrossRef] [PubMed]

139. Moniatte, M.; van der Goot, F.G.; Buckley, J.T.; Pattus, F.; van Dorsselaer, A. Characterisation of the heptameric pore-forming complex of the Aeromonas toxin aerolysin using MALDI-TOF mass spectrometry. *FEBS Lett.* **1996**, *384*, 269–272. [CrossRef]

140. Kitada, S.; Abe, Y.; Maeda, T.; Shimada, H. Parasporin-2 requires GPI-anchored proteins for the efficient cytocidal action to human hepatoma cells. *Toxicology* **2009**, *264*, 80–88. [CrossRef] [PubMed]

141. Sleytr, U.B.; Messner, P.; Pum, D.; Sára, M. Crystalline bacterial cell surface layers. *Mol. Microbiol.* **1993**, *10*, 911–916. [CrossRef]

142. Sleytr, U.B.; Shuster, B.; Egelseer, E.M.; Pum, D. S-layers: Principles and applications. *FEMS Microbiol. Rev.* **2014**, *38*, 823–864. [CrossRef]

143. Shi, Y.; Xu, W.; Yuan, M.; Tang, M.; Chen, J.; Pang, Y. Expression of *vip1/vip2* genes in *Escherichia coli* and *Bacillus thuringiensis* and the analysis of their signal peptides. *J. Appl. Microbiol.* **2004**, *97*, 757–765. [CrossRef]

144. Bi, Y.; Zhang, Y.; Shu, C.; Crickmore, N.; Wang, Q.; Du, L.; Song, F.; Zhang, J. Genomic sequencing identifies novel *Bacillus thuringiensis* Vip1/Vip2 binary and Cry8 toxins that have high toxicity to *Scarabaeoidea* larvae. *Appl. Microbiol. Biotechnol.* **2015**, *99*, 753–760. [CrossRef]

145. Warren, G.W. Vegetative insecticidal proteins: Novel proteins for control of corn pests. In *Advances in Insect Control, the Role of Transgenic Plants*; Carozzi, N.B., Koziel, M., Eds.; Taylor & Francis Ltd.: London, UK, 1997; pp. 109–121. [CrossRef]

146. Sattar, S.; Maiti, M.K. Molecular characterization of a novel vegetative insecticidal protein from *Bacillus thuringiensis* effective against sap-sucking insect pest. *J. Microbiol. Biotechnol.* **2011**, *21*, 937–946. [CrossRef]

147. Jucovic, M.; Walters, F.S.; Warren, G.W.; Palekar, N.V.; Chen, J.S. From enzyme to zymogen: Engineering Vip2, an ADP-ribosyltransferase from *Bacillus cereus*, for conditional toxicity. *Protein Eng. Des. Sel.* **2008**, *21*, 631–638. [CrossRef]

148. Leuber, M.; Orlik, F.; Schiffler, B.; Sickmann, A.; Benz, R. Vegetative insecticidal protein (Vip1Ac) of *Bacillus thuringiensis* HD201: Evidence for oligomer and channel formation. *Biochemistry* **2006**, *45*, 283–288. [CrossRef] [PubMed]

149. Barth, H.; Aktories, K.; Popoff, M.R.; Stiles, B.G. Binary bacterial toxins: Biochemistry, biology, and applications of common *Clostridium* and *Bacillus* proteins. *Microbiol. Mol. Biol. Rev.* **2004**, *68*, 373–402. [CrossRef] [PubMed]

150. Bel, Y.; Banyuls, N.; Chakroun, M.; Escriche, B.; Ferré, J. Insights into the Structure of the Vip3Aa Insecticidal Protein by Protease Digestion Analysis. *Toxins* **2017**, *4*, 131. [CrossRef] [PubMed]

151. Abdelkefi-Mesrati, L.; Boukedi, H.; Chakroun, M.; Kamoun, F.; Azzouz, H.; Tounsi, S.; Rouis, S.; Jaoua, S. Investigation of the steps involved in the difference of susceptibility of *Ephesitia kuehniella* and *Spodoptera littoralis* to the *Bacillus thuringiensis* Vip3Aa16 toxin. *J. Invertebr. Path.* **2011**, *107*, 198–201. [CrossRef] [PubMed]

152. Mendoza, G.; Portillo, A.; Arias, E.; Ribas, R.M.; Olmos, J. New Combinations of Cry Genes From *Bacillus Thuringiensis* Strains Isolated From Northwestern Mexico. *Int. Microbiol.* **2012**, *15*, 211–218. [PubMed]

153. Mendoza-Almanza, G.; Rocha-Zavaleta, L.; Aguilar-Zacarías, C.; Ayala-Luján, J.; Olmos, S.J. Cry1A Proteins are Cytotoxic to HeLa but not to SiHa Cervical Cancer Cells. *Curr. Pharm. Biotechnol.* **2019**, *20*, 1018. [CrossRef]

Review

Insecticidal Activity of *Bacillus thuringiensis* Proteins against Coleopteran Pests

Mikel Domínguez-Arrizabalaga [1], **Maite Villanueva** [1,2], **Baltasar Escriche** [3], **Carmen Ancín-Azpilicueta** [4] **and Primitivo Caballero** [1,*]

1 Institute for Multidisciplinary Research in Applied Biology-IMAB, Universidad Pública de Navarra, 31192 Mutilva, Navarra, Spain; mikel.dominguez@unavarra.es (M.D.-A.); maite.villanueva@unavarra.es (M.V.)

2 Bioinsectis SL, Avda Pamplona 123, 31192 Mutilva, Navarra, Spain

3 Departamento de Genética/ERI BioTecMed, Universitat de València, Burjassot, 46100 València, Spain; baltasar.escriche@uv.es

4 Departamento de Ciencias, Universidad Pública de Navarra, 31006 Pamplona, Spain; ancin@unavarra.es

* Correspondence: pcm92@unavarra.es; Tel.: +34-948-168-004

Received: 4 June 2020; Accepted: 25 June 2020; Published: 29 June 2020

Abstract: *Bacillus thuringiensis* is the most successful microbial insecticide agent and its proteins have been studied for many years due to its toxicity against insects mainly belonging to the orders Lepidoptera, Diptera and Coleoptera, which are pests of agro-forestry and medical-veterinary interest. However, studies on the interactions between this bacterium and the insect species classified in the order Coleoptera are more limited when compared to other insect orders. To date, 45 Cry proteins, 2 Cyt proteins, 11 Vip proteins, and 2 Sip proteins have been reported with activity against coleopteran species. A number of these proteins have been successfully used in some insecticidal formulations and in the construction of transgenic crops to provide protection against main beetle pests. In this review, we provide an update on the activity of Bt toxins against coleopteran insects, as well as specific information about the structure and mode of action of coleopteran Bt proteins.

Keywords: *Bacillus thuringiensis* proteins; coleopteran pests; insecticidal activity; structure; mode of action

Key Contribution: This contribution provide an update on the activity of Bt toxins against coleopteran insects.

---

1. Introduction

The use of entomopathogenic microorganisms as biological control agents has become one of the most effective alternatives to chemical pest control. Among all, the Gram-positive bacterium *Bacillus thuringiensis* (Bt) is the most important entomopathogenic microorganism used to date in crop protection. This bacterium is widely distributed in various ecological niches, such as water, soil, insects, and plants [1]. The feature that distinguishes *B. thuringiensis* from other members of the *Bacillus* group is the capacity to produce parasporal crystalline inclusions. These crystals are composed of proteins (Cry and Cyt) which are toxic against an increasing number of insect species from the orders Lepidoptera, Diptera, Coleoptera, Hymenoptera, and Hemiptera, among others, as well as against other organisms such as mites [2] and nematodes [3]. Bt also synthesizes insecticidal toxins associated with the vegetative growth phase, named Vip (vegetative insecticidal protein) and Sip (secreted insecticidal protein), which are secreted into the growth medium [4]. These toxins are uniquely specific, safe, and completely biodegradable, and have been used for more than 60 years as an alternative to chemical insecticides [5]. Products based on Bt isolates are the most successful microbial insecticides, with current worldwide benefits estimated at $8 billion annually [6]. However, not all Bt proteins are designated as toxins, for example, some parasporins do not have known insect targets, although they are toxic to human cancer cells [7]. The insecticidal activity of Bt toxins has also been transferred to crop plants through genetic engineering, providing very high protection levels against injurious pests and decreasing the use of chemical insecticides in many instances [8,9]. The success of these insecticidal proteins has fuelled the search for new Bt isolates and proteins that can render novel insecticidal agents with different specificities.

Since Schnepf and Whiteley cloned the first *cry* gene in the early 1980's [10], many others have been described and are now classified according to Bt Toxin Nomenclature, that consists of four ranks based on amino acid sequence identity [11]. To date, the Bt Toxin Nomenclature Committee [12] has reported at least 78 Cry protein groups, from Cry1 to Cry 78, divided into at least three phylogenetically non-related protein subfamilies that may have different modes of action: the three-domain Cry toxins (3-domain), the mosquitocidal Cry toxins (Etx_Mtx2), Toxin_10 proteins, and alpha-helical toxins (reviewed in [13,14]).

The largest group, with more than 53 Cry toxin subgroups, is the 3-domain Cry toxin group. Even though the sequence identity among these proteins is low, the overall structure of the three domains is quite similar, providing proteins with different specificities but with quite similar modes of action [15]. Thus, proteins such as Cry1Aa (lepidopteran specific) and Cry3Aa (coleopteran specific) have a 32.5% identity but a structural similarity as high as 98% [16]. Phylogenetic analysis shows that the great variability in the insecticidal activity of this 3-domain group has resulted from the independent evolution of the three structural domains as well as from the swapping of domain III between different toxins [15].

Due to their feeding habits, many species of coleoptera cause serious damage to both cultivated plants and stored products, leading to significant economic losses in all regions of the world [17,18]. Both larvae and adults have strong jaws, which enable them to feed on a wide variety of plant substrates, such as roots, stems, leaves, grains or wood [19]. Beetles represent the order of the Insecta class that includes the largest number of species. However, the studies carried out to identify toxins of *B. thuringiensis* active against beetles are far from being equal to those carried out in the order Lepidoptera. Thus, 45 Cry proteins, 2 Cyt proteins, 11 Vip proteins, and 2 Sip proteins have been reported with activity against coleopteran insects to date, of which the toxins of the Cry3 and Cry8 families have the largest host spectrum (Figure 1). In this review, we provide an update on the activity of Bt toxins against coleopteran pests.

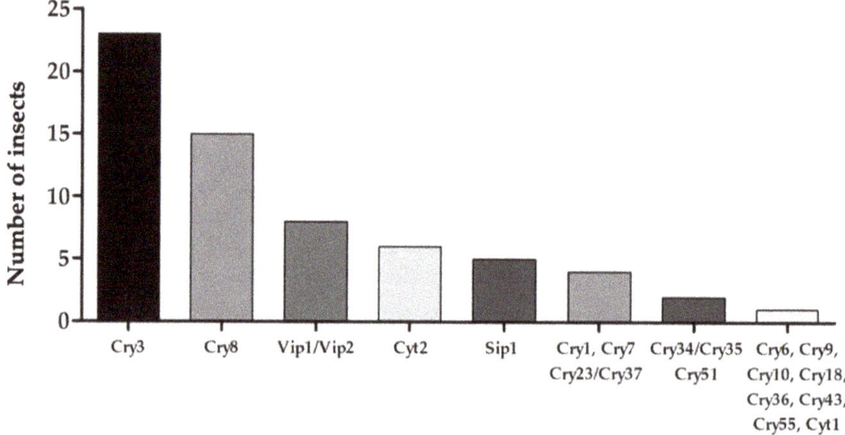

Figure 1. Number of susceptible coleopteran insects to Bt (*Bacillus thuringiensis*) proteins, grouped into protein families.

2. The Crystal Coleopteran-Active Proteins

Bt crystal proteins (δ-endotoxins) are produced during the stationary growth phase and have been isolated from a wide range of insect pests. These crystal inclusions are mainly formed by Cry and Cyt proteins that are toxic to a wide variety of insect species. Most of the information on the insecticidal properties has been obtained for the Cry3 family, and only a few data come from other Cry families. The Cyt proteins constitute a smaller group, mainly active against dipterans, although some Cyt proteins are toxic to coleopteran pests and increase the potential of certain Cry toxins [20].

2.1. Protein Structure

As mentioned above, Bt Cry proteins can be basically subdivided into three different groups according to their homology and molecular structure: the 3-domain group, Etx_Mtx2 proteins, Toxin_10 proteins, and alpha-helical toxins. The 3-domain Cry proteins constitute the largest and best-studied group, although there is increasing information on the 'non-3-domain' and Cyt proteins.

2.1.1. The 3-Domain Group Toxins

All 3-domain Cry proteins are produced as protoxins of two main sizes, a ~130 kDa protoxin and shorter one of approximately 70 kDa [16] (Figure 2). The 130 kDa proteins share a highly conserved C terminus containing 15-17 cysteine residues, which is dispensable for toxicity but necessary for the formation of intermolecular disulphide bonds during crystal formation [15,21]. This group has been mainly studied on lepidopteran toxins such as Cry1A, but also includes some coleopteran active toxins such as Cry7A and Cry8. The structure of the small protoxins is quite similar to the N-terminal half of the large toxin group. Since these do not contain the C-terminal extension, they require, in some cases, the presence of accessory proteins for crystallization [22,23]. This second group includes Cry2A, Cry11A, and some toxins active against Coleoptera, such as Cry3A or Cry3B. Proteolytic cleavage of the N-terminal peptide and the C-terminal extension (mainly in the long Cry protoxins) yields active ~60 kDa protease-resistant fragments [24]. The first crystal structure solved by X-ray crystallography was the coleopteran-specific Cry3Aa [25]. Since then, the tertiary structure of other six 3-domain Cry active proteins, including Cry1Aa, Cry2Aa, Cry3Bb, Cry4Aa, Cry4Aa, Cry4Ba and Cry8Ea, has been determined [26–31]. Among all, Cry3Aa, Cry3Bb and Cry8Ea have been defined as coleopteran-active proteins (Figure 3A,B). Using the FATCAT server [32], the structural alignment between these anti-coleopteran proteins is significantly similar, despite their low sequence identity. Pardo-López et al. [16] analyzed the structural similarity between Cry1Aa and the other 3-domain Cry proteins aforementioned, indicating the same structural likeness. The marked similarity in terms of the structure of the 3-domain Cry proteins, despite the low sequence identity and the differences in specificity, has rendered different proteins with similar modes of action.

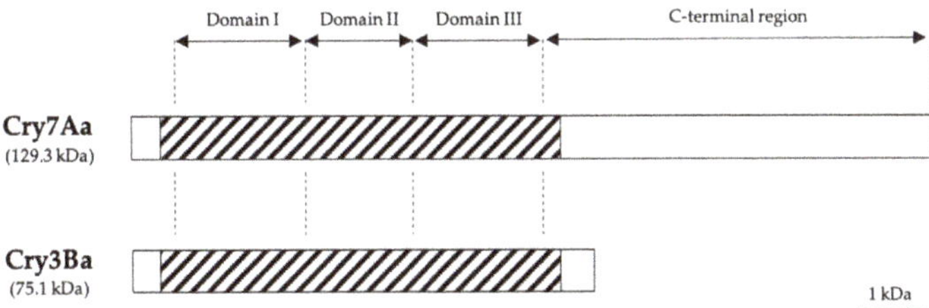

Figure 2. Relative length of 3-domain Cry proteins of *B. thuringiensis*, representing both main sizes of approximately 130 and 70 kDa. Dashed parts represent the activated toxin, while the white boxes represent the amino- and carboxy-terminal parts. Adapted from Bravo et al., 2007 [33].

Domain I consists of six α-helix surrounding a hydrophobic helix-α5. This domain, which shares strong similarities with the structure of the pore-forming domain of α+PFTs colicin A, might be responsible for membrane penetration and pore formation [23]. The binding domain II is constituted by three antiparallel β-sheets packing together and has an important role in receptor binding affinity. Finally, domain III is a two-twisted anti-parallel β-sheet and is also involved in receptor binding and pore formation [24,34]. Although it has been demonstrated that domains I and II have co-evolved over the years, swapping by homologous recombination of domain III has also been reported [15,35]. Local alignment of coleopteran-active Cry3, Cry7, and Cry8 showed that domain I was strongly conserved while domains II and III diversified [35]. Bt might use this mechanism to get adapted to a new insect host, which may explain the great variability in the biocidal activity of the 3-domain Cry proteins.

2.1.2. Non-3-Domain Cry Toxins

In addition to the 3-domain Cry proteins, some unrelated Cry proteins are also designated by the Cry nomenclature: Etx_Mtx2 proteins, Toxin_10 proteins and alpha-helical toxins [4]. The structure and function of Etx_Mtx2 proteins remains unclear, although the similarities with the *Clostridium perfringens* epsilon toxin (closely related to aerolysin) seem to indicate that they may have a β-sheet-based structure and a pore-forming activity [36]. It is important to notice that, while most of them have activity by themselves, some toxins are proposed as protein complexes to induce mortality, such as the Etx_Mtx2 protein Cry23 and the Cry37 protein [37]. The crystal structure of Cry23Aa reveals a single β-stranded domain protein, with structural similarity to several β-pore forming toxins as proaerolysins, produced by other bacterial species [38]. Cry37Aa conforms to a C2 β-sandwich fold, similar to the calcium phospholipid-binding domain observed in human cytosolic phospholipase A2 (Figure 3C) [38]. Moreover, the toxins Cry34 and Cry35 have been reported to have binary activity against coleopteran insects [39,40]. Crystal structures of Cry34Ab and Cry35Ab have been published (Figure 3D). Cry35Ab, a member of Toxin_10 proteins, shows an aerolysin-like fold, containing a β-trefoil N-terminal domain similar to the carbohydrate-binding domain in Mtx1. Cry34Ab is also a member of the aerolysin family with a β-sandwich fold, common among other cytolytic proteins [41].

2.1.3. Cyt Proteins

Similar to the Cry proteins, Cyt proteins are produced as protoxins with a proteolytically activated size of around 25 kDa [20]. As with some Cry proteins, the tertiary structure of some Cyt proteins has already been solved. Cyt1Aa [42], Cyt2Aa [43] and Cyt2Ba [44] show a similar structure composed of a single α–β-domain, with two outer layers of α-helix wrapped around a β-sheet (Figure 3E). Studies performed with peptides of Cyt1A show that α-helix peptides are major structural elements involved in membrane interaction [45] and also in the oligomerization process [46], while the β-strand forms an oligomeric pore with a β-barrel structure into the membrane [43].

2.2. Insecticidal Activity

The vast majority of Cry proteins described to date are toxic to lepidopteran pests, but there are also a few crystal proteins toxic to either coleopteran or dipteran insects, and a small number are toxic to nematodes [47]. Currently, 45 Bt crystal proteins, including Cry, Cyt or binary proteins, have been tested against different coleopteran insects (Table 1).

2.2.1. Host Range

Cry proteins are toxic to a large number of beetle pests. Mainly, the Cry3 group, the best-studied one, has been described with activity against most of the coleopteran species assayed. These Cry proteins, encoded by *cry3* genes, were first discovered in the subspecies *tenebrionis* [48] and *san diego* [49] although, years later, both strains turned out to be the same subsp. [50]. Since then, more isolates like Bt subsps. *tolworthi*, *kumamotoensis*, or *kurstaki* have been reported to encode a *cry3* gene [51,52]. Owing to the well-known activity in important coleopteran pests, such as *Leptinotarsa decemlineata* (Coleoptera: Chrysomelidae) or *Diabrotica* spp. (Coleoptera: Chrysomelidae), some of these isolates have been developed as bioinsecticides for beetle control [47]. Cry3Aa, Cry3Ba, Cry3Bb and Cry3Ca proteins have shown activity against most major coleopteran families, including *Chrysomelidae*, *Curculionidae*, *Scarabaeidae*, and *Tenebrionidae*, among others (Table 1). Although Cry3 proteins are the most effective Bt toxins against chrysomelid beetles, the widespread use of Cry3-based insecticides and Bt crops carries the risk of selecting insect biotypes tolerant to that proteins. The appearance of resistant populations of the chrysomelids *L. decemlineata*, *Chrysomela scripta* under laboratory conditions or *Diabrotica* spp. to Bt maize have been reported [53–55].

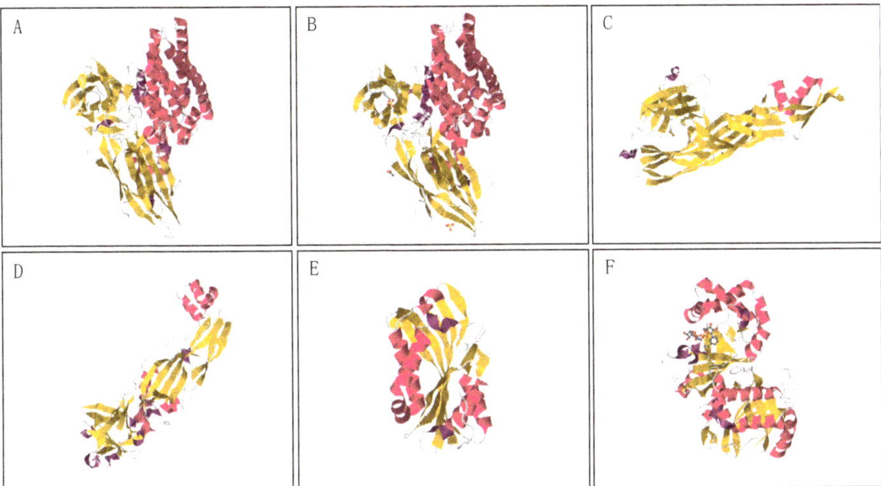

Figure 3. *Bacillus thuringiensis* proteins, with particular activity against coleopteran pests, for which three-dimensional structure has been predicted. (**A**) Cry3Aa (PBD accession number 4QX1); (**B**) Cry8Ea (PBD accession number 3EB7); (**C**) Protein complex Cry23Aa/Cry37Aa (PBD accession number 4RHZ); (**D**) Binary proteins Cry34Ab and Cry35Ab (PBD accession number 4JOX and 4JPO); (**E**) Cyt1Aa (PBD accession number 3RON); (**F**) Secretable protein Vip2Aa with a NAD complex (PBD accession number 1QS2).

Cry7 and Cry8 groups are comparatively less active on chrysomelids, but they represent a serious alternative to Cry3 proteins. Cry7Aa, formerly known as CryIIIC, is very toxic to Cylas species (Coleoptera: Brentidae) [56], even more than Cry3 protein, but it has no negative effects against *Anthonomus grandis* (Coleoptera: Curculionidae) or *D. undecimpuntata* [52]. Moreover, toxicity to Colorado potato beetle has been reported, but only after in vitro solubilization [52], which was countered by a recent report of a Cry7Aa-type protoxin which is active against *L. decemlineata* without any previous solubilization step [57]. Solubilized Cry7Ab is active against *Henosepilachna vigintiomaculata* (Coleoptera: Coccinellidae) and *Acanthoscelides obtectus* (Coleoptera: Chrysomelidae), but not against *Anomala corpulenta* (Coleoptera: Scarabaeidae) or *Pyrrhalta aenescens* (Coleoptera: Chrysomelidae) [57,58]. Cry8-type proteins are toxic to a large number of coleopteran pests, particularly against species in the Scarabaeidae family [59–61]. Furthermore, Cry8A and Cry8B proteins have shown activity against the chrysomelids *L. decemlineata* and *Diabrotica* spp., Cry8Ca against the tenebrionid *Alphitobius diaperinus* (Coleoptera: Tenebrionidae) [62] and Cry8Ka against the curculionid *A. grandis* [63]. Moreover, some Cry8 proteins, such as Cry8Ea, Cry8Ga or Cry8Na, are very specific, showing different activities against very closely related host species [64,65]. Cry6Aa and Cry6Ba are active against the curculionid beetles *Hypera postica* and *Hypera brunipennis*, two of the more important pests in alfalfa [66,67], as well as *D. virgifera*, which is susceptible to the activated toxin. Cry22 proteins also have activity to a wide spectrum of coleopteran insects. In particular, Cry22A and Cry22B proteins are toxic to coleopterans of the Brentidae, Chrysomelidae and Curculionidae families [56,68,69].

Generally, Bt protein groups are particularly toxic to a certain insect order. However, some proteins may be active against different orders [70]. Mainly lepidopteran proteins Cry1Ba and Cry1Ia have shown activity against the key coleopteran pests *A. grandis*, *A. obctetus*, *C. scripta* and *L. decemlineata* [71–76]. Dual activity against Lepidoptera-Coleoptera has also been demonstrated by Cry9-type proteins. Cry9 toxins exhibit strong activity against main lepidopteran pests, but Cry9Da is also toxic against the scarab *Anomala cuprea* [77]. Other example of cross-order toxicity is depicted by the dipteran toxin Cry10Aa, which can kill the Cotton boll weevil (*A. grandis*) [78]. Additionally, Cry51Aa is toxic against *Lygus* spp. (Hemiptera) and *L. decemlineata* [79] and Cry55Aa, a typical nematicidal protein, has been reported as toxic to the chrysomelid *Phyllotreta cruciferae* [80].

Binary toxins, structurally different from classical 3-domain Cry proteins [25], used to be considered as single toxins because both proteins are required to kill their target. To date, two binary complex toxins have been proposed to have activity against beetles. The coleopteran specific Cry23Aa has been assayed together with Cry37Aa protein to kill *Popillia japonica* (Coleoptera: Sacarabaeidae) and *Tribolium castaneum* (Coleoptera: Tenebrionidae) [37]. Furthermore, this protein mixture has been found to be active against *Cylas* spp. (Coleoptera: Brentidae) and *A. obtectus* [56,75]. On the other hand, Cry34 protein is only active in association with Cry35 protein [17]. Cry34 and Cry35 are closely related and are often encoded in the same operon, with coordinated function and appearance in crystals [40,81]. The Cry34/Cry35 binary proteins are mainly active against corn rootworms and have been developed for in-plant control in Bt maize [40,82].

B. thuringiensis Cyt proteins have an in vitro cytolytic (hemolytic) activity, hence their name, and show predominant dipteran specificity [24]. However, some of them are also toxic to coleopteran pests, such as Cyt1Aa to *C. scripta* [72] or Cyt2Ca to the chrysomelids *L. decemlineata* and *Diabrotica* spp. [83] and the curculionid *Diaprepes abbreviates* (Coleoptera: Curculionidae) [84,85]. Besides, Cyt proteins improve the activity of Cry proteins. For instance, Cyt1Aa is able to overcome high levels of resistance to Cry3Aa by *C. Scripta*, playing an important role in resistance management [72].

Table 1. Insecticidal activity of Cry and Cyt proteins against coleopteran pests.

Crystal Type Toxin	Target Insect		Activity [a]	LC$_{50}$ [b]	Reference
	Scientific Name	**Family**			
Cry1Aa	*Anoplophora glabripennis*	Cerambycidae	N		[86]
	Apriona germari	Cerambycidae	N		[87]
	Epilachna varivestis	Coccinellidae	A		[88]
	Tribolium castaneum	Tenebrionidae	LA		[89]
Cry1Ab	*Diabrotica undecimpuntata*	Chrysomelidae	N		[90]
	Leptinotarsa decemlineata	Chrysomelidae	N		[90]
	Phyllotreta armoraciae	Chrysomelidae	N		[90]
	Adalia bipunctata	Coccinellidae	N		[91]
	Atheta coriaria	Coccinellidae	N		[91]
	Cryptolaemus montrouzieri	Coccinellidae	N		[91]
	Harmonia axyridis	Coccinellidae	N		[92]
	Anthonomus grandis	Curculionidae	N		[90]
	Hypera postica	Curculionidae	N		[90]
	Popillia japonica	Scarabaeidae	N		[90]
Cry1Ac	*Diabrotica undecimpuntata*	Chrysomelidae	N		[90]
	Leptinotarsa decemlineata	Chrysomelidae	N		[90]
	Phyllotreta armoraciae	Chrysomelidae	N		[90]
	Hippodamia convergens	Coccinellidae	N		[93]
	Anthonomus grandis	Curculionidae	N		[90]
	Hypera postica	Curculionidae	N		[90,94]
	Haptoncus luteolus	Nitidulidae	N		[95]
	Tribolium castaneum	Tenebrionidae	N		[96]
	Popillia japonica	Scarabaeidae	N		[90]
Cry1Ah	*Propylea japónica*	Coccinellidae	N		[97]

Table 1. *Cont.*

Crystal Type Toxin	Target Insect		Activity [a]	LC$_{50}$ [b]	Reference
	Scientific Name	**Family**			
Cry1Aj	*Harmonia axyridis*	Coccinellidae	N		[92]
Cry1Ba	*Anoplophora glabripennis*	Cerambycidae	N		[86]
	Acanthoscelides obtectus	Chrysomelidae	A		[75]
	Chrysomela scripta F	Chrysomelidae	A	1.8 // 5.9	[71,72]
	Leptinotarsa decemlineata	Chrysomelidae	A	1050 // 142	[71,73]
	Phaedon cochleariae	Chrysomelidae	N		[98]
	Anthonomus grandis	Curculionidae	A	305.32	[74]
	Asymmathetes vulcanorum	Curculionidae	N		[99]
	Hypothenemus hampei	Curculionidae	A		[100]
	Tribolium castaneum	Tenebrionidae	N		[89]
Cry1Ca	*Tribolium castaneum*	Tenebrionidae	N		[89]
Cry1Da	*Tribolium castaneum*	Tenebrionidae	N		[89]
Cry1Ea	*Tribolium castaneum*	Tenebrionidae	N		[89]
Cry1Fa	*Cryptolestes pusillus*	Laemophloeidae	N		[17]
	Tribolium castaneum	Tenebrionidae	N		[17]
Cry1Fb	*Tribolium castaneum*	Tenebrionidae	N		[89]
Cry1Ia	*Acanthoscelides obtectus*	Chrysomelidae	A		[75]
	Agelastica coerulea	Chrysomelidae	N		[101,102]
	Diabrotica undecimpuntata	Chrysomelidae	N		[103]
	Leptinotarsa decemlineata	Chrysomelidae	A	33.7 // 10	[73,104]
	Phaedom brassicae	Chrysomelidae	N		[101]
	Anthonomus grandis	Curculionidae	A	21.5 // 230	[76,105]
	Asymmathetes vulcanorum	Curculionidae	N		[99]
	Tenebrio molitor	Tenebrionidae	N		[106]
	Tribolium castaneum	Tenebrionidae	N		[89]
Cry1Ib	*Phaedom brassicae*	Chrysomelidae	N		[101]
	Agelastica coerulea	Chrysomelidae	N		[101]
Cry1Id	*Agelastica coerulea*	Chrysomelidae	N		[102]
Cry1Ie	*Ceratoma trifurcata*	Chrysomelidae	N		[107]
	Pyrrhalta aenescens	Chrysomelidae	N		[108]
Cry1Jb	*Diabrotica undecimpuntata*	Chrysomelidae	N		[109]
	Leptinotarsa decemlineata	Chrysomelidae	N		[109]
Cry2Aa	*Diabrotica undecimpuntata*	Chrysomelidae	N		[93]
	Diabrotica virgifera	Chrysomelidae	N		[93]
	Leptinotarsa decemlineata	Chrysomelidae	N		[93]
	Hippodamia convergens	Coccinellidae	N		[93]
	Anthonomus grandis	Curculionidae	N		[93]
Cry2Ab	*Propylea japonica*	Coccinellidae	N		[97]
	Haptoncus luteolus	Nitidulidae	N		[95]
Cry3Aa	*Rhyzophertha dominica*	Bostrichidae	A	1.17 µg/mg	[110]
	Cylas brunneus	Brentidae	A	1.88 µg/g	[56]
	Cylas puncticollis	Brentidae	A	1.99 µg/g	[56]
	Apriona germari	Cerambycidae	A		[94,111]
	Acanthoscelides obtectus	Chrysomelidae	A		[75]
	Agelastica alni	Chrysomelidae	A		[112]
	Brontispa longissimi	Chrysomelidae	A	0.475 mg/mL	[113]
	Chrysomela tremulae	Chrysomelidae	A		[114]
	Chrysomela scripta F	Chrysomelidae	A		[115]
	Chrysomela scripta F	Chrysomelidae	A	2.22 // 1.8	[71,72]
	Colaphellus bowringi	Chrysomelidae	A	2.68 // 1.33	[116,117]
	Crioceris quaturdicerumpunctata	Chrysomelidae	A	3.82	[117]
	Diabrotica undecimpuntata	Chrysomelidae	N		[90,118]
	Diabrotica virgifera	Chrysomelidae	N		[118,119]
	Leptinotarsa decemlineata	Chrysomelidae	A	1.84 // 3.56	[73,118]
	Phaedom brassicae	Chrysomelidae	A	1.11	[117]
	Phaedon cochleariae	Chrysomelidae	A		[120]

Table 1. *Cont.*

| Crystal Type Toxin | Target Insect | | Activity [a] | LC$_{50}$ [b] | Reference |
	Scientific Name	Family			
	Phyllotreta armoraciae	Chrysomelidae	N		[90]
	Plagiodera versicolora	Chrysomelidae	A	1.13 // 3.09	[18]
	Pyrrhalta aenescens	Chrysomelidae	A	0.22 mg/mL	[121]
	Pyrrhalta luteola	Chrysomelidae	A	0.12 µg/cm^2	[49]
	Adalia bipunctata	Coccinellidae	N		[91]
	Atheta coriaria	Coccinellidae	N		[91]
	Cryptolaemus montrouzieri	Coccinellidae	N		[91]
	Epilachna varivestis	Coccinellidae	A		[88]
	Anthonomus grandis	Curculionidae	N		[90]
	Asymmathetes vulcanorum	Curculionidae	N		[99]
	Hypera postica	Curculionidae	N		[90]
	Hypothenemus hampei	Curculionidae	A		[100]
	Myllocerus undecimpustulatus	Curculionidae	A	152 ng/cm^2	[122]
	Premnotrypes vorax	Curculionidae	LA		[123]
	Sitophilus oryzae	Curculionidae	A		[124]
	Amphimallon solstitiale	Scarabaeidae	A		[112]
	Anomala corpulenta	Scarabaeidae	N		[116]
	Melontha melontha	Scarabaeidae	A		[112]
	Popillia japonica	Scarabaeidae	N		[90]
	Alphitobius diaperinus	Tenebrionidae	A	9.58 // 8 µg/cm^2	[62,125]
	Tribolium castaneum	Tenebrionidae	N		[96,110]
	Tribolium castaneum	Tenebrionidae	A	0.46 g/10 g	[89]
	Tenebrio molitor	Tenebrionidae	A	11.4 µg/larve	[126]
	Tenebrio molitor	Tenebrionidae	A		[110,127]
Cry3Ba	*Cylas brunneus*	Brentidae	A	1.304 µg/g	[56]
	Cylas puncticollis	Brentidae	A	1.273 µg/g	[56]
	Chrysomela scripta F	Chrysomelidae	A		[115]
	Diabrotica undecimpuntata	Chrysomelidae	A	107 ng/mm^2	[128]
	Leptinotarsa decemlineata	Chrysomelidae	A	1.35 ng/mm^2	[128]
	Epilachna varivestis	Coccinellidae	N		[88]
	Popillia japonica	Scarabaeidae	A	1	[37]
	Tribolium castaneum	Tenebrionidae	A	1.60 g/10 g	[89]
	Tribolium castaneum	Tenebrionidae	A	13.55 mg/mL	[37,96]
Cry3Bb	*Cylas brunneus*	Brentidae	A	1.83 µg/g	[56]
	Cylas puncticollis	Brentidae	A	1.82 µg/g	[56]
	Anoplophora glabripennis	Cerambycidae	N		[86]
	Diabrotica undecimpuntata	Chrysomelidae	A	9.49 // 1.18	[118,129]
	Diabrotica virgifera	Chrysomelidae	A	2.10 // 5.18	[118,129]
	Leptinotarsa decemlineata	Chrysomelidae	A	6.86 // 6.54	[118,129]
	Alphitobius diaperinus	Tenebrionidae	A	26.52 // 50 µg/cm^2	[62,125]
Cry3Ca	*Cylas brunneus*	Brentidae	A	0.69 µg/g	[56]
	Cylas puncticollis	Brentidae	A	0.57 µg/g	[56]
	Leptinotarsa decemlineata	Chrysomelidae	A	0.7 // 320.13	[130,131]
	Tribolium castaneum	Tenebrionidae	N		[96]
Cry6Aa	*Diabrotica virgifera*	Chrysomelidae	A	77 µg/cm^2	[66,119]
	Hypera brunneipennis	Curculionidae	A		[66]
	Hypera postica	Curculionidae	A		[66]
Cry6Ba	*Hypera postica*	Curculionidae	A	280 ng/µL	[94]
Cry7Aa	*Cylas brunneus*	Brentidae	A	0.44 µg/g	[56]
	Cylas puncticollis	Brentidae	A	0.34 µg/g	[56]
	Anoplophora glabripennis	Cerambycidae	N		[86]
	Diabrotica undecimpuntata	Chrysomelidae	N		[52]
	Leptinotarsa decemlineata	Chrysomelidae	A	13.1 // 18.8	[52,57]
	Anthonomus grandis	Curculionidae	N		[52]

Table 1. *Cont.*

Crystal Type Toxin	Target Insect		Activity [a]	LC$_{50}$ [b]	Reference
	Scientific Name	**Family**			
Cry7Ab	*Acanthoscelides obtectus*	Chrysomelidae	A		[75]
	Ceratoma trifurcata	Chrysomelidae	N		[107]
	Colaphellus bowringi	Chrysomelidae	A	293.79	[132]
	Pyrrhalta aenescens	Chrysomelidae	N		[58]
	Henosepilachna vigintioctomaculata	Coccinellidae	A	209	[58,133]
	Anomala corpulenta	Scarabaeidae	N		[58]
	Tribolium castaneum	Tenebrionidae	LA		[89]
Cry8Aa	*Leptinotarsa decemlineata*	Chrysomelidae	A		[134]
	Cotinis spp	Scarabaeidae	A		[135]
	Tribolium castaneum	Tenebrionidae	LA		[89]
Cry8Ab	*Holotrichia oblita*	Scarabaeidae	A	5.72 µg/g	[136]
	Holotrichia parallela	Scarabaeidae	A	2.00 µg/g	[136]
	Tenebrio molitor	Tenebrionidae	N		[136]
Cry8Ba	*Diabrotica virgifera*	Chrysomelidae	A		[119]
	Cotinis spp	Scarabaeidae	A		[137]
	Cyclocephala borealis	Scarabaeidae	A		[135]
	Cyclocephala pasadenae	Scarabaeidae	A		[135]
	Popillia japonica	Scarabaeidae	A		[135]
Cry8Bb	*Diabrotica undecimpuntata*	Chrysomelidae	A		[138]
	Diabrotica virgifera	Chrysomelidae	A		[138]
	Leptinotarsa decemlineata	Chrysomelidae	A		[138]
Cry8Ca	*Anoplophora glabripennis*	Cerambycidae	N		[86]
	Colaphellus bowringi	Chrysomelidae	N		[116]
	Leptinotarsa decemlineata	Chrysomelidae	N		[116]
	Epilachna varivestis	Coccinellidae	A		[88]
	Anomala corpulenta	Scarabaeidae	A	$1.75 \times 10 \times 10^8$ CFU/g	[116]
	Anomala corpulenta	Scarabaeidae	A	$1.6 \times 10 \times 10^8$ CFU/g	[139,140]
	Anomala cuprea	Scarabaeidae	A		[141]
	Anomala exoleta	Scarabaeidae	A		[142]
	Holotrichia parallela	Scarabaeidae	A	$9.24 \times 10 \times 10^8$ CFU/g	[140]
	Popillia japonica	Scarabaeidae	A	12.3 µg/g	[35]
	Alphitobius diaperinus	Tenebrionidae	A	7.71 // 10 µg/cm^2	[62,125]
	Tribolium castaneum	Tenebrionidae	N		[89]
Cry8Da	*Anomala cuprea*	Scarabaeidae	A		[143]
	Anomala orientalis	Scarabaeidae	A		[143]
	Popillia japonica	Scarabaeidae	A	17.0 µg/g	[35,143]
Cry8Db	*Popillia japonica*	Scarabaeidae	A	19.6 µg/g	[35]
Cry8Ea	*Plagiodera versicolora*	Chrysomelidae	A		[144]
	Anomala corpulenta	Scarabaeidae	A		[140]
	Holotrichia parallela	Scarabaeidae	A	$0.9 \times 10 \times 10^8$ CFU/mL	[59,64,144]
	Popillia japonica	Scarabaeidae	A		[144]
	Tenebrio molitor	Tenebrionidae	N		[64]
	Tribolium castaneum	Tenebrionidae	N		[64,89]
Cry8Fa	*Anomala corpulenta*	Scarabaeidae	N		[59]
	Holotrichia oblita	Scarabaeidae	N		[59]
	Holotrichia parallela	Scarabaeidae	N		[59]
	Tribolium castaneum	Tenebrionidae	N		[89]
Cry8Ga	*Holotrichia oblita*	Scarabaeidae	N		[60]
	Holotrichia parallela	Scarabaeidae	N		[60]
Cry8Ka	*Anthonomus grandis*	Curculionidae	A	2.83–8.93	[63]
Cry8Na	*Anomala corpulenta*	Scarabaeidae	N		[65]
	Holotrichia oblita	Scarabaeidae	N		[65]
	Holotrichia parallela	Scarabaeidae	A	$3.18 \times 10 \times 10^{10}$ CFU/g	[65]
Cry8Sa	*Holotrichia serrata* (F.)	Scarabaeidae	A		[145]
Cry9Bb	*Diabrotica undecimpuntata*	Chrysomelidae	N		[146]
	Diabrotica virgifera	Chrysomelidae	N		[146]
	Leptinotarsa decemlineata	Chrysomelidae	N		[146]
	Anthonomus grandis	Curculionidae	N		[146]

Table 1. *Cont.*

Crystal Type Toxin	Target Insect		Activity [a]	LC$_{50}$ [b]	Reference
	Scientific Name	**Family**			
Cry9Da	*Anomala cuprea*	Scarabaeidae	A		[77]
	Tribolium castaneum	Tenebrionidae	N		[89]
Cry10Aa	*Anthonomus grandis*	Curculionidae	A	7.12	[78]
Cry14Aa	*Tribolium castaneum*	Tenebrionidae	LA		[89]
Cry15Aa	*Leptinotarsa decemlineata*	Chrysomelidae	N		[147]
Cry18Aa1	*Melontha melontha*	Scarabaeidae	A		[148]
Cry22Aa	*Anthonomus grandis*	Curculionidae	A	0.75 µg/well	[68]
	Tribolium castaneum	Tenebrionidae	A	1.25 g/10 g	[89]
Cry22Ab	*Cylas brunneus*	Brentidae	A	1.01 µg/g	[56]
	Cylas puncticollis	Brentidae	A	0.78 µg/g	[56]
	Diabrotica virgifera	Chrysomelidae	A	39.4 µg/cm^2	[69]
	Diabrotica undecimpuntata	Chrysomelidae	N		[69]
	Leptinotarsa decemlineata	Chrysomelidae	N		[69]
	Anthonomus grandis	Curculionidae	A	3.12 µg/well	[68]
Cry22Ba	*Diabrotica virgifera*	Chrysomelidae	N		[68]
	Anthonomus grandis	Curculionidae	A		[68]
Cry23Aa/37Aa	*Cylas brunneus*	Brentidae	A	0.46 µg/g	[56]
	Cylas puncticollis	Brentidae	A	0.42 µg/g	[56]
	Acanthoscelides obtectus	Chrysomelidae	A		[75]
	Anthonomus grandis	Curculionidae	A		[149]
	Popillia japonica	Scarabaeidae	A		[37]
	Tribolium castaneum	Tenebrionidae	A	6.30 µg SC/µL	[37,61]
Cry34Aa	*Diabrotica virgifera*	Chrysomelidae	N		[40]
Cry34Ab	*Diabrotica undecimpuntata*	Chrysomelidae	LA		[150]
	Diabrotica virgifera	Chrysomelidae	N		[40,82]
Cry34Ac	*Diabrotica virgifera*	Chrysomelidae	N		[40]
Cry34Aa/35Aa	*Diabrotica undecimpuntata*	Chrysomelidae	A	34.1 µg/well	[151]
	Diabrotica virgifera	Chrysomelidae	A	34 µg/cm^2	[81,151]
Cry34Ab/35Ab	*Rhyzophertha dominica*	Bostrichidae	N		[17]
	Diabrotica undecimpuntata	Chrysomelidae	A		[150]
	Diabrotica virgifera	Chrysomelidae	A	3 µg/cm^2	[40,81]
	Oryzaephilus surinamensis	Cucujidae	LA		[17]
	Sitophilus oryzae	Curculionidae	LA		[17]
	Trogoderma variabile	Dermestidae	N		[17]
	Tenebrio molitor	Tenebrionidae	LA		[17]
	Tribolium castaneum	Tenebrionidae	LA		[17]
	Tribolium castaneum	Tenebrionidae	N		[96]
Cry34Ac/35Ac	*Diabrotica virgifera*	Chrysomelidae	A	7 µg/cm^2	[40,81]
Cry34Ba/35Ba	*Diabrotica virgifera*	Chrysomelidae	A		[39]
Cry35Aa	*Diabrotica virgifera*	Chrysomelidae	N		[40]
Cry35Ab	*Diabrotica virgifera*	Chrysomelidae	N		[40,82]
Cry35Ac	*Diabrotica virgifera*	Chrysomelidae	N		[40]
Cry36A	*Diabrotica virgifera*	Chrysomelidae	A	147.3 µg/well	[151]
Cry37Aa	*Tribolium castaneum*	Tenebrionidae	A	1.25 g/10 g	[89]
Cry38Aa	*Diabrotica virgifera*	Chrysomelidae	N		[39]
Cry43Aa	*Anomala cuprea*	Scarabaeidae	A		[152]
Cry43Ba	*Anomala cuprea*	Scarabaeidae	N		[152]
Cry51Aa	*Diabrotica undecimpuntata*	Chrysomelidae	N		[79]
	Diabrotica virgifera	Chrysomelidae	N		[79]
	Leptinotarsa decemlineata	Chrysomelidae	A		[79]
	Tribolium castaneum	Tenebrionidae	A	1.45 g/10 g	[89]

Table 1. *Cont.*

Crystal Type Toxin	Target Insect		Activity [a]	LC$_{50}$ [b]	Reference
	Scientific Name	**Family**			
Cry55Aa	*Phyllotreta cruciferae*	Chrysomelidae	A		[80]
	Tribolium castaneum	Tenebrionidae	N		[89]
Cyt1Aa	*Chrysomela scripta F*	Chrysomelidae	A	132.6	[72]
Cyt2Ca	*Diabrotica undecimpuntata*	Chrysomelidae	A	25 µg/well	[83]
	Diabrotica virgifera	Chrysomelidae	A	10.8 µg/well	[83]
	Leptinotarsa decemlineata	Chrysomelidae	A		[83]
	Diapepes abbreviatus	Curculionidae	A	50.7	[84,85]
	Popillia japonica	Scarabaeidae	A		[83]
	Tribolium castaneum	Tenebrionidae	A		[83]

[a] The parameter is mortality. A = active; N = not active; LA = low activity, with significant inhibition of growth;
[b] LC$_{50}$ = lethal concentration that causes 50% mortality of the insects. Data are expressed in µg/mL, unless otherwise stated. "//" separate two different values of the LC$_{50}$.

2.2.2. Genetically Engineered Cry Genes

Recent advances in next generation sequencing and genetic engineering technologies allow the construction of new synthetic *cry* genes that increase or amplify their toxicity. The domain regions of some lepidopteran-specific proteins have been modified in an attempt to improve their specific activity or broaden their host range [15,153]. The first coleopteran hybrid protein was made by fusing the sequences located in domain III of the *cry3A* and *cry1Aa* genes, although unfortunately, it caused the loss of activity against *L. decemlineata* [154]. Nonetheless, substituting domain III of Cry3Aa with the same domain from Cry1Ab induced activity against WCR (Western corn rootworm) larvae [155]. On a different approach, a *cry3Bb1* gene was engineered with five amino acid substitutions to produce the new Cry3Bb1.11098 protein, which increased the activity of the natural protein against WCR [156]. Similarly, a Cry3A variant (eCry3.1Ab) was designed to confer novel activity against rootworms by creating a cathepsin G protease recognition site [157]. This technology has been introduced successfully in the development of transgenic plants, mainly to overcome the appearance of resistance by WCR populations [158].

2.3. Mode of Action

The mode of action has been mostly studied in lepidopteran insects, although it is believed to be similar between different insect orders, with some peculiarities [8]. Briefly, it is widely accepted that the process begins once the target insect ingests the protein and reaches the insect midgut, where it is solubilized and proteolytically activated. Such an activation allows toxins to first bind to their specific receptors in the host cell membrane, then to their oligomerization and, eventually, to the formation of pores in the cell membrane (Figure 4). In this multi-step mode of action, several factors may contribute to protein specificity [159].

2.3.1. Solubilization and Proteolytic Processing

Once proteins reach the host midgut, they are released from their crystal package to initiate the pathogenic process. The crystals are stabilized by disulfide bridges among the C-terminal ends of the protoxins. More recently, the occurrence of 20 kbp DNA fragments with protoxins and 100–300 pb DNA fragments with in vitro proteolytic activated toxins has been established [160]. These DNA fragments have been observed to be associated with different Bt-toxins as Cry1A, Cry2A, etc., however, they have been more extensively studied on Cry8 toxins [21]. The sequence of the DNA fragments is not specific and they are located in plasmids and chromosomes [161]. Bioinformatics modelling suggests that two protoxin regions bind to major grooves and another one, combined with phosphoric acid, binds to the minor groove [162]. The associated DNA should be eliminated by the DNAses in the insect gut for the correct protein activation. In fact, DNA-protein association impairs the specific binding [163].

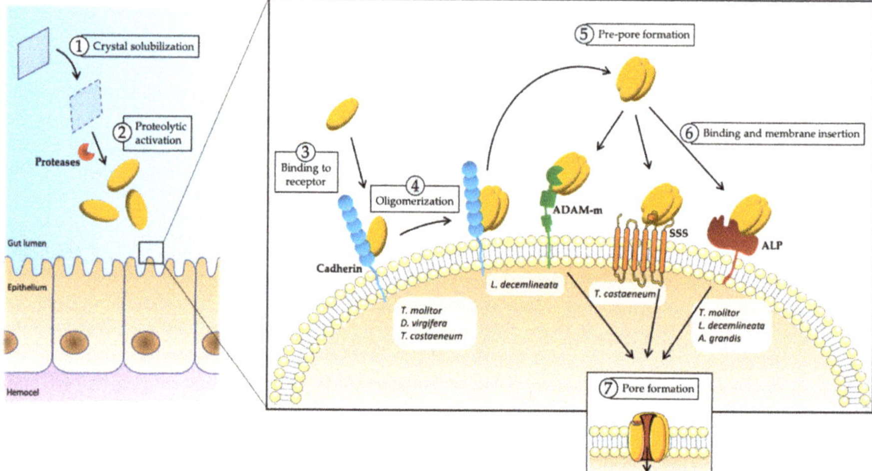

Figure 4. Schematic representation of the particularities in the mechanism of action of crystal proteins against coleopteran pests. (**1**) Crystal solubilizes in the acidic conditions of the coleopteran midgut lumen and (**2**) activates into toxin by proteolytic processing of the protoxin by the specific digestive enzymes, specially cysteine and aspartic proteases. (**3**) Toxins are able to bind to a first receptor (CADR), (**4**) oligomerizate and (**5**) form an oligomeric pre-pore structure that (**6**) is able to bind to a second specific receptor (ADAM metalloproteases/GPI-anchored alkaline phosphatases/sodium solute symporters). This event induces the insertion into the membrane, leading to (**7**) pore formation and finally to cell lysis.

It is well accepted that solubilization processes are due to the environmental conditions in the susceptible insect midgut, mainly to pH values. Of note, unlike the alkaline midgut of lepidopteran and dipteran insects, beetles have an acidic midgut, suggesting that different solubilization conditions are needed for each protein [164]. For instance, the midgut fluids of *L. decemlineata* and *D. virgifera* larvae do not seem to solubilize Cry1B and Cry7Aa1, and only after a previous in vitro solubilization, these proteins become active [52,71]. However, recent reports show that Cry7Ab2 and Cry7Aa2 proteins solubilize into midgut fluids of *H. vigintioctomaculata* and *L. decemlineata* larvae, respectively, suggesting that the lack of solubilization involves more factors than pH [57,58]. Cyt proteins dissolve readily under alkaline conditions, especially at pH 8 or higher, and they are harder to solubilize in neutral or slightly acidic pHs, which occurs in coleopteran midguts [72]. Another example of the importance of crystal solubilization was published by Galitsky et al. [28]. They related that differences in toxin solubility, oligomerization and binding for the Cry3-type toxins, in addition to differences in domain III, might explain the different specificities of Cry3A and Cry3B (e.g., WCR is susceptible to Cry3Bb1 but not to Cry3A). Solubilized proteins are proteolytically activated by gut proteases, which generate the toxic three-domain fragment of about 65 kDa [33]. In Lepidoptera and Diptera species, the main proteases present in the alkaline midgut juices are serine proteases, especially trypsin and chymotrypsin proteases [165]. However, the coleopteran species use digestive proteases belonging to cysteine and aspartic proteases and serine proteases are only present in some cases [166]. The presence of different proteases may be an important factor in toxin activation specificity, and improper processing of Bt toxins can involve the development of insect resistances. It has been reported that the combination of Cry3Aa protein and certain protease inhibitors enhances the toxicity against *Rhyzopertha dominica* (Coleoptera: Bostrichidae) larvae, evidencing that protease inhibitors may play an important role in resistant pests management [110]. Moreover, the relevance of a nicking in the N-terminal end, in the alpha 1–3 of Domain I in the activated Cry3A and Cry8Da toxins, has been shown, which rendered an 8 kDa fragment to obtain a functional 54 kDa toxin for receptor binding [167].

2.3.2. Binding to the Larval Epithelium

The activated toxin is able to bind to specific receptors located in the midgut epithelial cells to form an oligomeric pre-pore structure and alterations in the midgut receptors is a critical step for insect resistance appearance [159]. It has been demonstrated that Cry3Ba protein shares a binding receptor with Cry3Aa and Cry3Ca proteins, although heterologous-competition experiments show that both proteins may have other binding sites and only share one with Cry3Ba3 [168]. It has also been shown that Cry3Bb, Cry3Ca and Cry7Aa proteins competed for the same binding sites in *C. puncticollis*, so a mutation in the midgut receptor could render all three proteins ineffective [169]. To date, several specific coleopteran binding proteins have been identified. It has been shown that an ADAM metalloprotease can be considered as a Cry3Aa receptor in *L. decemlineata*, and this binding interaction improves Cry3Aa pore-formation [170]. GPI-anchored alkaline phosphatases (ALP) are also important for the Cry3Aa binding to *Tenebrio molitor* brush border membrane vesicles (BBMV) and are highly expressed when larvae are exposed to Cry3Aa [171]. In the same way, the Cry1Ba toxin binds to ALPs from *A. grandis* midgut cells [74]. Although some putative cadherines have been previously described [172,173], Fabrick et al. [127] were the first authors reporting a cadherin protein (TmCad1), cloned from *T. molitor* larval midgut as a Cry3Aa binding receptor. Furthermore, injection of *TmCad1* dsRNA into *T. molitor* larvae conferred resistance to Cry3Aa. Another truncated cadherin protein (DvCad1-CR8–10), isolated from the WCR, binds to activated Cry3Aa, Cry3Bb [118] and also Cry8Ca [62], enhancing the activity of *L. decemlineata*, *Diabrotica* spp. and *A. diaperinus*. Finally, in *T. castaeneum* larvae, a cadherine (TcCad1) and a sodium solute symporter (TcSSS) have been identified as putative Cry3Ba functional receptors, determinant for the specific Cry protein toxicity against coleopterans [174].

Studied Cry8-binding proteins revealed a difference from those confirmed previously as receptors for Cry1A or Cry3A proteins in lepidopteran and coleopteran insect species, such as aminopeptidases, cadherins or ABCC transporters [175,176]. A Cry8-like toxin without the C-terminal end has been described, which completely shared binding sites with Cry8Ga, despite only sharing 30% of the sequence, in *Holotrichia oblita*. Cry8Da tested on *Popillia japonica* BBMV, bound specifically with a 150 kDa membrane protein which shared homology with coleopteran β-glucosidases [177]. Cry8E and Cry8-like toxins showed, in *H. parallela* and *H. oblita*, binding to several different proteins. The most relevant for both insect species and Cry8 proteins were serine proteases, sodium/potassium-transporting proteins, and a transferrin-like protein [177,178].

There is evidence that some proteins work together to cause mortality in certain coleopteran species, although the mechanism of interaction between them remains unclear. In this way, it is hypothesized that Cry37 protein may facilitate linkage of channel-forming Cry23 toxin, given their homology to other binding proteins [24]. Moreover, the fact that Cry34Ab has some activity against the Western corn rootworm (WCR) on its own [150] seems to indicate that Cry35 has the role as a receptor of Cry34, which is mainly responsible for toxicity. Cyt proteins enhancing the insecticide potency of certain Cry toxins has been also observed. The Cyt1Aa protein, from Bt sub. *israelensis*, increases the activity of Cry11Aa toxin by acting as a membrane receptor [178]. Cyt1A also helps to overcome high levels of Cry3A resistance against *C. scripta* larvae [72]. Although this mechanism of action has not yet been elucidated, Cyt1A may act as a receptor of Cry3A to enhance the binding of this protein. This synergism between Cry and Cyt toxins is an excellent strategy to decrease the appearance of resistance to Cry proteins.

2.3.3. Oligomerization and Pore Formation

Although it remains unclear, some studies suggest that activated toxins need to form an oligomeric structure before insertion to the membrane as a result of binding to specific receptors [16]. In fact, Cry proteins that form oligomeric structures are related to a high pore activity [33]. Oligomerization of 3-domain Cry proteins has been described for toxins active against different insect orders, such as Cry3 proteins in coleopteran larvae. In the brush border membrane of *L. decemlineata*, Cry3A, Cry3B and Cry3C form an oligomer prior to membrane insertion, generating a pre-pore structure that can be inserted into the membrane [168]. Cry3Aa oligomeric structures have also been reported after incubation of Cry3Aa protoxin with *T. molitor* BBMV [127]. The oligomeric structure eventually leads to the lytic pore formation that disrupts the midgut insect cell by osmotic shock. However, oligomerization studies of Cry1Ab and Cry1Ia proteins incubated with lepidopteran and coleopteran BBMV, as well as culture insect cells, showed that Cry1Ia oligomerization may not be a requirement for toxicity [179]. Besides, the appearance of Cry1Ab oligomers when incubated with coleopteran BBMV could be due to an improper insertion of oligomers into the membrane or the inability to induce the post-pore events in the cells [179]. Either way, susceptible insects can withstand minor damage, but greater damage destroys the epithelium of the midgut, leading to a disruption in feeding and subsequent starvation death. Additional to the toxin action, spores may pass through the channel, to colonize and germinate in the hemolymph and contribute to insect death by septicemia [1].

3. The Secretable Coleopteran-Active Proteins

In addition to the δ-endotoxins produced during the stationary phase, other protein compounds have been found in the culture supernatant of certain entomopathogenic *Bacillus* isolates. These proteins, produced during the vegetative growth stage of the bacterium, were designated as vegetative insecticidal proteins (Vip) [180] and secreted insecticidal proteins (Sip) [181]. Within the Vip family, *vip1* and *vip2* genes are co-transcript in a single 4 kbp operon, which render proteins of about 100 kDa (Vip1) and 50 kDa (Vip2) [171]. The absence of toxicity of the proteins alone suggests that it is a binary toxin for some members of the coleopteran [180] and hemipteran [182] orders. In contrast, Vip3 proteins are single-chain toxins with insecticidal activity against a wide range of lepidopteran species [183]. While *B. thuringiensis* is a good source of Vip proteins, these proteins have also been found in other closely related bacteria, such as *Bacillus cereus*, *Lysinibacillus sphaericus*, or *Brevibacillus leterosporus*. Currently, two Sip proteins have been described, both active against several coleopteran pests. The fact that strains harboring *sip1Aa* and *sip1Ab* genes also contain *cry3* and *cry8* genes, respectively, suggests that Sip1 proteins may have a role in the insecticidal mechanism against coleopteran insects [184].

3.1. Protein Structure

Vip1 and Vip2 proteins are found in the culture supernatant before cell lysis due to specific secretion [181,185]. Both proteins have an N-terminal signal peptide for secretion, commonly cleaved after the secretion process is completed [24,181]. The Vip1/Vip2 homology with other bacterial binary toxins and the fact that these proteins are codified by two genes encoded in a single operon, suggest the presence of a typical "A+B" binary toxin [24,185]. It has been proposed that Vip1, with moderate sequence identity (30%) and structural similarity with the binding C2-II *Clostridium botulinum* toxin and the toxin "B" of *Clostridium difficile*, is the binding domain that translocates Vip2, with homology to the Rho-ADP-ribosylatin exotoxin C3 of *Clostridium* spp, to the host cell [186,187]. As occurs with other related "B" compounds, Vip1 is formed by four domains involved in docking to enzymatic components, binding to specific cell surface receptors, oligomerization, and channel formation in lipid membranes [188]. Coleopteran active Sip1Aa protein contains a predicted Gram-positive consensus secretion signal [4] and exhibits 46% similarity with Mtx3 mosquitocidal toxin of *Lysinibacillus sphaericus* [184]. This homology may indicate that Sip1Aa toxicity should be caused by pore formation.

3.2. Insecticidal Activity

The activity of the Bt secretable toxins against coleopterans is depicted in Table 2. Currently, four Vip protein families have been identified, but only Vip1/Vip2 showed activity against coleopteran pests [189]. Vip1/Vip2 proteins have been tested against different coleopteran families but they have shown active only against the Chrysomelidae, Curculionidae, and Scarabeidae families, being particularly toxic to corn rootworms. Single Vip1 or Vip2 showed no mortality, confirming that these proteins must act together to be toxic [185]. Vip1Aa was highly toxic against *Diabrotica* spp. when combined with Vip2Aa or Vip2Ab, but Vip1Ab/Vip2Ab (co-expressed in the same operon) and Vip1Ab/Vip2Aa were not active [185]. These data show the specificity of these proteins and suggest that the absence of toxicity is due to Vip1Ab. Moreover, Vip1Ba/Vip2Ba and Vip1Bb/Vip1Ba were toxic against *Diabrotica virgifera virgifera* [190] and binary Vip1Da/Vip1Ad had activity against the curculionid *A. grandis* and the chrysomelids *Diabrotica* spp and *L. decemlineata* [191]. These are the only Vip proteins active against the Colorado potato beetle. Vip1Ad/Vip2Ag binary proteins were the first report of demonstrated toxicity against any Scarabaeoidea larvae, being active against *Holotrichia parallela*, *H. oblita* and *Anomala corpulenta* [192]. Sip1Aa and Sip1Ab proteins have specific activity against coleopteran pests. Sip1Aa caused lethal toxicity for *L. decemlineata* larvae and stunting in *D. virgifera* and *D. undecimpunctata* larvae [181]. Sip1Ab was also toxic to *Colaphellus bowringi* Baly (Coleoptera: Chrysomelidae) but it did not harm *Hloltrichia diomphalia* (Coleoptera: Scarabaeidae) larvae [184], suggesting specific chrysomelid activity, although further studies are needed to determine its host range.

3.3. Mode of Action

The mode of action of coleopteran-specific Bt secretable proteins is poorly understood, but some information is available for this binary mechanism of action. The proposed multistep process begins with the ingestion of the two toxins by the susceptible larvae. Though the two encoded proteins are synthesized together, they are thought not to get associated in solution and reach the insect midgut as single proteins [188]. Then, the proteolytic processing by the trypsin-like proteases of the insect midgut juice of Vip1 allows the cell-bound "B" to bind to a specific membrane receptor, followed by the formation of oligomers containing seven Vip1 molecules [193]. It is at this stage when the docking between Vip1 and Vip2 translocates the toxic component (Vip2) into the cytoplasm though the "B" (Vip1) channel [188]. Recent studies in BBMVs of *H. parallela* evidenced that although Vip2Ag showed a low degree of binding on its own, the degree of binding increased when Vip1Ad was added, showing that Vip1Ad acted as a receptor to help Vip2 bind to BBMVs [194]. Once inside the cytosol, Vip2 destroys filamentous actin by blocking its polymerization and leading to cell death [195].

Sip1 proteins have no homology with Vip proteins, but Sip1A exhibits limited sequence similarity with the 36-kDa mosquitocidal Mtx3 protein of *B. sphaericus*, suggesting that toxicity is related with pore formation [181].

Table 2. Insecticidal activity of Vip and Sip proteins against coleopteran pests.

Crystal Type Toxin	Target Insect		Activity [a]	LC$_{50}$ [b]	Reference
	Scientific Name	Family			
Sip1Aa	*Diabrotica undecimpuntata*	Chrysomelidae	A		[181]
	Diabrotica virgifera	Chrysomelidae	A		[181]
	Colaphellus bowringi	Chrysomelidae	A	1.07	[184]
	Leptinotarsa decemlineata	Chrysomelidae	A	24	[181]
Sip1Ab	*Colaphellus bowringi*	Chrysomelidae	A	1.05	[184]
	Hloltrichia diomphalia	Scarabaeidae	N		[184]
Vip1Aa	*Diabrotica virgifera*	Chrysomelidae	N		[185]
Vip1Ac	*Holotrichia oblita*	Scarabaeidae	N		[196]
	Tenebrio molitor	Tenebrionidae	N		[195]
Vip1ad	*Anomala corpulenta*	Scarabaeidae	N		[192]
	Holotrichia oblita	Scarabaeidae	N		[192]
	Holotrichia parallela	Scarabaeidae	N		[192]
Vip1Da	*Diabrotica virgifera*	Chrysomelidae	N		[191]
Vip2Aa	*Diabrotica virgifera*	Chrysomelidae	N		[185]
Vip2Ac	*Tenebrio molitor*	Tenebrionidae	N		[195]
Vip2Ad	*Diabrotica virgifera*	Chrysomelidae	N		[191]

Table 2. *Cont.*

Crystal Type Toxin	Target Insect		Activity [a]	LC$_{50}$ [b]	Reference
	Scientific Name	Family			
Vip2Ae	*Holotrichia oblita*	Scarabaeidae	N		[196]
	Tenebrio molitor	Tenebrionidae	N		[196]
Vip2Ag	*Anomala corpulenta*	Scarabaeidae	N		[192]
	Holotrichia oblita	Scarabaeidae	N		[192]
	Holotrichia parallela	Scarabaeidae	N		[192]
Vip1Aa+Vip2Aa	*Diabrotica longicornis B.*	Chrysomelidae	A		[185]
	Diabrotica undecimpuntata	Chrysomelidae	A		[185]
	Diabrotica virgifera	Chrysomelidae	A		[185]
	Leptinotarsa decemlineata	Chrysomelidae	N		[185]
	Tenebrio molitor	Tenebrionidae	N		[185]
Vip1Aa+Vip2Ab	*Diabrotica virgifera*	Chrysomelidae	A		[185]
Vip1Ab+Vip2Aa	*Diabrotica virgifera*	Chrysomelidae	N		[185]
Vip1Ab+Vip2Ab	*Diabrotica virgifera*	Chrysomelidae	N		[185]
Vip1Ac+Vip2Ac	*Tenebrio molitor*	Tenebrionidae	N		[195]
Vip1Ac+Vip2Ae	*Holotrichia oblita*	Scarabaeidae	N		[196]
	Tenebrio molitor	Tenebrionidae	N		[196]
Vip1Ad+Vip2Ag	*Anomala corpulenta*	Scarabaeidae	A	220 ng/g soil	[192]
	Holotrichia oblita	Scarabaeidae	A	120 ng/g soil	[192]
	Holotrichia parallela	Scarabaeidae	A	80 // 2.33 ng/g soil	[195,197]
Vip1Ca+Vip2Aa	*Tenebrio molitor*	Tenebrionidae	N		[187]
Vip1Da+Vip2Ad	*Diabrotica longicornis B.*	Chrysomelidae	A	213	[191]
	Diabrotica undecimpuntata	Chrysomelidae	A	4.91	[191]
	Diabrotica virgifera	Chrysomelidae	A	437	[191]
	Leptinotarsa decemlineata	Chrysomelidae	A	37	[191]
	Anthonomus grandis	Curculionidae	A	207	[191]
Vip1Ba+Vip2Ba	*Diabrotica virgifera*	Chrysomelidae	A		[190]
Vip1Bb+Vip2Bb	*Diabrotica virgifera*	Chrysomelidae	A		[190]
Vip3Aa	*Tenebrio molitor*	Tenebrionidae	N		[197]

[a] The parameter is mortality. A = active; N = not active; [b] LC$_{50}$ = lethal concentration that causes 50% mortality of the insects. data are expressed in μg/mL, unless otherwise stated. "//" separate two different values of the LC$_{50}$.

4. Bt Based Insecticides

In 1938, the first insecticide based on *B. thuringiensis* was produced and marketed under the name *Sporéine* for the control of lepidopteran insect pests [47]. Since then, sporulated cultures of *B. thuringiensis* have been used widely as foliar sprays to protect crops from insect damage. Since *B. thuringiensis* subsp. *tenebrionis* was discovered [48], it was rapidly formulated as a bioinsecticide and commercialized against the Colorado potato beetle. Bt-based insecticides to control coleopteran pests are mainly developed against chrysomelid beetles [198]. Novodor® (Kenogard) uses the NB-176 strain of Bt subsp. *tenebrionis* as the active ingredient and is widely used for the control of *L. decemlineata*. However, the toxicity of this commercial product has been verified for other species of beetles, such as the chrysomelids *Chrysophtharta bimaculata, C. agricola* and *C. scripta* [199,200] under laboratory conditions. Furthermore, this product has been shown to be effective against *C. scripta* in field conditions [200], while the use of Novodor did not exert good control of the populations of *Lissorhoptrus oryzophilus* (Coleoptera: Curculionidae) [201].

To date, most of the Bt-based bioinsecticide products effectively use natural Bt strains for the control of foliar-feeding pests. However, several factors have limited their use. Usually, Bt strains have a narrow insecticidal spectrum compared with other insecticides, even when insects are closely related [202]. Advances in genetic manipulation technologies offer improvements in the efficiency of Bt-based formulates and reductions in their production costs. The development of new strains by conjugation or transduction has been used to confer natural strains with new insecticidal properties [203]. The natural Bt subsp. *kurstaki*, for example, has been modified to express several *cry3* genes and extend its host range to both lepidopteran and coleopteran pests [202]. The active ingredient in Foil® is the Bt strain EG2424, expressing both Cry1Ac and Cry3A proteins, the latter of which was transferred from a Cry3Aa-encoding plasmid belonging to the Bt subsp. *morrisoni* [204]. Similarly, the Cry3-overproducing strain, EG7673, was obtained by transforming a natural strain with a recombinant plasmid containing a *cry3Bb1* gene. A formulation with this strain as the active ingredient was commercialized as Raven® and was four-fold more active than the parental strain [205].

5. Bt-Crops

By expressing one or more Bt toxic genes in a target plant tissue transgenic insect-resistant crops, Bt crops, can be produced. Such cultivars need no further pest control measures. To date, the Bt crops extension has increased worldwide, particularly that of Bt cotton, Bt rice and Bt corn [9]. Bt plants have been created for the control of several insect pests, among others, Colorado potato beetle (*L. decemlineata*) and corn rootworms (*Diabrotica* spp.). The first human-modified pesticide-producing crop was potato, which expressed the *cry3A* gene from *B. thuringiensis* subsp. *tenebrionis* in their leaves [206]. The transgenic gene expression confers potato plants protection against the Colorado potato beetle and allows reducing insecticide applications [207]. A few years later, this Bt crop was complemented with another gene expression cassette that also provided protection against the Potato leafroll virus [208]. However, genetically modified potatoes were commercialized from 1995 to 2001, and eventually removed from the marketplace due to social concern for genetically modified crops [209].

A coleopteran-active Bt maize was designed for the control of corn rootworms, expressing a variant of the wild-type *cry3Bb1* gene from Bt subsp. *kumamotoensis* in the root tissue [210]. Currently, Bt maize hybrids express four different crystal proteins (Cry3Bb, mCry3A, Cry34Ab/35Ab and eCry3.1Ab), individually or co-expressing two toxins [211,212]. Vip1 and Vip2 proteins were also candidates to be expressed in maize plants, mainly due to the great toxicity against rootworms. However, the cytotoxic activity of the Vip2 protein has prevented the development of a Bt plant expressing this binary toxins [189]. The opportunity of expressing the toxin in a specific tissue allows minimization of the exposure of non-target fauna while increasing the control of tunneling and root pests, which are otherwise difficult to manage. However, Western corn rootworm has developed field resistance to all four currently available Bt toxins [212–214] as did *D. virgifera* in 2009 against Bt corn [55]. These facts show that although Bt crops have the potential to increase productivity while conserving biodiversity, resistance management programs and a better use of integrated pest management are necessary to delay resistance development as much as possible [215].

6. Resistance and Cross-Resistance

The widespread use of *B. thuringiensis* biopesticides, as well as the planting of millions of hectares of Bt plants to protect crops from pests, carry the risk of selecting insect biotypes that are tolerant or resistance to Bt toxins. The appearance of resistance may be due to alterations in any step involved their mode of action, from the solubilization and activation steps to the capacity of pore formation [159]. It is established that the lack of solubilization is favored by the physicochemical conditions of the midgut fluids, particularly the pH. The acidic midgut of the coleopteran insects seemed to be a limiting factor in the solubilization of Cry proteins, such as Cry1B and Cry7Aa [52,71], although recent reports seem to indicate that more factors are involved as Cry7Aa proteins are dissolved in *L. decemlineata* and *H. vigintioctomaculata* midgut fluids [57,58]. Once the Cry toxin is solubilized in the midgut, protoxins are proteolytically cleaved to activated toxins. This toxin processing depends on the presence of the right digestive enzymes in the host midgut fluid. As an example, it was observed in *D. virgifera* larvae that the Cry3Aa protein was poorly processed by its own proteases, which leads to low activity of Cry3Aa against rootworms [157]. Introduction of a chymotrypsin/cathepsin recognition site in domain I of Cry3A has been shown to enhance the bioactivity of this toxin against the western corn rootworm larvae [157].

Molecularly, the insect resistance basis is a modification or loss of the specific midgut cell membrane receptors or some mediator, which eliminates or reduces the capacity of the toxin to initiate a lethal pathway [216]. Cross-resistance between Cry toxins is often associated with sequence similarities in domains II and III, related to specific protein binding [217]. Under laboratory conditions, populations of *L. decemlineata* and *C. scripta* resistant to Cry3Aa have been described [53,54]. To date, the appearance of field resistance is still relatively low despite the extensive use of products based on the same protein, which increases the probability of resistance development.

Conversely, rootworm populations have developed resistance to all proteins used in transgenic corn. The intense selection pressure posed by the continuous exposure of insects to Bt toxins has increased the emergence of pest resistance. Since the first case of resistance to Cry3Bb1 Bt-maize in 2009, *Diabrotica* has developed resistance to Cry3Aa and Cry34/35Ab binary protein [211]. New strategies are being carried out to try to delay resistance, including a combined use of several proteins in the same Bt plant [218]. Pyramiding of two Bt proteins can delay resistance to those proteins because when insects become tolerant to one toxin, most will still be susceptible to the other toxin [211]. However, there is already evidence of cross-resistance to Cry3 proteins and even to Cry34/35, which may invalidate, in the long run, the use of all these proteins [212].

Author Contributions: Conceptualization, M.D.-A., C.A.-A. and P.C.; resources, C.A.-A., B.E., and P.C.; writing—original draft, M.D.-A.; writing and editing, M.D.-A., M.V., B.E., and P.C.; review and supervision, M.V. and P.C. All authors have read and agreed to the published version of the manuscript.

Funding: This research was funded by the Spanish Ministry of Science and Innovation (RTI2018-095204-B-C22).

Acknowledgments: M.D.-A. received a doctoral grant from Universidad Pública de Navarra, Pamplona, Spain.

References

1. Raymond, B.; Johnston, P.R.; Nielsen-LeRoux, C.; Lereclus, D.; Crickmore, N. *Bacillus thuringiensis*: An impotent pathogen? *Trends Microbiol.* **2010**, *18*, 189–194. [CrossRef]
2. Schnepf, E.; Crickmore, N.; Van Rie, J.; Lereclus, D.; Baum, J.; Feitelson, J.; Zeigler, D.R.; Dean, D.H. *Bacillus thuringiensis* and its pesticidal crystal proteins. *Microbiol. Mol. Biol. Rev.* **1998**, *62*, 775–806. [CrossRef] [PubMed]
3. Wei, J.-Z.; Hale, K.; Carta, L.; Platzer, E.; Wong, C.; Fang, S.-C.; Aroian, R.V. *Bacillus thuringiensis* crystal proteins that target nematodes. *Proc. Natl. Acad. Sci. USA* **2003**, *100*, 2760–2765. [CrossRef] [PubMed]
4. Palma, L.; Muñoz, D.; Berry, C.; Murillo, J.; Caballero, P. *Bacillus thuringiensis* Toxins: An Overview of Their Biocidal Activity. *Toxins* **2014**, *6*, 3296–3325. [CrossRef] [PubMed]
5. Nester, E.W.; Thomashow, L.S.; Metz, M.; Gordon, M. 100 Years of *Bacillus thuringiensis*: A critical scientific assessment. *Am. Soc. Microbiol.* **2002**.
6. Portela-Dussán, D.D.; Chaparro-Giralddo, A.; López-Pazos, S.A. La biotecnología de *Bacillus thuringiensis* en la agricultura. *Nova* **2013**, *11*, 87–96. [CrossRef]
7. Krishnan, V.; Domanska, B.; Elhigazi, A.; Afolabi, F.; West, M.J.; Crickmore, N. The human cancer cell active toxin Cry41Aa from *Bacillus thuringiensis* acts like its insecticidal counterparts. *Biochem. J.* **2017**, *474*, 1591–1602. [CrossRef]
8. Bravo, A.; Likitvivatanavong, S.; Gill, S.S.; Soberón, M. *Bacillus thuringiensis*: A story of a successful bioinsecticide. *Insect Biochem. Mol. Biol.* **2011**, *41*, 423–431. [CrossRef] [PubMed]
9. James, C. *Global Status of Commercialized Biotech/GM Crops in 2017: Biotech Crop Adoption Surges as Economic Benefits Accumulate in 22 Years*; ISAAA Br. No. 53; ISAAA: Ithaca, NY, USA, 2017.
10. Schnepf, H.E.; Whiteley, H.R. Cloning and expression of the *Bacillus thuringiensis* crystal protein gene in *Escherichia coli*. *Proc. Natl. Acad. Sci. USA* **1981**, *78*, 2893–2897. [CrossRef]
11. Crickmore, N.; Zeigler, D.R.; Feitelson, J.; Schnepf, E.; Van Rie, J.; Lereclus, D.; Baum, J.; Dean, D.H. Revision of the nomenclature for the *Bacillus thuringiensis* pesticidal crystal proteins. *Microbiol. Mol. Biol. Rev.* **1998**, *62*, 807–813. [CrossRef]

12. Crickmore, N.; Baum, J.A.; Bravo, A.; Lereclus, D.; Narva, K.E.; Sampson, K.; Schnepf, H.E.; Sun, M.; Zeigler, D.R. Bacillus thuringiensis Toxin Nomenclature. Available online: http://www.btnomenclature.info/ (accessed on 30 May 2020).

13. Bravo, A.; Soberón, M.; Gill, S.S. Bacillus thuringiensis: Mechanisms and Use. In *Insect Control: Biological and Synthetic Agents*; Gilbert, L.I., Gill, S.S., Eds.; Elsevier: Amsterdam, The Netherlands, 2005; pp. 175–205.

14. Berry, C.; Crickmore, N. Structural classification of insecticidal proteins—Towards an in silico characterisation of novel toxins. *J. Invertebr. Pathol.* **2017**, *142*, 16–22. [CrossRef] [PubMed]

15. de Maagd, R.A.; Bravo, A.; Crickmore, N. How *Bacillus thuringiensis* has evolved specific toxins to colonize the insect world. *Trends Genet.* **2001**, *17*, 193–199. [CrossRef]

16. Pardo-López, L.; Soberón, M.; Bravo, A. *Bacillus thuringiensis* insecticidal three-domain Cry toxins: Mode of action, insect resistance and consequences for crop protection. *FEMS Microbiol. Rev.* **2013**, *37*, 3–22. [CrossRef] [PubMed]

17. Oppert, B.; Ellis, R.T.; Babcock, J. Effects of Cry1F and Cry34Ab1/35Ab1 on storage pests. *J. Stored Prod. Res.* **2010**, *46*, 143–148. [CrossRef]

18. Yu, Y.; Yuan, Y.; Gao, M. Construction of an environmental safe *Bacillus thuringiensis* engineered strain against Coleoptera. *Appl. Microbiol. Biotechnol.* **2016**, *100*, 4027–4034. [CrossRef]

19. Jolivet, P. Food Habits and Food Selection of Chrysomelidae. Bionomic and Evolutionary Perspectives. In *Biology of Chrysomelidae*; Jolivet, P., Petitpierre, E., Hsiao, T.H., Eds.; Springer: Dordrecht, The Netherlands, 1988; pp. 1–24.

20. Soberón, M.; López-Díaz, J.A.; Bravo, A. Cyt toxins produced by *Bacillus thuringiensis*: A protein fold conserved in several pathogenic microorganisms. *Peptides* **2013**, *41*, 87–93. [CrossRef]

21. Bietlot, H.P.; Vishnubhatla, I.; Carey, P.R.; Pozsgay, M.; Kaplan, H. Characterization of the cysteine residues and disulphide linkages in the protein crystal of *Bacillus thuringiensis*. *Biochem. J.* **1990**, *267*, 309–315. [CrossRef]

22. Agaisse, H.; Lereclus, D. How does *Bacillus thuringiensis* produce so much insecticidal crystal protein? *J. Bacteriol.* **1995**, *177*, 6027–6032. [CrossRef]

23. Berry, C.; Ben-dov, E.; Jones, A.F.; Murphy, L.; Quail, M.A.; Holden, M.T.G.; Harris, D.; Zaritsky, A.; Parkhill, J. Complete sequence and organization of pBtoxis, the Toxin-Coding Plasmid of *Bacillus thuringiensis* subsp. israelensis. *Appl. Environ. Microbiol.* **2002**, *68*, 5082–5095. [CrossRef]

24. de Maagd, R.A.; Bravo, A.; Berry, C.; Crickmore, N.; Schnepf, H.E. Structure, diversity, and evolution of protein toxins from spore-forming entomopathogenic bacteria. *Annu. Rev. Genet.* **2003**, *37*, 409–433. [CrossRef]

25. Li, J.D.; Carroll, J.; Ellar, D.J. Crystal structure of insecticidal d-endotoxin from *Bacillus thuringiensis* at 2.5 A resolution. *Nature* **1991**, *353*, 815–821. [CrossRef] [PubMed]

26. Grochulski, P.; Masson, L.; Borisova, S.; Pusztai-Carey, M.; Schwartz, J.-L.; Brousseau, R.; Cygler, M. *Bacillus thuringiensis* CrylA(a) Insecticidal Toxin: Crystal structure and channel formation. *J. Mol. Biol.* **1995**, *254*, 447–464. [CrossRef] [PubMed]

27. Morse, R.J.; Yamamoto, T.; Stroud, R.M. Structure of Cry2Aa Suggests an Unexpected Receptor Binding Epitope. *Structure* **2001**, *9*, 409–417. [CrossRef]

28. Galitsky, N.; Cody, V.; Wojtczak, A.; Ghosh, D.; Luft, J.R.; Pangborn, W.; English, L. Structure of the insecticidal bacterial δ-endotoxin Cry3Bb1 of *Bacillus thuringiensis*. *Acta Crystallogr. Sect. D Biol. Crystallogr.* **2001**, *57*, 1101–1109. [CrossRef] [PubMed]

29. Boonserm, P.; Davis, P.; Ellar, D.J.; Li, J. Crystal structure of the mosquito-larvicidal toxin Cry4Ba and its biological implications. *J. Mol. Biol.* **2005**, *348*, 363–382. [CrossRef] [PubMed]

30. Boonserm, P.; Mo, M.; Angsuthanasombat, C.; Lescar, J. Structure of the functional form of the mosquito larvicidal Cry4Aa toxin from *Bacillus thuringiensis* at 2.8-Angstrom resolution. *J. Bacteriol.* **2006**, *188*, 3391–3401. [CrossRef]

31. Guo, S.; Ye, S.; Liu, Y.; Wei, L.; Xue, J.; Wu, H.; Song, F.; Zhang, J.; Wu, X.; Huang, D.; et al. Crystal structure of *Bacillus thuringiensis* Cry8Ea1: An insecticidal toxin toxic to underground pests, the larvae of Holotrichia parallela. *J. Struct. Biol.* **2009**, *168*, 259–266. [CrossRef]

32. Ye, Y.; Godzik, A. FATCAT: A web server for flexible structure comparison and structure similarity searching. *Nucleic Acids Res.* **2004**, *32*, W582–W585. [CrossRef]

33. Bravo, A.; Gill, S.S.; Soberón, M. Mode of action of *Bacillus thuringiensis* Cry and Cyt toxins and their potential for insect control. *Toxicon* **2007**, *49*, 423–435. [CrossRef]

34. Xu, C.; Wang, B.-C.; Yu, Z.; Sun, M. Structural Insights into *Bacillus thuringiensis* Cry, Cyt and Parasporin Toxins. *Toxins* **2014**, *6*, 2732–2770. [CrossRef]

35. Yamaguchi, T.; Sahara, K.; Bando, H.; Asano, S. Discovery of a novel *Bacillus thuringiensis* Cry8D protein and the unique toxicity of the Cry8D-class proteins against scarab beetles. *J. Invertebr. Pathol.* **2008**, *99*, 257–262. [CrossRef] [PubMed]

36. Bokori-Brown, M.; Savva, C.G.; Fernandes Da Costa, S.P.; Naylor, C.E.; Basak, A.K.; Titball, R.W. Molecular basis of toxicity of *Clostridium perfringens* epsilon toxin. *FEBS J.* **2011**, *278*, 4589–4601. [CrossRef] [PubMed]

37. Donovan, W.P.; Donovan, J.C.; Slaney, A.C. Bacillus thuringiensis CryET33 and CryET34 Compositions and Uses Therefor. U.S. Patent 6,063,756 A, 16 May 2000.

38. Rydel, T.; Sharamitaro, J.; Brown, G.R.; Gouzov, V.; Seale, J.; Sturman, E.; Thoma, R.; Gruys, K.; Isaac, B. The crystal structure of a coleopteran insect-active binary Bt protein toxin complex at 2.5 Å resolution. In Proceedings of the Annual Meeting of the American Crystallographic Association, Los Angeles, CA, USA, 21–26 July 2001.

39. Baum, J.A.; Chu, C.R.; Rupar, M.; Brown, G.R.; Donovan, W.P.; Huesing, J.E.; Ilagan, O.; Malvar, T.M.; Pleau, M.; Walters, M.; et al. Binary toxins from *Bacillus thuringiensis* active against the western corn rootworm, *Diabrotica virgifera virgifera* LeConte. *Appl. Environ. Microbiol.* **2004**, *70*, 4889–4898. [CrossRef]

40. Ellis, R.T.; Stockhoff, B.A.; Stamp, L.; Schnepf, H.E.; Schwab, G.E.; Knuth, M.; Russell, J.; Cardineau, G.A.; Narva, K.E. Novel *Bacillus thuringiensis* binary insecticidal Crystal proteins active on Western Corn Rootworm, *Diabrotica virgifera virgifera* LeConte. *Appl. Environ. Microbiol.* **2002**, *68*, 1137–1145. [CrossRef]

41. Kelker, M.S.; Berry, C.; Evans, S.L.; Pai, R.; McCaskill, D.G.; Wang, N.X.; Russell, J.C.; Baker, M.D.; Yang, C.; Pflugrath, J.W.; et al. Structural and Biophysical Characterization of *Bacillus thuringiensis* Insecticidal Proteins Cry34Ab1 and Cry35Ab1. *PLoS ONE* **2014**, *9*, e112555. [CrossRef]

42. Cohen, S.; Albeck, S.; Ben-Dov, E.; Cahan, R.; Firer, M.; Zaritsky, A.; Dym, O. Cyt1Aa Toxin: Crystal structure reveals implications for Its membrane-perforating function. *J. Mol. Biol.* **2011**, *413*, 804–814. [CrossRef] [PubMed]

43. Li, J.; Koni, P.A.; Ellar, D.J. Structure of the mosquitocidal δ-endotoxin CytB from *Bacillus thuringiensis* sp. kyushuensisand implications for membrane pore formation. *J. Mol. Biol.* **1996**, *257*, 129–152.

44. Cohen, S.; Dym, O.; Albeck, S.; Ben-Dov, E.; Cahan, R.; Firer, M.; Zaritsky, A. High-Resolution crystal structure of activated Cyt2Ba monomer from *Bacillus thuringiensis* subsp. *israelensis*. *J. Mol. Biol.* **2008**, *380*, 820–827. [CrossRef]

45. Gazit, E.; Burshtein, N.; Ellar, D.J.; Sawyer, T.; Shai, Y. *Bacillus thuringiensis* Cytolytic toxin associates specifically with Its synthetic Helices A and C in the membrane Bound State. Implications for the assembly of oligomeric transmembrane pores. *Biochemistry* **1997**, *36*, 15546–15554. [CrossRef]

46. Promdonkoy, B.; Rungrod, A.; Promdonkoy, P.; Pathaichindachote, W.; Krittanai, C.; Panyim, S. Amino acid substitutions in αA and αC of Cyt2Aa2 alter hemolytic activity and mosquito-larvicidal specificity. *J. Biotechnol.* **2008**, *133*, 287–293. [CrossRef]

47. Federici, B.A.; Park, H.-W.; Sakano, Y. Insecticidal Protein Crystals of *Bacillus thuringiensis*. In *Inclusions in Prokaryotes*; Springer: Berlin/Heidelberg, Germany, 2006; pp. 195–236.

48. Krieg, A.; Huger, A.M.; Langenbruch, G.A.; Schnetter, W. *Bacillus thuringiensis* var. tenebrionis: Ein neuer, gegenüber Larven von Coleopteren wirksamer Pathotyp. *Zeitschrift für Angew. Entomol.* **2009**, *96*, 500–508.

49. Herrnstadt, C.; Soares, G.G.; Wilcox, E.R.; Edwards, D.L. A new strain of *Bacillus thuringiensis* with activity against coleopteran insects. *Nat. Biotechnol.* **1986**, *4*, 305–308.

50. Barjac, H.; Frachon, E. Classification of *Bacillus thuringiensis* strains. *Entomophaga* **1990**, *35*, 233–240. [CrossRef]

51. Rupar, M.J.; Donovan, W.P.; Groat, R.G.; Slaney, A.C.; Mattison, J.W.; Johnson, T.B.; Charles, J.F.; Dumanoir, V.C.; de Barjac, H. Two novel strains of *Bacillus thuringiensis* toxic to coleopterans. *Appl. Environ. Microbiol.* **1991**, *57*, 3337–3344. [CrossRef] [PubMed]

52. Lambert, B.; Höfte, H.; Annys, K.; Jansens, S.; Soetaert, P.; Peferoen, M. Novel *Bacillus thuringiensis* insecticidal crystal protein with a silent activity against coleopteran larvae. *Appl. Environ. Microbiol.* **1992**, *58*, 2536–2542.

53. Whalon, M.E.; Miller, D.L.; Hollingworth, R.M.; Grafius, E.J.; Miller, J.R. Selection of a Colorado Potato Beetle (*Coleoptera: Chrysomelidae*) Strain Resistant to *Bacillus thuringiensis*. *J. Econ. Entomol.* **1993**, *86*, 226–233. [CrossRef]

54. Bauer, L.S. Resistance: A threat to the insecticidal crystal proteins of *Bacillus thuringiensis*. *Florida Entomol.* **1995**, *78*, 414–443. [CrossRef]

55. Gassmann, A.J.; Petzold-Maxwell, J.L.; Keweshan, R.S.; Dunbar, M.W. Field-Evolved Resistance to Bt Maize by Western Corn Rootworm. *PLoS ONE* **2011**, *6*, e22629. [CrossRef]

56. Ekobu, M.; Solera, M.; Kyamanywa, S.; Mwanga, R.O.M.; Odongo, B.; Ghislain, M.; Moar, W.J. Toxicity of seven *Bacillus thuringiensis* Cry proteins against *Cylas puncticollis* and *Cylas brunneus* (Coleoptera: Brentidae) using a novel artificial diet. *J. Econ. Entomol.* **2010**, *103*, 1493–1502. [CrossRef]

57. Domínguez-Arrizabalaga, M.; Villanueva, M.; Fernandez, A.B.; Caballero, P. A strain of *Bacillus thuringiensis* containing a novel *cry7Aa2* gene that Is toxic to *Leptinotarsa decemlineata* (Say) (Coleoptera: Chrysomelidae). *Insects* **2019**, *10*, 259. [CrossRef]

58. Song, P.; Wang, Q.; Nangong, Z.; Su, J.; Ge, D. Identification of *Henosepilachna vigintioctomaculata* (Coleoptera: Coccinellidae) midgut putative receptor for *Bacillus thuringiensis* insecticidal Cry7Ab3 toxin. *J. Invertebr. Pathol.* **2012**, *109*, 318–322. [CrossRef]

59. Shu, C.; Yu, H.; Wang, R.; Fen, S.; Su, X.; Huang, D.; Zhang, J.; Song, F. Characterization of Two Novel *cry8* genes from *Bacillus thuringiensis* Strain BT185. *Curr. Microbiol.* **2009**, *58*, 389–392. [CrossRef]

60. Shu, C.; Yan, G.; Wang, R.; Zhang, J.; Feng, S.; Huang, D.; Song, F. Characterization of a novel *cry8* gene specific to *Melolonthidae* pests: *Holotrichia oblita* and *Holotrichia parallela*. *Appl. Microbiol. Biotechnol.* **2009**, *84*, 701–707. [CrossRef] [PubMed]

61. Gindin, G.; Mendel, Z.; Levitin, B.; Kumar, P.; Levi, T.; Shahi, P.; Khasdan, V.; Weinthal, D.; Kuznetsova, T.; Einav, M.; et al. The basis for rootstock resilient to *Capnodis* species: Screening for genes encoding δ-endotoxins from *Bacillus thuringiensis*. *Pest Manag. Sci.* **2014**, *70*, 1283–1290. [CrossRef] [PubMed]

62. Park, Y.; Hua, G.; Taylor, M.D.; Adang, M.J. A coleopteran cadherin fragment synergizes toxicity of *Bacillus thuringiensis* toxins Cry3Aa, Cry3Bb, and Cry8Ca against lesser mealworm, *Alphitobius diaperinus* (Coleoptera: Tenebrionidae). *J. Invertebr. Pathol.* **2014**, *123*, 1–5. [PubMed]

63. Oliveira, G.R.; Silva, M.C.; Lucena, W.A.; Nakasu, E.Y.; Firmino, A.A.; Beneventi, M.A.; Souza, D.S.; Gomes, J.E.; de Souza, J.D.; Rigden, D.J.; et al. Improving Cry8Ka toxin activity towards the cotton boll weevil (*Anthonomus grandis*). *BMC Biotechnol.* **2011**, *11*, 85. [CrossRef] [PubMed]

64. Yu, H.; Zhang, J.; Huang, D.; Gao, J.; Song, F. Characterization of *Bacillus thuringiensis* strain Bt185 toxic to the Asian Cockchafer: *Holotrichia parallela*. *Curr. Microbiol.* **2006**, *53*, 13–17. [CrossRef] [PubMed]

65. Li, H.; Liu, R.; Shu, C.; Zhang, Q.; Zhao, S.; Shao, G.; Zhang, X.; Gao, J. Characterization of one novel *cry8* gene from *Bacillus thuringiensis* strain Q52-7. *World J. Microbiol. Biotechnol.* **2014**, *30*, 3075–3080. [CrossRef] [PubMed]

66. Thomson, M.; Knuth, M.; Cardineau, G. Bacillus thuringiensis Toxins with Improved Activity. U.S. Patent 5,874,288 A, 23 February 1999.

67. Shrestha, G.; Reddy, G.V.P.; Jaronski, S.T. Field efficacy of *Bacillus thuringiensis* galleriae strain SDS-502 for the management of alfalfa weevil and its impact on *Bathyplectes* spp. parasitization rate. *J. Invertebr. Pathol.* **2018**, *153*, 6–11. [CrossRef]

68. Isaac, B.; Krieger, E.K.; Mettus, A.-M.L.; Moshiri, F.; Sivasupramanian, S. Polypeptide Compositions Toxic to Anthonomus Insects and Methods of Use. U.S. Patent 6,541,448 B2, 1 August 2002.

69. Mettus, A.-M.L.; Baum, J.A. Polypeptide Compositions Toxic to Diabrotica Insects, Obtained from Bt; CryET70, and Methods of Use. Eur. Patent Application EP 1 129 198 B1, 11 May 2000. 29 September 1999.

70. van Frankenhuyzen, K. Cross-order and cross-phylum activity of *Bacillus thuringiensis* pesticidal proteins. *J. Invertebr. Pathol.* **2013**, *114*, 76–85. [CrossRef]

71. Bradley, D.; Harkey, M.A.; Kim, M.K.; Biever, K.D.; Bauer, L. The insecticidal CryIB Crystal protein of *Bacillus thuringiensis* ssp. thuringiensis has dual specificity to Coleopteran and Lepidopteran larvae. *J. Invertebr. Pathol.* **1995**, *65*, 162–173. [CrossRef]

72. Federici, B.A.; Bauer, L.S. Cyt1Aa protein of *Bacillus thuringiensis* Is toxic to the Cottonwood Leaf Beetle, *Chrysomela scripta*, and suppresses high levels of resistance to Cry3Aa. *Appl. Environ. Microbiol.* **1998**, *64*, 4368–4371. [CrossRef] [PubMed]

73. Naimov, S.; Weemen-Hendriks, M.; Dukiandjiev, S.; de Maagd, R. *Bacillus thuringiensis* Delta-Endotoxin Cry1 Hybrid Proteins with Increased Activity against the Colorado Potato Beetle. *Appl. Environ. Microbiol.* **2001**, *67*, 5328–5330. [CrossRef] [PubMed]

74. Martins, É.S.; Monnerat, R.G.; Queiroz, P.R.; Dumas, V.F.; Braz, S.V.; de Souza Aguiar, R.W.; Gomes, A.C.M.M.; Sánchez, J.; Bravo, A.; Ribeiro, B.M. Midgut GPI-anchored proteins with alkaline phosphatase activity from the cotton boll weevil (*Anthonomus grandis*) are putative receptors for the Cry1B protein of *Bacillus thuringiensis*. *Insect Biochem. Mol. Biol.* **2010**, *40*, 138–145. [CrossRef] [PubMed]

75. Rodríguez-González, Á.; Porteous-Álvarez, A.J.; Del Val, M.; Casquero, P.A.; Escriche, B. Toxicity of five Cry proteins against the insect pest *Acanthoscelides obtectus* (*Coleoptera: Chrisomelidae: Bruchinae*). *J. Invertebr. Pathol.* **2020**, *169*. [CrossRef] [PubMed]

76. Martins, É.S.; Praça, L.B.; Dumas, V.F.; Silva-Werneck, J.O.; Sone, E.H.; Waga, I.C.; Berry, C.; Monnerat, R.G. Characterization of *Bacillus thuringiensis* isolates toxic to cotton boll weevil (*Anthonomus grandis*). *Biol. Control* **2007**, *40*, 65–68. [CrossRef]

77. Asano, S. Identification of cry gene from *Bacillus thuringiensis* by PCR and isolation of unique insecticidal bacteria. *Mem. Fac. Agric.* **1996**, *19*, 529–563.

78. de Souza Aguiar, R.W.; Martins, É.S.; Ribeiro, B.M.; Monnerat, R.G. Cry10Aa Protein is highly toxic to *Anthonomus grandis* Boheman (*Coleoptera: Curculionidae*), an important insect pest in brazilian cotton crop fields. *Bt Res.* **2012**, *3*, 20–28.

79. Baum, J.A.; Sukuru, U.R.; Penn, S.R.; Meyer, S.E.; Subbarao, S.; Shi, X.; Flasinski, S.; Heck, G.R.; Brown, R.S.; Clark, T.L. Cotton Plants Expressing a Hemipteran-Active *Bacillus thuringiensis* Crystal Protein Impact the Development and Survival of *Lygus hesperus* (*Hemiptera: Miridae*) Nymphs. *J. Econ. Entomol.* **2012**, *105*, 616–624. [CrossRef]

80. Bradfisch, G.A.; Muller-Cohn, J.; Narva, K.E.; Fu, J.; Thomson, M. Bacillus thuringiensis toxins and Genes for Controlling Coleopteran Pests. U.S. Patent 6,710,027 B2, 23 March 2004.

81. Schnepf, H.E.; Lee, S.; Dojillo, J.; Burmeister, P.; Fencil, K.; Morera, L.; Nygaard, L.; Narva, K.E.; Wolt, J.D. Characterization of Cry34/Cry35 Binary Insecticidal Proteins from Diverse *Bacillus thuringiensis* Strain Collections. *Appl. Environ. Microbiol.* **2005**, *71*, 1765–1774. [CrossRef]

82. Moellenbeck, D.J.; Peters, M.L.; Bing, J.W.; Rouse, J.R.; Higgins, L.S.; Sims, L.; Nevshemal, T.; Marshall, L.; Ellis, R.T.; Bystrak, P.G.; et al. Insecticidal proteins from *Bacillus thuringiensis* protect corn from corn rootworms. *Nat. Biotechnol.* **2001**, *19*, 668–672. [CrossRef] [PubMed]

83. Rupar, M.J.; Donovan, W.P.; Tan, Y.; Slaney, A.C. Bacillus thuringiensis CryET29 Compositions Toxic to Coleopteran Insects and Ctenocephalides spp. U.S. Patent 6,093,695, 25 July 2000.

84. Weathersbee, A.A.; Lapointe, S.L.; Shatters, R.G. Activity of *Bacillus thuringiensis* isolates against *Diaprepes abbreviatus* (*Coleoptera: Curculionidae*). *Florida Entomol.* **2006**, *89*, 441–448. [CrossRef]

85. Mahmoud, S.B.; Ramos, J.E.; Shatters, R.G.; Hall, D.G.; Lapointe, S.L.; Niedz, R.P.; Rougé, P.; Cave, R.D.; Borovsky, D. Expression of *Bacillus thuringiensis* cytolytic toxin (Cyt2Ca1) in citrus roots to control *Diaprepes abbreviatus* larvae. *Pesticide Biochem. Physiol.* **2017**, *136*, 1–11. [CrossRef] [PubMed]

86. D'Amico, V.; Podgwaite, J.D.; Duke, S. Biological activity of *Bacillus thuringiensis* and associated toxins against the Asian Longhorned Beetle (*Coleoptera: Cerambycidae*). *J. Entomol. Sci.* **2004**, *39*, 318–324. [CrossRef]

87. Chen, J.; Dai, L.-Y.; Wang, X.-P.; Tian, Y.-C.; Lu, M.-Z. The *cry3Aa* gene of *Bacillus thuringiensis* Bt886 encodes a toxin against long-horned beetles. *Appl. Microbiol. Biotechnol.* **2005**, *67*, 351–356. [CrossRef] [PubMed]

88. Peña, G.; Miranda-Rios, J.; de la Riva, G.; Pardo-López, L.; Soberón, M.; Bravo, A. A *Bacillus thuringiensis* S-Layer Protein Involved in Toxicity against *Epilachna varivestis* (*Coleoptera: Coccinellidae*). *Appl. Environ. Microbiol.* **2006**, *127*, 353–360. [CrossRef]

89. Elgizawy, K.K.; Ashry, N.M. Efficiency of *Bacillus thuringiensis* strains and their Cry proteins against the Red Flour Beetle, *Tribolium castaneum* (Herbst.) (*Coleoptera: Tenebrionidae*). *Egypt. J. Biol. Pest Control* **2019**, *29*, 94. [CrossRef]

90. MacIntosh, S.C.; Stone, T.B.; Sims, S.R.; Hunst, P.L.; Greenplate, J.T.; Marrone, P.G.; Perlak, F.J.; Fischhoff, D.A.; Fuchs, R.L. Specificity and efficacy of purified *Bacillus thuringiensis* proteins against agronomically important insects. *J. Invertebr. Pathol.* **1990**, *56*, 258–266. [CrossRef]

91. Porcar, M.; García-Robles, I.; Domínguez-Escribà, L.; Latorre, A. Effects of *Bacillus thuringiensis* Cry1Ab and Cry3Aa endotoxins on predatory Coleoptera tested through artificial diet-incorporation bioassays. *Bull. Entomol. Res.* **2010**, *100*, 297–302. [CrossRef]

92. Chang, X.; Lu, Z.; Shen, Z.; Peng, Y.; Ye, G. Bitrophic and Tritrophic Effects of Transgenic *cry1Ab/cry2Aj* Maize on the Beneficial, Nontarget *Harmonia axyridis* (*Coleoptera: Coccinellidae*). *Environ. Entomol.* **2017**, *46*, 1171–1176. [CrossRef]

93. Sims, S.R. Host activity spectrum of the CryIIA *Bacillus thuringiensis* subsp. kurstaki protein. Effects on Lepidoptera, Diptera, and non-target arthropods. *Southwest. Entomol.* **1997**, *22*, 395–404.

94. Sharma, A.; Kumar, S.; Bhatnagar, R.K. *Bacillus thuringiensis* Protein Cry6B (BGSC ID 4D8) is Toxic to Larvae of *Hypera postica*. *Curr. Microbiol.* **2011**, *62*, 597–605. [CrossRef]

95. Niu, L.; Tian, Z.; Liu, H.; Zhou, H.; Ma, W.; Lei, C.; Chen, L. Transgenic Bt cotton expressing Cry1Ac/Cry2Ab or Cry1Ac/EPSPS does not affect the plant bug Adelphocoris suturalis or the pollinating beetle Haptoncus luteolus. *Environ. Pollut.* **2018**, *234*, 788–793. [CrossRef] [PubMed]

96. Contreras, E.; Rausell, C.; Real, M.D. Proteome response of *Tribolium castaneum* larvae to *Bacillus thuringiensis* toxin producing strains. *PLoS ONE* **2013**, *8*, e55330. [CrossRef]

97. Zhao, Y.; Zhang, S.; Luo, J.-Y.; Wang, C.-Y.; Lv, L.-M.; Wang, X.-P.; Cui, J.-J.; Lei, C.-L. Bt proteins Cry1Ah and Cry2Ab do not affect cotton aphid *Aphis gossypii* and ladybeetle *Propylea japonica*. *Sci. Rep.* **2016**, *6*, 20368. [CrossRef] [PubMed]

98. Zhong, C.; Ellar, D.J.; Bishop, A.; Johnson, C.; Lin, S.; Hart, E.R. Characterization of a *Bacillus thuringiensis* δ-Endotoxin which ia toxic to insects in three orders. *J. Invertebr. Pathol.* **2000**, *76*, 131–139. [CrossRef] [PubMed]

99. Gómez, J.E.C.; López-Pazos, S.A.; Cerón, J. Determination of Cry toxin activity and identification of an aminopeptidase N receptor-like gene in *Asymmathetes vulcanorum* (Coleoptera: Curculionidae). *J. Invertebr. Pathol.* **2012**, *111*, 94–98. [CrossRef] [PubMed]

100. López-Pazos, S.A.; Rojas Arias, A.C.; Ospina, S.A.; Cerón, J. Activity of *Bacillus thuringiensis* hybrid protein against a lepidopteran and a coleopteran pest. *FEMS Microbiol. Lett.* **2010**, *302*, 93–98. [CrossRef] [PubMed]

101. Shin, B.S.; Park, S.H.; Choi, S.K.; Koo, B.T.; Lee, S.T.; Kim, J.I. Distribution of *cryV*-type insecticidal protein genes in *Bacillus thuringiensis* and cloning of *cryV*-type genes from *Bacillus thuringiensis* subsp. kurstaki and *Bacillus thuringiensis* subsp. entomocidus. *Appl. Environ. Microbiol.* **1995**, *61*, 2402–2407. [CrossRef] [PubMed]

102. Choi, S.-K.; Shin, B.-S.; Kong, E.-M.; Rho, H.M.; Park, S.-H. Cloning of a new *Bacillus thuringiensis cryII* -Type crystal protein gene. *Curr. Microbiol.* **2000**, *41*, 65–69. [CrossRef]

103. Kostichka, K.; Warren, G.W.; Mullins, M.; Mullins, A.D.; Palekar, N.V.; Craig, J.A.; Koziel, M.G.; Estruch, J.J. Cloning of a *cryV*-type insecticidal protein gene from *Bacillus thuringiensis*: The *cryV*-encoded protein is expressed early in stationary phase. *J. Bacteriol.* **1996**, *178*, 2141–2144. [CrossRef] [PubMed]

104. Ruiz de Escudero, I.; Estela, A.; Porcar, M.; Martínez, C.; Oguiza, J.A.; Escriche, B.; Ferrè, J.; Caballero, P. Molecular and Insecticidal Characterization of a Cry1I Protein Toxic to Insects of the Families Noctuidae, Tortricidae, Plutellidae, and Chrysomelidae. *Appl. Environ. Microbiol.* **2006**, *72*, 4796–4804. [CrossRef] [PubMed]

105. Grossi-De-Sa, M.F.; De Magalhaes, M.Q.; Silva, M.S.; Silva, S.M.B.; Dias, S.C.; Nakasu, E.Y.T.; Brunetta, P.S.F.; Oliveira, G.R.; De Oliveira Neto, O.B.; De Oliveira, R.S.; et al. Susceptibility of *Anthonomus grandis* (Cotton Boll Weevil) and *Spodoptera frugiperda* (Fall Armyworm) to a Cry1Ia-type toxin from a Brazilian *Bacillus thuringiensis* strain. *BMB Rep.* **2007**, *40*, 773–782. [CrossRef] [PubMed]

106. Gleave, A.P.; Williams, R.; Hedges, R.J. Screening by polymerase chain reaction of *Bacillus thuringiensis* serotypes for the presence of *cryV*-like insecticidal protein genes and characterization of a *cryV* gene cloned from *B. thuringiensis* subsp. kurstaki. *Appl. Environ. Microbiol.* **1993**, *59*, 1683–1687. [CrossRef]

107. Mushtaq, R.; Behle, R.; Liu, R.; Niu, L.; Song, P.; Shakoori, A.R.; Jurat-Fuentes, J.L. Activity of *Bacillus thuringiensis* Cry1Ie2, Cry2Ac7, Vip3Aa11 and Cry7Ab3 proteins against *Anticarsia gemmatalis*, *Chrysodeixis includens* and *Ceratoma trifurcata*. *J. Invertebr. Pathol.* **2017**, *150*, 70–72. [CrossRef]

108. Liu, J.; Song, F.; Zhang, J.; Liu, R.; He, K.; Tan, J.; Huang, D. Identification of *vip3A*-type genes from *Bacillus thuringiensis* strains and characterization of a novel *vip3A*-type gene. *Lett. Appl. Microbiol.* **2007**, *45*, 432–438. [CrossRef]

109. von Tersch, M.A.; Gonzalez, J.M. Bacillus thuringiensis cryET1 Toxin Gene and Protein Toxic to Lepidopteran Insects. U.S. Patent 5,356,623 A, 18 October 1994.

110. Oppert, B.; Morgan, T.D.; Kramer, K.J. Efficacy of *Bacillus thuringiensis* Cry3Aa protoxin and protease inhibitors against coleopteran storage pests. *Pest Manag. Sci.* **2011**, *67*, 568–573. [CrossRef]

111. Zhongkamg, W.; He, W.; Peng, G.; Xia, Y. Transformation and expression of specific insecticide gene Bt cry3A in resident endogenetic bacteria isolated from *Apriona germari* (Hope) larvae intestines. *Acta Microbiol. Sin.* **2008**, *48*, 1168–1174.

112. Kurt, A.; Özkan, M.; Sezen, K.; Demirbağ, Z.; Özcengiz, G. Cry3Aa11: A new Cry3Aa δ-endotoxin from a local isolate of *Bacillus thuringiensis*. *Biotechnol. Lett.* **2005**, *27*, 1117–1121. [CrossRef]

113. Su, Z.; Deng, L.; Yi, X.; Xiao, S.; Zhang, C. The toxicity of Cry3Aa protein in *Brontispa longissima* by prokaryatic expression. *Genomics Appl. Biol.* **2009**, *28*, 691–694.

114. Génissel, A.; Leplé, J.C.; Millet, N.; Augustin, S.; Jouanin, L.; Pilate, G. High tolerance against *Chrysomela tremulae* of transgenic poplar plants expressing a synthetic *cry3Aa* gene from *Bacillus thuringiensis* ssp. *tenebrionis*. *Mol. Breed.* **2003**, *11*, 103–110. [CrossRef]

115. James, R.R.; Croft, B.A.; Strauss, S.H. Susceptibility of the Cottonwood Leaf Beetle (*Coleoptera: Chrysomelidae*) to Different Strains and Transgenic Toxins of *Bacillus thuringiensis*. *Environ. Entomol.* **1999**, *28*, 108–115. [CrossRef]

116. Yan, G.; Song, F.; Shu, C.; Liu, J.; Liu, C.; Huang, D.; Feng, S.; Zhang, J. An engineered *Bacillus thuringiensis* strain with insecticidal activity against Scarabaeidae (*Anomala corpulenta*) and Chrysomelidae (*Leptinotarsa decemlineata* and *Colaphellus bowringi*). *Biotechnol. Lett.* **2009**, *31*, 697–703. [CrossRef] [PubMed]

117. Gao, Y.; Jurat-Fuentes, J.L.; Oppert, B.; Fabrick, J.A.; Liu, C.; Gao, J.; Lei, Z. Increased toxicity of *Bacillus thuringiensis* Cry3Aa against *Crioceris quatuordecimpunctata*, *Phaedon brassicae* and *Colaphellus bowringi* by a *Tenebrio molitor* cadherin fragment. *Pest Manag. Sci.* **2011**, *67*, 1076–1081. [CrossRef]

118. Park, Y.; Abdullah, M.A.F.; Taylor, M.D.; Rahman, K.; Adang, M.J. Enhancement of *Bacillus thuringiensis* Cry3Aa and Cry3Bb Toxicities to Coleopteran Larvae by a Toxin-Binding Fragment of an Insect Cadherin. *Appl. Environ. Microbiol.* **2009**, *75*, 3086–3092. [CrossRef] [PubMed]

119. Li, H.; Olson, M.; Lin, G.; Hey, T.; Tan, S.Y.; Narva, K.E. *Bacillus thuringiensis* Cry34Ab1/Cry35Ab1 interactions with Western Corn Rootworm Midgut Membrane Binding Sites. *PLoS ONE* **2013**, *8*, e53079. [CrossRef] [PubMed]

120. Carroll, J.S.; Li, J.; Ellar, D.J. Proteolytic processing of a coleopteran-specific delta-endotoxin produced by *Bacillus thuringiensis* var. *tenebrionis*. *Biochem. J.* **1989**, *261*, 99–105. [CrossRef] [PubMed]

121. Wang, G.; Zhang, J.; Song, F.; Wu, J.; Feng, S.; Huang, D. Engineered *Bacillus thuringiensis* GO33A with broad insecticidal activity against lepidopteran and coleopteran pests. *Appl. Microbiol. Biotechnol.* **2006**, *72*, 924–930. [CrossRef] [PubMed]

122. Mahadeva Swamy, H.M.; Asokan, R.; Thimmegowda, G.G.; Mahmood, R. Expression of *cry3A* gene and its toxicity against Asian Gray Weevil *Myllocerus undecimpustulatus undatus* Marshall (*Coleoptera: Curculionidae*). *J. Basic Microbiol.* **2013**, *53*, 664–676. [CrossRef]

123. Gomez, S.; Mateus, A.C.; Hernandez, J.; Zimmermann, B.H. Recombinant Cry3Aa has insecticidal activity against the Andean potato weevil, *Premnotrypes vorax*. *Biochem. Biophys. Res. Commun.* **2000**, *279*, 653–656. [CrossRef]

124. Johnson, T.M.; Rishi, A.S.; Nayak, P.; Sen, S.K. Cloning of a *cryIIIA* endotoxin gene of *Bacillus thuringiensis* var. tenebrionis and its transient expression inindica rice. *J. Biosci.* **1996**, *21*, 673–685.

125. Adang, M.J. Enhancement of Bacillus thuringiensis Cry Toxicities to Lesser Mealworm Alphitobius diaperinus. U.S. Patent 0,201,549 A1, 18 August 2011.

126. Wu, S.-J.; Dean, D.H. Functional Significance of Loops in The Receptor Binding Domain of *Bacillus thuringiensis* CryIIIA δ-Endotoxin. *J. Mol. Biol.* **1996**, *255*, 628–640. [CrossRef] [PubMed]

127. Fabrick, J.; Oppert, C.; Lorenzen, M.D.; Morris, K.; Oppert, B.; Jurat-Fuentes, J.L. A Novel *Tenebrio molitor* Cadherin Is a functional receptor for *Bacillus thuringiensis* Cry3Aa toxin. *J. Biol. Chem.* **2009**, *284*, 18401–18410. [CrossRef]

128. Donovan, W.P.; Rupar, M.J.; Slaney, A.C.; Malvar, T.; Gawron-Burke, M.C.; Johnson, T.B. Characterization of two genes encoding *Bacillus thuringiensis* insecticidal crystal proteins toxic to Coleoptera species. *Appl. Environ. Microbiol.* **1992**, *58*, 3921–3927. [CrossRef]

129. Adang, M.J.; Adang, M.J.; Abdullah, M.A.F. Enhancement of Bacilus thuringiensis Cry Protein Toxicities to Coleopterans, and Novel Insect Cadhetin Fragments. U.S. Patent 8,486,887 B2, 16 July 2013.

130. Lambert, B.; Theunis, W.; Aguda, R.; Van Audenhove, K.; Decock, C.; Jansens, S.; Seurinck, J.; Peferoen, M. Nucleotide sequence of gene *cryIIID* encoding a novel coleopteran-active crystal protein from strain BTII09P of *Bacillus thuringiensis* subsp. *Kurstaki Gene* **1992**, *110*, 131–132. [CrossRef]

131. Haffani, Y.Z.; Cloutier, C.; Belzile, F.J. *Bacillus thuringiensis* cry3Ca1 protein is toxic to the Colorado Potato Beetle, *Leptinotarsa decemlineata* (Say). *Biotechnol. Prog.* **2001**, *17*, 211–216. [CrossRef] [PubMed]

132. Deng, S.; Shu, C.; Lin, Y.; Song, F.; Zhang, J. Identification, cloning and expression for novel *cry7Ab* gene and its insecticidal activity. *J. Agric. Biotechnol.* **2009**, *17*, 908–913.

133. Wang, L.; Guo, W.; Tan, J.; Sun, W.; Liu, T.; Sun, Y. Construction of an engineering strain expressing *cry7Ab7* gene cloned from *Bacillus thuringiensis*. *Front. Agric. China* **2010**, *4*, 328–333. [CrossRef]

134. Foncerrada, L.; Sick, A.J.; Payne, J.M. Novel Coleopteran-Active Bacillus thuringiensis Isolate and a Novel Gene Encoding a Coleopteran-Active Toxin. Patent Application EP0498537 A2, 16 January 1992. pp. 1–5.

135. Michaels, T.E.; Narva, K.E.; Foncerrada, L. Bacillus thuringiensis Toxins Active against Scarab Pests. U.S. Patent 5,554,534 A, 10 September 1996.

136. Zhang, Y.; Zheng, G.; Tan, J.; Li, C.; Cheng, L. Cloning and characterization of a novel *cry8Ab1* gene from *Bacillus thuringiensis* strain B-JJX with specific toxicity to scarabaeid (Coleoptera: Scarabaeidae) larvae. *Microbiol. Res.* **2013**, *168*, 512–517. [CrossRef]

137. Michaels, T.E.; Foncerrada, L.; Narva, K.E. Process for Controlling Scarab Pests with Bacillus thuringiensis Isolates. International Patent Application WO 93/15206, 5 August 1993.

138. Abad, A.; Duck, N.B.; Feng, X.; Flannagan, R.D.; Kahn, T.W.; Sims, L.E. Genes encoding Novel Proteins with Pesticidal Activity against Coleopterans. International Patent Application WO 02/034774 A2, 2 May 2002.

139. Liu, J.; Yan, G.; Shu, C.; Zhao, C.; Liu, C.; Song, F.; Zhou, L.; Ma, J.; Zhang, J.; Huang, D. Construction of a *Bacillus thuringiensis* engineered strain with high toxicity and broad pesticidal spectrum against coleopteran insects. *Appl. Microbiol. Biotechnol.* **2010**, *87*, 243–249. [CrossRef]

140. Jia, Y.; Zhao, C.; Wang, Q.; Shu, C.; Feng, X.; Song, F.; Zhang, J. A genetically modified broad-spectrum strain of *Bacillus thuringiensis* toxic against *Holotrichia parallela*, *Anomala corpulenta* and *Holotrichia oblita*. *World J. Microbiol. Biotechnol.* **2014**, *30*, 595–603. [CrossRef] [PubMed]

141. Sato, R.; Takeuchi, K.; Ogiwara, K.; Minami, M.; Kaji, Y.; Suzuki, N.; Hori, H.; Asano, S.; Ohba, M.; Iwahana, H. Cloning, heterologous expression, and localization of a novel crystal protein gene from *Bacillus thuringiensis* serovar *japonensis* strain Buibui toxic to scarabaeid insects. *Curr. Microbiol.* **1994**, *28*, 15–19. [CrossRef] [PubMed]

142. Huang, D.-F.; Zhang, J.; Song, F.-P.; Lang, Z.-H. Microbial control and biotechnology research on *Bacillus thuringiensis* in China. *J. Invertebr. Pathol.* **2007**, *95*, 175–180. [CrossRef] [PubMed]

143. Asano, S.; Yamashita, C.; Iizuka, T.; Takeuchi, K.; Yamanaka, S.; Cerf, D.; Yamamoto, T. A strain of *Bacillus thuringiensis* subsp. galleriae containing a novel cry8 gene highly toxic to Anomala cuprea (Coleoptera: Scarabaeidae). *Biol. Control* **2003**, *28*, 191–196. [CrossRef]

144. Liu, X.; Cheng, L.; Li, T.; Li, C.; Li, G. Cloning, expression and insecticidal activity of *cry8Ea2* gene from *Bacillus thuringiensis* strain B. *Acta Agric. Boreali-Sin.* **2008**, *23*, 1–4.

145. Singaravelu, B.; Crickmore, N.; Srikanth, J.; Hari, K.; Sankaranarayanan, C. Prospecting for Scarabid Specific *Bacillus thuringiensis* Crystal toxin *cry8* gene in sugarcane ecosystem of Tamil Nadu, India. *J. Sugarcane Res.* **2013**, *3*, 141–144.

146. Silva-Werneck, J.O.; Ellar, D.J. Characterization of a novel Cry9Bb δ-endotoxin from *Bacillus thuringiensis*. *J. Invertebr. Pathol.* **2008**, *98*, 320–328. [CrossRef]

147. Brown, K.L.; Whiteley, H.R. Molecular characterization of two novel crystal protein genes from *Bacillus thuringiensis* subsp. thompsoni. *J. Bacteriol.* **1992**, *174*, 549–557. [CrossRef]

148. Zhang, J.; Hodgman, T.C.; Krieger, L.; Schnetter, W.; Schairer, H.U. Cloning and analysis of the first *cry* gene from *Bacillus popilliae*. *J. Bacteriol.* **1997**, *179*, 4336–4341. [CrossRef]

149. Guzov, V.M.; Malvar, T.M.; Roberts, J.K.; Sivasupramanian, S. Insect Inhibitory Bacillus thuringiensis Proteins, Fusions, and Methods of Use Therefor. U.S. Patent 7,655,838 B2, 2 February 2010.

150. Herman, R.A.; Scherer, P.N.; Young, D.L.; Mihaliak, C.A.; Woodsworth, A.T.; Stockhoff, B.A.; Narva, K.E. Binary Insecticidal Crystal Protein from *Bacillus thuringiensis*, Strain PS149B1: Effects of Individual Protein Components and Mixtures in Laboratory Bioassays. *J. Econ. Entomol.* **2002**, *95*, 635–639. [CrossRef]

151. Rupar, M.J.; Donovan, W.P.; Chu, C.R.; Pease, E.; Tan, Y.; Slaney, A.C.; Malvar, T.M.; Baum, J. Coleopteran Toxin Polypeptide Compositions and Insect-Resistant Transgenic Plants. U.S. Patent 090,094,714 A1, 9 April 2009.

152. Yokoyama, T.; Tanaka, M.; Hasegawa, M. Novel *cry* gene from *Paenibacillus lentimorbus* strain Semadara inhibits ingestion and promotes insecticidal activity in *Anomala cuprea* larvae. *J. Invertebr. Pathol.* **2004**, *85*, 25–32. [CrossRef] [PubMed]

153. de Maagd, R.A.; Kwa, M.S.; van der Klei, H.; Yamamoto, T.; Schipper, B.; Vlak, J.M.; Stiekema, W.J.; Bosch, D. Domain III substitution in *Bacillus thuringiensis* delta-endotoxin CryIA(b) results in superior toxicity for *Spodoptera exigua* and altered membrane protein recognition. *Appl. Environ. Microbiol.* **1996**, *62*, 1537–1543. [CrossRef] [PubMed]

154. Shadenkov, A.A.; Kadyrov, R.M.; Uzbekova, S.V.; Kuz'min, E.V.; Osterman, A.L.; Chestukhina, G.G.; Shemiakin, M.F. Creation of a hybrid protein gene based on *Bacillus thuringiensis* delta endotoxins CryIIIA and CrIA(a) and expression of its derivatives in *Escherichia coli*. *Mol. Biol.* **1993**, *27*, 952–959.

155. Walters, F.S.; DeFontes, C.M.; Hart, H.; Warren, G.W.; Chen, J.S. Lepidopteran-Active Variable-Region Sequence Imparts Coleopteran Activity in eCry3.1Ab, an Engineered *Bacillus thuringiensis* Hybrid Insecticidal Protein. *Appl. Environ. Microbiol.* **2010**, *76*, 3082–3088. [CrossRef] [PubMed]

156. English, L.; Brussock, M.; Malvar, T.M.; Bryson, J.; Kulesza, C.; Walters, F.S.; Slatin, S.; von Tersch, M.A.; Romano, C. Nucleic Acid Segments Encoding Modified Bacillus thuringiensis Coleopteran-Toxic Crystal Proteins. U.S. Patent 6,060,594, 9 May 2000.

157. Walters, F.S.; Stacy, C.M.; Lee, M.K.; Palekar, N.; Chen, J.S. An engineered Chymotrypsin/Cathepsin G site in Domain I renders *Bacillus thuringiensis* Cry3A active against Western Corn Rootworm larvae. *Appl. Environ. Microbiol.* **2008**, *74*, 367–374. [CrossRef]

158. Jouzani, G.S.; Valijanian, E.; Sharafi, R. *Bacillus thuringiensis*: A successful insecticide with new environmental features and tidings. *Appl. Microbiol. Biotechnol.* **2017**, *101*, 2691–2711. [CrossRef]

159. Jurat-Fuentes, J.L.; Crickmore, N. Specificity determinants for Cry insecticidal proteins: Insights from their mode of action. *J. Invertebr. Pathol.* **2017**, *142*, 5–10. [CrossRef]

160. Guo, S.; Li, J.; Liu, Y.; Song, F.; Zhang, J. The role of DNA binding with the Cry8Ea1 toxin of *Bacillus thuringiensis*. *FEMS Microbiol. Lett.* **2011**, *317*, 203–210. [CrossRef]

161. Xia, L.; Sun, Y.; Ding, X.; Fu, Z.; Mo, X.; Zhang, H.; Yuan, Z. Identification of cry-Type Genes on 20-kb DNA Associated with Cry1 Crystal Proteins from *Bacillus thuringiensis*. *Curr. Microbiol.* **2005**, *51*, 53–58. [CrossRef]

162. Wu, F.; Zhao, X.; Sun, Y.; Li, W.; Xia, L.; Ding, X.; Yin, J.; Hu, S.; Yu, Z.; Tang, Y. Construction of gene library of 20 kb DNAs from parasporal Crystal in *Bacillus thuringiensis* strain 4.0718: Phylogenetic analysis and molecular docking. *Curr. Microbiol.* **2012**, *64*, 106–111. [CrossRef]

163. Ai, B.; Li, J.; Feng, D.; Li, F.; Guo, S. The Elimination of DNA from the Cry Toxin-DNA Complex Is a Necessary Step in the Mode of Action of the Cry8 Toxin. *PLoS ONE* **2013**, *8*, e81335. [CrossRef] [PubMed]

164. Koller, C.N.; Bauer, L.S.; Hollingworth, R.M. Characterization of the pH-mediated solubility of *Bacillus thuringiensis* var. san diego native δ-endotoxin crystals. *Biochem. Biophys. Res. Commun.* **1992**, *184*, 692–699. [CrossRef]

165. Chougule, N.P.; Doyle, E.; Fitches, E.; Gatehouse, J.A. Biochemical characterization of midgut digestive proteases from *Mamestra brassicae* (cabbage moth; Lepidoptera: Noctuidae) and effect of soybean Kunitz inhibitor (SKTI) in feeding assays. *J. Insect Physiol.* **2008**, *54*, 563–572. [CrossRef] [PubMed]

166. Michaud, D.; Bernier-Vadnais, N.; Overney, S.; Yelle, S. Constitutive expression of digestive cysteine proteinase forms during development of the colorado potato beetle, *Leptinotarsa decemlineata* Say (Coleoptera: Chrysomelidae). *Insect Biochem. Mol. Biol.* **1995**, *25*, 1041–1048. [CrossRef]

167. Yamaguchi, T.; Sahara, K.; Bando, H.; Asano, S. Intramolecular proteolytic nicking and binding of *Bacillus thuringiensis* Cry8Da toxin in BBMVs of Japanese beetle. *J. Invertebr. Pathol.* **2010**, *105*, 243–247. [CrossRef]

168. Rausell, C.; García-Robles, I.; Sánchez, J.; Muñoz-Garay, C.; Martínez-Ramírez, A.C.; Real, M.D.; Bravo, A. Role of toxin activation on binding and pore formation activity of the *Bacillus thuringiensis* Cry3 toxins in membranes of *Leptinotarsa decemlineata* (Say). *Biochim. Biophys. Acta Biomembr.* **2004**, *1660*, 99–105. [CrossRef]

169. Hernández-Martínez, P.; Vera-Velasco, N.M.; Martínez-Solís, M.; Ghislain, M.; Ferré, J.; Escriche, B. Shared binding sites for the *Bacillus thuringiensis* proteins Cry3Bb, Cry3Ca, and Cry7Aa in the African Sweet Potato Pest *Cylas puncticollis* (Brentidae). *Appl. Environ. Microbiol.* **2014**, *80*, 7545–7550. [CrossRef]

170. Ochoa-Campuzano, C.; Real, M.D.; Martínez-Ramírez, A.C.; Bravo, A.; Rausell, C. An ADAM metalloprotease is a Cry3Aa *Bacillus thuringiensis* toxin receptor. *Biochem. Biophys. Res. Commun.* **2007**, *362*, 437–442. [CrossRef]

171. Zúñiga-Navarrete, F.; Gómez, I.; Peña, G.; Bravo, A.; Soberón, M. A Tenebrio molitor GPI-anchored alkaline phosphatase is involved in binding of *Bacillus thuringiensis* Cry3Aa to brush border membrane vesicles. *Peptides* **2013**, *41*, 81–86. [CrossRef]

172. Siegfried, B.D.; Waterfield, N.; Ffrench-Constant, R.H. Expressed sequence tags from *Diabrotica virgifera virgifera* midgut identify a coleopteran cadherin and a diversity of cathepsins. *Insect Mol. Biol.* **2005**, *14*, 137–143. [CrossRef]

173. Sayed, A.; Nekl, E.R.; Siqueira, H.A.A.; Wang, H.-C.; Ffrench-Constant, R.H.; Bagley, M.; Siegfried, B.D. A novel cadherin-like gene from western corn rootworm, *Diabrotica virgifera virgifera* (Coleoptera: Chrysomelidae), larval midgut tissue. *Insect Mol. Biol.* **2007**, *16*. [CrossRef] [PubMed]

174. Contreras, E.; Schoppmeier, M.; Real, M.D.; Rausell, C. Sodium Solute Symporter and Cadherin proteins Act as *Bacillus thuringiensis* Cry3Ba toxin functional receptors in *Tribolium castaneum*. *J. Biol. Chem.* **2013**, *288*, 18013–18021. [CrossRef] [PubMed]

175. Shu, C.; Tan, S.; Yin, J.; Soberón, M.; Bravo, A.; Liu, C.; Geng, L.; Song, F.; Li, K.; Zhang, J. Assembling of *Holotrichia parallela* (dark black chafer) midgut tissue transcriptome and identification of midgut proteins that bind to Cry8Ea toxin from *Bacillus thuringiensis*. *Appl. Microbiol. Biotechnol.* **2015**, *99*, 7209–7218. [CrossRef] [PubMed]

176. Jiang, J.; Huang, Y.; Shu, C.; Soberón, M.; Bravo, A.; Liu, C.; Song, F.; Lai, J.; Zhang, J. *Holotrichia oblita* Midgut Proteins That Bind to *Bacillus thuringiensis* Cry8-Like Toxin and Assembly of the *H. oblita* Midgut Tissue Transcriptome. *Appl. Environ. Microbiol.* **2017**, *83*, 1–11. [CrossRef]

177. Yamaguchi, T.; Bando, H.; Asano, S. Identification of a *Bacillus thuringiensis* Cry8Da toxin-binding glucosidase from the adult Japanese beetle, *Popillia japonica*. *J. Invertebr. Pathol.* **2013**, *113*, 123–128. [CrossRef]

178. Pérez, C.; Muñoz-Garay, C.; Portugal, L.C.; Sánchez, J.; Gill, S.S.; Soberón, M.; Bravo, A. *Bacillus thuringiensis* ssp. israelensis Cyt1Aa enhances activity of Cry11Aa toxin by facilitating the formation of a pre-pore oligomeric structure. *Cell. Microbiol.* **2007**, *9*, 2931–2937.

179. Khorramnejad, A.; Domínguez-Arrizabalaga, M.; Caballero, P.; Escriche, B.; Bel, Y. Study of the *Bacillus thuringiensis* Cry1Ia protein oligomerization promoted by midgut Brush Border Membrane Vesicles of lepidopteran and coleopteran insects, or cultured insect cells. *Toxins* **2020**, *12*, 133. [CrossRef]

180. Warren, G.W.; Koziel, M.G.; Mullins, M.A.; Kostichka, K.; Estruch, J.J. Novel Pesticidal Proteins and Strains. Patent Coop. Treaty WO 96/10083 A1, 4 April 1996.

181. Donovan, W.P.; Engleman, J.T.; Donovan, J.C.; Baum, J.A.; Bunkers, G.J.; Chi, D.J.; Clinton, W.P.; English, L.; Heck, G.R.; Ilagan, O.M.; et al. Discovery and characterization of Sip1A: A novel secreted protein from *Bacillus thuringiensis* with activity against coleopteran larvae. *Appl. Microbiol. Biotechnol.* **2006**, *72*, 713–719. [CrossRef]

182. Sattar, S.; Maiti, M.K. Molecular Characterization of a Novel Vegetative Insecticidal Protein from *Bacillus thuringiensis* Effective Against Sap-Sucking Insect Pest. *J. Microbiol. Biotechnol.* **2011**, *21*, 937–946. [CrossRef]

183. Estruch, J.J.; Warren, G.W.; Mullins, M.A.; Nye, G.J.; Craig, J.A.; Koziel, M.G. Vip3A, a novel *Bacillus thuringiensis* vegetative insecticidal protein with a wide spectrum of activities against lepidopteran insects. *Proc. Natl. Acad. Sci. USA* **1996**, *93*, 5389–5394. [CrossRef]

184. Sha, J.; Zhang, J.; Chi, B.; Liu, R.; Li, H.; Gao, J. *sip1Ab* gene from a native *Bacillus thuringiensis* strain QZL38 and its insecticidal activity against *Colaphellus bowringi* Baly. *Biocontrol Sci. Technol.* **2018**, *28*, 459–467. [CrossRef]

185. Warren, G.W. Vegetative Insecticidal Proteins. In *Advances in Insect Control*; Taylor & Francis: Abingdon, UK, 1997; pp. 109–122.

186. Han, S.; Craig, J.A.; Putnam, C.D.; Carozzi, N.B.; Tainer, J.A. Evolution and mechanism from structures of an ADP-ribosylating toxin and NAD complex. *Nat. Struct. Biol.* **1999**, *6*, 932–936.

187. Shi, Y.; Ma, W.; Yuan, M.; Sun, F.; Pang, Y. Cloning of *vip1/vip2* genes and expression of Vip1Ca/Vip2Ac proteins in *Bacillus thuringiensis*. *World J. Microbiol. Biotechnol.* **2007**, *23*, 501–507. [CrossRef]

188. Barth, H.; Aktories, K.; Popoff, M.R.; Stiles, B.G. Binary bacterial toxins: Biochemistry, biology, and applications of common *Clostridium* and *Bacillus* proteins. *Microbiol. Mol. Biol. Rev.* **2004**, *68*, 373–402. [CrossRef] [PubMed]

189. Chakroun, M.; Banyuls, N.; Bel, Y.; Escriche, B.; Ferré, J. Bacterial Vegetative Insecticidal Proteins (Vip) from Entomopathogenic Bacteria. *Microbiol. Mol. Biol. Rev.* **2016**, *80*, 329–350. [CrossRef]

190. Feitelson, J.S.; Schnepf, H.E.; Narva, K.E.; Stockhoff, B.A.; Schmeits, J.; Loewer, D.; Dullum, C.J.; Muller-Cohn, J.; Stamp, L. Pesticidal Toxins and Nucleotide Sequences Which Encode These Toxins. U.S. Patent 6,204,435 B1, 20 March 2001.

191. Boets, A.; Arnaut, G.; Van Rie, J.; Damme, N. Toxins. U.S. Patent 7,919,609 B2, 5 April 2011.

192. Bi, Y.; Zhang, Y.; Shu, C.; Crickmore, N.; Wang, Q.; Du, L.; Song, F.; Zhang, J. Genomic sequencing identifies novel *Bacillus thuringiensis* Vip1/Vip2 binary and Cry8 toxins that have high toxicity to Scarabaeoidea larvae. *Appl. Microbiol. Biotechnol.* **2015**, *99*, 753–760. [CrossRef]

193. Leuber, M.; Orlik, F.; Schiffler, B.; Sickmann, A.; Benz, R. Vegetative Insecticidal Protein (Vip1Ac) of *Bacillus thuringiensis* HD201: Evidence for oligomer and channel formation. *Biochemistry* **2006**, *45*, 283–288. [CrossRef]

194. Geng, J.; Jiang, J.; Shu, C.; Wang, Z.; Song, F.; Geng, L.; Duan, J.; Zhang, J. *Bacillus thuringiensis* Vip1 functions as a receptor of Vip2 toxin for binary insecticidal activity against *Holotrichia parallela*. *Toxins* **2019**, *11*, 440. [CrossRef]

195. Shi, Y.; Xu, W.; Yuan, M.; Tang, M.; Chen, J.; Pang, Y. Expression of vip1/vip2 genes in *Escherichia coli* and *Bacillus thuringiensis* and the analysis of their signal peptides. *J. Appl. Microbiol.* **2004**, *97*, 757–765. [CrossRef]

196. Yu, X.; Liu, T.; Liang, X.; Tang, C.; Zhu, J.; Wang, S.; Li, S.; Deng, Q.; Wang, L.; Zheng, A.; et al. Rapid detection of *vip1*-type genes from *Bacillus cereus* and characterization of a novel *vip* binary toxin gene. *FEMS Microbiol. Lett.* **2011**, *325*, 30–36. [CrossRef]

197. Baranek, J.; Kaznowski, A.; Konecka, E.; Naimov, S. Activity of vegetative insecticidal proteins Vip3Aa58 and Vip3Aa59 of *Bacillus thuringiensis* against lepidopteran pests. *J. Invertebr. Pathol.* **2015**, *130*, 72–81. [CrossRef] [PubMed]

198. Wraight, S.P.; Lacey, L.A.; Kabaluk, J.T.; Gottel, M.S. Potential for Microbial Biological Control of Coleopteran and Hemipteran Pests of Potato—Agriculture and Agri-Food Canada (AAFC). *Fruit Veg. Cereal Sci. Biotechnol.* **2009**, *3*, 25–38.

199. Beveridge, N.; Elek, J.A. Insect and host-tree species influence the effectiveness of a *Bacillus thuringiensis* ssp. tenebrionis-based insecticide for controlling chrysomelid leaf beetles. *Aust. J. Entomol.* **2001**, *40*, 386–390. [CrossRef]

200. Coyle, D.R.; McMillin, J.D.; Krause, S.C.; Hart, E.R. Laboratory and field evaluations of two *Bacillus thuringiensis* formulations, Novodor and Raven, for control of Cottonwood Leaf Beetle (Coleoptera: Chrysomelidae). *J. Econ. Entomol.* **2000**, *93*, 713–720. [CrossRef] [PubMed]

201. Way, M.O.; Wallace, R.G.; Harper, H.B.; Landry, C.L. Control of Rice Water Weevil with Novodor 3, 1999. *Arthropod Manag. Tests* **2000**, *25*, 133. [CrossRef]

202. Baum, J.A.; Johnson, T.B. Bacillus thuringiensis: Natural and recombinant bioinsecticide products. In *Methods in Biotechnology, Vol5: Biopesticides: Use and Deliver*; Humana Press Inc.: Totowa, NJ, USA, 1999; pp. 189–209.

203. Sanchis, V. From microbial sprays to insect-resistant transgenic plants: History of the biospesticide *Bacillus thuringiensis*. A review. *Agron. Sustain. Dev.* **2011**, *31*, 217–231. [CrossRef]

204. Gawron-Burke, C.; Baum, J.A. Genetic Manipulation of *Bacillus thuringiensis* Insecticidal Crystal Protein Genes in Bacteria. In *Genetic Engineering*; Springer US: Boston, MA, USA, 1991; Volume 13, pp. 237–263.

205. Baum, J.A.; Kakefuda, M.; Gawron-burke, C. Engineering *Bacillus thuringiensis* Bioinsecticides with an Indigenous Site-Specific Recombination System. *Appl. Environ. Microbiol.* **1996**, *62*, 4367–4373. [CrossRef]

206. Perlak, F.; Stone, T.; Muskopf, Y.; Petersen, L.; Parker, G.; McPherson, S.; Wyman, J.; Love, S.; Reed, G.; Biever, D. Genetically improved potatoes: Protection from damage by Colorado potato beetles. *Plant Mol. Biol.* **1993**, *22*, 313–321. [CrossRef]

207. Duncan, D.R.; Hammond, D.; Zalewski, J.; Cudnohufsky, J.; Kaniewski, W.; Thornton, M.; Bookout, J.T.; Lavrik, P.; Rogan, G.J.; Feldman-Riebe, J. Field performance of "Transgenic" potato, with resistance to Colorado Potato Beetle and Viruses. *HortScience* **2002**, *37*, 556E–557. [CrossRef]

208. Thomas, P.E.; Kaniewski, W.K.; Lawson, E.C. Reduced Field Spread of Potato Leafroll Virus in Potatoes Transformed with the Potato Leafroll Virus Coat Protein Gene. *Plant Dis.* **1997**, *81*, 1447–1453. [CrossRef]

209. Thornton, M. The Rise and Fall of NewLeaf Potatoes. *NABC Rep.* **2004**, *15*, 235–243.

210. Vaughn, T.; Cavato, T.; Brar, G.; Coombe, T.; DeGooyer, T.; Ford, S.; Groth, M.; Howe, A.; Johnson, S.; Kolacz, K.; et al. A Method of controlling corn rootworm feeding using a *Bacillus thuringiensis* protein expressed in transgenic maize. *Crop Sci.* **2005**, *45*, 931–938. [CrossRef]

211. Gassmann, A.J.; Shrestha, R.B.; Jakka, S.R.K.; Dunbar, M.W.; Clifton, E.H.; Paolino, A.R.; Ingber, D.A.; French, B.W.; Masloski, K.E.; Dounda, J.W.; et al. Evidence of resistance to Cry34/35Ab1 Corn by Western corn rootworm (Coleoptera: Chrysomelidae): Root Injury in the Field and Larval Survival in Plant-Based Bioassays. *J. Econ. Entomol.* **2016**, *109*, 1872–1880. [CrossRef]

212. Zukoff, S.N.; Ostlie, K.R.; Potter, B.; Meihls, L.N.; Zukoff, A.L.; French, L.; Ellersieck, M.R.; French, B.W.; Hibbard, B.E. Multiple assays indicate varying levels of cross resistance in Cry3Bb1-selected field populations of the western corn rootworm to mCry3A, eCry3.1Ab, and Cry34/35Ab1. *J. Econ. Entomol.* **2016**, *109*, 1387–1398. [CrossRef] [PubMed]

213. Wangila, D.S.; Gassmann, A.J.; Petzold-Maxwell, J.L.; French, B.W.; Meinke, L.J. Susceptibility of Nebraska Western Corn Rootworm (*Coleoptera: Chrysomelidae*) populations to Bt corn events. *J. Econ. Entomol.* **2015**, *108*, 742–751. [CrossRef]

214. Jakka, S.R.K.; Shrestha, R.B.; Gassmann, A.J. Broad-spectrum resistance to *Bacillus thuringiensis* toxins by western corn rootworm (*Diabrotica virgifera virgifera*). *Sci. Rep.* **2016**, *6*, 27860. [CrossRef]

215. Shrestha, R.B.; Dunbar, M.W.; French, B.W.; Gassmann, A.J. Effects of field history on resistance to Bt maize by western corn rootworm, *Diabrotica virgifera virgifera* LeConte. *PLoS ONE* **2018**, *13*, e0200156. [CrossRef]

216. Ferré, J.; Van Rie, J.; Macintosh, S.C. Insecticidal Genetically Modified Crops and Insect Resistance Management (IRM). In *Integration of Insect-Resistant Genetically Modified Cropts within IPM Programs*; Springer Science + Business Media: Dordrecht, The Netherlands, 2008.

217. Carrière, Y.; Crickmore, N.; Tabashnik, B.E. Optimizing yramided transgenic Bt crops for sustainable pest management. *Nat. Biotechnol.* **2015**, *33*, 61–168. [CrossRef] [PubMed]

218. Onstad, D.W.; Meinke, L.J. Modeling evolution of *Diabrotica virgifera virgifera* to transgenic corn with two insecticidal traits. *J. Econ. Entomol.* **2010**, *103*, 849–860. [CrossRef] [PubMed]

Article

The Cadherin Protein Is Not Involved in Susceptibility to *Bacillus thuringiensis* Cry1Ab or Cry1Fa Toxins in *Spodoptera frugiperda*

Jianfeng Zhang [1,†], Minghui Jin [2,†], Yanchao Yang [1], Leilei Liu [1], Yongbo Yang [1], Isabel Gómez [3], Alejandra Bravo [3], Mario Soberón [3], Yutao Xiao [2] and Kaiyu Liu [1,*]

[1] Institute of Entomology, School of Life Sciences, Central China Normal University, Wuhan 430079, China; jianfengzhang@mails.ccnu.edu.cn (J.Z.); yangyanchao@mails.ccnu.edu.cn (Y.Y.); liulei@mails.ccnu.edu.cn (L.L.); yongboyang@mail.ccnu.edu.cn (Y.Y.)
[2] Agricultural Genomics Institute at Shenzhen, Chinese Academy of Agricultural Sciences, Shenzhen 518120, China; jinminghui@caas.cn (M.J.); xiaoyutao@caas.cn (Y.X.)
[3] Instituto de Biotecnología, Universidad Nacional Autónoma de México, Apdo. Postal 510-3, Cuernavaca 62250, Morelos, Mexico; isabelg@ibt.unam.mx (I.G.); bravo@ibt.unam.mx (A.B.); mario@ibt.unam.mx (M.S.)
* Correspondence: liukaiyu@mail.ccnu.edu.cn; Tel.: +86-27-67867221
† These authors equally contributed to this work.

Received: 4 May 2020; Accepted: 3 June 2020; Published: 6 June 2020

Abstract: It is well known that insect larval midgut cadherin protein serves as a receptor of *Bacillus thuringiensis* (Bt) crystal Cry1Ac or Cry1Ab toxins, since structural mutations and downregulation of *cad* gene expression are linked with resistance to Cry1Ac toxin in several lepidopteran insects. However, the role of *Spodoptera frugiperda* cadherin protein (SfCad) in the mode of action of Bt toxins remains elusive. Here, we investigated whether SfCad is involved in susceptibility to Cry1Ab or Cry1Fa toxins. *In vivo*, knockout of the *SfCad* gene by CRISPR/Cas 9 did not increase tolerance to either of these toxins in *S. frugiperda* larvae. *In vitro* cytotoxicity assays demonstrated that cultured insect TnHi5 cells expressing GFP-tagged SfCad did not increase susceptibility to activated Cry1Ab or Cry1Fa toxins. In contrast, expression of another well recognized Cry1A receptor in this cell line, the ABCC2 transporter, increased the toxicity of both Cry1Ab and Cry1Fa toxins, suggesting that SfABCC2 functions as a receptor of these toxins. Finally, we showed that the toxin-binding region of SfCad did not bind to activated Cry1Ab, Cry1Ac, nor Cry1Fa. All these results support that SfCad is not involved in the mode of action of Cry1Ab or Cry1Fa toxins in *S. frugiperda*.

Keywords: *Bacillus thuringiensis*; *Spodoptera frugiperda*; cadherin; Cry1Ab; Cry1Fa; mode of action of Cry toxin

Key Contribution: The CRISPR/Cas 9 gene editing, cytotoxicity assessment and biochemical analysis demonstrate the mutations of *cadherin* gene does not result in the development of resistance to Cry1Ab or Cry1Fa toxins in *S. frugiperda*.

---

1. Introduction

The crystal (Cry) toxins and vegetative insecticidal proteins (Vip) produced by *Bacillus thuringiensis* (Bt) bacteria are important biological tools for the control of insect pests and provide good protection for plants growth [1]. During sporulation, Bt bacteria accumulate Cry toxins in crystal inclusion bodies inside the mother cell, while the Vip proteins are secreted in the vegetative phase of growth [2,3]. The Bt toxin receptors, located on the larval midgut cells, play important roles in the toxicity of these Bt toxins. After ingestion of Bt crystal inclusions or Vip protein by the larvae, these proteins are dissolved under the alkaline conditions of the gut lumen, releasing protoxins that are activated by midgut proteases. The activated toxins bind to receptors, forming oligomers that insert into the cell membrane leading to pore formation, which results in death of the larvae [2,3]. The mode of action of Vip3Aa might be different from crystal toxins, since receptors for Vip3Aa are not shared with the Cry toxins [4–7].

In several lepidopteran insects, mutations in the *cadherin* gene (*cad*) are associated with resistance to Cry1Ac or Cry1Ab toxins [8–12]. The Cry1Ac toxin-binding region of *Helicoverpa armigera* cadherin (HaCad) and the membrane-proximal region of HaCad are required for Cry1Ac toxicity [13,14]. The downregulated expression of the *cadherin* gene has also been associated with resistance against the Bt Cry1Ac toxin in *Pectinophora gossypiella* [15]. Besides cadherin, the ATP-binding cassette sub-family C member 2 (ABCC2) is also recognized as an important insect molecule involved in the mode of action of Cry1A toxins [16]. Furthermore, it is known that HaCad and *Heliothis virescens* cadherin (HvCad) have a synergistic effect with ABCC2 on toxicity of Cry1A in cultured insect cells, since co-expression of cadherin receptors or the toxin-binding region of HaCad with the ABCC2 protein induced a synergistic effect on the cytotoxicity of Cry1Ac [14,16].

Even though cadherin has been shown to be an important Cry1A receptor in different Lepidopteran species, this is not always the case for some other lepidopteran insects. For instance, it has been reported that the cadherin from *Plutella xylostella* (PxCad) is not associated with resistance in *P. xylostella* to Cry1Ac [17]. However, other reports suggest that PxCad is a functional receptor of Cry1Ac, since PxCad can increase cytotoxicity of Cry1Ac when expressed in the Sf9 cell line [18–20]. In addition, we reported that *Spodoptera litura* cadherin (SlCad), in contrast to HaCad, cannot increase cytotoxicity of Cry1Ac when expressed in Hi5 cells, suggesting that SlCad is not a functional receptor of Cry1Ac in *S. litura* [14].

Although *S. frugiperda* is susceptible to Cry1Ab, Cry2Ab, Cry1Fa, and Vip3Aa toxins [21–30], there are no reports regarding whether *S. frugiperda* cadherin (SfCad) is involved in the mode of action of these Bt toxins. It has been shown that resistance to Cry1Fa in *S. frugiperda* is linked to different ABCC2 mutant alleles [27,28,31]. In addition, most *S. frugiperda* populations show low susceptibility to Cry1Ab or Cry1Ac toxins, in contrast to Cry1Fa that is highly active against this pest [21,32]. Here, we investigated whether SfCad is involved in the toxicity of Cry1Ab and Cry1Fa using both CRISPR/Cas 9 genome editing technology and cytotoxicity assays of Bt toxins in an insect cell line expressing SfCad. Our results suggest that *S. frugiperda* cadherin is not involved in the mode of action of Cry1Ab or Cry1Fa toxins.

2. Results

2.1. Construction of SfCad Gene Deleted Mutant by CRISPR/Cas 9 Genome Editing

To construct an *S. frugiperda cadherin* gene knockout mutant strain, we made use of the CRISPR/Cas 9 system to produce a large fragment deletion by designing two sgRNAs targeting different exons of the *SfCad* gene (Figure 1A). Freshly laid eggs were co-injected with the two *in vitro* transcribed sgRNAs, that are complementary to 20 bp DNA sequences from the fourth or fifth exons of *SfCad*, respectively, along with Cas 9 protein (Figure 1A). The results show that 22.5% (45/200) of the injected eggs hatched, and 71.1% (32) of the 45 neonates, raised in diet, survived into adults (F$_0$). The F$_0$ male and female moths were mass backcrossed with the DH19 strain to produce the next generation in single pair matings (F$_1$).

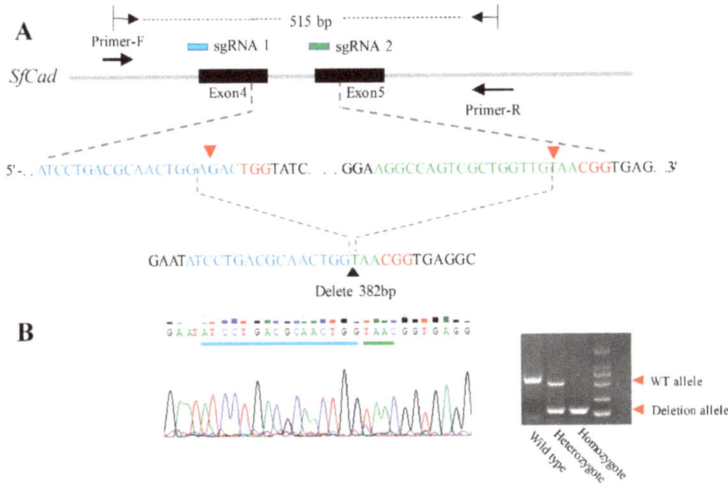

Figure 1. The knock-out of the *SfCad* gene. (**A**). Deleted fragment of the *SfCad* gene by the CRISPR/Cas 9 system between the two red arrow heads. (**B**). Sequencing and agarose gel electrophoresis of DNA confirming the knock-out of the *SfCad* gene.

After enough eggs were collected, genomic DNA from individual F_1 moths was prepared. Deletion events were detected by PCR using primers across the two target site regions (Figure 1A). Fragment deletions were initially screened by agarose gel electrophoresis and then those samples that showed multiple bands were cloned using T vector, and the DNA was sequenced to identify their mutations. We found that 25% (8/32) of the examined individuals showed deletions in the *cad* gene.

From the detected mutations, we selected a 382-bp deletion to generate a homozygous knockout strain (Figure 1B). The F_1 larvae (progeny crosses of the 382-bp deletion F_0 moth and strain DH19) were reared to pupation, and 96 exuviates of the final instar larvae were used to prepare genomic DNA. The DNA fragments flanking the two target sites were amplified by PCR, which were 515 bp in the wild type and 133 bp in the mutant (Figure 1B).

Among the 96 pupae screened, 30 carried the 382-bp deletion allele. Adults from these pupae were mass-crossed to obtain the F_2 generation. The genotypes of more than 100 F_2 individuals were visualized by agarose gel electrophoresis using gDNA samples from randomly selected exuviates of final-instar larvae. The agarose gel electrophoresis results showed that 21.4% (30/140) were homozygous for the 382-bp deletion. The 30 individuals were further sequenced and verified to be homozygotes. Finally, these homozygous individuals were pooled and mass-crossed to establish the *SfCad* knockout strain (Cad-KO).

2.2. Susceptibility of Cad-KO Strain to the Bt Toxins

To determine whether *SfCad* is involved in Cry1Ab or Cry1Fa resistance in *S. frugiperda*, we performed bioassays using the Cad-KO (knockout strain) and the progenitor DH19 *S. frugiperda* strains. Bioassay results showed that the knockout strain Cad-KO did not decrease susceptibility to these two Bt toxins (Figure 2). The LC_{50} values of Cry1Ab and Cry1Fa to the Cad-KO strain were not significantly different from the control DH19 strain because their 95% fiducial limit (FL) values overlapped (Table 1), suggesting that SfCad is not a functional receptor of Cry1Ab and Cry1Fa. The bioassay data also showed that Cry1Ab was at least 20- to 40-fold less toxic to both *S. frugiperda* strains analyzed compared to Cry1Fa toxin.

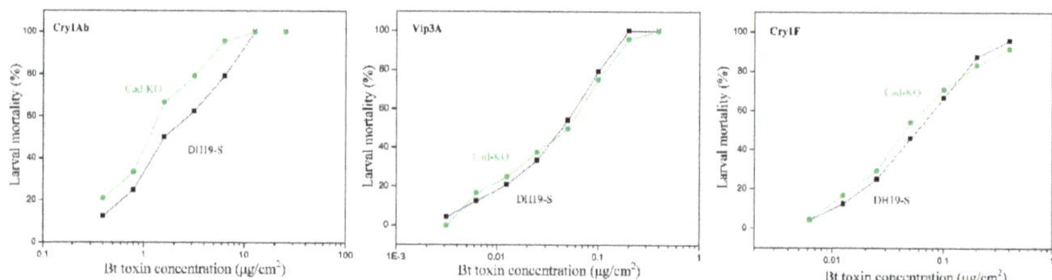

Figure 2. Influences of the knockout of the *SfCad* gene on susceptibility of the first instar *S. frugiperda* larvae to Cry1Ab, Vip3Aa, and Cry1Fa, respectively. Cad-KO (knockout strain), *S. frugiperda* larvae with the knockout of the *SfCad* gene. DH19-S, Bt toxin-susceptible *S. frugiperda* without knockout of the *SfCad* gene.

Table 1. Comparison of the susceptibility of first instar *S. frugiperda* larvae from the Cad-KO and DH19-S strains to different Bt toxins.

Strain.	Cry1Ab LC_{50} in $\mu g/cm^2$ (95% of FL)	Cry1Fa LC_{50} in $\mu g/cm^2$ (95% of FL)	Vip3Aa LC_{50} in $\mu g/cm^2$ (95% of FL)
Cad-KO	1.103 (0.798–1.453)	0.05 (0.037–0.069)	0.035 (0.026–0.047)
DH19-S	1.797 (1.311–2.458)	0.054 (0.041–0.073)	0.033 (0.025–0.044)

As an additional control, we also tested toxicity of the Vip3Aa protein; we found that the knockout strain Cad-KO did not decrease susceptibility to the Vip3Aa toxin (Table 1 and Figure 2), indicating that SfCad does not participate in Vip3Aa toxicity.

A total of 24 larvae in each group were tested with the indicated concentrations of Bt toxins, and the values of LC_{50} were calculated after day 7 of oral feeding. Assays were done in triplicate. The 95% fiducial limits (FL) values, shown inside the parenthesis, indicate that there are no significant differences between Cad-KO and DH19-S strains in each column, since these values overlap. Cad-KO are *S. frugiperda* larvae with the knockout of the *SfCad* gene. DH19-S are Bt toxin-susceptible *S. frugiperda* larvae without knockout of the *SfCad* gene.

2.3. SfCAD Expression Did Not Increase Susceptibility of Hi5 Insect Cells to Cry1Ab or Cry1Fa Toxins

The plasmid pIE2-SfCad-GFP was used to transiently express the fusion protein SfCad-GFP in Hi5 cells. As a control, Hi5 cells were also transfected with pIE2-SfABCC2-GFP that was previously shown to confer susceptibility to Hi5 cells to Cry1Ac toxin [33]. After transfection, cells were observed under the confocal fluorescent microscope, and the results revealed that SfCad-GFP was mainly localized on the cell membrane, suggesting proper expression and folding of the recombinant protein (Figure 3).

The transfection efficiency of the plasmid pIE2-SfCad-GFP was around 45–50% in Hi5 cells. Bioassay data showed that Hi5 cells expressing SfCad-GFP were still tolerant to Cry1Ab or Cry1Fa toxins, since the toxin-treated cells did not swell even at the highest concentration, 20 $\mu g/mL$, of these toxins (Figure 4 and Table 2). In contrast, Hi5 cells that were transfected with plasmid pIE2-SfABCC2 were susceptible to Cry1Ab or Cry1Fa toxins (Table 2 and Figure 4). The EC_{50} values of Bt toxins mediated by SfCad could not be calculated because there were no swollen cells after treatment with the Bt toxins for 1 h. These results also confirmed that SfCad could not mediate cytotoxicity of Cry1Ab and Cry1Fa in Hi5 cells.

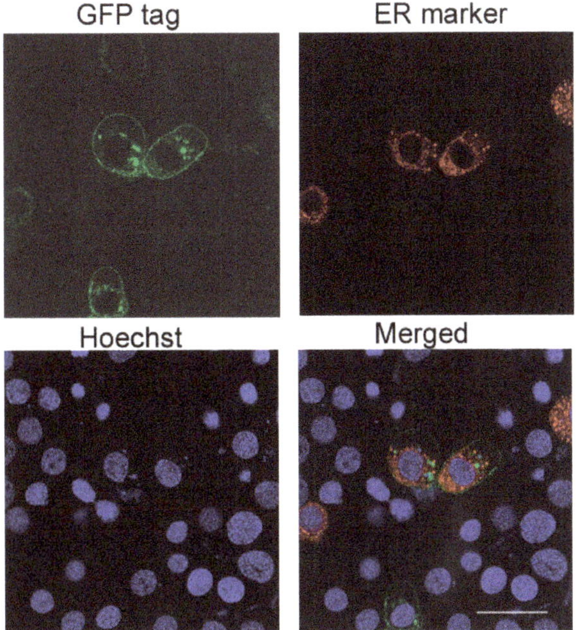

Figure 3. Subcellular localization of SfCad-GFP in *Trichoplusia ni* Hi5 cells. Green, SfCad-GFP (GFP tag); red, endoplasmic reticulum marker (ER marker); blue, nucleus (Hoechst). Scale bar, 50 μm.

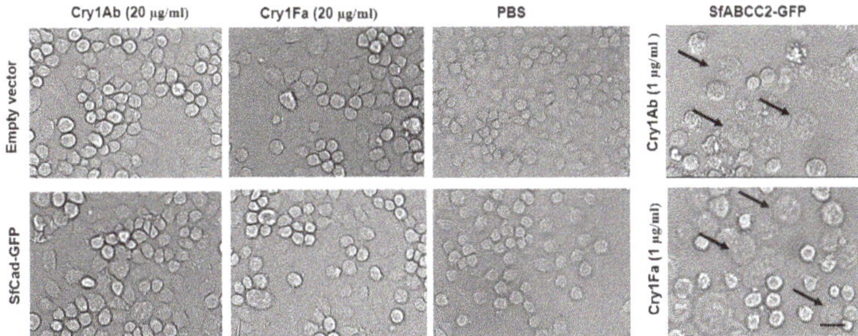

Figure 4. Susceptibility of Hi5 cells expressing SfCad-GFP to activated Cry1Ab and Cry1Fa toxins. The cells were transfected with plasmids pIE2-SfCad-GFP or pIE2-GFP (empty vector), respectively, and cultured for 24 h. Then, they were treated with activated toxins at 20 μg/mL for 1 h. A negative control of PBS-treated cells, treated with buffer, is included in the figure. A positive control of cells expressing SfABCC2-GFP is also shown in the figure. The susceptible cells pointed by arrow heads would become swollen, as shown in the positive control. Cells expressing SfCad-GFP or transfected with empty vector showed no swelling of the cells, similar to the negative control. Scale bar, 50 μm.

Table 2. Effect of SfCad or SfABCC2 on the cytotoxicity of activated Cry1Ab and Cry1Fa toxins in Hi5 cells.

Toxin	Putative Receptor	EC$_{50}$ (µg/mL)	95% FL	Slope ± SE	χ^2	df
Cry1Ab	SfCAD-GFP	— *	—	—	—	—
Cry1Ab	SfABCC2-GFP	0.06$_a$ **	0.04–0.08	1.55 ± 0.08	7.20	3
Cry1Fa	SfCAD-GFP	—	—	—	—	—
Cry1Fa	SfABCC2-GFP	0.23$_b$	0.19–0.27	2.18 ± 0.09	6.77	3

* indicates that the cells expressing the putative receptors are not susceptible to the indicated toxins; ** the different lowercase letters indicate that there are significant differences between EC$_{50}$ values of Cry1Ab and Cry1Fa in the same column.

2.4. Cry1Ac, Cry1Ab, and Cry1Fa Did Not Bind to the Toxin-Binding Region (TBR) of SfCad

The phylogenetic tree constructed with cadherin protein sequences from different Lepidopteran insects showed that the cadherin proteins of three *Spodoptera* species (*S. frugiperda*, *S. exigua*, and *S. litura*) cluster together and share high amino acid sequence identities (around 84%). The *Spodoptera* cadherin cluster is far away from *Helicoverpa armigera* cadherin (Figure 5). It is known that HaCad can mediate toxicity of Cry1Ac in larvae and also induces susceptibility to Cry1Ac when expressed in Hi5 cells [13,14,34].

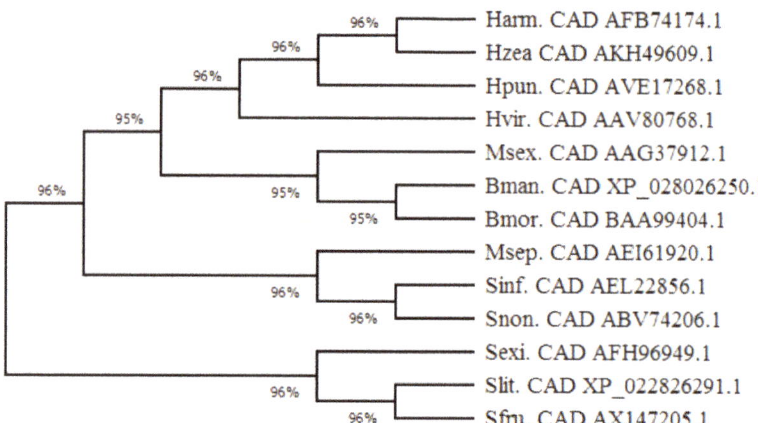

Figure 5. Phylogenetic analysis of cadherin protein in Lepidoptera insects. Harm, *Helicoverpa armigera*; Hzea, *Helicoverpa zea*; Hpun, *Helicoverpa punctigera*; Hvir, *Heliothis virescens*; Msex, *Manduca sexta*; Bman, *Bombyx mandarina*; Bmor, *Bombyx mori*; Msep, *Mythimna separata*; Sinf, *Sesamia inferens*; Snon, *Sesamia nonagrioides*; Sexi, *S. exigua*; Slit, *S. litura*. The Genbank accession numbers of the Cad proteins sequences used in this phylogenetic analysis are indicated in the graph.

Finally, we performed ligand blot binding assays confirming that the toxin-binding regions (TBR) of SfCad, SeCad, and SlCad did not bind to the activated Cry1Ac, Cry1Ab, and Cry1Fa toxins, in contrast to the positive control HaCad that clearly bound to Cry1Ac- and Cry1Ab-activated toxins (Figure 6). Cry1Fa did not bind to any of the TBR regions analyzed, including the TBR from the HaCad protein (Figure 6).

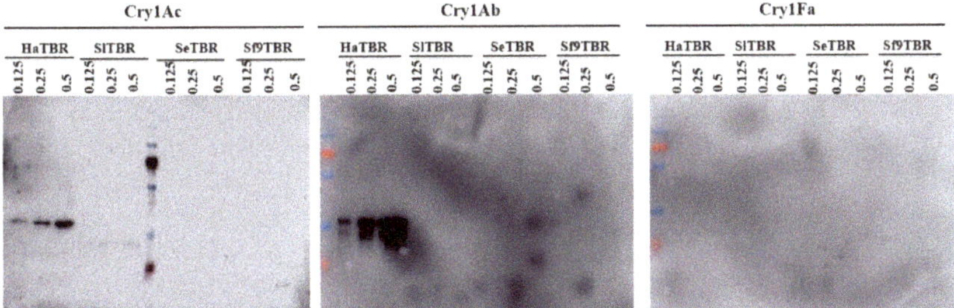

Figure 6. Ligand blot analysis of His-tagged toxin-binding regions (TBRs) binding to activated Cry1Ac, Cry1Ab, or Cry1Fa. The His-tagged TBRs of HaCad, SlCad, SeCad, and SlCad were used at different concentrations (0.125, 0.250, and 0.50 µg/mL) with one lane for each His-TBR. Binding of all toxins was assayed at 10 nM. Bound toxin was revealed with the corresponding polyclonal antibody (rabbit anti-Cry1Ac, rabbit anti-Cry1Ab, or rabbit anti-Cry1Fa antibody) as indicated in the Materials and Methods Section.

3. Discussion

The cadherin proteins from some Lepidopteran insects are involved in susceptibility of those larvae to Cry1A toxins. Mutations or reduced expression of *cadherin* genes in *H. virescens*, *H. armigera*, or *P. gossypiella* are associated with resistance to Cry1Ac [8,11,13,35]. In addition, *Bombyx mori* cadherin was shown to be involved in toxicity of Cry1Aa and Cry1Ab toxins [36,37]. However, *S. litura* and *Trichoplusia ni* cadherins do not function as Cry1Ac receptors [14,38]. A previous study demonstrated that cadherin protein from *H. virescens* functions as a receptor for Cry1A toxins, but not for Cry1Fa, when expressed in *Drosophila* S2 cells, suggesting that Cry1A and Cry1Fa toxins may rely on different receptor molecules [39]. In the present study, both the knockout in *S. frugiperda* insect larvae and over-expression of the *SfCad* gene in cultured Hi5 insect cells indicated that SfCad is not involved in toxicity of Cry1Ab or Cry1Fa in *S. frugiperda*. As described above, it has been shown that Vip3Aa does not share receptors with Cry1A or Cry1Fa toxins [4–7]. Thus, we also performed bioassays of the Cad-KO and DH19 *S. frugiperda* strains with Vip3Aa and showed that there was also no difference in the toxicity of Vip3Aa in the two *S. frugiperda* strains (Figure 2 and Table 1). These results also show that SfCad is not a functional receptor of Vip3Aa in *S. frugiperda*.

Interestingly, we showed that expression of the *SfABCC2* transporter gene in Hi5 cells greatly increased the susceptibility to Cry1Ab or Cry1Fa toxins (Table 2), supporting that ABCC2 is a functional receptor for both Cry1Ab and Cry1Fa toxins in *S. frugiperda*, as previously reported [27]. In the case of Cry1Fa, our results agree with the fact that resistance to Cry1Fa in different populations is linked to mutant alleles of *ABCC2* [28,31]. However, the toxicity of Cry1Ab to Hi5 cells expressing SfABCC2 was 3.5-fold higher than that of Cry1Fa. These results do not correlate with the toxicities of both toxins to wild type DH19 *S. frugiperda* larvae, where Cry1Fa showed 20- to 40-fold higher toxicity than Cry1Ab (Figure 2). These results indicate that Cry1Ab toxicity is limited by some additional mechanisms, other than receptor binding, in the wild type DH19 larvae. It was reported that the lack of toxicity of Cry1Ab to an *S. frugiperda* population from México correlated with enhanced toxin degradation by midgut proteases and also with reduced receptor binding [32]. In addition, it is still possible that an additional Cry1Ab receptor is expressed in the Hi5 cells but not in *S. frugiperda* larvae. Thus, different toxin susceptibility to midgut proteases or lower binding to brush border membrane vesicles (BBMV) could explain the differences in the larval susceptibility to Cry1Fa and Cry1Ab. These hypotheses remain to be analyzed.

As mentioned above, *S. frugiperda ABCC2* mutations are linked with resistance to Cry1Fa [27,28,31]. Interestingly, some Cry1Fa-resistant *S. frugiperda* strains showed cross-resistance to Cry1Ab or Cry1Ac toxins but not to Cry2Ab or Vip3Aa toxins [20,23]. In the present study, the knockout and over-expression of the *SfCad* gene revealed that SfCad is not involved in susceptibility of *S. frugiperda* to Cry1Ab,

Cry1Fa, nor Vip3Aa toxins. Nevertheless, RNAi silencing experiments of *SeCad* showed that cadherin from *S. exigua* might be involved in the toxicity of Cry1Ac and Cry2Aa to some degree [34]. Previously, we reported that SlCad did not increase the toxicity of Cry1Ac when expressed in Hi5 cells, indicating that SlCad is not a functional receptor of Cry1Ac [14]. These data agree with the lack of binding of the TBR from SlCad, SeCad, or SfCad to the Cry1Ac toxin (Figure 6) and support that cadherin proteins from the *Spodoptera* cluster species are not involved in the toxicity of Cry1Ab or Cry1Ac toxins. In the future, we will investigate whether SfCad is involved in Cry2A toxicity in *S. frugiperda*.

Even though some cadherins are not functional receptors of Cry1Ab or Cry1Ac toxins in different insect species, the cadherin protein could be a target for the evolution of Cry1Ab or Cry1Ac variants that bind to this receptor and increase their toxicity to susceptible or resistant insects where cadherin is not a functional receptor of the wild type of Cry1Ab or Cry1Ac. In the case of *T. ni*, Cry1Ac variants that could bind to the TnCad protein were selected by continuous evolution, and it was found that the Cry1Ac variants that were able to bind to TnCad were also able to counter resistance of *T. ni* insects linked to *ABCC2* mutations [40]. In addition, Cry1Ab domain III mutants were shown to increase the toxicity of Cry1Ab to different *S. frugiperda* strains, which was correlated with their increased binding to SfCad receptor [32]. Overall our results show that *S. frugiperda* cadherin is not a functional receptor of Cry1Fa and Cry1Ab toxins. Defining the structural basis for the lack of binding between Cry1Ab or Cry1Fa with SfCad could provide strategies for improving binding and toxicity of these Cry proteins to this invasive pest.

4. Materials and Methods

4.1. S. frugiperda Strain and Insect Cell Cultures

The *S. frugiperda* strain DH19 was established from individual moths collected from Dehong, Yunnan Province of China in January 2019 and reared in laboratory conditions on artificial diet without exposure to any insecticide or Bt toxin. Insects were reared at 27 ± 2 °C and 75% ± 10% relative humidity (RH) with a photoperiod of 14L:10D. For adults, 10% sugar solution was supplied as a food source.

The *Trichoplusia ni* BTI-Tn-5B1 cell line (Hi5) was established from insect ovaries [41] and kept in our laboratory. The cell line was cultured in Grace's insect cell culture medium (Life Technologies Co., Gand Island, NY, USA) supplemented with 10% fetal bovine serum (Life Technologies Inc.), 100 U/mL penicillin and 100 µg/mL streptomycin (Life technologies Inc.) at 28 °C under normal atmospheric conditions.

4.2. Preparation of sgRNAs

A pair of sgRNAs against the *S. frugiperda cadherin* gene (*SfCad*) (Genbank accession no.: AX147205.1) was designed using the sgRNAcas9 design tool [42]. The sgRNA1 target sequence (5′-ATC CTG ACG CAA CTG GAG ACT GG-3′) and sgRNA2 target sequence (5′-AGG CCA GTC GCT GGT TGT AAC GG-3′) were selected in exons 4 and 5 of the *SfCad* gene, respectively, (Figure 1A). The selected sgRNAs were analyzed in the *S. frugiperda* genome (https://bipaa.genouest.org), and no potential off-target sites were found. The DNA template for *in vitro* transcription of these sgRNAs was constructed by using PCR-based fusion of two oligonucleotides with a T7 promoter (Target F: TAA TAC GAC TCA CTA TAG + the target sequence; Target R: TTC TAG CTC TAA AAC + the reverse complementary sequence of the target). The PCR conditions were as reported by Jin et al. [43]. The sgRNAs were synthesized using an *in vitro* transcription GeneArt Precision gRNA Synthesis Kit (Thermo Fisher Scientific, Shanghai, China), according to the manufacturer's instructions.

4.3. Cas 9 Protein

Cas 9 protein (GeneArt Platinum Cas 9 Nuclease) was purchased from Thermo Fisher Scientific (Shanghai, China).

4.4. Egg Collection and Microinjection

Freshly laid eggs (within 2 h of oviposition) were washed with distilled water. Then, the eggs were placed on a microscope slide and fixed with double-sided adhesive tape. We injected each egg with 1–2 nL of a mixture solution containing two sgRNAs (150 ng/μL for each) and Cas 9 protein (50 ng/μL) using Nanoject III (Drummond, Broomall, PA, USA). The injected eggs were incubated at 25 °C and 65% RH for hatching.

4.5. Identification of SfCad Mutations Mediated by CRISPR/Cas 9 System

To identify the mutations, a specific pair of primers (Cad-F: CCT CCT CAA ATA AGA TTA CC; Cad-R: ATG ATG GGC GCA TTG TCG T) were designed that flanked the target sites, and genomic DNA samples of individual insects were used as the template. The genomic DNA of the larvae was extracted using a Multisource Genomic DNA Miniprep Kit (Axygen, New York, NY, USA) according to the manufacturer's instructions. The PCR conditions were as reported by Jin et al. [42]. Then, 10 μL PCR products were analyzed by agarose gel electrophoresis. Multiple bands indicated that double nicking had occurred. To analyze the exact type of mutation (insertion or deletion), the bands were recovered, cloned, and sequenced by Sangon Biotech (Shanghai, China).

4.6. Bt Toxins and Bioassay

The activated Cry1Ab toxin and Vip3Aa protoxins used in the *in vivo* bioassay were provided by the institute of Plant Protection, Chinese Academy of Agricultural Science (CAAS), Beijing, China. The other purified activated and lyophilized Cry1Ac, Cry1Ab, and Cry1Fa toxins were kindly donated by Dr. Marianne Pusztai-Carey from Case Western Reserve University, USA. Toxicity of each Bt toxin to DH19 and SfCad knockout strain was determined with diet overlay bioassays. Gradient concentrations of Bt toxin solution were prepared by diluting the stock suspensions in PBS (pH 7.0) solution. Artificial diet (900 μL) was dispensed into a 24-well plate (surface area per well = 2 cm^2) and after the diet cooled down, 50 μL Bt toxin solution was applied on the surface in each well. A single 1st-instar larva was put in each well after the toxin solution was dried at room temperature, and mortality was recorded after 7 days. The LC$_{50}$ (median lethal concentration that killed 50% of the tested larvae) and the corresponding 95% fiducial limits were calculated through Probit analysis of the mortality data using SPSS. Control wells were treated with buffer solution.

4.7. Plasmids, Transfection, and Fluorescence Observation

The *SfCad* gene was synthesized by Genescript Company (Nanjing, China) and inserted into a pGEM-T easy vector (Promega Inc., Madison, WI, USA). Then, the gene was amplified by PCR and inserted into pIE2-GFP-N1, and the new construct was named as pIE2-SfCad-GFP [14]. The plasmid purified from the transformed *E. coli* DH5α was transfected into Hi5 cells as previously reported [14]. Briefly, Hi5 cells were grown overnight in 48-well cell culture plates (Corning Inc., New York, NY, USA) at 1.2 × 10^5 cells/well. Then, the transfection was performed using the mixture of the plasmid (250 ng/well) with a transfection kit Genefusion HD (1 μL/well) (Promega Inc., Madison, WI, USA). Plasmids pIE2-GFP-N1, pIE2-SfCad-GFP, pIE2-SfABCC2-GFP, and pIE2-dsRED-ER were previously reported [33,44]. At 24 h post transfection, the cells were fixed using 4% paraformaldehyde for 10 min, and stained by Hoechst 33,342 (1 μg/mL) for 10 min. Then, the cells were observed and photographed under a laser confocal scanning microscope (Carl Zeiss, Jena Deutschland, Germany). The transfection efficiency was calculated: A = the number of cells emitting green fluorescence (SfCad-GFP) divided by the number of cells emitting blue fluorescence (nucleus stained by Hoechst 33,342) × 100%. Three biological replicates were performed.

4.8. Cytotoxicity Assay

Hi5 cells were transfected using single plasmids (pIE2-SfCad-GFP or pIE2-SfABCC2-GFP) as described above. At 24 h post transfection, the cells were treated with the indicated toxin concentrations (at least five different concentrations, two-fold serial dilution) of activated Bt toxins (Cry1Ab or Cry1Fa) for 1 h, and they were photographed under an inverted confocal microscope (Nikon, Tokyo, Japan). The cells transfected with the empty vector (pIE2-GFP-N1) were used as a control group and were also treated with the Cry toxins. An additional negative control of cells treated with phosphate buffer solution (PBS) was included in these assays. The percentage of swollen cells resulting from these toxin treatments was calculated as follows: B = the number of the swollen cells divided by the number of the total cells × 100%. The percentage of the swollen cells expressing SfCad-GFP or SfABCC2-GFP was calculated as follows: C = B/A × 100%. The transfection efficiency (A) was described above in Section 4.7. The effective concentration inducing 50% mortality value (EC_{50}) was obtained by regression analysis using SPPS 16.0 software. For two particular populations, the EC_{50} values were considered as significantly different if their 95% fiducial limits (FL) did not overlap [45].

4.9. Construction of a Lepidoptera Cadherin Phylogenetic Tree

The sequences of Lepidopteran insect cadherin proteins were selected for constructing a cadherin evolutionary tree by analyzing their phylogeny. GenBank accession numbers of the sequences of these cadherin proteins are as follow. Harm: *Helicoverpa armigera* cadherin, AFB74174.1; Hzea: *Helicoverpa zea* cadherin, AKH49609.1; Hpun: *Helicoverpa punctigera* cadherin, AVE17268.1; Hvir: *Heliothis virescens* cadherin, AAV80768.1; Msex: *Manduca sexta* cadherin, AAG37912.1; Bman: *Bombyx mandarina* cadherin, XP_028026250.1; Bmor: *Bombyx mori* cadherin, BAA99404.1; Msep: *Mythimna separata* cadherin, AEI61920.1; Sinf: *Sesamia inferens* cadherin, AEL22856.1; Snon: *Sesamia nonagrioides* cadherin, ABV74206.1; Sexi: *S. exigua* cadherin, AFH96949.1; Slit: *S. litura* cadherin, XP_022826291.1. The phylogeny of these sequences was analyzed using the neighbor-joining tree method with MEGA 5.0 software (https://mega.software.informer.com/5.0/).

4.10. Purification of Proteins Expressed in Bacteria

The coding DNA of toxin-binding regions of *SfCad, SlCad, SeCad,* and *HaCad* were amplified by PCR from the corresponding plasmids containing these genes or the cDNA obtained from midgut tissue of these insects [14]. The primers are listed in Table 3. The amplified fragments were purified and digested with restriction nucleases and cloned into the cleaved plasmids listed in Table 3. The constructs were transformed into *Escherichia coli* BL21 cells and the His-tagged proteins were purified with Ni-NTA affinity column (GE Healthcare Bioscience, Piscataway, NJ, USA) according to the manufacturer's manual. Detailed protocols were described previously [14]. All the purified proteins were stored at −80 °C until use.

Table 3. Primers used for expression of the different fragments of proteins.

Fragment	Forward Primer (5′–3′)	Reverse Primer (5′–3′)	Vector
His-HaTBR	CCGGAATTCTACGATTC GTGCTACGGACGGT	CCCAAGCTTCAGGTACA CCTTCACTTCCGT-3	pET22b (Novagen, Madison, WI, USA)
His-SeTBR	CCGGAATTCTGTTATCC GAGCTACTGATGG	CCCAAGCTTCATGAAGA TTGTCACTTCAGCTCGATC	pET22b
His-SlTBR	CCGGAATTCTGTTATTCG TGCCACGGATGGT	CCCAAGCTTCATGTAGA TTATAACTTTTGCTCG	pET22b
His-SfTBR	CCGGAATTCTGGAGGCG GTGGAGGCGGTGTTATT CGGGCCACGGACGGCG	CCCAAGCTTCATGTAAA TTGACACTTTTGCTCGAT CACTCGC	pET22b

The underlined letters indicate restriction sites of endonucleases.

4.11. Ligand Blot Assays

The 6 × His-tagged HaCad, SfCad, SlCad, or SeCad TBRs were separated on 12% SDS-PAGE gel and transferred onto Polyvinylidene fluoride (PVDF) membrane (three different membranes were prepared). The loading of proteins on these membranes was checked by Ponceau S staining 0.2% (*w/v*) in 3% (*v/v*) acetic acid and followed by complete destaining by washing with water. These membranes were blocked with 2% BSA in PBS-Tween (0.2%) for 3 h, then each membrane was incubated with a different activated Cry toxin (Cry1Ab, Cry1Ac or Cry1Fa toxin) at 10 nM for 2 h. After washing three times with PBS-Tween (0.2%), the membranes were further incubated with the corresponding polyclonal antibody (rabbit anti-Cry1Ac, rabbit anti-Cry1Ab, or rabbit anti-Cry1Fa antibody) diluted in PBS-Tween (0.2%) at 1:1000 dilution for 3 h. Then, the membranes were incubated with horseradish peroxidase (HRP)-conjugated goat anti-rabbit secondary antibody (Abbkine) diluted in PBS at 1:10,000. Finally, the membranes were incubated with the enhanced chemiluminescence (ECL) reagent (GE Healthcare Biosciences, Piscataway, NJ, USA) and then covered with X-ray film for exposure for a few minutes, and the film was developed and fixed as previously described [46].

Author Contributions: Conceptualization, K.L. and A.B.; Formal analysis, J.Z., M.J., Y.Y. (Yanchao Yang), Y.X., L.L., I.G., and M.S.; Funding acquisition, K.L.; Investigation, J.Z., M.J., Y.Y. (Yangchao Yang), and Y.Y. (Yongbo Yang); Writing original draft, J.Z. and M.J.; Writing review and editing, K.L., A.B., and M.S.; Supervision, K.L. All authors have read and agreed to the published version of the manuscript.

Funding: This work was supported by the National Key R&D Program of China (grant number: 2017YFD0200400).

Conflicts of Interest: M.S. and A.B. are coauthors of a patent on modified Bt toxins. "Suppression of resistance in insects to *Bacillus thuringiensis* Cry toxins, using toxins that do not require the cadherin receptor" (patent numbers: CA2690188A1, CN101730712A, EP2184293A2, EP2184293A4, EP2184293B1, WO2008150150A2, WO2008150150A3).

References

1. Wu, K.M.; Lu, Y.H.; Feng, H.Q.; Jiang, Y.Y.; Zhao, J.Z. Suppression of cotton bollworm in multiple crops in China in areas with Bt toxin–containing cotton. *Science* **2008**, *321*, 1676–1678. [CrossRef] [PubMed]

2. Gill, S.S.; Cowles, E.A.; Pietrantonio, P.V. The mode of action of *Bacillus thuringiensis* endotoxins. *Annu. Rev. Entomol.* **1992**, *37*, 615–636. [CrossRef] [PubMed]

3. Bravo, A.; Likitvivatanavong, S.; Gill, S.S.; Soberón, M. *Bacillus thuringiensis*: A story of a successful bioinsecticide. *Insect Biochem. Mol. Biol.* **2011**, *41*, 423–431. [CrossRef] [PubMed]

4. Estruch, J.J.; Warren, G.W.; Mullins, M.A.; Nye, G.J.; Craig, J.A.; Koziel, M.G. Vip3A, a novel *Bacillus thuringiensis* vegetative insecticidal protein with a wide spectrum of activities against lepidopteran insects. *Proc. Natl. Acad. Sci. USA* **1996**, *93*, 5389–5394. [CrossRef] [PubMed]

5. Lee, M.K.; Walters, F.S.; Hart, H.; Palekar, N.; Chen, J.S. The mode of action of the *Bacillus thuringiensis* vegetative insecticidal protein Vip3A differs from that of Cry1Ab delta-endotoxin. *Appl. Environ. Microbiol.* **2003**, *69*, 4648–4657. [CrossRef]

6. Jiang, K.; Hou, X.; Han, L.; Tan, T.; Cao, Z.; Cai, J. Fibroblast growth factor receptor, a novel receptor for vegetative insecticidal protein Vip3Aa. *Toxins* **2018**, *10*, 546. [CrossRef]

7. Jiang, K.; Hou, X.Y.; Tan, T.T.; Cao, Z.L.; Mei, S.Q.; Yan, B.; Chang, J.; Han, L.; Zhao, D.; Cai, J. Scavenger receptor-C acts as a receptor for *Bacillus thuringiensis* vegetative insecticidal protein Vip3Aa and mediates the internalization of Vip3Aa via endocytosis. *PLoS Pathog.* **2018**, *14*. [CrossRef]

8. Gahan, L.J.; Gould, F.; Heckel, D.G. Identification of a gene associated with Bt resistance in *Heliothis virescens*. *Science* **2001**, *293*, 857–860. [CrossRef]

9. Morin, S.; Biggs, R.W.; Sisterson, M.S.; Shriver, L.; Ellers-Kirk, C.; Higginson, D.; Holley, D.; Gahan, L.J.; Heckel, D.G.; Carrière, Y.; et al. Three cadherin alleles associated with resistance to *Bacillus thuringiensis* in pink bollworm. *Proc. Natl. Acad. Sci. USA* **2003**, *100*, 5004–5009. [CrossRef]

10. Yang, Y.; Yang, Y.; Gao, W.; Guo, J.; Wu, Y.; Wu, Y. Introgression of a disrupted cadherin gene enables susceptible *Helicoverpa armigera* to obtain resistance to *Bacillus thuringiensis* toxin Cry1Ac. *Bull. Entomol. Res.* **2009**, *99*, 175–181. [CrossRef]

11. Wang, J.; Zhang, H.; Wang, H.; Zhao, S.; Zuo, Y.; Yang, Y.; Wu, Y. Functional validation of cadherin as a receptor of Bt toxin Cry1Ac in *Helicoverpa armigera* utilizing the CRISPR/Cas9 system. *Insect Biochem. Mol. Biol.* **2016**, *76*, 11–17. [CrossRef] [PubMed]

12. Wang, L.; Wang, J.; Ma, Y.; Wan, P.; Liu, K.; Cong, S.; Xiao, Y.; Xu, D.; Wu, K.; Fabrick, J.A.; et al. Transposon insertion causes *cadherin* mis-splicing and confers resistance to Bt cotton in pink bollworm from China. *Sci. Rep.* **2019**, *9*. [CrossRef] [PubMed]

13. Wang, G.; Wu, K.; Liang, G.; Guo, Y. Gene cloning and expression of cadherin in midgut of *Helicoverpa armigera* and its Cry1A binding region. *Sci. China C Life Sci.* **2005**, *48*, 346–356. [CrossRef] [PubMed]

14. Ma, Y.; Zhang, J.; Xiao, Y.; Yang, Y.; Liu, C.; Peng, R.; Yang, Y.; Bravo, A.; Soberón, M.; Liu, K. The cadherin Cry1Ac binding-region is necessary for the cooperative effect with ABCC2 transporter enhancing insecticidal activity of *Bacillus thuringiensis* Cry1Ac toxin. *Toxins* **2019**, *11*. [CrossRef]

15. Fabrick, J.A.; Mathew, L.G.; LeRoy, D.M.; Hull, J.J.; Unnithan, G.C.; Yelich, A.J.; Carrière, Y.; Li, X.; Tabashnik, B.E. Reduced cadherin expression associated with resistance to Bt toxin Cry1Ac in pink bollworm. *Pest Manag. Sci.* **2020**, *76*, 67–74. [CrossRef]

16. Brestschneider, A.; Heckel, D.G.; Pauchet, Y. Three toxins, two receptors, one mechanism: Mode of action of Cry1A toxins from *Bacillus thuringiensis* in *Heliothis virescens*. *Insect Biochem. Mol. Biol.* **2016**, *76*, 109–117. [CrossRef]

17. Guo, Z.; Kang, S.; Zhu, X.; Wu, Q.; Wang, S.; Xie, W.; Zhang, Y. The midgut cadherin-like gene is not associated with resistance to *Bacillus thuringiensis* toxin Cry1Ac in *Plutella xylostella* (L.). *J. Invertebr. Pathol.* **2015**, *126*, 21–30. [CrossRef]

18. Park, Y.; Herrero, S.; Kim, Y. A single type of cadherin is involved in *Bacillus thuringiensis* toxicity in *Plutella xylostella*. *Insect Mol. Biol.* **2015**, *24*, 624–633. [CrossRef]

19. Gao, M.; Dong, S.; Hu, X.; Zhang, X.; Liu, Y.; Zhong, J.; Lu, L.; Wang, Y.; Chen, L.; Liu, X. Roles of midgut cadherin from two moths in different *Bacillus thuringiensis* action mechanisms: Correlation among toxin binding, cellular toxicity, and synergism. *J. Agric. Food Chem.* **2019**, *67*, 13237–13246. [CrossRef]

20. Hu, X.; Zhang, X.; Zhong, J.; Liu, Y.; Zhang, C.; Xie, Y.; Lin, M.; Xu, C.; Lu, L.; Zhu, Q.; et al. Expression of Cry1Ac toxin-binding region in *Plutella xylostella* cadherin-like receptor and studying their interaction mode by molecular docking and site-directed mutagenesis. *Int. J. Biol. Macromol.* **2018**, *111*, 822–831. [CrossRef]

21. Ingber, D.A.; Mason, C.E.; Flexner, L. Cry1 Bt susceptibilities of fall armyworm (Lepidoptera: Noctuidae) host strains. *J. Econ. Entomol.* **2018**, *111*, 361–368. [CrossRef] [PubMed]

22. Yang, F.; Morsello, S.; Head, G.P.; Sansone, C.; Huang, F.; Gilreath, R.T.; Kerns, D.L. F_2 screen, inheritance and cross-resistance of field-derived Vip3A resistance in *Spodoptera frugiperda* (Lepidoptera: Noctuidae) collected from Louisiana, USA. *Pest Manag. Sci.* **2018**, *74*, 1769–1778. [CrossRef] [PubMed]

23. Vélez, A.M.; Spencer, T.A.; Alves, A.P.; Moellenbeck, D.; Meagher, R.L.; Chirakkal, H.; Siegfried, B.D. Inheritance of Cry1F resistance, cross-resistance and frequency of resistant alleles in *Spodoptera frugiperda* (Lepidoptera: Noctuidae). *Bull. Entomol. Res.* **2013**, *103*, 700–713. [CrossRef] [PubMed]

24. Tay, W.T.; Mahon, R.J.; Heckel, D.G.; Walsh, T.K.; Downes, S.; James, W.J.; Lee, S.F.; Reineke, A.; Williams, A.K.; Gordon, K.H. Insect resistance to *Bacillus thuringiensis* toxin Cry2Ab is conferred by mutations in an ABC transporter subfamily A protein. *PLoS Genet.* **2015**, *11*. [CrossRef] [PubMed]

25. Storer, N.P.; Babcock, J.M.; Schlenz, M.; Meade, T.; Thompson, G.D.; Bing, J.W.; Huckaba, R.M. Discovery and characterization of field resistance to Bt maize: *Spodoptera frugiperda* (Lepidoptera: Noctuidae) in Puerto Rico. *J. Econ. Entomol.* **2010**, *103*, 1031–1038. [CrossRef]

26. Monnerat, R.; Martins, E.; Macedo, C.; Queiroz, P.; Praça, L.; Soares, C.M.; Moreira, H.; Grisi, I.; Silva, J.; Soberón, M.; et al. Evidence of field-evolved resistance of *Spodoptera frugiperda* to Bt corn expressing Cry1F in Brazil that is still sensitive to modified Bt toxins. *PLoS ONE* **2015**, *10*. [CrossRef]

27. Banerjee, R.; Hasler, J.; Meagher, R.; Nagoshi, R.; Hietala, L.; Huang, F.; Narva, K.; Jurat-Fuentes, J.L. Mechanism and DNA/based detection of field-evolved resistance to transgenic Bt corn in fall armyworm (*Spodoptera frugiperda*). *Sci. Rep.* **2017**, *7*. [CrossRef]

28. Boaventura, D.; Ulrich, J.; Lueke, B.; Bolzan, A.; Okuma, D.; Gutbrod, O.; Geibel, S.; Zeng, Q.; Dourado, P.M.; Martinelli, S.; et al. Molecular characterization of Cry1F resistance in fall armyworm, *Spodoptera frugiperda* from Brazil. *Insect Biochem. Mol. Biol.* **2020**, *116*. [CrossRef]

29. Bernardi, O.; Bernardi, D.; Horikoshi, R.J.; Okuma, D.M.; Miraldo, L.L.; Fatoretto, J.; Medeiros, F.C.; Burd, T.; Omoto, C. Selection and characterization of resistance to the Vip3Aa20 protein from *Bacillus thuringiensis* in *Spodoptera frugiperda*. *Pest Manag. Sci.* **2016**, *72*, 1794–1802. [CrossRef]

30. Kahn, T.W.; Chakroun, M.; Williams, J.; Walsh, T.; James, B.; Monserrate, J.; Ferré, J. Efficacy and resistance management potential of a modified Vip3C protein for control of *Spodoptera frugiperda* in Maize. *Sci. Rep.* **2018**, *8*. [CrossRef]

31. Flagel, L.; Lee, Y.W.; Wanjugi, H.; Swarup, S.; Brown, A.; Wang, J.; Kraft, E.; Greenplate, J.; Simmons, J.; Adams, N.; et al. Mutational disruption of the *ABCC2* gene in fall armyworm, *Spodoptera frugiperda*, confers resistance to the Cry1Fa and Cry1A.105 insecticidal proteins. *Sci. Rep.* **2018**, *8*. [CrossRef] [PubMed]

32. Gómez, I.; Ocelotl, J.; Jorge Sánchez, J.; Lima, C.; Martins, E.; Rosales-Juárez, A.; Aguilar-Medel, S.; Abad, A.; Dong, H.; Monnerat, R.; et al. Enhancement of *Bacillus thuringiensis* Cry1Ab and Cry1Fa toxicity to *Spodoptera frugiperda* by domain III mutations indicates two limiting steps in toxicity as defined by receptor binding and protein stability. *Appl. Environ. Microbiol.* **2018**, *84*. [CrossRef] [PubMed]

33. Liu, L.; Chen, Z.; Yang, Y.; Xiao, Y.; Liu, C.; Ma, Y.; Soberón, M.; Bravo, A.; Yang, Y.; Liu, K. A single amino acid polymorphism in ABCC2 loop 1 is responsible for differential toxicity of *Bacillus thuringiensis* Cry1Ac toxin in different *Spodoptera* (Noctuidae) species. *Insect Biochem. Mol. Biol.* **2018**, *100*, 59–65. [CrossRef] [PubMed]

34. Xu, X.; Yu, L.; Wu, Y. Disruption of a cadherin gene associated with resistance to Cry1Ac endotoxin of *Bacillus thuringiensis* in *Helicoverpa armigera*. *Appl. Environ. Microbiol.* **2005**, *71*, 948–954. [CrossRef] [PubMed]

35. Qiu, L.; Hou, L.; Zhang, B.; Liu, L.; Li, B.; Deng, P.; Ma, W.; Wang, X.; Fabrick, J.A.; Chen, L.; et al. Cadherin is involved in the action of *Bacillus thuringiensis* toxins Cry1Ac and Cry2Aa in the beet armyworm, *Spodoptera exigua*. *J. Invertebr. Pathol.* **2015**, *127*, 47–53. [CrossRef] [PubMed]

36. Adegawa, S.; Nakama, Y.; Endo, H.; Shinkawa, N.; Kikuta, S.; Sato, R. The domain II loops of *Bacillus thuringiensis* Cry1Aa form an overlapping interaction site for two *Bombyx mori* larvae functional receptors, ABC transporter C2 and cadherin-like receptor. *Biochim. Biophys. Acta Proteins Proteom.* **2017**, *1865*, 220–231. [CrossRef] [PubMed]

37. Endo, H.; Adegawa, S.; Kikuta, S.; Sato, R. The intracellular region of silkworm cadherin-like protein is not necessary to mediate the toxicity of *Bacillus thuringiensis* Cry1Aa and Cry1Ab toxins. *Insect Biochem. Mol. Biol.* **2018**, *94*, 36–41. [CrossRef]

38. Wang, S.; Kain, W.; Wang, P. *Bacillus thuringiensis* Cry1A toxins exert toxicity by multiple pathways in insects. *Insect Biochem. Mol. Biol.* **2018**, *102*, 59–66. [CrossRef]

39. Jurat-Fuentes, J.L.; Adang, M.J. The *Heliothis virescens* cadherin protein expressed in Drosophila S2 cells functions as receptor for *Bacillus thuringiensis* Cry1A but not Cry1Fa toxins. *Biochemistry* **2006**, *45*, 9688–9695. [CrossRef]

40. Badran, A.H.; Guzov, V.M.; Huai, Q.; Kemp, M.M.; Vishwanath, P.; Kain, W.; Nance, A.M.; Evdokimov, A.; Moshiri, F.; Turner, K.H.; et al. Continuous evolution of *Bacillus thuringiensis* toxins overcomes insect resistance. *Nature* **2016**, *533*, 58–63. [CrossRef]

41. Davis, T.R.; Wickham, T.J.; McKenna, K.A.; Granados, R.R.; Shuler, M.L.; Wood, H.A. Comparative recombinant production of eight insect lines. *In Vitro Cell. Dev. Biol. Anim.* **1993**, *29A*, 388–390. [CrossRef] [PubMed]

42. Xie, S.; Shen, B.; Zhang, C.; Huang, X.; Zhang, Y. sgRNAcas9: A software package for designing CRISPR sgRNA and evaluating potential off-target cleavage sites. *PLoS ONE* **2014**, *9*. [CrossRef] [PubMed]

43. Jin, M.; Tao, J.; Li, Q.; Cheng, Y.; Sun, X.; Wu, K.; Xiao, Y. Genome editing of the *SfABCC2* gene confers resistance to Cry1F toxin from *Bacillus thuringiensis* in *Spodoptera frugiperda*. *J. Integr. Agric.* **2019**, *18*, 2–7.

44. Li, J.; Ma, Y.; Yuan, W.; Xiao, Y.; Liu, C.; Wang, J.; Peng, J.; Peng, R.; Soberón, M.; Bravo, A.; et al. FOXA transcriptional factor modulates insect susceptibility to *Bacillus thuringiensis* Cry1Ac toxin by regulating the expression of toxin-receptor *ABCC2* and *ABCC3* genes. *Insect Biochem. Mol. Biol.* **2017**, *88*, 1–11. [CrossRef] [PubMed]

45. Tabashnik, B.E.; Cushing, N.L.; Johnson, M.W. Diamondback moth (Lepidoptera: Plutellidae) resistance to insecticides in Hawaii: Intra-island variation and cross-resistance. *J. Econ. Entomol.* **1987**, *80*, 1091–1099. [CrossRef]

46. Xie, R.; Zhuang, M.; Ross, L.S.; Gomez, I.; Oltean, D.I.; Bravo, A.; Soberón, M.; Gill, S.S. Single amino acid mutations in the cadherin receptor from *Heliothis virescens* affect its toxin binding ability to Cry1A toxins. *J. Biol. Chem.* **2005**, *280*, 8416–8425. [CrossRef]

Article

CRISPR-Mediated Knockout of the *ABCC2* Gene in *Ostrinia furnacalis* Confers High-Level Resistance to the *Bacillus thuringiensis* Cry1Fa Toxin

Xingliang Wang, Yanjun Xu, Jianlei Huang, Wenzhong Jin, Yihua Yang and Yidong Wu *

College of Plant Protection, Nanjing Agricultural University, Nanjing 210095, China; wxl@njau.edu.cn (X.W.);
2016102098@njau.edu.cn (Y.X.); 2016202024@njau.edu.cn (J.H.); 2018102105@njau.edu.cn (W.J.);
yhyang@njau.edu.cn (Y.Y.)
* Correspondence: wyd@njau.edu.cn; Tel.: +86-25-8439-6062

Received: 16 March 2020; Accepted: 9 April 2020; Published: 11 April 2020

Abstract: The adoption of transgenic crops expressing *Bacillus thuringiensis* (Bt) insecticidal crystalline (Cry) proteins has reduced insecticide application, increased yields, and contributed to food safety worldwide. However, the efficacy of transgenic Bt crops is put at risk by the adaptive resistance evolution of target pests. Previous studies indicate that resistance to *Bacillus thuringiensis* Cry1A and Cry1F toxins was genetically linked with mutations of ATP-binding cassette (ABC) transporter subfamily C gene *ABCC2* in at least seven lepidopteran insects. Several strains selected in the laboratory of the Asian corn borer, *Ostrinia furnacalis*, a destructive pest of corn in Asian Western Pacific countries, developed high levels of resistance to Cry1A and Cry1F toxins. The causality between the *O. furnacalis ABCC2* (*OfABCC2*) gene and resistance to Cry1A and Cry1F toxins remains unknown. Here, we successfully generated a homozygous strain (OfC2-KO) of *O. furnacalis* with an 8-bp deletion mutation of *ABCC2* by the CRISPR/Cas9 approach. The 8-bp deletion mutation results in a frame shift in the open reading frame of transcripts, which produced a predicted protein truncated in the TM4-TM5 loop region. The knockout strain OfC2-KO showed much more than a 300-fold resistance to Cry1Fa, and low levels of resistance to Cry1Ab and Cry1Ac (<10-fold), but no significant effects on the toxicities of Cry1Aa and two chemical insecticides (abamectin and chlorantraniliprole), compared to the background NJ-S strain. Furthermore, we found that the Cry1Fa resistance was autosomal, recessive, and significantly linked with the 8-bp deletion mutation of *OfABCC2* in the OfC2-KO strain. In conclusion, *in vivo* functional investigation demonstrates the causality of the *OfABCC2* truncating mutation with high-level resistance to the Cry1Fa toxin in *O. furnacalis*. Our results suggest that the *OfABCC2* protein might be a functional receptor for Cry1Fa and reinforces the association of this gene to the mode of action of the Cry1Fa toxin.

Keywords: Asian corn borer; *ABCC2*; CRISPR/Cas9; Cry1Fa; resistance

Key Contribution: In vivo functional investigation demonstrates the causality of the *ABCC2* truncating mutation with high level of resistance to the Cry1Fa toxin in *Ostrinia furnacalis*. Our results suggest that *O. furnacalis* ABCC2 might be a functional receptor for Cry1Fa and reinforces the association of this gene to the mode of action of the Cry1Fa toxin.

---

1. Introduction

Transgenic crops expressing *Bacillus thuringiensis* (Bt) insecticidal crystalline (Cry) proteins have been commercialized worldwide since 1996. The global planting area of Bt crops reached 104 million hectares in 2018 [1]. The widespread Bt crop adoption has suppressed pest populations, reduced insecticide usage, promoted biocontrol services, and economically benefited growers [2].

However, the efficacy of Bt crops is put at risk from the adaptive evolution of resistance by the target pests, and practical resistance to Bt crops has been documented at least in nine pest species in six countries [3–5].

The European corn borer *Ostrinia nubilalis* (Hübner) and the Asian corn borer *Ostrinia furnacalis* (Guenée) are two sibling species, both of which are economically important insect pests of corn, *Zea mays* (L.) [6]. *O. nubilalis* is present in Europe, North Africa, Central Asia, and North America [7], while *O. furnacalis* is distributed widely in East and Southeast Asia, Australia, and the Western Pacific Islands [8]. Bt corn expressing Cry1Ab has been widely planted for the control of some lepidopteran pests, including *O. nubilalis*, in North America since 1996, resulting in the suppression of target pest populations and reduced insecticide applications in both Bt and non-Bt corn [9,10]. No practical resistance to Cry1Ab has been identified in *O. nubilalis* field populations from North America for more than 20 years [5]. Bt corn expressing Cry1F has been used commercially in North America since 2003, and the frequency of alleles conferring Cry1F resistance did not increase in field populations of *O. nubilalis* sampled during 2003 to 2009 from the US corn belt [11]. However, practical resistance to Cry1F was discovered in 2018 from *O. nubilalis* populations from Nova Scotia of Canada [4]. It indicates that Bt resistane has already become a real threat to the long-term effectiveness of Bt corn for the control of *O. nubilalis*.

China is a major corn producer in the world and its corn acreage was 41.5 million hectares in 2018 [12]. *O. furnacalis* is the domiant pest and widely distributed in most of the corn-growing regions of China, while *O. nubilalis* is limited to some regions of northwestern China [13]. Although the commercial planting of Bt corn has not yet been approved in China, two Bt corn events (DBN9936 and Shuangkang12-5) obtained biosafety certificates in 2019 (MARA, 2020) [14], which is considered a prerequisite and landmark for commercial production. To be prepared for the switch to the adoption of Bt corn in the near future, a number of investigations have been conducted in China to assess resistance risk and cross-resistance by laboratory selection of *O. furnacalis* with Bt proteins. Under laboratory selection conditions, *O. furnacalis* developed high levels of resistance to various Cry1 toxins, including Cry1Ab, Cry1Ac, Cry1Ah, Cry1F, and Cry1Ie [15–18], proving its potential to develop Bt resistance in the field. Symmetrical cross-resistance was found among Cry1Ab, Cry1Ac, Cry1Ah, and Cry1F [15–18]. Asymmetrical cross-resistance was observed between Cry1Ie and other Cry1 toxins. Selection with Cry1Ab, Cry1Ac, Cry1Ah, or Cry1F did not confer cross-resistance to Cry1Ie, but selection with Cry1Ie resulted in high-level cross-resistance to Cry1F [15–19]. These results are valuable for the future designing of resistance management strategies for Bt corn in China. However, the resistance mechanisms underlying Bt resistance of *O. furnacalis* remain elusive.

Several proteins have been identified and characterized as receptors for Cry toxins, including cadherins, aminopeptidase *N* (APN), alkaline phosphatases (ALP), and ATP-binding cassette (ABC) transporters [20]. One of the major mechanisms of resistance to Cry toxin is reduced toxin binding to their specific larval midgut receptors through the disruption of the receptor genes [21]. Since the disruption of the ABC transporter subfamily C2 (*ABCC2*) gene was first identified to confer Cry1Ac resistance in *Heliothis virescens* [22], mutations of the homologous *ABCC2* genes associated with Cry1A and/or Cry1F resistance have been found in several lepidopteran insects, including *Plutella xylostella*, *Trichoplusia ni* [23], *Bombyx mori* [24], *Helicoverpa armigera* [25], *Spodoptera exigua* [26], and *Spodoptera frugiperda* [27–29]. Recently, the CRISPR/Cas9 system has been applied to investigate the *in vivo* role of insect *ABCC2* in the mode of action and resistance mechanisms of Bt toxins. The causal relationship between *ABCC2* knockout and Cry1A/Cry1F resistance has been confirmed in *P. xylostella* [30], *S. frugiperda* [31], and *S. exigua* [32]. Interestingly, the knockout of either *ABCC2* or *ABCC3* of *H. armigera* did not confer Cry1Ac resistance, whereas the knockout of *ABCC2* and *ABCC3* together resulted in extremely high levels of resistance to Cry1Ac [33]. However, until now, whether or not the *ABCC2* gene of *O. furnacalis* (*OfABCC2*) is involved in Bt resistance development remains unknown.

In this study, we knocked out the *OfABCC2* employing the CRISPR/Cas9 system and constructed a homozygous mutant strain (OfC2-KO). Next, we performed toxicity bioassays and found that the

OfABCC2 knockout obtained a resistance to Cry1Fa greater than 300-fold compared to the wild-type control strain. Finally, we accessed the inheritance mode of the acquired resistance and confirmed the linkage between manipulated gene deletion and high-level resistance to Cry1Fa in the OfC2-KO strain.

2. Results

2.1. CRISPR/Cas9-Mediated Mutagenesis of OfABCC2 in O. Furnacalis

A total of 572 newly laid eggs (< 2 h) were injected with a mixture of the synthesized sgRNA and Cas9 protein. A total of 150 injected embryos (~26%) hatched and developed to 5th instar larvae. In order to obtain individuals with edited genomes as quickly as possible, the genomic DNA of 90 exuviates of the final instar larvae were isolated, and *OfABCC2* genotypes were identified by the direct sequencing of PCR products flanking the target site. Sequencing chromatograms revealed that 7.8% (7/90) of the examined G_0 individuals were mutagenized at the target site with a stretch of double peaks. Then, the seven chimeras (six females and one male) were single crossed with the wild-type NJ-S moth, respectively (G_0, Figure 1).

After G_0 oviposited, the genomic DNA of randomly selected eight–nine exuviates from each single-pair progeny were prepared, and their *OfABCC2* genotype was identified as described above. Among the 60 exuviates genotyped, 46 samples were wild-type homozygotes, seven individuals were heterozygotes harboring a wild-type allele and an 8-bp deletion allele, three samples were heterozygotes carrying a wild-type allele and a 1-bp insertion allele, and the genotype of the rest of the four individuals could not be identified by visual checks based on the sequencing chromatogram. We therefore confirmed efficient mutagenesis induced by CRISPR/Cas9 system had occurred in *OfABCC2* and the genome-edited alleles were transmitted to G_1.

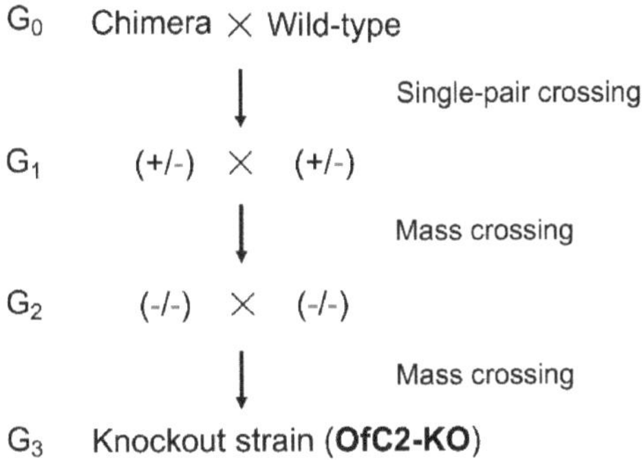

Figure 1. Diagram of the crossing strategy used to obtain the knockout strain homozygous for the 8-bp deletion mutation in exon 4 of *OfABCC2*. (+/-) means heterozygote (0/-8), (-/-) means mutant homozygote (-8/-8).

2.2. Construction of a Homozygous Strain with OfABCC2 Knocked Out

The mass mating was made among the above seven heterozygotes (three males and four females) that were harboring a wild-type allele and an 8-bp deletion allele (+/-) in G_1 (Figure 1). The genomic DNA of 30 exuviates from G_2 progeny were isolated and the genotype of *OfABCC2* was screened, and 21, five, and four samples were respectively identified as wild-type homozygotes (+/+), heterozygotes (+/-), and mutant homozygotes (-/-). The four moths (three females and one male) harboring both the 8-bp deletion alleles were mass crossed and their progeny (G_3) was reared to form a homozygous knockout strain named OfC2-KO (Figure 1). Subsequently, the TA-clone sequencing of the PCR products using both gDNA and cDNA from the OfC2-KO larvae were performed, and confirmed the *OfABCC2* carrying the 8-bp deletion at the desired site (data not shown).

Based on the deduced peptide sequences, the 8-bp deletion at exon 4 caused a pre-mature stop codon (Figure 2a,b). The consequence of this 8-bp deletion is predicted to lose TM5-TM12 transmembrane segments and two nucleotide-binding domains (NBDs) (Figure 2c). In view of the absence of about 70% of the protein structure, the *ABCC2* gene in the OfC2-KO strain is predicted to be defective and most likely non-functional.

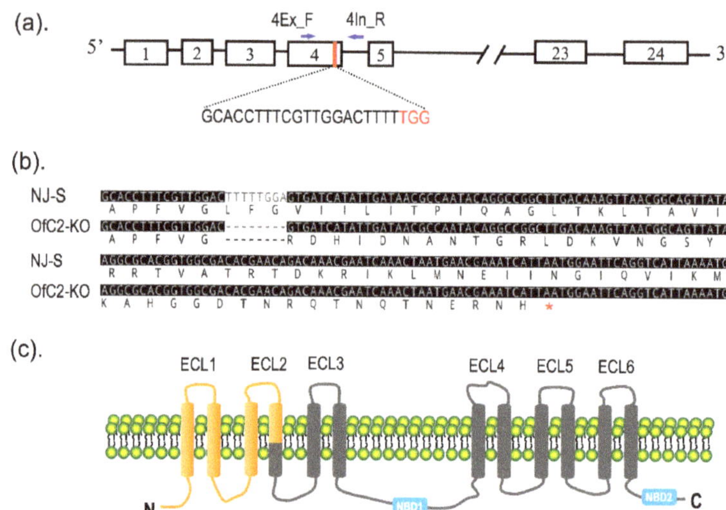

Figure 2. CRISPR/Cas9-mediated editing of the *OfABCC2* gene. (**a**) A diagram of the *OfABCC2* gene and sgRNA targeting site. The white boxes represent predicted exons through sequence alignment with *ABCC2*s from *Heliothis virescens* and *Plutella xylostella*. The sgRNA targeting site was located at exon 4, containing a proto spacer and a protospacer adjacent motif (PAM) sequence (TGG, in red). (**b**) The deduced peptide sequences from partial exon 4 to exon 6 of *OfABCC2*. The stop code is shown by a red asterisk. (**c**) A schematic diagram of the 12 transmembrane domains (TM1–TM12). The cleaved site induced by CRISPR/Cas9 is located at TM4, resulting in a frame shift of the transcript. The predicted protein produced from this mutant allele would be truncated in the intracellular TM4–TM5 loop of OfABCC2.

2.3. Impact of OfABCC2 Disruption on the Susceptibility of O. Furnacalis to Bt Toxins and Chemical Insecticides

Toxicity assays with larvae from the mutagenesis OfC2-KO strain and the background NJ-S strain against four Bt Cry toxins and two insecticides were carried out with the aim of assessing the impact of disrupted *OfABCC2* on larvae's susceptibility. The bioassay results indicate that the OfC2-KO strain showed low levels of resistance to Cry1Ac (8.1-fold) and Cry1Ab (3.6-fold), but no significant resistance to Cry1Aa (1.4-fold) compared to the NJ-S strain based on LC_{50} values (Table 1). However, because the susceptibility of the OfC2-KO strain to Cry1Fa was reduced to a large extent, the LC_{50} cannot be obtained by bioassay. The mortality of the OfC2-KO larvae was only 4% when treated by 120 µg/g Cry1Fa, and the estimated resistance ratio was much more than 300-fold. In contrast, the two strains exhibited approximately equal susceptibility to two chemical insecticides (abamectin and chlorantraniliprole) with 0.6- and 1.3-fold difference of LC_{50}s. Our findings provide strong reverse genetics evidence for *OfABCC2* involved in the toxicity and mode of action of Cry1Fa.

Table 1. Toxicity response to four Bt toxins and two chemical insecticides of larvae from the original NJ-S and OfC2-KO strains of *O. furnacalis*.

Toxin/Insecticide	Strain	N [1]	Slope ± SE	LC$_{50}$ (µg/g)	95% Fiducial Limits	RR [2]
Cry1Aa	NJ-S	312	3.714 ± 0.519	0.391	0.320-0.455	1
	OfC2-KO	384	2.583 ± 0.386	0.527	0.359-0.737	1.4
Cry1Ab	NJ-S	360	2.978 ± 0.362	0.116	0.074-0.177	1
	OfC2-KO	384	2.339 ± 0.286	0.414	0.259-0.585	3.6
Cry1Ac	NJ-S	720	3.248 ± 0.427	0.100	0.069-0.136	1
	OfC2-KO	384	3.531 ± 0.572	0.808	0.676-0.947	8.1
Cry1Fa [3]	NJ-S	408	4.488 ± 0.505	0.411	0.349-0.466	1
	OfC2-KO	48	-	-	-	>300
Abamectin	NJ-S	192	2.221 ± 0.227	0.118	0.090-0.153	1
	OfC2-KO	432	1.937 ± 0.171	0.153	0.122-0.188	1.3
Chlorantraniliprole	NJ-S	432	2.106 ± 0.217	0.031	0.025-0.037	1
	OfC2-KO	432	1.387 ± 0.137	0.018	0.013-0.023	0.6

[1] Numbers of larvae used in bioassay; [2] RR (resistance ratio) = LC$_{50}$ (OfC2-KO)/LC$_{50}$ (NJ-S); [3] LC$_{50}$ for OfC2-KO could not be determined because of an insufficient dose response (only 4% mortality at 120 µg/g of Cry1Fa treatment).

2.4. Dominance of Cry1Fa and Cry1Ac Resistance in the OfC2-KO Strain

To investigate the inheritance of different levels of resistance to Cry1Fa (high) and Cry1Ac (low) in the OfC2-KO strain, it was crossed with the susceptible NJ-S strain, and the responses of the two strains and their F_1 progeny were determined at a diagnostic concentration of Cry1Fa (2 µg/g) and Cry1Ac (1 µg/g), respectively (Table 2). For Cry1Fa, the F_{1a} and F_{1b} progeny had a high mortality (100% and 98.3%) at 2 µg/g, and the dominance parameters (h) were 0 and 0.02. Similarly, for Cry1Ac, the corresponding mortality was 100%, and both of the two h values were 0. The results indicated that either a high level of resistance to Cry1Fa or a low level of resistance to Cry1Ac in OfC2-KO strain was inherited as a recessive mode.

Table 2. Mortality and dominance of the susceptible NJ-S strain, OfC2-KO strain, and their F_1 progeny from reciprocal crosses to the diagnostic concentration of Cry1Fa and Cry1Ac, respectively.

Strain/cross	Treatment	N [1]	Survival Number	h [2]
NJ-S	Cry1Fa	72	0	
	Cry1Ac	48	0	
OfC2-KO	Cry1Fa	72	67	
	Cry1Ac	96	37	
F_{1a} (OfC2-KO♀×NJ-S♂)	Cry1Fa	120	0	0
	Cry1Ac	120	0	0
F_{1b} (OfC2-KO♂×NJ-S♀)	Cry1Fa	120	2	0.02
	Cry1Ac	120	0	0

[1] Numbers of larvae measured at the Cry1Fa (2 µg/g) or Cry1Ac (1 µg/g) diagnostic concentration; [2] The degree of dominance (h) = (survival of F_1 - survival of NJ-S)/(survival of OfC2-KO - survival of NJ-S). h = 0, completely recessive; h = 1, completely dominant.

2.5. Genetic Association between the 8-bp Deletion of OfABCC2 and Cry1Fa Resistance

To clarify the causal relationship of the 8-bp deletion in exon 4 of *OfABCC2* with high levels of Cry1Fa resistance, a set of genetic crosses was performed (Figure 3a). By using direct-sequencing analysis of the target PCR products (see typical chromatogram in Figure 3b), the genotype of 25 individuals from NJ-S were homozygous for the wild-type (*ss*) and that of 30 individuals from the OfC2-KO strain were homozygous for the 8-bp deletion of *OfABCC2* (*rr*) (Table 3). When treated with 2 µg/g of Cry1Fa in F_2 progeny, 22.6% (38/168) of the larvae survived after 7 days of treatment. All the detected 21 survivors randomly selected from the F_2-treated group were homozygous for the 8-bp deletion of *OfABCC2* (*rr*), and the F_2-untreated individuals were separated into wild-type homozygous (*ss*: 7), heterozygous (*rs*: 13), and 8-bp deletion homozygous (*rr*: 9). Our results clearly demonstrated that the 8-bp deletion of *OfABCC2* is significantly linked (Fisher's exact test, $p < 0.0001$) with Cry1Fa resistance in the manipulated OfC2-KO strain.

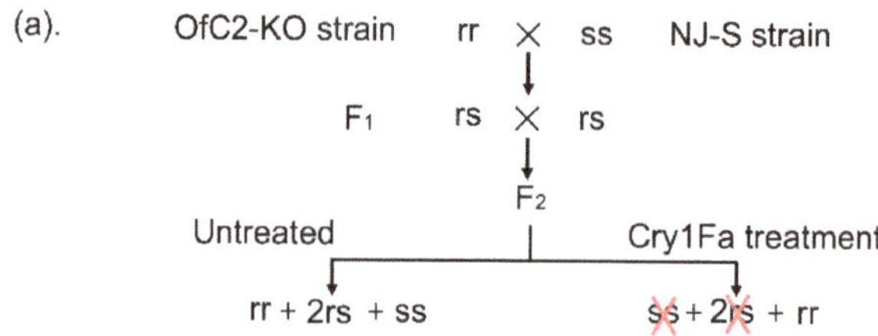

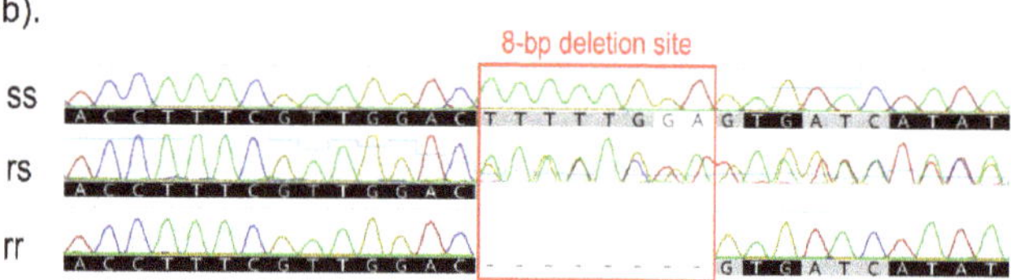

Figure 3. Linkage analysis of Cry1Fa resistance in the OfC2-KO strain of *O. furnacalis*. *OfABCC2* genotypes: *ss* = wild type; *rs* = heterozygous mutant; *rr* = homozygous mutant (8-bp deletion). (**a**) The crossing design used to generate F$_2$ progeny (1*ss*: 2*rs*:1*rr*). (**b**) The direct sequencing chromatograms of PCR products amplified from a fragment of gDNA flanking the 8-bp deletion site (red box) of *OfABCC2*.

Table 3. Genetic linkage between the 8-bp deletion of *OfABCC2* and resistance to Cry1Fa in *O. furnacalis*.

F$_2$ Progeny [1]	Number of Individuals for Each Genotype [2]		
	ss	*rs*	*rr*
NJ-S	25	0	0
OfC2-KO	0	0	30
F$_2$-untreated larvae (n = 29)	7	13	9
F$_2$-treated survivors (n = 21)	0	0	21

[1] F$_1$ progeny between the susceptible NJ-S and Cry1Fa-resistant *OfABCC2* strains were crossed to produce F$_2$ progeny. 168 larvae from the F$_2$ progeny were treated with 2 µg/g of Cry1Fa toxin. 21 of 38 survivors and 30 untreated larvae were genotyped individually by direct sequencing of the PCR products; [2] *ss* represent homozygous for the wild type *OfABCC2*, while *rs* means heterozygous for the 8-bp deletion allele of *OfABCC2*, and *rr* represent homozygous for the 8-bp deletion allele of *OfABCC2*.

3. Discussion

In the current study, we successfully induced a deletion mutation of 8-bp into the *OfABCC2* gene of *O. furnacalis* by the CRISPR/Cas9 genome editing system, and characterized Bt resistance properties of the knockout OfC2-KO strain. We found that the OfC2-KO strain acquired a high level of resistance to Cry1Fa (>300-fold) and low levels of resistance to Cry1Ab and Cry1Ac (< 10-fold). We also confirmed the genetic association between the 8-bp deletion of *OfABCC2* and the obtained resistance to Cry1Fa in the knockout strain. These findings provide strong evidence that OfABCC2 plays a major role in meditating the toxicity of Cry1Fa in *O. furnacalis*. Moreover, the cross-resistance and inheritance pattern results provide helpful information for designing of resistance management strategies for future adoption of Bt corn in China. Furthermore, the OfC2-KO strain can be employed in an F$_1$ screen program to investigate the diversity and frequency of the *OfABCC2* mutant alleles in field populations of *O. furnacalis*.

ABCC2 proteins have been identified as receptors for Bt toxins Cry1A and/or Cry1F in a number of lepidopteran insects, but they have differential contributions to the toxicities for individual Cry1 toxins. The CRISPR-mediated knockout of *P. xylostella ABCC2* conferred high levels of resistance to Cry1Aa, Cry1Ab, and Cry1Ac [30]. The double knockout of *ABCC2* and *ABCC3* of *H. armigera* confers a >15,000-fold resistance to Cry1Ac [33]. A point mutation in the *ABCC2* of *B. mori* resulted in high levels of resistance to Cry1Ab and Cry1Ac, but not to Cry1Aa [24]. The CRISPR-mediated knockout of

S. frugiperda ABCC2 conferred a 118-fold resistance to Cry1F [31], and the knockout of *S. exigua ABCC2* resulted in high levels of resistance to both Cry1Fa and Cry1Ac [32]. In our study, the knockout of *ABCC2* in *O. furnacalis* produced high-level resistance to Cry1Fa, low levels of resistance to Cry1Ab and Cry1Ac, and no resistance to Cry1Aa. The present study provides a new case for the investigation of the interaction between lepidopteran ABCC2 receptors and Bt Cry1 toxins.

A laboratory-selected strain of *O. nubilalis* developed a >1200-fold resistance to Cry1F, and the Cry1F resistance trait is controlled by a single quantitative trait locus (QTL) on linkage group 12 [34]. The subsequent fine mapping of the Cry1F resistance QTL identified a genomic region containing the *ABCC2* locus tightly linked with Cry1F resistance [35]. Practical resistance to Cry1F was recently documented in some field populations of Canadian *O. nubilalis* [4]. It will be interesting to check whether there are mutations in the *ABCC2* gene in both the laboratory-selected strain and field-derived resistant populations of *O. nubilalis*. The identification of the specific gene for Cry1F resistance of *O. nubilalis* is urgently needed for developing molecular tools to monitor the spreading of the practical resistance in the field.

Several studies reported the potential mechanisms of Cry1Ab and Cry1Ac resistance in the laboratory-selected strains of *O. furnacalis*, such as the up-regulation of the V-ATPase catalytic subunit A and heat shock 70 kDa proteins [36], the down-regulation and mutation of a cadherin gene [37], the differential expression of the miRNAs targeting potential Bt receptors [38], and the reduced expression of APN and ABC subfamily G transcripts [39]. The CRISPR-mediated knockout approach established for *O. furnacalis* in the present study can be used to evaluate the functional role of the candidate genes relating to Bt resistance.

The characterization of the inheritance of Bt resistance will provide important information for evaluating the risks of evolution of resistance and will make it possible to formulate effective resistance management strategies. Based on previous reports, resistance to Cry1-type toxins mediated by *ABCC2* mutations was recessive or incompletely recessive [22–24,27,28,30,32,33]. Consistent with these results, both the high-level resistance to Cry1Fa (>300-fold) and low-level resistance to Cry1Ac (~8-fold) were inherited as a recessive mode in the knockout OfC2-KO strain of *O. furnacalis*.

In the present work, the obtained Cry1Fa resistance was confirmed to be genetically associated with the 8-bp deletion of *OfABCC2*, which excludes the CRISPR-mediated off-target effects on resistance phenotype. We analyzed 18 research cases that employed the CRISPR/Cas9 system to manipulate the resistance genes to Bt toxins or insecticides, and found that only five of them performed linkage analysis between acquired resistance and the introduced mutation, including the knockout of the cadherin gene in *H. armigera* [40], nicotinic acetylcholine receptor α6 subunit in *P. xylostella* and *S. exigua* [41,42], the ryanodine receptor G4946E mutation in *Drosophila melanogaster* [43], and a *CYP9M10* gene in *Culex quinquefasciatus* [44]. We therefore recommend that when CRISPR-based gene editing is conducted to verify the function of a candidate gene, it is necessary to perform a genetic linkage analysis in order to clarify whether there are off-target effects.

4. Materials and Methods

4.1. Insect Strains and Rearing

The susceptible NJ-S strain was originally collected from Nanjing, China, in May 2010, and has been maintained in the laboratory without exposure to any insecticides or Bt toxins. By using the CRISPR/Cas9 genome-editing system, the *OfABCC2* gene in the background strain NJ-S was knocked out to construct a manipulated strain denoted as OfC2-KO. The genome-edited OfC2-KO strain is homozygous for the 8-bp deletion in exon 4 of *OfABCC2*, which was predicted to produce a truncated and loss-of-function protein.

The larvae of *O. furnacalis* were reared on an artificial diet with corn and soybean powder as major ingredients at 27 ± 1 °C, 80% relative humidity (RH), and a photoperiod of 16 h light:8 h dark. The pupae were transferred to mating cages with more than 80% RH and a photoperiod of 16:8 h (L: D).

Adults were supplied with 10% sugar solution to replenish energy. About 5–6 pieces of waxed papers, as substrate for oviposition, were placed on the top of the cage, and the bottom sheet was collected daily. Egg masses were incubated in plastic boxes lined with moistened filter paper until hatching.

4.2. Diet Bioassay

The activated Cry1Aa, Cry1Ab, Cry1Ac, and Cry1Fa toxins were purchased from Marianne Pusztai-Carey (Case Western Reserve University, Cleveland, OH, USA). Abamectin (20 g/L EC) was obtained from the Institute of Plant Protection, Guangdong Academy of Agricultural Sciences, Guangzhou, Guangdong Province, China. Chlorantraniliprole (200 g/L SC) was purchased from DuPont Agricultural Chemicals Ltd. (Wilmington, DE, USA).

We used the diet incorporation method to evaluate the toxicity of Cry toxin or chemical insecticide to *O. furnacalis*. Briefly, 5 to 7 concentrations of Bt or insecticide test solutions were first diluted in distilled water. Then, we added the solution (or distilled water for control) to a proper amount diet in a clean mixing bowl and thoroughly mixed all the ingredients together until a soft, smooth dough was obtained. Next, we dispensed the toxin-incorporated diet into each well of a 24-well plate. After the diet cooled and solidified, one *O. furnacalis* larva (neonate for Cry toxin susceptibility test and 2nd instar larva for chemical insecticide bioassay) was placed in each well. All the plates were kept at an illumination incubator set at 27 ± 1 °C, 80% RH, and a photoperiod of 16:8 h (L:D). For Cry toxin, the mortality was recorded after 7 days of treatment, and the larvae were considered dead if they died or weighed less than 5 mg. For abamectin and chlorantraniliprole, the mortality was recorded after 2 days of treatment, and the larvae were considered dead if they did not move after gentle prodding with a brush. The data were analyzed with PoloPlus (LeOra Software) [45] to estimate the LC_{50} with 95% fiducial limits (FL), as well as the slopes of the concentration–mortality lines. Resistance ratios (RR) were calculated by dividing the LC_{50} for a particular strain by the LC_{50} for the susceptible NJ-S strain.

4.3. Design and Preparation of sgRNA

In our previous work, the full-length sequences of *OfABCC2* mRNA (GenBank no. MN783372) had been obtained from the susceptible NJ-S strain of *O. furnacalis*. By scanning the $GN_{19}NGG$ motifs, we identified a sgRNA target site (5′-GCACCTTTCGTTGGACTTTT<u>TGG</u>-3′) in predicted exon 4 of *OfABCC2* (Figure 2a). A PCR-based approach was employed to prepare sgRNA according to the instructions [46]. In brief, a forward oligonucleotide encoding a T7 polymerase-binding site and the sgRNA target sequences GN_{19} (OfC2_sgF, Table 4) and a universal oligonucleotide encoding the remaining sgRNA sequences (OfC2_sgR, Table 4) were designed at first. The OfC2_sgF and OfC2_sgR were fused by PCR to generate a sgRNA DNA template. The PCR reaction mixture (50 μL) contained 10 μL of 5 × PCR buffer, 4 μL of 2.5 mM dNTP, 4 μL of 10 μM OfC2_sgF, 4 μL of 10 μM OfC2_sgR, 0.5 μL of PrimeSTAR polymerase (TaKaRa, Dalian, China), and 27.5 μL of Nuclease-free water. PCR was performed at 98 °C for 30 s, 30 cycles (98 °C 5 s, 60 °C 30 s, 72 °C 15 s), 72 °C for 10 min, and holding at 4 °C. After identification by electrophoresis, the PCR products were purified by a QIAprep® Spin Miniprep Kit (QIAGEN, Hilden, Germany). A MEGAshortscript™ T7 High Yield Transcription Kit (Ambion, Foster City, CA, USA) was used for sgRNA in vitro transcription according to the manufacturer's protocol.

Table 4. Primers used in this study.

Primer Name	Primer Sequences (5′ to 3′)
OfC2_sgF	GAAATTAATACGACTCACTATAGCACCTTTCGTTGGACTTTTGTTTTAGAGCTAGAAATAGC
OfC2_sgR	AAAAGCACCGACTCGGTGCCACTTTTTCAAGTTGATAACGGACTAGCCTTATTTTAACTTGCTATTTCTAGCTCTAAAAC
4Ex_F	TAAACCAAGTGTCCATAGGAGACG
5Ex_R	TTCGTTTGTCTGTTCGTGTCGC
4In_R	GCTGACTATGACATCCACAAAGACAA

4.4. Egg Collection and Microinjection

Mated female moths of *O. furnacalis* were allowed to lay egg masses on pieces of wax paper previously placed on the top of the mating cage. Fresh egg masses (within 2 h after oviposition) were immediately collected by cutting the wax paper. Then, the eggs were lined on double-sided adhesive tape on a microscope slide. About 1 nL of mixture containing 150 ng/μL of Cas9 protein (GeneArt™ Platinum™ Cas9 Nuclease, Thermo Fisher Scientific, Shanghai, China) and 300 ng/μL of sgRNA were injected into each egg using a FemtoJet and InjectMan NI 2 microinjection system (Eppendorf, Hamburg, Germany). After injection, the embryos were incubated in an incubator as described above. The hatched larvae were fed an artificial diet without any toxin.

4.5. Generation of OfABCC2 Mutation Mediated by CRISPR/Cas9

After embryo injection, the genomic DNAs of exuviate from 90 5th-instar larva were isolated individually using an AxyPrep Multisource Genomic DNA Miniprep Kit (Axygen, Hangzhou, China) following the manufacturer's instruction. To identify the indel mutations at predicted exon 4 of *OfABCC2*, the intron 4 sequences was first amplified by a specific pair of primers (4Ex_F and 5Ex_R, Table 4) and then by using the primers of 4Ex_F and 4In_R (Table 4) to amplify a 280-bp region flanking the desired cleavage site. The second PCR reaction mixture contained 1 μL of template, 1 μL of each of the 4Ex_F or 4In_R primer, 12.5 μL of 2× Taq Master Mix (TaKaRa, Dalian, China), and 9.5 μL of PCR-grade water in a final volume of 25 μL. PCR was performed at 95 °C 3 min, 35 cycles (95 °C 30 s, 55°C 30 s, 72 °C 1 min), 72 °C for 10 min, and 4 °C forever, and then the PCR products were directly sequenced with 4Ex_F (sequencing primer) by TSINGKE Biological Technology (Nanjing, China). Direct sequencing chromatograms of mutant *OfABCC2* have double peaks around the cutting site at G_0 generation. To detect the detailed deletion information of G_2 genomic DNAs, the 280-bp PCR products were subcloned into pTOPO-T vector (Aidlab Biotechnologies, Beijing, China) and sequenced by TSINGKE Biological Technology. The acquired 8-bp deletion in *OfABCC2* was reconfirmed by clone sequencing using genomic DNA and mRNA templates from the knockout strain OfC2-KO.

4.6. Inheritance Model Determination and Genetic Association Analysis

The sex of *O. furnacalis* was visually determined based on the bottom characters of the pupa. Male adults (30 moths) from the original strain NJ-S were mass crossed with virgin female adults (30 moths) of the knockout strain OfC2-KO and vice versa. The degree of dominance (*h*) was estimated using the formula: $h = (Srs − Sss)/(Srr − Sss)$, where Srs, Sss, and Srr are the survival rate for F_1 hybrids, the NJ-S strain, and the OfC2-KO strains, respectively. The *h* varies from 0 for completely recessive resistance to 1 for completely dominant resistance [47].

For genetic association analysis between the 8-bp deletion of *OfABCC2* and Cry1Fa resistance phenotype, the F_1 progeny from the reciprocal crosses were pooled and mass crossed to produce F_2 progeny (Figure 3a). A total of 168 newly hatched larvae of the F_2 progeny were treated with 2 μg/g of Cry1Fa toxin. The survivors (F_2-treated) were collected after 7 days of treatment. The DNA from random selected parents (NJ-S and OfC2-KO), F_2-treated survivors, and F_2-untreated individuals were extracted for *OfABCC2* genotyping.

Author Contributions: Conceptualization, Y.W. and Y.Y.; methodology, X.W., Y.Y. and Y.W.; Investigation, X.W., Y.X., J.H. and W.J.; visualization, X.W., Y.X. and Y.Y. Funding acquisition, Y.W.; writing-original draft preparation, X.W.; writing-review and editing, Y.W. All authors have read and agreed to the published version of the manuscript.

Funding: This work was supported by grants from National Science and Technology Major Project of China (2019ZX08012004-005) and Fundamental Research Funds for the Central Universities of China (KYT201803).

References

1. ISAAA. *Global Status of Commercialized Biotech/GM Crops in 2018, ISAAA Brief no. 54*; ISAAA: Ithaca, NY, USA, 2018.
2. NASEM. Genetically Engineered Crops: Experiences and Prospects. In *National Academies of Sciences, Engineering, and Medicine*; National Academies Press: Washington, DC, USA, 2016.
3. Calles-Torrez, V.; Knodel, J.J.; Boetel, M.A.; French, B.W.; Fuller, B.W.; Ransom, J.K. Field-evolved resistance of northern and western corn rootworm (Coleoptera: Chrysomelidae) populations to corn hybrids expressing single and pyramided Cry3Bb1 and Cry34/35Ab1 Bt proteins in North Dakota. *J. Econ. Entomol.* **2019**, *112*, 1875–1886. [CrossRef] [PubMed]
4. Smith, J.L.; Farhan, Y.; Schaafsma, A.W. Practical resistance of *Ostrinia nubilalis* (Lepidoptera: Crambidae) to Cry1F *Bacillus thuringiensis* maize discovered in Nova Scotia, Canada. *Sci. Rep.* **2019**, *9*, 18247. [CrossRef] [PubMed]
5. Tabashnik, B.E.; Carrière, Y. Global patterns of resistance to Bt crops highlighting pink bollworm in the United States, China, and India. *J. Econ. Entomol.* **2019**, *112*, 2513–2523. [CrossRef] [PubMed]
6. Mutuura, A.; Munroe, E. Taxonomy and distribution of the European corn borer and allied species: Genus *ostrinia* (lepidoptera: Pyralidae). *Mem. Entomol. Soc. Can.* **1970**, *102*, 1–112. [CrossRef]
7. Frolov, A.; Bourguet, D.; Ponsard, S. Reconsidering the taxomony of several *Ostrinia* species in the light of reproductive isolation: a tale for Ernst Mayr. *Biol. J. Lin. Soc.* **2007**, *91*, 49–72. [CrossRef]
8. Nafus, D.M.; Schreiner, I.H. Review of the biology and control of the Asian corn borer, *Ostrinia furnacalis* (Lep, Pyralidae). *Trop. Pest Manag.* **1991**, *37*, 41–56. [CrossRef]
9. Hutchison, W.D.; Burkness, E.C.; Mitchell, P.D.; Moon, R.D.; Leslie, T.W.; Fleischer, S.J.; Abrahamson, M.; Hamilton, K.L.; Steffey, K.L.; Gray, M.E.; et al. Areawide suppression of European corn borer with Bt maize reaps savings to non-Bt maize growers. *Science* **2010**, *330*, 222–225. [CrossRef]
10. Dively, G.P.; Venugopal, P.D.; Bean, D.; Whalen, J.; Holmstrom, K.; Kuhar, T.P.; Doughty, H.B.; Patton, T.; Cissel, W.; Hutchison, W.D. Regional pest suppression associated with widespread Bt maize adoption benefits vegetable growers. *Proc. Natl. Acad. Sci. USA* **2018**, *115*, 3320–3325. [CrossRef]
11. Siegfried, B.D.; Rangasamy, M.; Wang, H.; Spencer, T.; Haridas, C.V.; Tenhumberg, B.; Sumerford, D.V.; Storer, N.P. Estimating the frequency of Cry1F resistance in field populations of the European corn borer (Lepidoptera: Crambidae). *Pest Manag. Sci.* **2014**, *70*, 725–733. [CrossRef]
12. CSY. *China Statistical Yearbook*; National Bureau of Statistics of China, China Statistics Press: Beijing, China, 2018.
13. Zhou, D.R.; Wang, Y.S.; Li, W.D. Studies on the identification of the dominant corn borer species in China. *Acta. Phytophy. Sin.* **1988**, *15*, 145–152.
14. MARA. *Approved List of Agricultural GMO Safety Certificates*; Ministry of Agriculture and Rural Affairs of the People's Republic of China: Beijing, China, 2020.
15. Xu, L.; Wang, Z.; Zhang, J.; He, K.; Ferry, N.; Gatehouse, A.M.R. Cross-resistance of Cry1Ab-selected Asian corn borer to other Cry toxins. *J. Appl. Entomol.* **2010**, *134*, 429–438. [CrossRef]
16. Zhang, T.; He, M.; Gatehouse, A.M.; Wang, Z.; Edwards, M.G.; Li, Q.; He, K. Inheritance patterns, dominance and cross-resistance of Cry1Ab- and Cry1Ac-selected *Ostrinia furnacalis* (Guenée). *Toxins* **2014**, *6*, 2694–2707. [CrossRef] [PubMed]
17. Wang, Y.; Wang, Y.; Wang, Z.; Bravo, A.; Soberón, M.; He, K. Genetic basis of Cry1F resistance in a laboratory selected Asian corn borer strain and its cross-resistance to other *Bacillus thuringiensis* toxins. *PLoS ONE* **2016**, *11*, e0161189. [CrossRef] [PubMed]
18. Shabbir, M.Z.; Quan, Y.; Wang, Z.; Bravo, A.; Soberón, M.; He, K. Characterization of the Cry1Ah resistance in Asian corn borer and its cross-resistance to other *Bacillus thuringiensis* toxins. *Sci. Rep.* **2018**, *8*, 234. [CrossRef]
19. Wang, Y.Q.; Quan, Y.D.; Yang, J.; Shu, C.L.; Wang, Z.Y.; Zhang, J.; Gatehouse, A.M.R.; Tabashnik, B.E.; He, K.L. Evolution of Asian corn borer resistance to Bt toxins used singly or in pairs. *Toxins* **2019**, *11*, 461. [CrossRef]
20. Wu, Y.D. Detection and mechanisms of resistance evolved in insects to Cry toxins from *Bacillus thuringiensis*. *Adv. Insect Physiol.* **2014**, *47*, 297–342.
21. Pardo-López, L.; Soberón, M.; Bravo, A. *Bacillus thuringiensis* insecticidal three-domain Cry toxins: mode of action, insect resistance and consequences for crop protection. *FEMS Microbiol. Rev.* **2013**, *37*, 3–22. [CrossRef]

22. Gahan, L.J.; Pauchet, Y.; Vogel, H.; Heckel, D.G. An ABC transporter mutation is correlated with insect resistance to *Bacillus thuringiensis* Cry1Ac toxin. *PLoS Genet.* **2010**, *6*, e1001248. [CrossRef]

23. Baxter, S.W.; Badenes-Perez, F.R.; Morrison, A.; Vogel, H.; Crickmore, N.; Kain, W.; Wang, P.; Heckel, D.G.; Jiggins, C.D. Parallel evolution of *Bacillus thuringiensis* toxin resistance in lepidoptera. *Genetics* **2011**, *189*, 675–679. [CrossRef]

24. Atsumi, S.; Miyamoto, K.; Yamamoto, K.; Narukawa, J.; Kawai, S.; Sezutsu, H.; Kobayashi, I.; Uchino, K.; Tamura, T.; Mita, K.; et al. Single amino acid mutation in an ATP-binding cassette transporter gene causes resistance to Bt toxin Cry1Ab in the silkworm, *Bombyx mori. Proc. Natl. Acad. Sci. USA* **2012**, *109*, 1591–1598. [CrossRef]

25. Xiao, Y.; Zhang, T.; Liu, C.; Heckel, D.G.; Li, X.; Tabashnik, B.E.; Wu, K. Mis-splicing of the *ABCC2* gene linked with Bt toxin resistance in *Helicoverpa armigera. Sci. Rep.* **2014**, *4*, 6184. [CrossRef] [PubMed]

26. Park, Y.; Gonzalez-Martinez, R.M.; Navarro-Cerrillo, G.; Chakroun, M.; Kim, Y.; Ziarsolo, P.; Blanca, J.; Canizares, J.; Ferre, J.; Herrero, S. ABCC transporters mediate insect resistance to multiple Bt toxins revealed by bulk segregant analysis. *BMC Biol.* **2014**, *12*, 46. [CrossRef] [PubMed]

27. Banerjee, R.; Hasler, J.; Meagher, R.; Nagoshi, R.; Hietala, L.; Huang, F.; Narva, K.; Jurat-Fuentes, J.L. Mechanism and DNA-based detection of field-evolved resistance to transgenic Bt corn in fall armyworm (*Spodoptera frugiperda*). *Sci. Rep.* **2017**, *7*, 10877. [CrossRef]

28. Flagel, L.; Lee, Y.W.; Wanjugi, H.; Swarup, S.; Brown, A.; Wang, J.L.; Kraft, E.; Greenplate, J.; Simmons, J.; Adams, N.; et al. Mutational disruption of the *ABCC2* gene in fall armyworm, *Spodoptera frugiperda*, confers resistance to the Cry1Fa and Cry1A.105 insecticidal proteins. *Sci. Rep.* **2018**, *8*, 7255. [CrossRef] [PubMed]

29. Boaventura, D.; Ulrich, J.; Lueke, B.; Bolzan, A.; Okuma, D.; Gutbrod, O.; Geibel, S.; Zeng, Q.; Dourado, P.M.; Martinelli, S.; et al. Molecular characterization of Cry1F resistance in fall armyworm, *Spodoptera frugiperda* from Brazil. *Insect Biochem. Mol. Biol.* **2020**, *116*, 103280. [CrossRef] [PubMed]

30. Guo, Z.J.; Sun, D.; Kang, S.; Zhou, J.L.; Gong, L.J.; Qin, J.Y.; Guo, L.; Zhu, L.H.; Bai, Y.; Luo, L.; et al. CRISPR/Cas9-mediated knockout of both the *PxABCC2* and *PxABCC3* genes confers high-level resistance to *Bacillus thuringiensis* Cry1Ac toxin in the diamondback moth, *Plutella xylostella* (L.). *Insect Biochem. Mol. Biol.* **2019**, *107*, 31–38. [CrossRef]

31. Jin, M.; Tao, J.; Li, Q.; Cheng, Y.; Sun, X.; Wu, K.; Xiao, Y. Genome editing of the *SfABCC2* gene confers resistance to Cry1F toxin from *Bacillus thuringiensis* in *Spodoptera frugiperda. J. Integr. Agr.* **2020**. [CrossRef]

32. Huang, J.L.; Xu, Y.J.; Zuo, Y.Y.; Yang, Y.H.; Tabashnik, B.E.; Wu, Y.D. Evaluation of five candidate Bt toxin receptors in the beet armyworm using CRISPR mediated gene knockouts. *Insect Biochem. Mol. Biol.* **2020**, *121*, 103361. [CrossRef]

33. Wang, J.; Ma, H.H.; Zhao, S.; Huang, J.L.; Yang, Y.H.; Tabashnik, B.E.; Wu, Y.D. Functional redundancy of two transporter proteins in mediating toxicity of *Bacillus thuringiensis* to cotton bollworm. *PLoS Pathog.* **2020**, *16*, e1008427. [CrossRef]

34. Coates, B.S.; Sumerford, D.V.; Lopez, M.D.; Wang, H.; Fraser, L.; Kroemer, M.J.A.; Spencer, T.; Kim, K.S.; Abel, C.A.; Hellmich, R.L.; et al. A single major QTL controls expression of larval Cry1F resistance trait in *Ostrinia nubilalis* (Lepidoptera: Crambidae) and is independent of midgut receptor genes. *Genetica* **2011**, *139*, 961. [CrossRef]

35. Coates, B.S.; Siegfried, B.D. Linkage of an ABCC transporter to a single QTL that controls *Ostrinia nubilalis* larval resistance to the *Bacillus thuringiensis* Cry1Fa toxin. *Insect Biochem. Mol. Biol.* **2015**, *63*, 86–96. [CrossRef] [PubMed]

36. Xu, L.N.; Ferry, N.; Wang, Z.Y.; Zhang, J.; Edwards, M.G.; Gatehouse, A.M.R.; He, K.L. A proteomic approach to study the mechanism of tolerance to Bt toxins in *Ostrinia furnacalis* larvae selected for resistance to Cry1Ab. *Transgenic. Res.* **2013**, *22*, 1155–1166. [CrossRef] [PubMed]

37. Jin, T.; Chang, X.; Gatehouse, A.M.; Wang, Z.; Edwards, M.G.; He, K. Downregulation and mutation of a cadherin gene associated with Cry1Ac resistance in the Asian corn borer, *Ostrinia furnacal* is (Guenée). *Toxins* **2014**, *6*, 2676–2693. [CrossRef] [PubMed]

38. Xu, L.N.; Ling, Y.H.; Wang, Y.Q.; Wang, Z.Y.; Hu, B.J.; Zhou, Z.Y.; Hu, F.; He, K.L. Identification of differentially expressed microRNAs between *Bacillus thuringiensis* Cry1Ab-resistant and -susceptible strains of *Ostrinia furnacalis. Sci. Rep.* **2015**, *5*, 15461. [CrossRef]

39. Zhang, T.T.; Coates, B.S.; Wang, Y.Q.; Wang, Y.D.; Bai, S.X.; Wang, Z.Y.; He, K.L. Down-regulation of aminopeptidase N and ABC transporter subfamily G transcripts in Cry1Ab and Cry1Ac resistant Asian corn borer, *Ostrinia furnacalis* (Lepidoptera: Crambidae). *Int. J. Biol. Sci.* **2017**, *13*, 835–851. [CrossRef]

40. Wang, J.; Zhang, H.; Wang, H.D.; Zhao, S.; Zuo, Y.; Yang, Y.H.; Wu, Y.D. Functional validation of cadherin as a receptor of Bt toxin Cry1Ac in *Helicoverpa armigera* utilizing the CRISPR/Cas9 system. *Insect Biochem. Mol. Biol.* **2016**, *76*, 11–17. [CrossRef]

41. Wang, X.L.; Ma, Y.M.; Wang, F.L.; Yang, Y.H.; Wu, S.W.; Wu, Y.D. Disruption of nicotinic acetylcholine receptor alpha 6 mediated by CRISPR/Cas9 confers resistance to spinosyns in *Plutella xylostella*. *Pest Manag. Sci.* **2020**, *76*, 1618–1625. [CrossRef]

42. Zuo, Y.; Xue, Y.; Lu, W.; Ma, H.; Chen, M.; Wu, Y.; Yang, Y.; Hu, Z. Functional validation of nicotinic acetylcholine receptor (nAChR) alpha6 as a target of spinosyns in *Spodoptera exigua* utilizing the CRISPR/Cas9 system. *Pest Manag. Sci.* **2020**. [CrossRef]

43. Douris, V.; Papapostolou, K.M.; Ilias, A.; Roditakis, E.; Kounadi, S.; Riga, M.; Nauen, R.; Vontas, J. Investigation of the contribution of RyR target-site mutations in diamide resistance by CRISPR/Cas9 genome modification in *Drosophila*. *Insect Biochem. Mol. Biol.* **2017**, *87*, 127–135. [CrossRef]

44. Itokawa, K.; Komagata, O.; Kasai, S.; Ogawa, K.; Tomita, T. Testing the causality between CYP9M10 and pyrethroid resistance using the TALEN and CRISPR/Cas9 technologies. *Sci. Rep.* **2016**, *6*, 24652. [CrossRef]

45. LeOra Software. *Polo Plus: A User's Guide to Probit and Logit Analysis*; LeOra Software: Berkeley, CA, USA, 2002.

46. Bassett, A.R.; Tibbit, C.; Ponting, C.P.; Liu, J.L. Highly efficient targeted mutagenesis of *Drosophila* with the CRISPR/Cas9 System. *Cell Rep.* **2013**, *4*, 220–228. [CrossRef] [PubMed]

47. Liu, Y.B.; Tabashnik, B.E. Inheritance of resistance to the *Bacillus thuringiensis* toxin Cry1C in the diamondback moth. *Appl. Environ. Microb.* **1997**, *63*, 2218–2223. [CrossRef]

 *toxins*

Article

ATP-Binding Cassette Subfamily a Member 2 Is a Functional Receptor for *Bacillus thuringiensis* Cry2A Toxins in *Bombyx mori*, But Not for Cry1A, Cry1C, Cry1D, Cry1F, or Cry9A Toxins

Xiaoyi Li [1], Kazuhisa Miyamoto [2], Yoko Takasu [2], Sanae Wada [2], Tetsuya Iizuka [2], Satomi Adegawa [1], Ryoichi Sato [1,*] and Kenji Watanabe [2,*]

[1] Graduate School of Bio-Applications and Systems Engineering, Tokyo University of Agriculture and Technology, Naka 2-24-16, Koganei, Tokyo 184-8588, Japan; s159217x@st.go.tuat.ac.jp (X.L.); s174092v@st.go.tuat.ac.jp (S.A.)

[2] Institute of Agrobiological Sciences, NARO, 1-2 Ohwashi, Tsukuba, Ibaraki 305-8634, Japan; miyamoto401606@gmail.com (K.M.); takasu@affrc.go.jp (Y.T.); sunny@affrc.go.jp (S.W.); tiizuka@affrc.go.jp (T.I.)

* Correspondence: ryoichi@cc.tuat.ac.jp (R.S.); nabek@affrc.go.jp (K.W.); Tel.: +81-42-388-7277 (R.S.); +81-29-838-6081 (K.W.)

Received: 29 December 2019; Accepted: 3 February 2020; Published: 6 February 2020

Abstract: Cry toxins are insecticidal proteins produced by *Bacillus thuringiensis* (Bt). They are used commercially to control insect pests since they are very active in specific insects and are harmless to the environment and human health. The gene encoding ATP-binding cassette subfamily A member 2 (ABCA2) was identified in an analysis of Cry2A toxin resistance genes. However, we do not have direct evidence for the role of ABCA2 for Cry2A toxins or why Cry2A toxin resistance does not cross to other Cry toxins. Therefore, we performed two experiments. First, we edited the *ABCA2* sequence in *Bombyx mori* using transcription activator-like effector-nucleases (TALENs) and confirmed the susceptibility-determining ability in a diet overlay bioassay. Strains with C-terminal half-deleted BmABCA2 showed strong and specific resistance to Cry2A toxins; even strains carrying a deletion of 1 to 3 amino acids showed resistance. However, the C-terminal half-deleted strains did not show cross-resistance to other toxins. Second, we conducted a cell swelling assay and confirmed the specific ability of BmABCA2 to Cry2A toxins in HEK239T cells. Those demonstrated that BmABCA2 is a functional receptor for Cry2A toxins and that BmABCA2 deficiency-dependent Cry2A resistance does not confer cross-resistance to Cry1A, Cry1Ca, Cry1Da, Cry1Fa or Cry9Aa toxins.

Keywords: Cry2Ab toxin; *Bombyx mori*; ATP-binding cassette subfamily a member 2 (ABCA2); genome editing; transcription activator-like effector-nucleases (TALENs); HEK293T cell; functional receptor

Key Contribution: To know Cry2A-susceptibility determining role of BmABCA2, *B. mori* strains were created by genome editing using TALEN. Larvae of genome edited strains showed high level resistance to both Cry2Aa and Cry2Ab but not to Cry1A, Cry1Ca, Cry1Da, Cry1Fa, and Cry9Aa, indicating that BmABCA2 is a susceptibility determinant specific to Cry2A toxins. The heterologous expressing of BmABCA2 in HEK293T cell showed high susceptibility to Cry2Ab toxin, but not to Cry1A and Cry9Aa toxins, suggesting that BmABCA2 is a functional receptor specific to Cry2A toxins.

1. Introduction

Cry toxins are insecticidal crystal proteins and pore-forming toxins produced by *Bacillus thuringiensis* (Bt) [1,2]. Proteases activate Cry toxins in the midgut of the host insect; the toxins then interact with specific receptors on the columnar cell membrane [3]. This interaction drives the toxins to insert partial structures into the membrane, forming ion channels [4]. A cation influx triggers the influx of water [5], resulting in cell swelling and lysis [2,6]. Given their strong toxicity in specific species and inability to harm the environment and human health, Cry toxins are used widely in pest control [7]. As a gene source, Bt toxin genes have been used efficiently to make transgenic crops (Bt crops) that resist pests [8]. However, resistant insect strains have been found in these crops [9].

To delay the selection and evolution of resistance in exposed insect populations, current commercial insecticidal systems combine two or more Cry toxins that bind to different receptors in the target pests [10]. Nevertheless, in traditional insecticidal systems that use a single Cry toxin, the generation of resistant insects and cross-resistance to other Cry toxins are still problems, reducing the value of commercial Bt crops. Regarding cross-resistance, in *Ostrinia nubilalis* and *Spodoptera frugiperda*, Cry1Ab, Cry1Ac, and Cry1Fa compete for the same binding sites with high affinity; however, Cry2Ab does not compete for the binding sites of Cry1 proteins [10]. Heterologous competition binding assays in *Helicoverpa armigera* and *Helicoverpa zea* midguts showed a common binding site for three Cry2A toxins (Cry2Aa, Cry2Ab, and Cry2Ae), but this binding site was not shared with Cry1Ac [11]. Cry 1Ab-resistant *Pectinophora gossypiella* was also reported to have strong cross-resistance to Cry1Aa, but little or no cross-resistance to Cry1Ca, Cry1Da, Cry2Aa, and Cry9Aa [12]. In the diamondback moth (*Plutella xylostella*), the Cry1C-resistant strain had strong cross-resistance to Cry1Ab, Cry1Ac, and Cry1F, but low cross-resistance to Cry1Aa and no cross-resistance to Cry2Aa [13]. Therefore, it is necessary to clarify the reason for cross-resistance or the receptors used by each Cry toxin to devise new strategies to defeat cross-resistance.

ATP-binding cassette (ABC) transporters, a class of transmembrane proteins, are found widely in organisms and are involved in Bt toxin activities [14]. Many studies seeking to identify functional receptors of target Cry toxins have focused on ABC transporters [15]. When BmABCC2 was expressed in *S. frugiperda* (Sf9) cells and *Drosophila* tissues that were originally insensitive, they became sensitive to Cry1 toxins, and when BmABCC2 was expressed in *Xenopus* oocytes the Cry toxins made pores in the membrane and cations flowed into the cells through these pores [16–18]. These results indicate that BmABCC2 is a receptor for Cry1A toxins. Using heterologous expression in Sf9 cells, ABCB1 was found to be a functional Cry3Aa receptor [19]. In comparison, an *H. armigera* strain with 6000-fold resistance to Cry2Ab [15] had a mutation in ABC subfamily A member 2 (HaABCA2), suggesting that ABCA2 is linked to Cry2Ab resistance [20]. This was confirmed using CRISPR/Cas9-mediated genome editing, leading to the conclusion that HaABCA2 determines the susceptibility of *H. armigera* to Cry2Aa and Cry2Ab [21]. The knockout of ABCA2 using CRISPR/Cas9 conferred resistance to Cry2Ab on *Trichoplusia ni* [22]. Furthermore, in Cry2Ab-resistant *P. gossypiella* strains, PgABCA2 was disrupted in several ways, indicating that the Cry2Ab-resistance of *P. gossypiella* is associated with an ABCA2 deficiency [23]. Therefore, ABCA2s seem to cause Cry2 resistance in most insects. In addition, since ABCC2 and ABCB1 function as receptors for Cry1A and Cry3A toxins, respectively, ABCA2s likely function as receptors for Cry2A toxins. However, to confirm whether ABCAs are really functional receptors for Cry2 toxins, it is necessary to show the receptor function of ABCAs using a heterologous expression system.

In this study, we used transcription activator-like effector nucleases (TALEN) to edit the genome at the *BmABCA2* locus and created mutant strains of *Bombyx mori* as a model system to clarify susceptibility determining activity. Then, we performed Cry2A toxin-contaminated leaf disc feeding assays using these mutant strains to determine whether BmABCA2 is really a susceptibility determinant for Cry2A toxins in *B. mori*. Using HEK293T cells expressing BmABCA2, we conducted cell swelling assays to demonstrate the Cry2A toxin-specific receptor function of BmABCA2.

2. Results

2.1. Creation of Silkworm Strains with C-Terminal Half-Deleted BmABCA2s and Mutants with C-Terminal Deletions in TM7 by TALEN-Mediated Mutagenesis

First, 200 eggs were injected individually with TALEN mRNAs and a donor oligonucleotide prepared to mutate the C-terminus of transmembrane domain 7 (TM7) in exon 15 (Figure 1A,B). The 14 neonates that hatched were allowed to develop (G_0). Two female moths survived and were crossed singly with the wild-type strain to produce the next generation (G_1). The G_1 larvae were allowed to develop into adults. Genomic DNA samples of individual G_1 moths were prepared, and insertion/deletion mutations at TM7 in exon 15 of *BmABCA2* were identified. One mutant allele (named A2T01) was presumably derived from homology-directed repair (HDR), and 12 mutant alleles (named A2T03–A2T14) obtained via non-homologous end joining (NHEJ) were detected (Figure 1C). The mutations in A2T01, A2T05, A2T06, A2T09, A2T10, A2T12, A2T13, and A2T14 led to truncation of the C-terminus of TM7 in BmABCA2. The mutations in A2T03, A2T04, A2T07, A2T08, and A2T11 were predicted to cause 3, 4, 1, 1, and 2 amino-acid deletions from the C-terminus of TM7, respectively.

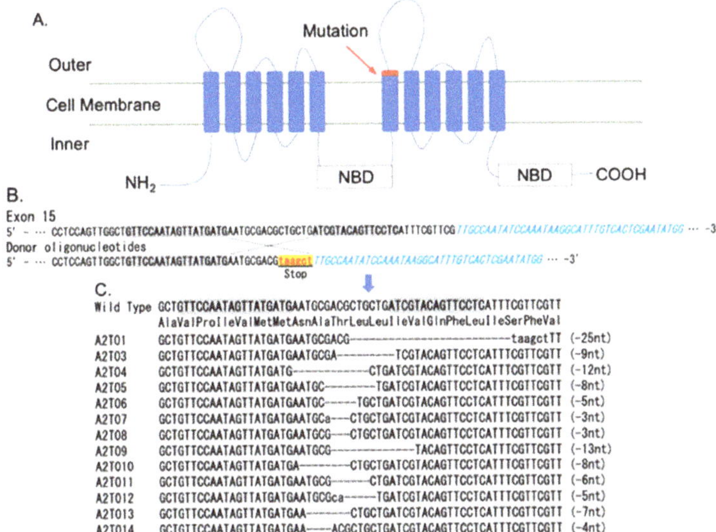

Figure 1. TALEN-induced mutations generated at the C-terminus of TM7 in BmABCA2. (**A**) Schematic structure of BmABCA2 and the mutation sites created with the TALEN system in the wild-type (Ringetsu) *B. mori* strain. The transmembrane topology of BmABCA2 was predicted by Phobius (http://phobius.sbc.su.se/). The approximate position of the mutation is indicated by the red rectangle located at the C-terminus of TM7 in BmABCA2. (**B**) The TALEN-binding site to TM7 in exon 15 used to introduce an insertion/deletion mutation in BmABCA2 and donor oligonucleotides for HDR. (**C**) The detected G_1 mutant alleles of the mutant strains. A2T01 was obtained as a result of HDR, while A2T03–A2T14 were obtained as the result of NHEJ.

To construct homozygous strains, heterozygous G_1 strains were selected after confirming the PCR products with direct sequencing and were mated with individuals of the same genotype. The homozygous individuals of the next generation (G_2) were screened by genotyping, resulting in the establishment of three strains with C-terminal half-deleted BmABCA2 (A2T01, A2T06, and A2T14) and three strains (A2T03, A2T08, and A2T11) with amino acids deleted from the C-terminus of TM7 (Figure S1).

2.2. BmABCA2 Activity against Cry2A Toxins

To test the susceptibility of the homozygous BmABCA2 mutant strains to Cry2Ab, larvae were reared on leaf disks contaminated with Cry2Ab for 2 days, and then on leaf disks without toxin for an additional 2 days. Larvae of all of the strains with C-terminal half-deleted BmABCA2s (A2T01, A2T06, and A2T14; Figure 1) showed strong resistance to Cry2Ab (Figure 2A). The median lethal dose (LC$_{50}$) of Cry2Ab on A2T14 was >9990-fold higher than that of the wild-type strain (Table 1). Surprisingly, the larvae of strains with BmABCA2s in which 1–3 amino acids had been deleted from the C-terminus of TM7 (A2T03, A2T11, and A2T08; Figure S1), were also resistant to Cry2Ab (Figure 2B,C). However, in association with the Cry2Ab concentration, those strains were slightly susceptible to Cry2Ab, i.e., the resistance of those strains was not as high as that of strains with C-terminal half-deleted BmABCA2s.

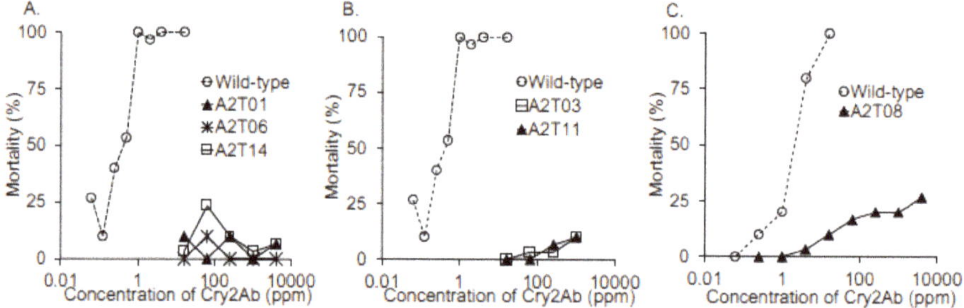

Figure 2. Toxin feeding assay to evaluate the susceptibility of the larvae with mutations in BmABCA2 to Cry2Ab. (**A**) Evaluation of strains with C-terminal half-deleted BmABCA2s (A2T01, A2T06, and A2T14; indicated in Figure 1C. The wild-type strain (Ringetsu), which is susceptible to Cry2A toxins, was used as a control. (**B,C**) The susceptibility of strains with BmABCA2s carrying 1–3 amino acid deletions at the C-terminus of TM7 [A2T03, A2T11 (**B**), and A2T08 (**C**); indicated in Figure 1A,C] to Cry2Ab. The wild-type strain was used as a control. Cry2Ab was spread on the leaf disk at 10 µL/cm^2. The toxin feeding assays were performed as described in the Materials and Methods.

Table 1. Responses of the knock-out (A2T14) and wild-type strains to Cry toxins.

	B. mori Strains	N [1]	Cry Toxin LC$_{50}$ (ppm, 95%CI) [2]	Slope [3]	RR [4]
2Aa	A2T14	180	>332	-	>9.182
	Wild-type	210	36.155 (26.431–49.518)	2.173	1.000
2Ab	A2T14	180	>4096	-	>9990.244
	Wild-type	270	0.410 (0.302–0.799)	2.114	1.000
1Aa	A2T14	240	1.382 (1.152–1.730)	4.009	1.544
	Wild-type	240	0.895 (0.705–1.128)	2.410	1.000
1Ab	A2T14	210	4.808 (3.818–6.344)	1.994	1.057
	Wild-type	210	4.616 (3.515–6.771)	2.119	1.000
1Ac	A2T14	210	22.078 (17.150–28.664)	2.116	4.212
	Wild-type	210	5.242 (3.579–9.043)	1.746	1.000
1Ca	A2T14	240	0.450 (0.365–0.567)	2.916	0.569
	Wild-type	210	0.791 (0.530–1.627)	1.478	1.000
1Da	A2T14	210	0.836 (0.600–1.301)	1.498	1.101
	Wild-type	210	0.759 (0.596–0.981)	2.497	1.000
1Fa	A2T14	210	57.116 (47.210–73.042)	3.801	2.514
	Wild-type	240	22.719 (17.757–30.446)	1.816	1.000
9Aa	A2T14	180	0.264 (0.183–0.467)	1.492	0.708
	Wild-type	180	0.373 (0.311–0.475)	3.985	1.000

[1] Number of larvae tested. [2] Concentration of toxins killing 50% of larvae and its 95% confidence interval (CI). [3] Slope of the concentration-mortality line. [4] Resistance ratio (RR) = LC50 of knock-out strain divided by LC50 of the same toxin for wild-type.

The susceptibility of *B. mori* strains carrying mutant BmABCA2 to Cry2Aa was also investigated. All of the strains with C-terminal half-deleted BmABCA2s (A2T01 and A2T14) and the strains with BmABCA2s in which 1 or 2 amino acids were deleted from the C-terminus of TM7 (A2T08 and A2T11) were resistant to Cry2Aa (Figure 3). With increasing Cry2Aa concentrations, A2T08 showed slight susceptibility to Cry2Aa (Figure 3B).

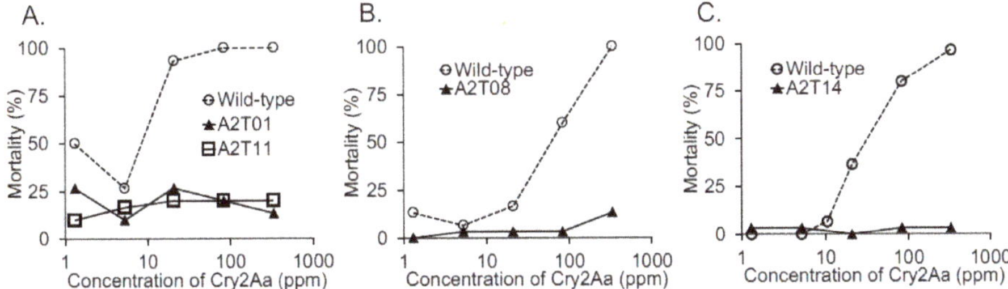

Figure 3. Toxin feeding assay to evaluate the susceptibility of larvae with mutations in BmABCA2 to Cry2Aa. Strains with C-terminal half-deleted BmABCA2s [A2T01 (**A**) and A2T14 (**C**)] and strains with BmABCA2s carrying 1 or 2 amino acid deletions at the C-terminal end of TM7 [see Figure 1A,C; A2T08 (**B**) and A2T11 (**A**)] were evaluated using the toxin feeding assay as described in Figure 2 and the Materials and Methods. The wild-type strain (Ringetsu) was used as a control.

The susceptibility of a *B. mori* strain with C-terminal half-deleted BmABCA2s (A2T14) to Cry1A, Cry1Ca, Cry1Da, Cry1Fa, and Cry9Aa was investigated further. We found that the susceptibility of strain A2T14 to Cry1Aa, Cry1Ab, Cry1Ac, Cry1Ca, Cry1Da, Cry1Fa, and Cry9Aa was similar to that of the wild-type strain (Figure 4). Moreover, the LC_{50} with the 95% confidence interval of each toxin was calculated (Table 1). The LC_{50} of every toxin did not differ between the wild-type and A2T14 strains, but A2T14 strains had no or very limited resistance to Cry1Aa (<2-fold), Cry1Ac (<5-fold) and Cry1Fa (<3-fold), indicating that the knockout of BmABCA2 did not affect susceptibility to Cry1A, Cry1Ca, Cry1Da, Cry1Fa, or Cry9Aa.

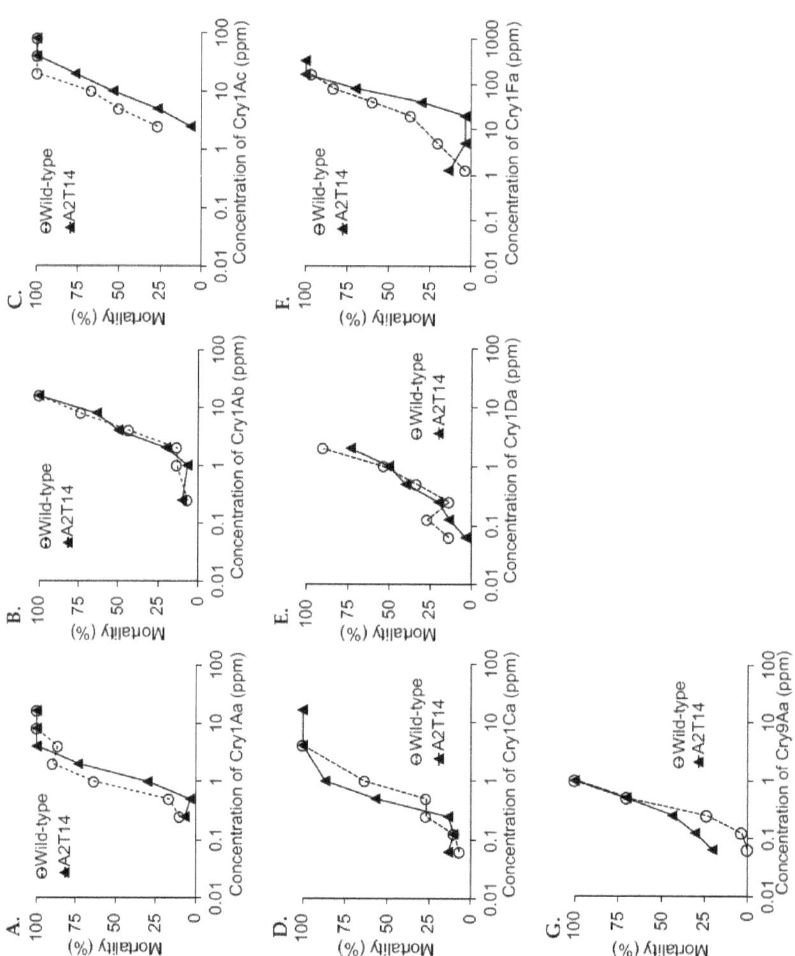

Figure 4. Toxin feeding assay to evaluate the susceptibility of the *B. mori* strain with a C-terminal half-deleted BmABCA2 (A2T14) to several Cry toxins. The strain with C-terminal half-deleted BmABCA2 (A2T14) and the wild-type strain (Ringetsu) were fed leaf disks contaminated with Cry1Aa (**A**), Cry1Ab (**B**), Cry1Ac (**C**), Cry1Ca (**D**), Cry1Da (**E**), Cry1Fa (**F**), or Cry9Aa (**G**) as described in the Materials and Methods.

2.3. BmABCA2-Dependent Cry2A Toxins Induce Cell Swelling

To examine whether BmABCA2 is a functional Cry2A toxin receptor, we used cell swelling assays. Enhanced green fluorescent protein (EGFP) cDNA was equipped with BmABCA2 cDNA in the same vector and transiently expressed in HEK293T cells showing green fluorescence on transfection. Cry2Ab was administered to those transient expression cells. Only the EGFP-positive cells swelled when they were treated with more than 40 nM Cry2Ab (Figure 5). The cells swelled in a Cry2Ab toxin concentration-dependent manner (Figure 5). By contrast, even 1.1 µM Cry2Ab did not induce swelling in cells not transfected with the BmABCA2 expression vector (Figure 5). Furthermore, cells that were transfected with the BmABCA2 expression vector did not swell when they were incubated in buffer lacking Cry2Ab.

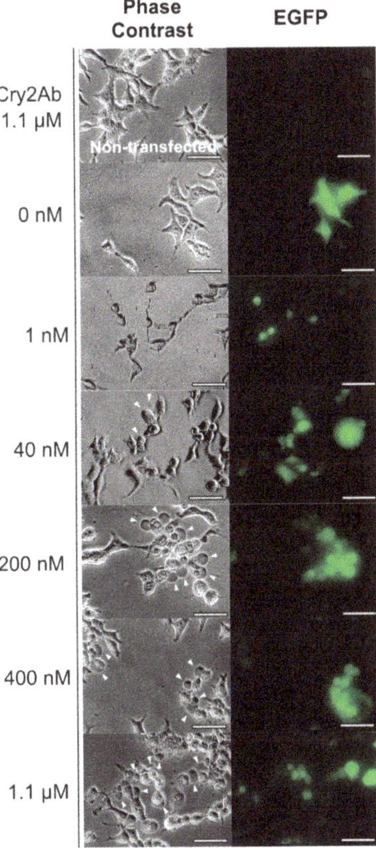

Figure 5. Cell swelling assay against Cry2Ab using BmABCA2-expressing HEK292T cells. The cells were attached to coverslips set on the six plates. BmABCA2 was transiently expressed on the surface of HEK293T cells via the transfection of an expression vector with attached EGFP. The coverslips were set on the wells of glass slides filled with the Cry2Ab test solutions, incubated for 1 h, and observed under a microscope. Arrowheads indicate swollen cells. (Scale bar = 50 µm).

To examine whether BmABCA2 acts as a functional receptor for other Cry toxins, BmABCA2-expressing cells were treated with Cry1Aa, Cry1Ac, and Cry9Aa. However, no cells were swollen after treatment with up to 1.5 µM Cry1Aa, 1.1 µM Cry1Ac, and 3.3 µM Cry9Aa (Figure 6).

To clarify whether the BmABCC2 receptor for Cry1A toxins can function as a Cry2Ab receptor, BmABCC2-expressing cells were administered Cry2Ab. However, no cells swelled with up to 1.1 µM Cry2Ab (Figure 7). By contrast, Cry 1Aa induced swelling in BmABCC2-expressing cells at 15 nM, indicating that the level of BmABCC2 expression was sufficient to assess Cry2Ab receptor function (Figure 7). Cry1Ac induced swelling of the BmABCC2-expressing cells at 500 nM (Figure 7).

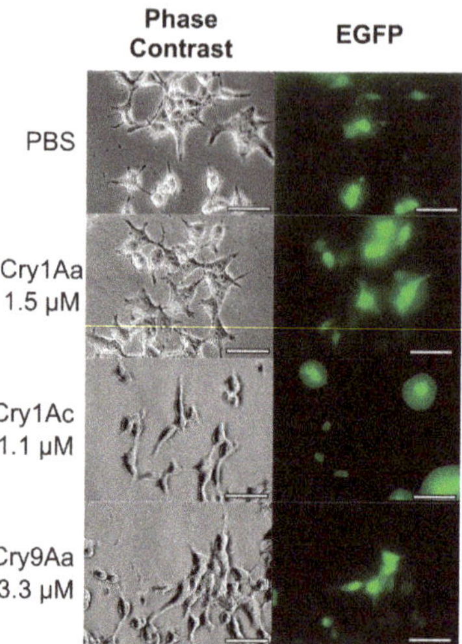

Figure 6. Cell swelling assay against Cry1A toxins and Cry9Aa using BmABCA2-expressing HEK293T cells. The cells were attached to coverslips set on the six plates. BmABCA2 was transiently expressed on the surface of HEK293T cells via the transfection of an expression vector with attached EGFP. Then, the cells were incubated for 1 h with Cry toxins on glass slides and observed under a microscope, as described in Figure 5. (Scale bar = 50 µm).

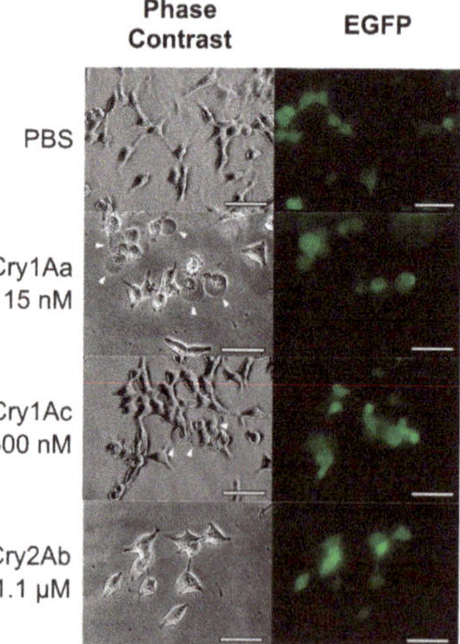

Figure 7. Cell swelling assay against Cry1A toxins and Cry2Ab using BmABCC2-expressing HEK293T cells. The cells were attached to coverslips set on the six plates. BmABCC2 co-expressed with EGFP was transiently expressed on the surface of HEK293T cells via transfection of the expression vector. Then, the cells were incubated for 1 h with Cry toxins on glass slides and observed under a microscope, as described in Figure 5. Arrowheads indicate swollen cells. (Scale bar = 50 µm).

3. Discussion

Toxicity tests of the strains with C-terminal half-deleted BmABCA2 (A2T01, A2T06, and A2T14) showed that they were highly resistant to Cry2Ab (Figure 2), indicating that BmABCA2 plays an essential role in determining the susceptibility of *B. mori* to Cry2Ab. ABCA2 was first suggested to be linked to Cry2Ab resistance in *H. armigera* [20]. This was confirmed by generating an ABCA2 knockout strain via CRISPR/Cas9 mutagenesis in which the resistance of *H. armigera* and *T. ni* larvae to Cry2Ab was linked to defects in ABCA2 [21,22], indicating that ABCA2 plays an important role in the mode of action of Cry2A toxins. Although the resistance was slightly lower than that of the C-terminal half-deleted strains (Figure 2B,C), toxicity testing of the strains with BmABCA2s that had 1–3 amino acids deleted from the C-terminal end of TM7 (A2T03, A2T08, and A2T11) showed that they were also resistant to Cry2Ab (Figure 2B,C). Thus, BmABCA2 also plays an essential role in determining the susceptibility to Cry2Ab in *B. mori*. In *B. mori*, even the partial deletion and alanine replacement of several amino acid residues at the N-terminus of extracellular loop 4 (ECL4) decreased the susceptibility-conferring activity of BmABCC2 to Cry1Aa [17]. This suggests that ECL4 of BmABCC2 is part of the site that interacts with Cry1Aa. ECL4 of BmABCA2 might also play a role in the interaction with Cry2Ab (Figure 1A), and the 1–3 amino acids that were deleted from the C-terminal end of TM7 might affect the interaction of BmABCA2 with Cry2Ab by changing the three-dimensional structure of the binding epitopes.

It is still not known how ABCA2 is involved in the susceptibility of larvae to Cry2A toxins. ABCA2 is a membrane protein and the ABC transporter ABCC2 is a functional Cry1A toxin receptor [5,16,18,24,25]. Therefore, we examined whether BmABCA2 transiently expressed in HEK293T cells (Figure 5) could function as a Cry2Ab receptor and whether HEK293T cells would swell like the columnar cells in a Cry toxin-intoxicated midgut. The BmABCA2-expressing cells started swelling on exposure to 40 nM Cry2Ab (Figure 5). When BmABCC2, a highly functional receptor for a single Cry1Aa molecule [16], was expressed in HEK293T cells using a very similar expression system, the HEK293T cells started swelling in response to 1 nM Cry1Aa [26]. By contrast, when BmABCC3, a less functional receptor for a single Cry1Aa molecule, was expressed in HEK293T cells, the cells started to swell in response to 100 nM Cry1Aa [26]. Based on our cell swelling assays using HEK293T cells, BmABCA2 has sufficient functional receptor activity for Cry2Ab. This is consistent with the strong resistance to Cry2Ab that was generated by the TALEN-induced BmABCA2 mutation (Figure 2).

In *H. armigera*, an HaABCA2 knockout strain showed resistance to Cry2Ab, but not to Cry1Ac [21], suggesting that this receptor is highly tuned to Cry2A toxins. Regarding Cry1Aa, the BmABCC2 binding site was thought to be the pocket made by loops 2 and 3 [27]. However, we could not find any amino acids near loops 2 and 3 that were conserved in Cry1Aa and Cry2Ab. This might explain the receptor specificity difference between Cry1Aa and Cry2Ab. The BmABCA2 knockout strains were susceptible to Cry1Aa (Figure 4). In addition, the BmABCA2 knockout strains were susceptible to all of the Cry1 and Cry9 toxins tested (Figure 4). This implies that Cry2A toxin-resistant insects lack cross-resistance to Cry1Ca, Cry1Da, Cry1Fa, or Cry9Aa. Actually, a Cry1Ca-resistant diamondback moth was reported to lack cross-resistance to Cry2Aa [13]. Furthermore, BmABCA2-expressing HEK293T cells were susceptible to Cry2Ab, but not to Cry1A or Cry9A toxins (Figures 5 and 6). Therefore, our results suggest that the specificity of BmABCA2 as a Cry toxin receptor is narrowly tuned to Cry2A toxins. By contrast, BmABCC2-expressing HEK293T cells were susceptible to Cry1A toxins, but not to Cry2Ab (Figure 7). This also suggests that ABC transporters are highly tuned to a narrow group of Cry toxins.

4. Materials and Methods

4.1. Silkworm Strains and Rearing

The wild-type silkworm strain was distributed from the Genetics Resources Center, National Agriculture and Food Research Organization (NARO) and was reared on mulberry leaves or artificial diet (Nihon Nosan Kogyo, Yokohama, Japan) at 25 °C.

4.2. DNA Target Site Selection and Preparation of TALEN mRNA

The DNA target site was selected in the fifteenth exon of the BmABCA2 (KP219767) gene that encodes the extracellular region between the 7th and 8th transmembrane regions. Two TALEN half sites were designed, as shown in Figure 1. TALEN-encoding genes were constructed by Golden Gate assembly, as described previously [28]. To prepare mRNAs for microinjection, TALEN-encoding plasmids were linearized by *Xba*I (TaKaRa Bio, Kusatsu, Japan) and transcribed using an mMESSAGE mMACHINE T7 transcription kit (Thermo Fisher Scientific, Waltham, MA, USA) according to the protocols of manufacturer.

4.3. Egg Microinjection

The poly(A)-tailed TALEN mRNAs (0.2 µg/µL) were dissolved in injection buffer (5 mM KCl, 0.5 mM phosphate buffer, pH 7.0) together with donor oligonucleotides (0.2–0.4 µg/µL), as described previously [29], and injected to the silkworm eggs at the syncytial preblastodermal stage [30]. The embryos were incubated at 25 °C in a humidified atmosphere.

4.4. Identification of BmABCA2 Mutation Induced by TALENs

To extract genomic DNA, a leg of each G_1 moth was homogenized in 50 µL of DNAzol® Direct (Molecular Research Center, Cincinnati, OH, USA). After 10 min of incubation at room temperature, the homogenate was mixed vigorously and separated by centrifugation. The supernatant containing genomic DNA was used as a template for PCR. The target region of the BmABCA2 gene was amplified using a specific primer set (forward: 5′-GTGTCAGGAGCAAGTCTGGTC-3′, reverse: 5′-AGACGTGTTAAATATCTCGTCTCG-3′). Direct sequencing of the PCR products was performed using the reverse primer as a sequencing primer. Mutations induced by TALENs were identified according to the sequencing results.

4.5. Cry Toxins Preparation

The DNAs of Cry2Ab (AAA22342), and Cry1Fa (AAA22348) genes which were optimized for expression of heterologous proteins in *Escherichia coli* was synthesized by Strings DNA Fragments service (Thermo Fisher Scientific) and subcloned into between *Bam*HI and *Xho*I sites of pGEX6P-3 (GE healthcare lifesciences, Amersham, UK) using In-fusion HD Cloning kit (TaKaRa Bio, Kusatsu, Japan). The DNA clones were used to produce the Cry2Ab, and Cry1Fa toxins. For the production of the Cry1Aa, Cry1Ab, Cry1Ac, Cry1Da, and Cry9Aa toxins, the genes of these toxins were subcloned into pGEX4T-3 and then *E. coli* cells were transformed with those as described previously [16]. The transformed cells were cultured in LB liquid medium with ampicillin at 37 °C and gene expression was induced by isopropyl thio-b-d-galactoside. The inclusion bodies were harvested and washed as describe as previously [31]. Inclusion bodies of Cry2Aa toxin was prepared as described elsewhere [32]. The Cry1Ca toxin was produced by a *B. thuringiensis* recombinant stain, and inclusion bodies of Cry1Ca were washed as described elsewhere [33]. The Cry1Aa, Cry1Ac, and Cry9Aa were solubilized and activated as described previously [27]. The inclusion bodies of Cry2Ab was activated with a High-Performance Liquid Chromatography method (HPLC) that the Cry2Ab was solubilized as same as described above. After the solution, the pro-toxin of Cry2Ab was dialyzed with 20 mM Tris-HCl pH 9 and passed through a 0.45 µM filter (Millipore Millex-HP Hydrophilic PVDF, Millipore,

Burlington, MA, USA) to remove bacteria. Then, the pro-toxin was applied to an HPLC system equipped with a Shodex IEC DEAE-825 column (0.8 × 7–5cm, Showa Denko Co.) and equilibrated with 250 mM Tris-HCl, pH 9. The pro-toxin was bound with the column, and non-absorbent pellets were washing by 100 mM Tris-HCl pH 9. Then, the bound pro-toxin was treated with 0.0625 mg/mL Trypsin (Sigma-Aldrich, St. Louis, MO, USA) for 40 min at 37 °C. The treatment Cry2Ab was eluted by Elix water 5 min later, with a linear gradient of 0~250 mM Tris-HCl pH 9 buffer eluted over 55 min with a 0.5 mL/min flow rate, and the activated toxin was dialyzed by phosphate-buffered saline (PBS, 137 mM sodium chloride, 2.7 mM potassium chloride, 10 mM sodium phosphate dibasic, 1.8 mM potassium dihydrogen phosphate, pH 7.4) for cell swelling assay.

4.6. Diet Overlay Bioassays

Susceptibility to Cry toxins of the BmABCA2 genome edited mutants and wild-type strains were evaluated with diet overlay bioassays. To make leaf disks, sixth open leaves from the top of each branch of the mulberry leaves were picked up. Cry toxins solutions were diluted with Silwet® L-77 (Momentive Performance Materials, Waterford, NY, USA), and the suspensions were spread to be 10 µL/cm^2 on the leaf disks. After Cry toxins suspension were dried at room temperature, two leaf disks were put in each petri dish with 10 larvae of 2nd instar of the genome edited, and wild-type strains were respectively reared on each disk for 2 days. After 2 days, the larvae were moved to non-toxins leaf disks and mortality was recorded at 4 days after feeding initiation.

The median lethal dose (LC$_{50}$) values and the 95% confidence interval were calculated based on Probit Analysis.

4.7. cDNA Cloning of BmABCA2 and Construction of Expression Vectors for HEK293T Cells

Total RNA was isolated from midgut tissue of 5th instar larvae of the wild-type silkworm strain, using TRIzol (Thermo Fisher Scientific, Waltham, MA, USA) according to the protocols of manufacturer, and used for cDNA synthesis as a template. The BmABCA2 cDNA was amplified by one-step RT-PCR using PrimeScript High Fidelity RT-PCR Kit (TaKaRa Bio). The amplified cDNA using primers (forward: 5′-CCACCCGGATCCGATATGAGACCTCAGAGAAAAGAAGCC-3′, reverse: 5′-GTCTTTGTAGTCGATCAAGCCTTCCCTTTGATATTTCGT-3′) was cloned into *EcoRV* site of the pcDNA3.1 (Thermo Fisher Scientific, Waltham, MA, USA) using In-Fusion® HD Cloning Kit (TaKaRa Bio, Kusatsu, Japan). The neomycin resistance gene of pcDNA3.1 was replaced by Enhanced green fluorescent protein (EGFP) - *Streptoalloteichus hindustanus* ble (Sh ble) fusion gene by In-fusion cloning method described below. The linearized vector was generated from pcDNA3.1 plasmid as a template by PCR using primers (forward: 5′-GCCCTTGCTCACCATGCGAACGATCCTCATCCTGTC-3′, reverse: 5′- GAGGAGCAGGACTGAGCGGGACTCTGGGGTTCG-3′). The EGFP-ble fusion gene was amplified using primers (forward: 5′-ATGGTGAGCAAGGGCGAGGAG-3′, reverse: 5′-TCAGTCCTGCTCCTCGGCCAC-3′) and cloned into the linearized vector.

4.8. Expression of BmABCA2 and BmABCC2 in HEK293T Cells and Cell Swelling Assay with Cry Toxins

HEK293T cells were cultured and transfected, as described previously [5]. The HEK293T cells were cultured on the cover glass in a 6-well plate (Truesline; Nippon Genetics, Tokyo, Japan). Until the HEK293T cells grow up to 70 ~ 80% confluence, expression vectors for BmABCA2 and BmABCC2 were transfected in Opti-MEM® (Thermo Fisher Scientific, Waltham, MA, USA) with polyethylenimine (PEI Max, Polysciences, Warrington, PA, USA) for 2 h incubation at 37 °C. Then, the media were changed to a fresh Dulbecco's modified Eagle medium (D-MEM) and incubated at 37 °C for 48 h in a CO$_2$ incubator. After that, the cover glass was taken out and covered onto Cry toxins solutions in a hole of the 2-hole glass slide. Cry toxins were diluted with PBS (pH 7.4). After 1 h incubation at 37 °C, the cells were observed by phase-contrast and fluorescent microscopy.

Supplementary Materials: The following are available online at http://www.mdpi.com/2072-6651/12/2/104/s1, Figure S1: Mutations in BmABCA2 in mutant strains produced by genome-editing method using TALENs.

Author Contributions: K.W. and R.S. conceived and designed the experiments; X.L., K.W., S.A., and K.M. performed the experiments and analyzed the data; Y.T., S.W. and T.I. helped to prepare the mutant silkworm strains used in the study; X.L., K.W. and R.S. wrote the manuscript. All authors have read and agreed to the published version of the manuscript.

Funding: This research was partially supported by a Grant-in-Aid for Scientific Research (B) (18H03397) from the Ministry of Education, Culture, Sports, Science, and Technology of Japan.

Acknowledgments: We thank Ritsuko Murakami for rearing the silkworms and providing technical assistance. We also thank Shin-ichiro Asano for providing the *Bacillus thuringiensis* strain Bt51(pHS2). The silkworm strain "Ringetsu" was provided from The Genetic Resources Center, NARO. The work was funded by a Grant-in-Aid for Scientific Research (B) (18H03397) from the Ministry of Education, Culture, Sports, Science, and Technology of Japan.

Conflicts of Interest: We declare no conflict of interest, and the funders had no role in the design of the study.

References

1. Palma, L.; Muñoz, D.; Berry, C.; Murillo, J.; Caballero, P. Bacillus thuringiensis Toxins: An Overview of Their Biocidal Activity. *Toxins* **2014**, *6*, 3296–3325. [CrossRef] [PubMed]

2. Soberón, M.; Pardo, L.; Muñóz-Garay, C.; Sánchez, J.; Gómez, I.; Porta, H.; Bravo, A. Pore formation by Cry toxins. In *Proteins Membrane Binding and Pore Formation*; Springer: New York, NY, USA, 2010; pp. 127–142.

3. Pigott, C.R.; Ellar, D.J. Role of receptors in *Bacillus thuringiensis* crystal toxin activity. *Microbiol. Mol. Biol. Rev.* **2007**, *71*, 255–281. [CrossRef] [PubMed]

4. Villalon, M.; Vachon, V.; Brousseau, R.; Schwartz, J.L.; Laprade, R. Video imaging analysis of the plasma membrane permeabilizing effects of *Bacillus thuringiensis* insecticidal toxins in Sf9 cells. *Biochim. Biophys. Acta (BBA) Biomembr.* **1998**, *1368*, 27–34. [CrossRef]

5. Endo, H.; Tanaka, S.; Imamura, K.; Adegawa, S.; Kikuta, S.; Sato, R. Cry toxin specificities of insect ABCC transporters closely related to lepidopteran ABCC2 transporters. *Peptides* **2017**, *98*, 86–92. [CrossRef] [PubMed]

6. Melo, A.L.D.A.; Soccol, V.T.; Soccol, C.R. Bacillus thuringiensis: Mechanism of action, resistance, and new applications: A review. *Crit. Rev. Biotechnol.* **2016**, *36*, 317–326. [CrossRef]

7. Roh, J.Y.; Choi, J.Y.; Li, M.S.; Jin, B.R.; Je, Y.H. Bacillus thuringiensis as a specific, safe, and effective tool for insect pest control. *J. Microbiol. Biotechnol.* **2007**, *17*, 547.

8. Jouzani, G.S.; Valijanian, E.; Sharafi, R. Bacillus thuringiensis: A successful insecticide with new environmental features and tidings. *Appl. Microbiol. Biotechnol.* **2017**, *101*, 2691–2711. [CrossRef]

9. Tabashnik, B.E.; Carrière, Y. Surge in insect resistance to transgenic crops and prospects for sustainability. *Nat. Biotechnol.* **2017**, *35*, 926. [CrossRef]

10. Hernández-Rodríguez, C.S.; Hernández-Martínez, P.; Van Rie, J.; Escriche, B.; Ferré, J. Shared midgut binding sites for Cry1A. 105, Cry1Aa, Cry1Ab, Cry1Ac and Cry1Fa proteins from *Bacillus thuringiensis* in two important corn pests, *Ostrinia nubilalis* and *Spodoptera frugiperda*. *PLoS ONE* **2013**, *8*, e68164. [CrossRef]

11. Hernández-Rodríguez, C.S.; Van Vliet, A.; Bautsoens, N.; Van Rie, J.; Ferré, J. Specific binding of *Bacillus thuringiensis* Cry2A insecticidal proteins to a common site in the midgut of *Helicoverpa* species. *Appl. Environ. Microbiol.* **2008**, *74*, 7654–7659. [CrossRef]

12. Tabashnik, B.E.; Liu, Y.B.; de Maagd, R.A.; Dennehy, T.J. Cross-resistance of pink bollworm (Pectinophora gossypiella) to *Bacillus thuringiensis* toxins. *Appl. Environ. Microbiol.* **2000**, *66*, 4582–4584. [CrossRef] [PubMed]

13. Liu, Y.B.; Tabashnik, B.E.; Meyer, S.K.; Crickmore, N. Cross-resistance and stability of resistance to *Bacillus thuringiensis* toxin Cry1C in diamondback moth. *Appl. Environ. Microbiol.* **2001**, *67*, 3216–3219. [CrossRef] [PubMed]

14. Wu, C.; Chakrabarty, S.; Jin, M.; Liu, K.; Xiao, Y. Insect ATP-Binding Cassette (ABC) Transporters: Roles in Xenobiotic Detoxification and Bt Insecticidal Activity. *Int. J. Mol. Sci.* **2019**, *20*, 2829. [CrossRef] [PubMed]

15. Sato, R.; Adegawa, S.; Li, X.; Tanaka, S.; Endo, H. Function and role of ATP-binding cassette transporters as receptors for 3D-Cry toxins. *Toxins* **2019**, *11*, 124. [CrossRef]

16. Tanaka, S.; Miyamoto, K.; Noda, H.; Jurat-Fuentes, J.L.; Yoshizawa, Y.; Endo, H.; Sato, R. The ATP-binding cassette transporter subfamily C member 2 in B ombyx mori larvae is a functional receptor for C ry toxins from B acillus thuringiensis. *FEBS J.* **2013**, *280*, 1782–1794. [CrossRef]

17. Tanaka, S.; Endo, H.; Adegawa, S.; Kikuta, S.; Sato, R. Functional characterization of *Bacillus thuringiensis* Cry toxin receptors explains resistance in insects. *FEBS J.* **2016**, *283*, 4474–4490. [CrossRef]

18. Obata, F.; Tanaka, S.; Kashio, S.; Tsujimura, H.; Sato, R.; Miura, M. Induction of rapid and selective cell necrosis in Drosophila using *Bacillus thuringiensis* Cry toxin and its silkworm receptor. *BMC Biol.* **2015**, *13*, 48. [CrossRef]

19. Pauchet, Y.; Bretschneider, A.; Augustin, S.; Heckel, D.G. A P-glycoprotein is linked to resistance to the *Bacillus thuringiensis* Cry3Aa toxin in a leaf beetle. *Toxins* **2016**, *8*, 362. [CrossRef]

20. Tay, W.T.; Mahon, R.J.; Heckel, D.G.; Walsh, T.K.; Downes, S.; James, W.J.; Lee, S.F.; Reineke, A.; Williams, A.K.; Gordon, K.H. Insect resistance to *Bacillus thuringiensis* toxin Cry2Ab is conferred by mutations in an ABC transporter subfamily a protein. *PLoS Genet.* **2015**, *11*, e1005534. [CrossRef]

21. Wang, J.; Wang, H.; Liu, S.; Liu, L.; Tay, W.T.; Walsh, T.K.; Yang, Y.; Wu, Y. CRISPR/Cas9 mediated genome editing of *Helicoverpa armigera* with mutations of an ABC transporter gene HaABCA2 confers resistance to *Bacillus thuringiensis* Cry2A toxins. *Insect Biochem. Mol. Biol.* **2017**, *87*, 147–153. [CrossRef]

22. Yang, X.; Chen, W.; Song, X.; Ma, X.; Cotto-Rivera, R.O.; Kain, W.; Chu, H.; Chen, Y.R.; Fei, Z.; Wang, P. Mutation of ABC transporter ABCA2 confers resistance to Bt toxin Cry2Ab in Trichoplusia ni. *Insect Biochem. Mol. Biol.* **2019**, *112*, 103209. [CrossRef] [PubMed]

23. Mathew, L.G.; Ponnuraj, J.; Mallappa, B.; Chowdary, L.R.; Zhang, J.; Tay, W.T.; Walsh, T.K.; Gordon, K.H.; Heckel, D.G.; Carrière, Y. ABC transporter mis-splicing associated with resistance to Bt toxin Cry2Ab in laboratory-and field-selected pink bollworm. *Sci. Rep.* **2018**, *8*, 13531. [CrossRef] [PubMed]

24. Bretschneider, A.; Heckel, D.G.; Pauchet, Y. Three toxins, two receptors, one mechanism: Mode of action of Cry1A toxins from *Bacillus thuringiensis* in Heliothis virescens. *Insect Biochem. Mol. Biol.* **2016**, *76*, 109–117. [CrossRef] [PubMed]

25. Stevens, T.; Song, S.; Bruning, J.B.; Choo, A.; Baxter, S.W. Expressing a moth abcc2 gene in transgenic Drosophila causes susceptibility to Bt Cry1Ac without requiring a cadherin-like protein receptor. *Insect Biochem. Mol. Biol.* **2017**, *80*, 61–70. [CrossRef]

26. Endo, H.; Tanaka, S.; Adegawa, S.; Ichino, F.; Tabunoki, H.; Kikuta, S.; Sato, R. Extracellular loop structures in silkworm ABCC transporters determine their specificities for *Bacillus thuringiensis* Cry toxins. *J. Biol. Chem.* **2018**, *293*, 8569–8577. [CrossRef]

27. Adegawa, S.; Nakama, Y.; Endo, H.; Shinkawa, N.; Kikuta, S.; Sato, R. The domain II loops of *Bacillus thuringiensis* Cry1Aa form an overlapping interaction site for two *Bombyx mori* larvae functional receptors, ABC transporter C2 and cadherin-like receptor. *Biochim. Biophys. Acta (BBA) Proteins Proteom.* **2017**, *1865*, 220–231. [CrossRef]

28. Takasu, Y.; Sajwan, S.; Daimon, T.; Osanai-Futahashi, M.; Uchino, K.; Sezutsu, H.; Tamura, T.; Zurovec, M. Efficient TALEN construction for *Bombyx mori* gene targeting. *PLoS ONE* **2013**, *8*, e73458. [CrossRef]

29. Takasu, Y.; Kobayashi, I.; Tamura, T.; Uchino, K.; Sezutsu, H.; Zurovec, M. Precise genome editing in the silkworm *Bombyx mori* using TALENs and ds-and ssDNA donors—A practical approach. *Insect Biochem. Mol. Biol.* **2016**, *78*, 29–38. [CrossRef]

30. Tamura, T.; Thibert, C.; Royer, C.; Kanda, T.; Eappen, A.; Kamba, M.; Kômoto, N.; Thomas, J.L.; Mauchamp, B.; Shirk, P. Germline transformation of the silkworm *Bombyx mori* L. using a piggyBac transposon-derived vector. *Nat. Biotechnol.* **2000**, *18*, 81.

31. Atsumi, S.; Mizuno, E.; Hara, H.; Nakanishi, K.; Kitami, M.; Miura, N.; Tabunoki, H.; Watanabe, A.; Sato, R. Location of the *Bombyx mori* aminopeptidase N type 1 binding site on *Bacillus thuringiensis* Cry1Aa toxin. *Appl. Environ. Microbiol.* **2005**, *71*, 3966–3977. [CrossRef]

32. Sasaki, J.; Asano, S.; Hashimoto, N.; Lay, B.W.; Hastowo, S.; Bando, H.; Iizuka, T. Characterization of a cry2A Gene Cloned from an Isolate of *Bacillus thuringiensis* serovar sotto. *Curr. Microbiol.* **1997**, *35*, 1–8. [CrossRef] [PubMed]
33. Masson, L.; Mazza, A.; Sangadala, S.; Adang, M.J.; Brousseau, R. Polydispersity of *Bacillus thuringiensis* Cry1 toxins in solution and its effect on receptor binding kinetics. *Biochim. Biophys. Acta (BBA) Protein Struct. Mol. Enzymol.* **2002**, *1594*, 266–275. [CrossRef]

Article

Reduced Expression of a Novel Midgut Trypsin Gene Involved in Protoxin Activation Correlates with Cry1Ac Resistance in a Laboratory-Selected Strain of *Plutella xylostella* (L.)

Lijun Gong [1,2], Shi Kang [2], Junlei Zhou [2], Dan Sun [2], Le Guo [1,2], Jianying Qin [2], Liuhong Zhu [2], Yang Bai [2], Fan Ye [1,2], Mazarin Akami [2], Qingjun Wu [2], Shaoli Wang [2], Baoyun Xu [2], Zhongxia Yang [1], Alejandra Bravo [3], Mario Soberón [3], Zhaojiang Guo [2,*], Lizhang Wen [1,*] and Youjun Zhang [2,*]

[1] College of Plant Protection, Hunan Agricultural University, Changsha 410125, China; gonglijun025@163.com (L.G.); guole930323@163.com (L.G.); yefan9605@163.com (F.Y.); yzxmichelle@aliyun.com (Z.Y.)

[2] Department of Plant Protection, Institute of Vegetables and Flowers, Chinese Academy of Agricultural Sciences, Beijing 100081, China; kangshi0718@163.com (S.K.); zhoujunlei2006@126.com (J.Z.); sun13804560684@163.com (D.S.); qinjianying0203@163.com (J.Q.); liuhongzhu1992@126.com (L.Z.); baiyang15765533722@163.com (Y.B.); makami1987@gmail.com (M.A.); wuqingjun@caas.cn (Q.W.); wangshaoli@caas.cn (S.W.); xubaoyun@caas.cn (B.X.)

[3] Departamento de Microbiología Molecular, Instituto de Biotecnología, Universidad Nacional Autónoma de México, Apdo. Postal 510-3, Cuernavaca, Morelos 62250, Mexico; bravo@ibt.unam.mx (A.B.); mario@ibt.unam.mx (M.S.)

[*] Correspondence: guozhaojiang@caas.cn (Z.G.); weninsect123@aliyun.com (L.W.); zhangyoujun@caas.cn (Y.Z.); Tel.: +86-10-82109518 (Z.G.); +86-0731-84618163 (L.W.); +86-10-62152945 (Y.Z.)

Received: 7 January 2020; Accepted: 21 January 2020; Published: 23 January 2020

Abstract: *Bacillus thuringiensis* (Bt) produce diverse insecticidal proteins to kill insect pests. Nevertheless, evolution of resistance to Bt toxins hampers the sustainable use of this technology. Previously, we identified down-regulation of a trypsin-like serine protease gene *PxTryp_SPc1* in the midgut transcriptome and RNA-Seq data of a laboratory-selected Cry1Ac-resistant *Plutella xylostella* strain, SZ-R. We show here that reduced *PxTryp_SPc1* expression significantly reduced caseinolytic and trypsin protease activities affecting Cry1Ac protoxin activation, thereby conferring higher resistance to Cry1Ac protoxin than activated toxin in SZ-R strain. Herein, the full-length cDNA sequence of *PxTryp_SPc1* gene was cloned, and we found that it was mainly expressed in midgut tissue in all larval instars. Subsequently, we confirmed that the *PxTryp_SPc1* gene was significantly decreased in SZ-R larval midgut and was further reduced when selected with high dose of Cry1Ac protoxin. Moreover, down-regulation of the *PxTryp_SPc1* gene was genetically linked to resistance to Cry1Ac in the SZ-R strain. Finally, RNAi-mediated silencing of *PxTryp_SPc1* gene expression decreased larval susceptibility to Cry1Ac protoxin in the susceptible DBM1Ac-S strain, supporting that low expression of *PxTryp_SPc1* gene is involved in Cry1Ac resistance in *P. xylostella*. These findings contribute to understanding the role of midgut proteases in the mechanisms underlying insect resistance to Bt toxins.

Keywords: *Bacillus thuringiensis*; *Plutella xylostella*; Cry1Ac resistance; trypsin-like midgut protease; protoxin activation

Key Contribution: Low expression of a novel midgut trypsin-like protease gene *PxTryp_SPc1* involved in protoxin activation contributes to Cry1Ac resistance in a laboratory-selected *P. xylostella* strain.

---

1. Introduction

Bacillus thuringiensis (Bt) are gram-positive entomopathogenic bacteria most widely used as a biopesticide worldwide, and transgenic crops expressing insecticidal toxins produced by these bacteria (transgenic Bt crops) have been planted in 104 million hectares globally in 2018, which has a central role in pest control and global food security [1,2]. However, field-evolved resistance to Bt crops soared from three cases in 2005 to 16 in 2016, documenting an accelerated evolution of practical resistance [3]. Due to the commercial application of Bt proteins, such as Cry proteins for the control of insect pests, it is necessary to probe the resistance mechanism to Cry proteins in order to propose effective strategies to delay the resistance evolution.

Bt Cry proteins are produced as inactive and insoluble crystals formed by protoxins [4]. Cry protoxins are solubilized in the alkaline environment of the midgut and are further processed into activated toxins by midgut proteases when ingested by susceptible larvae [5]. Activated toxins then interact with specific midgut receptor proteins, such as aminopeptidase N (APN), alkaline phosphatase (ALP), cadherin (CAD) and ABC transporters, located in the brush border membrane (BBM) of the midgut epithelium cells from the larvae [6,7] Receptor binding leads to the formation of lytic pores in the membrane that burst cells and finally kill the insects [8]. However, it has been shown that protoxins also bind to specific receptors, and then they are activated by midgut proteases inducing also toxin oligomerization and pore-formation [8,9]. Whether protoxins are activated before or after receptor binding is an important step, transforming the 130 kDa protoxin into a 55–65 kDa activated toxin [4,7]. Trypsin proteases are important midgut proteinases, which participate in Bt Cry protein degradation and protoxin activation [10,11]. It has been reported that alteration of the midgut trypsin genes or trypsin proteolytic activities are linked to Bt resistance in *Plodia interpunctella* (Hübner) (Lepidoptera: Pyralidae) [10], *Ostrinia nubilalis* (Hübner) (Lepidoptera: Pyralidae) [12–14], *Spodoptera frugiperda* (JE Smith) (Lepidoptera: Noctuidae) [15], *Helicoverpa armigera* (Hübner) (Lepidoptera: Noctuidae) [16,17], *Aedes aegypti* (L.) (Diptera: Culicidae) [18], *Mythimna unipuncta* (Haworth) (Lepidoptera: Noctuidae) [19], and *Helicoverpa zea* (Boddie) (Lepidoptera: Noctuidae) [20].

The diamondback moth, *Plutella xylostella* (L.), is a cosmopolitan insect pest of cruciferous crops that was the first example of resistance to Bt sprays in the field [21]. The economic damage produced by *P. xylostella* was estimated to be up to USD 5 billion every year [22]. Moreover, since *P. xylostella* was the first documented insect developing field-evolved Bt resistance, it is a good model to understand insect resistance mechanisms to Bt toxins. Previous studies showed that in *P. xylostella*, resistance to the Cry1Ac toxin was not associated with alterations of the *PxABCH1* and *PxCAD* genes [23–25]. In contrast, Cry1Ac resistance rather correlated with a mutation in the *PxABCC2* gene [26] or with the differential expression of the *PxmALP*, *PxABCB1*, *PxABCC1*, *PxABCC2*, *PxABCC3*, and *PxABCG1* genes, which were shown to be down-regulated by an enhanced MAPK signaling pathway [27–29]. In the MAPK-mediated *trans*-regulatory mechanism, we reported that over-expression of the *PxMAP4K4* gene resulted in down-regulation of diverse midgut genes, thereby conferring a Cry1Ac resistance phenotype. Different *P. xylostella* strains that are resistant to Cry1Ac showed different induction levels of *PxMAP4K4*, resulting in different resistance levels. For example, the SZ-R resistant strain showed moderate resistance levels (662-fold to Cry1Ac) in contrast with the near-isogenic strain (NIL-R) that was highly resistant to Cry1Ac (>3900-fold). The relative expression of *PxMAP4K4* in SZ-R is slightly higher than that of the susceptible DBM1Ac-S strain; thus, expression of some Cry toxin receptors are also down-regulated in the SZ-R strain although at a lower level than that of the NIL-R, which showed the highest constitutive expression of the *PxMAP4K4* gene [27]. In particular, the high Cry1Ac resistance levels in the NIL-R strain does not involve the Cry1Ac protoxin activation mechanism [30]. However, the relationship between the protoxin activation mechanisms in other resistant strains of *P. xylostella* remains unclear.

Here, we compared data from midgut transcriptome and RNA-Seq analyses that were previously done, showing a significant decrease expression of a novel trypsin-like protease in a *P. xylostella* strain SZ-R that shows resistance to Cry1Ac toxin [31,32]. Then, we cloned and characterized the midgut trypsin protease gene of *P. xylostella* (*PxTryp_SPc1*). Finally, we demonstrated that down-regulation of the *PxTryp_SPc1* gene in the midgut tissue of SZ-R strain is related to Cry1Ac resistance by using different genomic, molecular, biochemical, and genetic tools. The conclusions of this study provide a new insight into the Bt resistance mechanism that could give hints for the control of insect pests.

2. Results

2.1. Comparison of Midgut Protease Activities and Cry1Ac Protoxin Activation between Susceptible DBM1Ac-S and Resistant SZ-R Strains

Previously, differential expression of midgut trypsin-like serine protease (*Tryp_SP*) genes was identified in the Cry1Ac-resistant strain SZ-R in contrast to the susceptible *P. xylostella* strain DBM1Ac-S [31,32]. To determine whether the potential altered *PxTryp_SP* gene expression can change the midgut protease activities and affect Cry1Ac protoxin activation in SZ-R strain, we first compared the midgut protease activities and Cry1Ac protoxin activation in both susceptible DBM1Ac-S and resistant SZ-R strains. The resistant SZ-R strain showed significantly lower caseinolytic protease activity in the midgut extracts than the susceptible DBM1Ac-S larvae ($p < 0.05$; Duncan's tests; n = 3), likewise, the trypsin activity of resistant SZ-R was also significantly lower than the susceptible DBM1Ac-S larvae, but the chymotrypsin activity was similar between these two *P. xylostella* strains ($p < 0.05$; Duncan's tests; n = 3) (Figure 1A). Subsequently, the incubation of Cry1Ac protoxin with midgut protease extracts from susceptible DBM1Ac-S or resistant SZ-R larvae were compared (Figure 1B). After 1-h incubation, a strong single band of about 65 kDa was produced in DBM1Ac-S gut extracts, corresponding to the processed Cry1Ac protein, having a similar band size as produced by the control treatment with bovine trypsin. In contrast, two bands were observed in the activation produced by SZ-R gut extracts: one similar to that produced by the control bovine trypsin or DBM1Ac-S and the other of higher molecular weight. These results confirmed that the potential altered *PxTryp_SP* gene expression might be involved in Bt Cry1Ac resistance in SZ-R strain via decreasing midgut protease activities and protoxin activation.

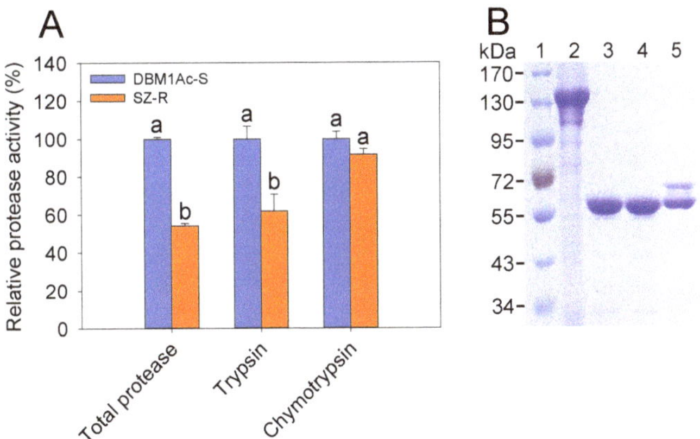

Figure 1. Midgut protease activities (caseinolytic proteases, trypsin, and chymotrypsin) and the activation differences between the *P. xylostella* DBM1Ac-S and SZ-R strains. (**A**). Protease activities were calculated relative to the activities shown by the susceptible DBM1Ac-S strain (100%). Different letters stand for statistically significant differences within the three replicates and four technical repeats ($p < 0.05$; Duncan's test; n = 3). (**B**) Activation of Cry1Ac protoxin with protease midgut extracts from control (DBM1Ac-S) or resistant (SZ-R) strains. Lane 1: protein maker; Lane 2: Cry1Ac protoxin; Lane 3: Cry1Ac incubated with bovine trypsin (positive control); Lane 4: Cry1Ac incubated with protease midgut extracts from DBM1Ac-S strain; Lane 5: Cry1Ac incubated with midgut extracts from SZ-R strain.

2.2. Bioassay Analyses of Cry1Ac Protoxin and Activated Toxin

To further validate the influence of altered *PxTryp_SP* gene expression on the resistance to Cry1Ac toxin in SZ-R strain, bioassays with Cry1Ac (protoxin and activated toxin) were further conducted. Bioassays revealed that the resistance levels of Cry1Ac protoxin or activated toxin by trypsin were different in the SZ-R strain (Table 1). The resistance ratios (RR) were 662 for Cry1Ac protoxin and 422 for the activated Cry1Ac toxin in the SZ-R strain. The LC_{50} values of activated toxin and protoxin showed slightly but significant differences since the LC_{50} value of activated toxin was less than two fold lower to that of Cry1Ac protoxin (Table 1). These data correlated with the partial activation of Cry1Ac protoxin as shown above (Figure 1B). Furthermore, to verify whether Cry1Ac protoxin is about half as potent as the activated toxin due to their different molecular weights (130 vs. 65 kDa), we estimated the potency of Cry1Ac protoxin relative to activated toxin as reported before [33]. The potency ratios (PR) were 0.83 for the DBM1Ac-S strain and 0.53 for the SZ-R strain, which did not differ significantly from the predicted value of 0.50 (one sample t-test, df = 1, t = 3.25, p = 0.19), implying that the Cry1Ac protoxin was no more effective than the activated toxin in both strains analyzed (Table 1).

Table 1. Bioassays of Cry1Ac protoxin and activated toxin in DBM1Ac-S and resistant SZ-R larvae.

Strains	Treatments	Slope (± SEM) [a]	LC_{50} (95% FL) [b]	RR [c]	PR [d]
DBM1Ac-S	Protoxin	2.008 (±0.232)	0.83 (0.64–1.06)	1.0	
SZ-R	Protoxin	1.815 (±0.252)	549.62 (411.60–797.83) *	662	
DBM1Ac-S	Activated toxin	2.156 (±0.252)	0.69 (0.54–0.87)	1.0	0.83
SZ-R	Activated toxin	1.561 (±0.200)	291.04 (216.61–408.95) *	422	0.53

[a] Slope of the dose response-mortality. SEM stands for standard error of the mean. [b] LC_{50} (95% FL): Toxin concentration (mg/L) killing 50% of larvae and its 95% fiducial limits (lower-upper). [c] RR: Resistance ratio calculated by the ratio between the LC_{50} value of SZ-R by the LC_{50} of DBM1Ac-S. [d] PR: Potency was calculated as the ratio of LC_{50} value of activated toxin by the LC_{50} of protoxin as reported [33,34]. Potency values < 1 indicate the activated Cry1Ac toxin is more potent than protoxin, while potency values > 1 indicate that Cry1Ac protoxin is more potent than activated toxin. * Asterisks represent significantly different LC_{50} value by the conservative criterion of non-overlapped 95% CL value.

2.3. Cloning, Characterization, and Phylogenetic Analyses of the PxTryp_SPc1 Gene

During the previous characterization of differentially altered genes in SZ-R, we identified that a *PxTryp_SPc1* gene was possibly down-regulated [31,32]. Thus, we further explored this gene in *P. xylostella*. Based on the unigene sequences from the midgut transcriptome database of *P. xylostella* [32], the full-length cDNA sequence of *PxTryp_SPc1* gene (GenBank accession no. MN422356) was cloned from fourth-instar *P. xylostella* larval midgut tissue using specific primers (Supplementary Materials Table S1). The cDNA sequence of the *PxTryp_SPc1* gene (799 bp) contains an ORF of 768 nucleotides encoding 222 amino acid residues. The genomic DNA (gDNA) sequence of this gene can be found in the *P. xylostella* genome (DBM-DB: http://iae.fafu.edu.cn/DBM, Gene ID: Px016056). The genomic analysis revealed that it contains four exons (Figure 2A). The amino acid sequence of the *PxTryp_SPc1* showed structural features characteristic of members of the trypsin family, as three catalytic residues His (^{70}H), Ser (^{116}S), and Asp (^{211}D) (Figure 2B).

The PxTryp_SPc1 protein shares sequence identity from 17% to 52% with other insect trypsin orthologs, as revealed by the BLASTp homology search of the GenBank database (Supplementary Materials Figure S1). Moreover, phylogenetic analysis of different insect trypsin orthologs showed that trypsin proteins from different insect orders are clustered in independent branches and are evolutionarily conserved (Figure 2C). Additionally, the phylogenetic tree revealed close relationship among trypsin proteins from Lepidoptera and PxTryp_SPc1, which indicated that these trypsin proteins are homologous. Moreover, those trypsin proteins that were reported to be related to Bt resistance were not identified as PxTryp_SPc1 orthologs and were not included in this phylogenetic tree.

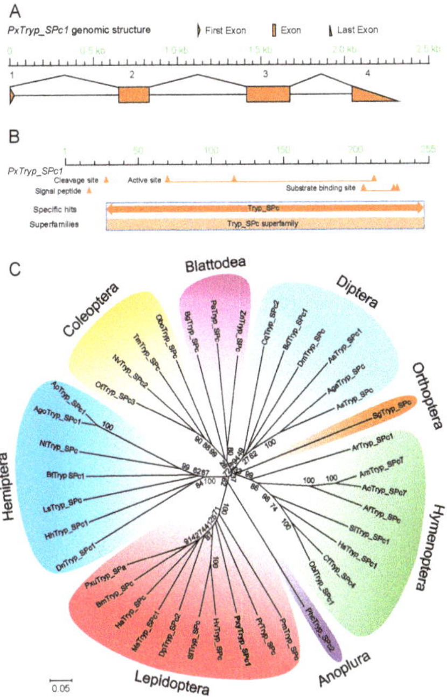

Figure 2. Structural and phylogenetic relationship analyses of *PxTryp_SPc1* gene. (**A**) Genomic structure of *PxTryp_SPc1* gene. Orange boxes represent exons, and the spaces between two boxes represent introns. The figure is drawn to scale. (**B**) Conserved domain annotation obtained from NCBI annotation of the PxTryp_SPc1 protein sequence. The protein sequence was considered as a characteristic member of the trypsin family. The location of the signal peptide, cleavage site, active sites, and substrate binding sites are indicated by orange triangles. (**C**) Phylogenetic analysis of the PxTryp_SPc1 protein and its orthologs in diverse insects by the neighbor-joining (NJ) method. The unrooted phylogenetic tree was constructed by ClustalW alignment of amino acid sequences in MEGA-X. The bootstrap values with 1000 replications are shown on branches. The amino acid sequences of these trypsins were retrieved from the GenBank database (GenBank accession numbers are listed below). The scale bar shows the evolutionary distances. Abbreviations: 1. Lepidoptera (Pm (*Papilio machaon*, KPJ14943); Pr (*Pieris rapae*, XP_022118678); Pxy (*Plutella xylostella*, MN422356); Hv (*Heliothis virescens*, AFO68329); Sl (*Spodoptera litura*, XP_022815738); Dp (*Danaus plexippus*, OWR45697); Ms (*Manduca sexta*, CAM84320); Ha (*Helicoverpa armigera*, ABU98624); Bm (*Bombyx mori*, XP_004923288); Pxu (*Papilio xuthus*, KPJ03461)); 2. Hemiptera (Dn (*Diuraphis noxia*, XP_015367971); Hh (*Halyomorpha halys*, XP_024219146); Ls (*Laodelphax striatellus*, RZF38227); Bt (*Bemisia tabaci*, XP_018896298); Nl (*Nilaparvata lugens*, XP_022184709)); Ago (*Aphis gossypii*, XP_027850262); Ap (*Acyrthosiphon pisum*, XP_001943273)); 3. Coleoptera (Obo (*Oryctes borbonicus*, KRT83696); Tm (*Tenebrio molitor*, AFB81537); Nv (*Nicrophorus vespilloides*, XP_017773892); Ot (*Onthophagus taurus*, XP_022900611)); 4. Blattodea (Bg, (*Blattella germanica*, AAZ78212); Pa (*Periplaneta americana*, AIA09342); Zn (*Zootermopsis nevadensis*, XP_021914447)); 5. Diptera (As (*Anopheles sinensis*, KFB42846); Aga (*Anopheles gambiae*, XP_317171.2); Aa (*Aedes aegypti*, XP_001657786); Bd (*Bactrocera dorsalis*, XP_011214086); Cq (*Culex quinquefasciatus*, XP_001847028); Dm (*Drosophila melanogaster*, NP_001285772)); 6. Orthoptera (Sg, (*Schistocerca gregaria*, CAA70820)); 7. Hymenoptera (Ar (*Athalia rosae*, XP_020711972); Am (*Apis mellifera*, XP_623564); Ac (*Apis cerana*, XP_016922703); Af (*Apis florea*, XP_012344846); Si (*Solenopsis invicta*, XP_011166798); Hs (*Harpegnathos saltator*, EFN81462); Cf (*Camponotus floridanus*, XP_011266670); Ob (*Ooceraea biroi*, XP_011336015)); 8. Anoplura (Phc (*Pediculus humanus corporis*, AAV48634)).

2.4. Tissue Expression Profiles of the PxTryp_SPc1 Gene

Expression analysis of the *PxTryp_SPc1* gene by qPCR in the different tissues of the fourth-instar larvae indicated that it was specifically expressed in the midgut (MG) tissue, in contrast to its expression in the head, integument, testis, and Malpighian tubules ($p < 0.05$; Duncan's test; n = 3) (Figure 3A). Moreover, expression analysis of *PxTryp_SPc1* gene in various developmental stages showed that its expression levels gradually raised from egg (EG) into the larval stages and reached the highest peak in the fourth-instar larvae (L4), while it showed low expression in pre-pupae, pupae, female, and male adults (Figure 3B). The expression of *PxTryp_SPc1* gene was high in midgut and larval stages of *P. xylostella*.

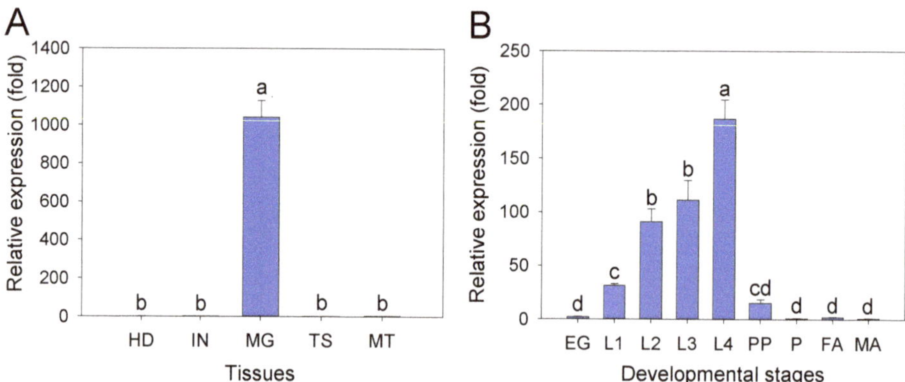

Figure 3. qPCR expression profile of the *P. xylostella PxTryp_SPc1* gene in different tissues and developmental stages. (**A**) Relative expression levels of *PxTryp_SPc1* in different tissues including head (HD), integument (IN), midgut (MG), testis (TS), and Malpighian tubules (MT) of fourth-instar larvae. (**B**) Expression profile of *PxTryp_SPc1* in different developmental stages: eggs (EG), first-instar larvae (L1), second-instar larvae (L2), third-instar larvae (L3), fourth-instar larvae (L4), prepupae (PP), pupae (P), male adults (MA), and female adults (FA). *RPL32* gene expression was used as the internal reference gene to normalize and calculate the gene expression levels. Expression level was calculated according to the value of the lowest expression identified (Tissues: HD; developmental stages: P), which was given an arbitrary value of 1. The means and the corresponding standard errors are shown. Different letters stand for statistically significant differences within the three replicates and four technical repeats ($p < 0.05$; Duncan's test; n = 3).

2.5. The Expression of PxTryp_SPc1 Gene in Susceptible DBM1Ac-S and Resistant SZ-R Strains

Expression difference of the *PxTryp_SPc1* gene by qPCR was compared in the resistant SZ-R and susceptible DBM1Ac-S strains (Figure 4). In general, the transcript levels of *PxTryp_SPc1* showed a significantly reduced expression (about 2.8-fold down) in the SZ-R strain compared to the DBM1Ac-S strain ($p < 0.05$; Duncan's test; n = 3). Furthermore, treatment of third-instar SZ-R larvae with a high concentration of Cry1Ac protoxin (2000 mg/L), showed that the transcript level of *PxTryp_SPc1* gene was further down-regulated ($p < 0.05$; Duncan's test; n = 3), showing a ratio of ~5.1-fold down compared to the DBM1Ac-S strain (Figure 4).

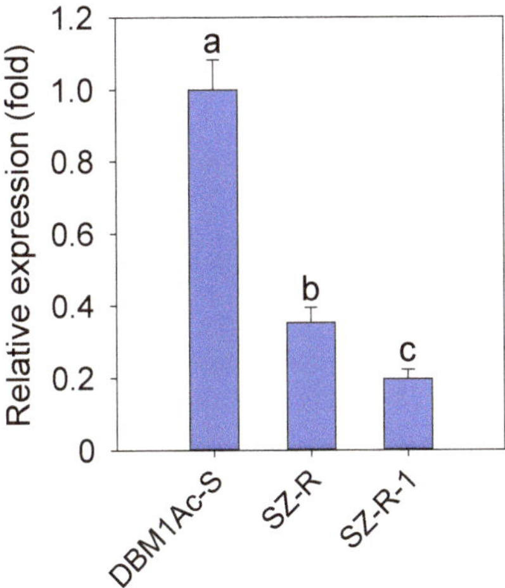

Figure 4. The expression differences of *PxTryp_SPc1* gene between susceptible and resistant strains. Expression levels of *PxTryp_SPc1* by qPCR in fourth-instar larval midgut tissue from susceptible and resistant strains. Lane 1: DBM1Ac-S; Lane 2: SZ-R; Lane 3: SZ-R intoxicated with 2000 mg/L Cry1Ac protoxin. *RPL32* gene was considered as a reference gene to normalize and calculate the level of gene expression. The expression level was calculated based on the value of the highest expression (DBM1Ac-S, arbitrary value of 1). The means and standard errors are shown. Different letters represent statistically significant differences with three independent repeats and four technical replications ($p < 0.05$; Duncan's test, n = 3).

2.6. Linkage between Decreased PxTryp_SPc1 Gene Expression and Cry1Ac Resistance in SZ-R Strain

To determine the genetic linkage of decreased *PxTryp_SPc1* expression with Cry1Ac resistance in the SZ-R strain, a single-pair cross between a male SZ-R larva and a female DBM1Ac-S larva was performed to obtain F1 progeny. Subsequently, backcross family a or b generated from reciprocal crosses between SZ-R moths and F1 progeny were selected and fed on cabbage leaves without or with a diagnostic dose of Cry1Ac protoxin (20 mg/L), and the midgut samples from fourth-instar *P. xylostella* larvae were subjected to qPCR analysis. The qPCR results indicated that *PxTryp_SPc1* gene expression levels in individual fourth-instar larval midguts from F1 generation resemble those in their susceptible DBM1Ac-S strain (Figure 5), implying that the resistance trait in SZ-R is recessive. Nevertheless, the expression levels of *PxTryp_SPc1* in midgut tissue from two backcross families (backcross a and b) showed two different groups; one displayed notable decreased expression levels of *PxTryp_SPc1* (< ~2.8-fold), but another group demonstrated similar expression levels to those of larvae midgut tissue from the original susceptible DBM1Ac-S strain or the F1 generation from the DBM1Ac-S and SZ-R strains cross (Figure 5). The ratios between the two families of individuals were found to be 10:8 (backcross a) and 9:9 (backcross b), following the calculated random assortment ratio 1:1 basically ($p > 0.1$ or $p = 1.0$; χ^2 test). On the contrary, all of the survivals from Cry1Ac exposure in the two backcross families showed decreased expression levels of *PxTryp_SPc1* (<~2.8-fold) compared to larvae of the DBM1Ac-S strain or the F1 progenies, testifying a co-segregation (linkage) with resistance to Cry1Ac in SZ-R ($p < 0.05$, χ^2 test) (Figure 5). Thus, the decreased expression level of the *PxTryp_SPc1* gene was tightly linked to Cry1Ac resistance in the *P. xylostella* SZ-R strain.

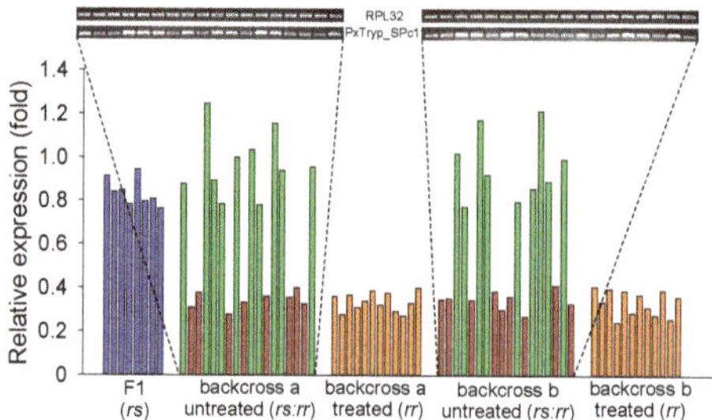

Figure 5. Genetic linkage analysis of the decreased *PxTryp_SPc1* expression level in the SZ-R strain of *P. xylostella* and resistance to Cry1Ac. The expression levels of *PxTryp_SPc1* in F1 larvae, Cry1Ac-treated backcross families (family a and b), and non-selected (untreated) are shown in relation to the levels in the DBM1Ac-S strain. Corresponding intensity of PCR bands for the *PxTryp_SPc1* and the reference *RPL32* gene are exhibited (Upper).

2.7. RNAi-Mediated Functional Assay of the PxTryp_SPc1 Gene

The *PxTryp_SPc1* gene expression was silenced by microinjection of *P. xylostella* susceptible larvae with *PxTryp_SPc1* dsRNA to determine the potential role of the *PxTryp_SPc1* gene in Cry1Ac resistance. The expression levels were statistically reduced after 24 h post dsRNA injection, and the reduction effect lasted almost 96 h. In contrast, controls treated with buffer or dsEGFP, did not show any silencing effect on *PxTryp_SPc1* expression (Figure 6A). The subsequent bioassays revealed that silencing of *PxTryp_SPc1* gene reduced larval susceptibility to Cry1Ac protoxin at 1 mg/L (the LC_{50} value) or at 2 mg/L (the LC_{90} value) after 48 h post-injection compared to control larvae injected with buffer or dsEGFP (Figure 6B), suggesting that the reduced expression of *PxTryp_SPc1* gene correlated with higher tolerance of *P. xylostella* to Cry1Ac. We also determined the effect of *PxTryp_SPc1* gene silencing on Cry1Ac protoxin activation using larval midgut extracts from different RNAi treatments. After 1 h of Cry1Ac incubation from larvae injected with nonspecific dsEGFP and buffer-only, a single band of ~65 kDa was observed (Figure 6C, lanes 4 and 6). In contrast, larvae treated with dsPxTryp_SPc1 showed the two bands of Cry1Ac as previously observed with midgut extract from the SZ-R strain (Figure 6C, lane 5 compared to Figure 6C lane 5).

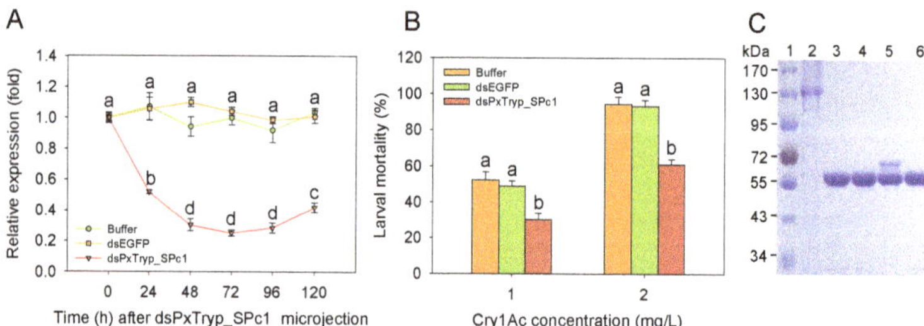

Figure 6. Influences on Cry1Ac toxicity of *PxTryp_SPc1* expression in larval midgut after RNAi silencing. (**A**) Impacts on injection of larvae with buffer, dsEGFP, or dsPxTryp_SPc1 on *PxTryp_SPc1* expression after 120 h RNAi silencing from *P. xylostella*. Different letters represent statistically significant differences within three repeats and four technical replications ($p < 0.05$; Duncan's test, n = 3). (**B**) Mortality of *P. xylostella* larvae after treatment with two concentrations of Cry1Ac protoxin; larvae were injected with buffer, dsEGFP, or dsPxTryp_SPc1. Within each group, different letters denote statistically significant differences between treatments ($p < 0.05$; Duncan's test, n = 3). (**C**) Activation analysis of Cry1Ac protoxin by *P. xylostella* midgut extracts from larvae injected with: dsEGFP (lane 4), dsPxTryp_SPc1 (lane 5), and buffer only (lane 6). Lane1: protein maker; Lane 2: Cry1Ac protoxin; Lane 3: Cry1Ac incubated with bovine trypsin (positive control).

3. Discussion

The role and function of the insect proteases present in midgut tissue in mediating Bt resistance has been analyzed in different *P. xylostella* resistant strains. It was shown before that decreased activation of protoxin to toxin could be a major Bt resistance mechanism in the Cry1Ac-resistant Cry1Ac-SEL strain of *P. xylostella* [35] and a Bt resistance strain of *P. xylostella* [36], but the specific midgut protease gene involved was not identified. Here, we show that the reduced expression of the *PxTryp_SPc1* trypsin gene contributes to Cry1Ac resistance in the Cry1Ac-resistant SZ-R strain. The activities of caseinolytic and trypsin proteases in midgut extracts significantly decreased in contrast to the susceptible strain, suggesting that reduced trypsin protease activity is associated with a resistant phenotype of the SZ-R strain. Based on previously published transcriptome, RNA-Seq, and proteomics-based studies [31,32,37], we identified a new trypsin gene, *PxTryp_SPc1* (GenBank accession no. MN422356, DBM-DB gene ID: Px016056), which is mainly expressed in the midgut and that is down-regulated in the SZ-R resistant strain. Previous identified trypsin proteins involved in Bt resistance in different lepidopteran insects including OnT23 (~50%, GenBank accession no. AAR98919), SfT6 (~43%, GenBank accession no. ACR25157), HaSP2 (~33%, GenBank accession no. ABP96915), and HaTryR (~28%, GenBank accession no. AHL46496) have been reported to be associated with Bt resistance in *O. nubilalis* [14], *S. frugiperda* [15], and *H. armigera* [16,17] (the percentage in parentheses means the identity between the amino acid sequences of these trypsins and PxTryp_SPc1). Although the protein identity between PxTryp_SPc1 and OnT23 or SfT6 is as high as 50% and 43%, respectively, another *P. xylostella* trypsin that shared 57% protein similarity with PxTryp_SPc1 was identified in the *P. xylostella* genome database (DBM-DB gene ID: Px015403). This *P. xylostella* trypsin has higher protein identity (55% and 63%, respectively) with OnT23 or SfT6, indicating that PxTryp_SPc1 and OnT23 or SfT6 are actually not orthologs. Some of the orthologs of PxTryp_SPc1 in other lepidopteran insects are shown in the phylogenetic tree constructed in this study (Figure 2C). The role of other orthologs from different lepidopteran insects in Cry toxin activation still remains to be identified. These results indicated that PxTryp_SPc1 is a novel trypsin member related to Bt resistance, which enriches the Bt-responsive midgut trypsin gene repertoire in insects.

The most common mechanism of high resistance levels to Bt Cry toxins in lepidopteran insects is related to reduced toxin binding to midgut receptors [7,38], while an altered protease expression mechanism has been associated with low or moderate resistance in lepidopteran insects [39]. Indeed, altered processing of Cry1Ac protoxin by midgut proteases is not related to high-level field-evolved Cry1Ac resistance in the *P. xylostella* NIL-R strain [30]. The laboratory-selected strain SZ-R shows moderate Cry1Ac resistance to Cry1Ac activated toxin or protoxin (Table 1). We reported that the *PxCAD* and *PxABCH1* genes are not associated with Cry1Ac resistance in the SZ-R strain [24], but differential expression of the *PxmALP*, *PxABCB1*, *PxABCC1*, *PxABCC2*, *PxABCC3*, and *PxABCG1* genes was shown to be associated with Cry1Ac resistance in the SZ-R strain [27–29], suggesting that reduction in toxin binding is associated with Cry1Ac resistance in the SZ-R strain. Thus, the resistance mechanism related to reduced expression of the trypsin gene *PxTryp_SPc1* identified in this study is an additional mechanism in this *P. xylostella* strain. The reduced expression of the *PxTryp_SPc1* gene correlated well with altered Cry1Ac protoxin activation, suggesting that incomplete activation of protoxin is conducive to developing the resistant phenotype (Figure 1B). However, our data showed that protoxin activation was not completely blocked in the SZ-R strain since treatment of Cry1Ac protoxin with midgut juice from the resistant population resulted in two protein bands; one correlated with the 65 kDa Cry1Ac activated toxin, implying that other midgut proteases may still participate in the activation of Cry1Ac protoxin in the resistant SZ-R strain. These data correlated with the toxicity bioassays revealed that SZ-R was only two-fold more susceptible to the Cry1Ac activated toxin compared with the protoxin (Table 1). Nevertheless, RNAi analysis showed that silencing *PxTryp_SPc1* gene did reduce the larval susceptibility to Cry1Ac toxin supporting that this protein contributes to the Cry1Ac resistance phenotype of SZ-R strain (Figure 6). Moreover, reduced expression of *PxTryp_SPc1* was linked to Cry1Ac resistance in SZ-R strain (Figure 5).

The regulation mechanism involved in the reduced expression of *PxTryp_SPc1* in the SZ-R strain still remains to be determined. Interestingly, down-regulation of the trypsin gene *HaTryR* in *H. armigera* is caused by a promoter sequence mutation mediated *cis*-regulatory mechanism [17]. Our previous studies demonstrated that different expression of the *PxmALP*, *PxABCB1*, *PxABCC1–3*, and *PxABCG1* genes can be modulated by the MAPK signaling pathway [27,29]. Thus, whether reduced expression of *PxTryp_SPc1* in SZ-R strain is conferred by the promoter mutation-induced *cis*- or MAPK-induced *trans*-regulatory mechanism warrants further study. In addition, considering that the *PxABCC2* and *PxABCC3* genes were successfully knocked out by a novel CRISPR/Cas9 genome editing tool confirming their involvement in Bt Cry1Ac resistance [40], we will further utilize the CRISPR/Cas9 system to edit the *PxTryp_SPc1* gene to offer in vivo reverse genetic evidence of its involvement in Cry1Ac resistance, which thus could help us in the future to determine what is the function of *PxTryp_SPc1* gene during Bt Cry1Ac toxin activation processing in *P. xylostella*. Moreover, the *P. xylostella* genome contains many different trypsin genes [41]; some other trypsin genes displaying marked differential expression levels between Cry1Ac-susceptible and -resistant *P. xylostella* strains have also been found previously by transcriptome, RNA-Seq, and proteomics analyses, and further investigations of their potential functions in Bt Cry1Ac resistance in *P. xylostella* are also needed.

Overall, our data confirmed that down-regulation of a novel trypsin gene (*PxTryp_SPc1*) is associated with Cry1Ac resistance in the SZ-R strain of *P. xylostella*. This study is helpful to elucidate the complex causes of Bt Cry1Ac resistance in *P. xylostella*. The deeper understanding that we have of these mechanisms, the stronger and better strategies we will be able to propose to cope with the evolution of insect resistance to Bt toxins.

4. Materials and Methods

4.1. Insect Strains

The *P. xylostella* susceptible DBM1Ac-S and resistant SZ-R strains that were used in this study were previously described [24,28]. The SZ-R strain was originated from field collected moths at Shenzhen in China (2003), and it was constructed by constant selection with a concentration of Cry1Ac protoxin that generally kills 50%–70% of larvae in the laboratory for more than 200 generations. Both *P. xylostella* strains were reared on Chinese cabbage, JingFeng No. 1 (*Brassica oleracea* var. *capitata*), at 65% RH, 25 °C, with a photoperiod of 16 h light:8 h dark, and adults were fed with a 10% sucrose solution.

4.2. Midgut Protease Activity Assays

The caseinolytic protease was measured at 28 °C using the substrate azocasein (Sigma, St. Louis, MO, USA), as previously reported [42]. In brief, midgut extracts (20 µL) were mixed with 1% azocasein in 50 mM $NaHCO_3$-Na_2CO_3 buffer (150 µL) and incubated for 2 h at 28 °C. Then 10% trichloroacetic acid (TCA) (170 µL) was used to stop the reaction. The solution was incubated at 25 °C for 1 h and centrifuged for 15 min at $16,000\times g$ at room temperature to remove the debris. Then, 1 M NaOH (340 µL) was mixed and the optical density (OD) of collected supernatant was measured at 450 nm in a SpectraMaxM2^e microplate reader (Molecular Devices, Sunnyvale, CA, USA).

Chymotrypsin and trypsin activities were detected using 1 mM Nα-benzoyl-L-arginine-*p*-nitroanilide (BApNA, Sigma) and 1 mM *N*-succinyl-Ala-Ala-Pro-phenylalanine *p*-nitroanilide (SAApFpNA, Sigma) as respective specific substrates. For chymotrypsin activity determination, 5 µL midgut extract was mixed with 3 mL of 1 mM SAApFpNA in 50 mM $NaHCO_3$-Na_2CO_3 buffer. For trypsin activity examination, 10 µL midgut extract was mixed with 3 mL of 1 mM BApNA in 50 mM $NaHCO_3$-Na_2CO_3 buffer. The peptidolytic reaction was tested immediately by recording continuously the optic density (OD) value at 405 nm every 15 s at 28 °C for 30 min. The enzyme activities are exhibited as relative activities of the DBM1Ac-S midgut extract protease activities, which were considered as 100%. Biological assay was performed in triplicate and four technical repetitions each were used to confirm the protease activities. For analysis of statistical differences among samples, one-way ANOVA with Duncan's tests ($p < 0.05$) was used.

4.3. Bioassays

The Cry1Ac protoxin and trypsin-activated toxin were obtained as previously described [30]. The Cry1Ac toxin was finally dissolved in 50 mM Na_2CO_3 (pH 9.6) and stored at $-20\,°C$. The respective toxicity of the Cry1Ac protoxin and trypsin-activated toxin was determined by 72-h leaf-dip bioassays using a total of 280 third-instar *P. xylostella* larvae per bioassay as described before [27,28]. In short, ten larvae that were exposed to seven different concentrations of Cry1Ac toxin in each group, and four repeats were performed for all bioassays. The control mortality did not exceed 5%. We used the POLO Plus 2.0 software (LeOra Software, Berkeley, CA, USA) to calculate the LC_{50} values (median lethal concentrations killing 50% of the tested larvae) and 95% CL (95% confidence limits of the LC_{50}) values by Probit analysis.

4.4. RNA Extraction and cDNA Synthesis

The methods of RNA extraction and cDNA synthesis from *P. xylostella* were previously described [27]. The midgut samples were extracted in TRIzol Reagent (Invitrogen, Carlsbad, CA, USA); then the concentration of RNA was quantified by a NanoDrop 2000c spectrophotometer (Thermo Fisher Scientific Inc., Waltham, MA, USA). PrimeScript II 1st strand cDNA Synthesis Kit (TaKaRa, Dalian, China) was used to synthetize first-strand DNA. For qPCR analysis, 1 µg total RNA was used to perform the first-strand cDNA with the PrimeScript RT kit (containing gDNA Eraser, Perfect Real Time) (TaKaRa, Dalian, China) following the manufacturer's instructions. The synthesized cDNA was immediately used or stored at $-20\,°C$ until used.

4.5. Gene Identification and Cloning

The candidate cDNA sequence of *PxTryp_SPc1* gene was identified in our previous midgut transcriptome database of *P. xylostella* [32] and was further in silico corrected by the *P. xylostella* genome database (DBM-DB: http://iae.fafu.edu.cn/DBM, Gene ID: Px016056); then the specific primers were designed (Supplementary Materials Table S1) and were used in subsequent PCR amplification assays. The full-length cDNA sequence *of PxTryp_SPc1* gene was finally obtained and deposited in the GenBank database (accession no. MN422356).

As described previously [27], the PCR reaction (25 µL total volume) contained 0.2 µL LA Taq HS polymerase (TaKaRa, Dalian, China) in an C1000 Thermal Cycler PCR system (BioRad, Philadelphia, PA, USA) for 35 cycles using LA Taq polymerase (TaKaRa, Dalian, China). A gel extraction kit (CWBIO, Beijing, China) was used for purification of the PCR products of *PxTryp_SPc1*, which were further cloned into the pEASY-T1 vector (TransGen, Beijing, China). For gene sequencing, *Escherichia coli* TOP10 competent cells (TransGen, Beijing, China) were transformed with candidate plasmids.

4.6. Gene Sequence Analysis

DANMAN 8.0 (Lynnon BioSoft, San Ramon, CA, USA) software was used for gene sequence assembly, exon-intron analysis, and multiple sequence alignment. The open reading frame (ORF) was identified by the ORF finder tool at the NCBI (https://www.ncbi.nlm.nih.gov/orffinder/), and predicted amino acid sequences were achieved by ExPASy online tool to translate (https://web.expasy.org/translate/). The BLAST tool at the GenBank database (https://blast.ncbi.nlm.nih.gov) was used for the sequence-similarity analyses. The protein-specific motifs and active sites were found and annotated at the GenBank database (https://www.ncbi.nlm.nih.gov/). The signal peptide was predicted by SignalP-5.0 Server online (http://www.cbs.dtu.dk/services/SignalP/).

4.7. Phylogenetic Tree Construction

To verify the classification of the *PxTryp_SPc1* gene, phylogenetic analysis of the PxTryp_SPc1 protein was done by using the full-length amino acid sequences of its orthologs from other insects. MEGA-X software (https://www.megasoftware.net/) with ClustalW algorithm was used to construct the phylogenetic tree. An unrooted neighbor-joining (NJ) phylogenetic tree was done choosing the "p-distance" as the amino acid substitution model; the bootstrap value was determined from 1000 replicates.

4.8. Sample Preparation

Samples from different developmental stages were collected, and different tissues were also dissected from the fourth-instar DBM1Ac-S larvae to characterize the spatio-temporal expression patterns of the *PxTryp_SPc1* gene. Moreover, in order to resolve whether the expression level of *PxTryp_SPc1* was related to Cry1Ac resistance, third-instar SZ-R larvae were treated with a high concentration of Cry1Ac protoxin (2000 mg/L). After the midguts from the survivors were dissected, the extraction of total RNA and cDNA was synthesized as mentioned above. Data were obtained from three biological replications performed in all samples.

4.9. Gene Expression Analysis

Gene expression differences were determined by real-time quantitative PCR (qPCR) as described before with slight modification [27,28]. Briefly, Primer Premier 5.0 (PREMIER Biosoft international, Palo Alto, CA, USA) was used for defining specific *PxTryp_SPc1* gene primers (Supplementary Materials Table S1). PCR reactions (20 μL) contained 7.4 μL RNase-Free ddH$_2$O, 10 μL of 2 × FastFire qPCR PreMix Plus (TIANGEN, Beijing, China), 5 μM of each specific primer, 1 μL of first-strand cDNA template, and 0.4 μL 50 × ROX Reference Dye (TIANGEN, Beijing, China). The running program consisted of a denaturation at 95 °C for 10 min followed by 40 denaturalized cycles at 95 °C for 15 s, annealing at 57 °C for 30 s, and extension at 72 °C for 30 s. All reactions were performed in an Applied Biosystems QuantStudio 3 Real-Time PCR System (Applied Biosystems, Forster City, CA, USA). As an internal control for relative quantification, the *ribosomal protein L32 (RPL32)* gene (GenBank accession no. AB180441) was used in qPCR data analysis. Three biological repetitions and four technical repetitions were conducted for each sample. To define the statistically differences, one-way ANOVAs with Duncan's test (overall significance level $p < 0.05$) were used.

4.10. Linkage Analysis

Genetic linkage analysis was performed as previously described [27,28]. F1 progeny was generated by a single-pair mating between a SZ-R male and a DBM1Ac-S female. A diagnostic Cry1Ac protoxin diagnostic dose (20 mg/L) killed all the F1 (heterozygous) larvae was determined in a toxicity bioassay. Reciprocal crosses between SZ-R moths and F1 progeny were made to generate backcross family a and b. Forty larvae from each backcross families of progeny were fed on cabbage (non-Cry1Ac-selected) or cabbage with 20 mg/L of Cry1Ac protoxin (Cry1Ac-selected), and midguts tissues from the survived fourth-instar *P. xylostella* larvae were dissected for qPCR analysis as mentioned above.

4.11. RNA Interference (RNAi)

To determine the impact of *PxTryp_SPc1* gene expression in *P. xylostella* resistance to Cry1Ac, RNAi-mediated down expression of *PxTryp_SPc1* gene was performed. Early third-instar *P. xylostella* larvae were microinjected with specific dsRNA targeting *PxTryp_SPc1* gene (dsPxTryp_SPc1), as described previously [24]. Briefly, Primer Premier 5.0 (PREMIER Biosoft international, Palo Alto, CA, USA) was used to design the dsRNA primers containing the T7 promoter on the 5′ end targeting gene-specific region of *PxTryp_SPc1* (GenBank accession no. MN422356) or *EGFP* gene (GenBank accession no. KC896843) (Supplementary Materials Table S1). To further validate the specificity of these dsRNAs, BLASTn searches of the GenBank database (https://www.ncbi.nlm.nih.gov/) and the *P. xylostella* genome database (DBM-DB: http://iae.fafu.edu.cn/DBM) were performed showing no unspecific hit diminishing potential off-target effects. The amplicons (389 bp for dsPxTryp_SPc1 and 469 bp for dsEGFP) were used as templates for in vitro transcription reactions to produce dsRNAs using the T7 RiboMAX Express RNAi System (Promega, Madison, WI, USA). The synthesized dsRNA was dissolved in 10 mM Tris–HCl (pH 7.0), and 1 mM EDTA was used as injection buffer and mixed with Metafectene PRO transfection reagent (Biontex, Planegg, Germany). A total of 300 ng (70 nL) dsEGFP or dsPxTryp_SPc1 were injected into the hemocoel of DBM1Ac-S larvae, resulting in less than 20% larval mortality determined after 5 days. Finally, to determine the silencing efficiency at 48 h post-injection, midgut tissue was dissected from injected larvae. The control group was injected with equal volumes of buffer alone. At least 30 larvae were analyzed for each treatment, and three replicate experiments were conducted. The bioassay data were processed as mentioned above. Statistically significant differences between qPCR and bioassay analyses were determined by one-way ANOVAs with Duncan's tests (overall significance level $p < 0.05$).

Supplementary Materials: The following are available online at http://www.mdpi.com/2072-6651/12/2/76/s1, Figure S1. Pairwise comparisons of protein sequence identities among nine trypsin orthologs of PxTryp_SPc1 from different insect species. Table S1. List of primers used in this study.

Author Contributions: L.G. (Lijun Gong), Z.G., L.W., and Y.Z. conceived and designed the experiments. L.G. (Lijun Gong), Z.G., S.K., J.Z., D.S., L.G. (Le Guo), J.Q., L.Z., Y.B., F.Y., M.A., Q.W., S.W., B.X., and Z.Y. performed the experiments. L.G. and Z.G. analyzed the data. L.G. (Lijun Gong), A.B., M.S., Z.G., L.W., and Y.Z. wrote the paper. All authors have read and agreed to the published version of the manuscript.

Funding: This research was funded by the National Natural Science Foundation of China (31630059; 31701813), the Beijing Key Laboratory for Pest Control and Sustainable Cultivation of Vegetables, and the Science and Technology Innovation Program of the Chinese Academy of Agricultural Sciences (CAAS-ASTIP-IVFCAAS).

Conflicts of Interest: The authors declare no conflict of interest.

References

1. ISAAA. *Global Status of Commercialized Biotech/GM Crops in 2018: Biotech Crops Continue to Help Meet the Challenges of Increased Population and Climate Change*; ISAAA Brief No. 54.; ISAAA: Ithaca, NY, USA, 2018.

2. Bravo, A.; Likitvivatanavong, S.; Gill, S.S.; Soberón, M. *Bacillus thuringiensis*: A story of a successful bioinsecticide. *Insect Biochem. Mol. Biol.* **2011**, *41*, 423–431. [CrossRef]

3. Tabashnik, B.E.; Carrière, Y. Surge in insect resistance to transgenic crops and prospects for sustainability. *Nat. Biotechnol.* **2017**, *35*, 926–935. [CrossRef] [PubMed]

4. Pardo-López, L.; Soberón, M.; Bravo, A. *Bacillus thuringiensis* insecticidal three-domain Cry toxins: Mode of action, insect resistance and consequences for crop protection. *FEMS Microbiol. Rev.* **2013**, *37*, 3–22. [CrossRef] [PubMed]

5. Palma, L.; Muñoz, D.; Berry, C.; Murillo, J.; Caballero, P. *Bacillus thuringiensis* toxins: An overview of their biocidal activity. *Toxins* **2014**, *6*, 3296–3325. [CrossRef] [PubMed]

6. Pigott, C.R.; Ellar, D.J. Role of receptors in *Bacillus thuringiensis* crystal toxin activity. *Microbiol. Mol. Biol. Rev.* **2007**, *71*, 255–281. [CrossRef]

7. Adang, M.J.; Crickmore, N.; Jurat-Fuentes, J.L. Diversity of *Bacillus thuringiensis* crystal toxins and mechanism of action. *Adv. Insect Physiol.* **2014**, *47*, 39–87.

8. Gómez, I.; Sánchez, J.; Muñoz-Garay, C.; Matus, V.; Gill, S.S.; Soberón, M.; Bravo, A. *Bacillus thuringiensis* Cry1A toxins are versatile proteins with multiple modes of action: Two distinct pre-pores are involved in toxicity. *Biochem. J.* **2014**, *459*, 383–396. [CrossRef]

9. Peña-Cardeña, A.; Grande, R.; Sánchez, J.; Tabashnik, B.E.; Bravo, A.; Soberón, M.; Gómez, I. The C-terminal protoxin domain of *Bacillus thuringiensis* Cry1Ab toxin has a functional role in binding to GPI-anchored receptors in the insect midgut. *J. Biol. Chem.* **2018**, *293*, 20263–20272. [CrossRef]

10. Oppert, B.; Kramer, K.J.; Beeman, R.W.; Johnson, D.; McGaughey, W.H. Proteinase-mediated insect resistance to *Bacillus thuringiensis* toxins. *J. Biol. Chem.* **1997**, *272*, 23473–23476. [CrossRef]

11. Lightwood, D.J.; Ellar, D.J.; Jarrett, P. Role of proteolysis in determining potency of *Bacillus thuringiensis* Cry1Ac δ-endotoxin. *Appl. Environ. Microbiol.* **2000**, *66*, 5174–5181. [CrossRef]

12. Huang, F.; Zhu, K.Y.; Buschman, L.L.; Higgins, R.A.; Oppert, B. Comparison of midgut proteinases in *Bacillus thuringiensis*-susceptible and -resistant European corn borer, *Ostrinia nubilalis* (Lepidoptera; Pyralidae). *Pestic. Biochem. Physiol.* **1999**, *65*, 132–139. [CrossRef]

13. Li, H.; Oppert, B.; Higgins, R.A.; Huang, F.; Zhu, K.Y.; Buschman, L.L. Comparative analysis of proteinase activities of *Bacillus thuringiensis*-resistant and -susceptible *Ostrinia nubilalis* (Lepidoptera: Crambidae). *Insect Biochem. Mol. Biol.* **2004**, *34*, 753–762. [CrossRef] [PubMed]

14. Li, H.; Oppert, B.; Higgins, R.A.; Huang, F.; Buschman, L.L.; Gao, J.R.; Zhu, K.Y. Characterization of cDNAs encoding three trypsin-like proteinases and mRNA quantitative analysis in Bt-resistant and -susceptible strains of *Ostrinia nubilalis*. *Insect Biochem. Mol. Biol.* **2005**, *35*, 847–860. [CrossRef] [PubMed]

15. Rodríguez-Cabrera, L.; Trujillo-Bacallao, D.; Borrás-Hidalgo, O.; Wright, D.J.; Ayra-Pardo, C. RNAi-mediated knockdown of a *Spodoptera frugiperda* trypsin-like serine-protease gene reduces susceptibility to a *Bacillus thuringiensis* Cry1Ca1 protoxin. *Environ. Microbiol.* **2010**, *12*, 2894–2903. [CrossRef]

16. Rajagopal, R.; Arora, N.; Sivakumar, S.; Rao, N.G.V.; Nimbalkar, S.A.; Bhatnagar, R.K. Resistance of *Helicoverpa armigera* to Cry1Ac toxin from *Bacillus thuringiensis* is due to improper processing of the protoxin. *Biochem. J.* **2009**, *419*, 309–316. [CrossRef]

17. Liu, C.; Xiao, Y.; Li, X.; Oppert, B.; Tabashnik, B.E.; Wu, K. *Cis*-mediated down-regulation of a trypsin gene associated with Bt resistance in cotton bollworm. *Sci. Rep.* **2014**, *4*, 7219. [CrossRef]

18. Tetreau, G.; Stalinski, R.; David, J.P.; Despres, L. Increase in larval gut proteolytic activities and Bti resistance in the dengue fever mosquito. *Arch. Insect Biochem. Physiol.* **2013**, *82*, 71–83. [CrossRef]

19. González-Cabrera, J.; García, M.; Hernández-Crespo, P.; Farinós, G.P.; Ortego, F.; Castañera, P. Resistance to Bt maize in *Mythimna unipuncta* (Lepidoptera: Noctuidae) is mediated by alteration in Cry1Ab protein activation. *Insect Biochem. Mol. Biol.* **2013**, *43*, 635–643. [CrossRef]

20. Zhang, M.; Wei, J.; Ni, X.; Zhang, J.; Jurat-Fuentes, J.L.; Fabrick, J.A.; Carrière, Y.; Tabashnik, B.E.; Li, X. Decreased Cry1Ac activation by midgut proteases associated with Cry1Ac resistance in *Helicoverpa zea*. *Pest Manag. Sci.* **2019**, *75*, 1099–1106. [CrossRef]

21. Tabashnik, B.E.; Cushing, N.L.; Finson, N.; Johnson, M.W. Field development of resistance to *Bacillus thuringiensis* in diamondback moth (Lepidoptera: Plutellidae). *J. Econ. Entomol.* **1990**, *83*, 1671–1676. [CrossRef]

22. Furlong, M.J.; Wright, D.J.; Dosdall, L.M. Diamondback moth ecology and management: Problems, progress and prospects. *Annu. Rev. Entomol.* **2013**, *58*, 517–541. [CrossRef] [PubMed]

23. Baxter, S.W.; Zhao, J.Z.; Gahan, L.J.; Shelton, A.M.; Tabashnik, B.E.; Heckel, D.G. Novel genetic basis of field-evolved resistance to Bt toxins in *Plutella xylostella*. *Insect Mol. Biol.* **2005**, *14*, 327–334. [CrossRef] [PubMed]

24. Guo, Z.; Kang, S.; Zhu, X.; Wu, Q.; Wang, S.; Xie, W.; Zhang, Y. The midgut cadherin-like gene is not associated with resistance to *Bacillus thuringiensis* toxin Cry1Ac in *Plutella xylostella* (L.). *J. Invertebr. Pathol.* **2015**, *126*, 21–30. [CrossRef] [PubMed]

25. Guo, Z.; Kang, S.; Zhu, X.; Xia, J.; Wu, Q.; Wang, S.; Xie, W.; Zhang, Y. The novel ABC transporter ABCH1 is a potential target for RNAi-based insect pest control and resistance management. *Sci. Rep.* **2015**, *5*, 13728. [CrossRef]

26. Baxter, S.W.; Badenes-Pérez, F.R.; Morrison, A.; Vogel, H.; Crickmore, N.; Kain, W.; Wang, P.; Heckel, D.G.; Jiggins, C.D. Parallel evolution of *Bacillus thuringiensis* toxin resistance in Lepidoptera. *Genetics* **2011**, *189*, 675–679. [CrossRef]

27. Guo, Z.; Kang, S.; Chen, D.; Wu, Q.; Wang, S.; Xie, W.; Zhu, X.; Baxter, S.W.; Zhou, X.; Jurat-Fuentes, J.L.; et al. MAPK signaling pathway alters expression of midgut ALP and ABCC genes and causes resistance to *Bacillus thuringiensis* Cry1Ac toxin in diamondback moth. *PLoS Genet.* **2015**, *11*, e1005124. [CrossRef]

28. Guo, Z.; Kang, S.; Zhu, X.; Xia, J.; Wu, Q.; Wang, S.; Xie, W.; Zhang, Y. Down-regulation of a novel ABC transporter gene (*Pxwhite*) is associated with Cry1Ac resistance in the diamondback moth, *Plutella xylostella* (L.). *Insect Biochem. Mol. Biol.* **2015**, *59*, 30–40. [CrossRef]

29. Zhou, J.; Guo, Z.; Kang, S.; Qin, J.; Gong, L.; Sun, D.; Guo, L.; Zhu, L.; Bai, Y.; Zhang, Z.; et al. Reduced expression of the P-glycoprotein gene *PxABCB1* is linked to resistance to *Bacillus thuringiensis* Cry1Ac toxin in *Plutella xylostella* (L.). *Pest Manag. Sci.* **2020**, *76*, 712–720. [CrossRef]

30. Guo, Z.; Gong, L.; Kang, S.; Zhou, J.; Sun, D.; Qin, J.; Guo, L.; Zhu, L.; Bai, Y.; Bravo, A.; et al. Comprehensive analysis of Cry1Ac protoxin activation mediated by midgut proteases in susceptible and resistant *Plutella xylostella* (L.). *Pestic. Biochem. Physiol.* **2020**, *163*, 23–30. [CrossRef]

31. Lei, Y.; Zhu, X.; Xie, W.; Wu, Q.; Wang, S.; Guo, Z.; Xu, B.; Li, X.; Zhou, X.; Zhang, Y. Midgut transcriptome response to a Cry toxin in the diamondback moth, *Plutella xylostella* (Lepidoptera: Plutellidae). *Gene* **2014**, *533*, 180–187. [CrossRef]

32. Xie, W.; Lei, Y.; Fu, W.; Yang, Z.; Zhu, X.; Guo, Z.; Wu, Q.; Wang, S.; Xu, B.; Zhou, X.; et al. Tissue-specific transcriptome profiling of *Plutella xylostella* third instar larval midgut. *Int. J. Biol. Sci.* **2012**, *8*, 1142–1155. [CrossRef] [PubMed]

33. Tabashnik, B.E.; Zhang, M.; Fabrick, J.A.; Wu, Y.; Gao, M.; Huang, F.; Wei, J.; Zhang, J.; Yelich, A.; Unnithan, G.C.; et al. Dual mode of action of Bt proteins: Protoxin efficacy against resistant insects. *Sci. Rep.* **2015**, *5*, 15107. [CrossRef] [PubMed]

34. Tabashnik, B.E.; Huang, F.; Ghimire, M.N.; Leonard, B.R.; Siegfried, B.D.; Rangasamy, M.; Yang, Y.; Wu, Y.; Gahan, L.J.; Heckel, D.G.; et al. Efficacy of genetically modified Bt toxins against insects with different genetic mechanisms of resistance. *Nat. Biotechnol.* **2011**, *29*, 1128–1131. [CrossRef] [PubMed]

35. Sayyed, A.H.; Gatsi, R.; Kouskoura, T.; Wright, D.J.; Crickmore, N. Susceptibility of a field-derived, *Bacillus thuringiensis*-resistant strain of diamondback moth to in vitro-activated Cry1Ac toxin. *Appl. Environ. Microbiol.* **2001**, *67*, 4372–4373. [CrossRef] [PubMed]

36. Talaei-Hassanloui, R.; Bakhshaei, R.; Hosseininaveh, V.; Khorramnezhad, A. Effect of midgut proteolytic activity on susceptibility of lepidopteran larvae to *Bacillus thuringiensis* subsp. *Kurstaki. Front. Physiol.* **2013**, *4*, 406. [CrossRef] [PubMed]

37. Xia, J.; Guo, Z.; Yang, Z.; Zhu, X.; Kang, S.; Yang, X.; Yang, F.; Wu, Q.; Wang, S.; Xie, W.; et al. Proteomics-based identification of midgut proteins correlated with Cry1Ac resistance in *Plutella xylostella* (L.). *Pestic. Biochem. Physiol.* **2016**, *132*, 108–117. [CrossRef]

38. Ferré, J.; Van Rie, J. Biochemistry and genetics of insect resistance to *Bacillus thuringiensis. Annu. Rev. Entomol.* **2002**, *47*, 501–533. [CrossRef]

39. De Bortoli, C.P.; Jurat-Fuentes, J.L. Mechanisms of resistance to commercially relevant entomopathogenic bacteria. *Curr. Opin. Insect Sci.* **2019**, *33*, 56–62. [CrossRef]

40. Guo, Z.; Sun, D.; Kang, S.; Zhou, J.; Gong, L.; Qin, J.; Guo, L.; Zhu, L.; Bai, Y.; Luo, L.; et al. CRISPR/Cas9-mediated knockout of both the *PxABCC2* and *PxABCC3* genes confers high-level resistance to *Bacillus thuringiensis* Cry1Ac toxin in the diamondback moth, *Plutella xylostella* (L.). *Insect Biochem. Mol. Biol.* **2019**, *107*, 31–38. [CrossRef]

41. Lin, H.; Lin, X.; Zhu, J.; Yu, X.Q.; Xia, X.; Yao, F.; Yang, G.; You, M. Characterization and expression profiling of serine protease inhibitors in the diamondback moth, *Plutella xylostella* (Lepidoptera: Plutellidae). *BMC Genom.* **2017**, *18*, 162. [CrossRef]

42. Wang, P.; Zhao, J.Z.; Rodrigo-Simón, A.; Kain, W.; Janmaat, A.F.; Shelton, A.M.; Ferré, J.; Myers, J. Mechanism of resistance to *Bacillus thuringiensis* toxin Cry1Ac in a greenhouse population of the cabbage looper, *Trichoplusia ni. Appl. Environ. Microbiol.* **2007**, *73*, 1199–1207. [CrossRef] [PubMed]

Communication

Study of the *Bacillus thuringiensis* Cry1Ia Protein Oligomerization Promoted by Midgut Brush Border Membrane Vesicles of Lepidopteran and Coleopteran Insects, or Cultured Insect Cells

Ayda Khorramnejad [1,2], Mikel Domínguez-Arrizabalaga [3], Primitivo Caballero [3], Baltasar Escriche [1] and Yolanda Bel [1,*]

[1] Departamento de Genética/ERI BioTecMed, Universitat de València, Burjassot, 46100 València, Spain; ayda.khoramnejad@uv.es (A.K.); baltasar.escriche@uv.es (B.E.)

[2] Department of Plant Protection, College of Agriculture and Natural Resources, University of Tehran, Karaj 31578-77871, Alborz, Iran

[3] Departamento de Agronomía, Biotecnología y Alimentación, Universidad Pública de Navarra, Pamplona, 31006 Navarra, Spain; mikel.dominguez@unavarra.es (M.D.-A.); pcm92@unavarra.es (P.C.)

* Correspondence: yolanda.bel@uv.es

Received: 16 December 2019; Accepted: 19 February 2020; Published: 21 February 2020

Abstract: *Bacillus thuringiensis* (Bt) produces insecticidal proteins that are either secreted during the vegetative growth phase or accumulated in the crystal inclusions (Cry proteins) in the stationary phase. Cry1I proteins share the three domain (3D) structure typical of crystal proteins but are secreted to the media early in the stationary growth phase. In the generally accepted mode of action of 3D Cry proteins (sequential binding model), the formation of an oligomer (tetramer) has been described as a major step, necessary for pore formation and subsequent toxicity. To know if this could be extended to Cry1I proteins, the formation of Cry1Ia oligomers was studied by Western blot, after the incubation of trypsin activated Cry1Ia with insect brush border membrane vesicles (BBMV) or insect cultured cells, using Cry1Ab as control. Our results showed that Cry1Ia oligomers were observed only after incubation with susceptible coleopteran BBMV, but not following incubation with susceptible lepidopteran BBMV or non-susceptible Sf21 insect cells, while Cry1Ab oligomers were persistently detected after incubation with all insect tissues tested, regardless of its host susceptibility. The data suggested oligomerization may not necessarily be a requirement for the toxicity of Cry1I proteins.

Keywords: Cry1Ab; oligomer formation; Sf21 cell line; *Ostrinia nubilalis*; *Lobesia botrana*; *Leptinotarsa decemlineata*; bioassay

Key Contribution: The paper studies the oligomer formation of trypsin activated Cry1I in vitro, after incubation with insect BBMV or insect cultured cells. The results show that Cry1Ia oligomers are only visualized after incubation with coleopteran susceptible BBMV. This could suggest that oligomerization may not be a limiting step in the mode of action of Cry1I protein.

---

1. Introduction

The entomopathogenic Gram-positive bacterium, *Bacillus thuringiensis* (Bt), is the most successful bioinsecticide commercialized to date. It generates a wide variety of insecticidal proteins that can be produced at different growth stages. Cry, Cyt and parasporin proteins are synthetized during the stationary growth phase, Cry1I proteins are secreted in the initial phase of sporulation and Vip and Sip proteins are secreted during the vegetative phase of bacterial growth [1–5]. The success of Bt-based insecticides is due to its narrow spectrum of activity, environmental safety and because it is harmless to animals and plants [6]. To date, the most studied Bt entomopathogenic proteins are the three domain (3D) Cry proteins such as Cry1 and Cry2 toxins. Their mode of action is not completely known, but it is commonly accepted that specific binding to insect midgut receptors is essential to exert their toxicity [7,8]. The current models of Cry toxin action include the "signal transduction model" that claims that the toxicity is mediated by intracellular pathways [9], and the "sequential mode of action"—the most accepted model so far, that is based on the sequential binding of Bt toxin to several midgut receptors, the promotion of a pre-pore oligomer, the insertion of pre-pore oligomer into the midgut membrane, pore formation, osmotic imbalance, midgut epithelium disruption, septicemia and insect death [10]. Regarding this model, some authors have stated that the formation of the oligomer prior to toxin insertion into membrane is a major step in the toxicity process, and that the oligomeric (tetrameric) structure, is necessary for the final pore formation; also, it has been claimed that the oligomer formation is a conserved mechanism in the mode of action of the Cry proteins [11–16].

Oligomerization has been studied in several wild type and mutant toxins such as Cry1Aa, Cry1Ab, Cry1Ac, Cry1Ca, Cry1Da, Cry1Ea, Cry1Fa, Cry2Ab, Cry3Aa, Cry3Ba, Cry3Ca, Cry4Ba, Cry11Aa and Cry46Aa1 [15,17–25]. Oligomerization has also been detected in other Bt toxins such as Cyt [26] and Vip [27]. However, regarding the Cry1I protein family, only one study has described the possible oligomerization of this protein in solution, in the absence of insect midgut proteins [28]. So far, the promotion of Cry1I oligomer formation after incubation with insect tissues or insect cell-derived models has never been experimentally addressed.

Cry1I proteins are included within the Cry1 family (3D Cry proteins) because of their sequence similarities and structural characteristics. However, Cry1I proteins display several unique, specific and remarkable characteristics, that include their secretion early in the stationary growth phase of Bt (instead of forming part of the Bt crystals), their unusual protoxin molecular weight (80 kDa), and the dual insecticidal activity against lepidopteran and coleopteran pests [29,30]. This, together with the lack of cross-resistance with other Cry1A and Cry1F insecticidal proteins [31–33], make Cry1I proteins interesting for developing new insecticidal products and indeed, they have been recently introduced in Bt crops [34,35].

Understanding the mode of action of Bt toxins is critical for enhancing and sustaining their efficacy against pests. In this context, in the present work, for the first time, the oligomer formation of Cry1Ia toxins has been examined after incubation with midgut insect tissues, trying to clarify its relevance in the Cry1I mode of action. Cry1Ab, for which oligomerization has been reported in several studies [36–39], has been used as control. In this study, midgut brush border membrane vesicles (BBMV) of different insect species from Lepidoptera (*Ostrinia nubilalis* and *Lobesia botrana*) and Coleoptera (*Leptinotarsa decemlineata*), as well as a cell line derived from *Spodoptera frugiperda* (Sf21) were used to promote the Cry oligomer formation. This selection covered both susceptible and tolerant insect species for Cry1Ia and Cry1Ab. Based on this, the oligomer formation of Cry1Ia toxin compared to Cry1Ab, and its possible correlation with insecticidal activity, have been examined.

2. Results

2.1. Toxicity of Cry1Ia and Cry1Ab against Lepidopteran and Coleopteran Species

The toxicity of Cry1Ia and Cry1Ab protoxins towards the insects' species used in the present study were assessed by performing bioassays. The results are summarized in Table 1. The Cry1Ia protein was found to be toxic for the two selected lepidopteran species, *L. botrana*, and *O. nubilalis*, with LC_{50} values of 80 and 273 ng/cm^2 respectively, as well as for the coleopteran *L. decemlineata* (LC_{50} = 22 µg/mL). On the other hand, Cry1Ab was only toxic for the lepidopteran insect species.

Table 1. Toxicity parameters of Cry1Ab and Cry1Ia protoxins.

Insect Species	Bt Toxins	LC$_{50}$	Fiducial Limits (95%)		Regression Line	
			Lower	Upper	Slope $\pm$ SE	a $\pm$ SE
L. botrana	Cry1Ab	153	106	217	1.15 $\pm$ 0.13	2.48 $\pm$ 0.3
	Cry1Ia	80	56	108	1.28 $\pm$ 0.13	2.56 $\pm$ 0.31
O. nubilalis	Cry1Ab	69	47	97	1.06 $\pm$ 0.11	3.04 $\pm$ 0.23
	Cry1Ia	273	88	1011	1.50 $\pm$ 0.15	1.34 $\pm$ 0.14
L. decemlineata	Cry1Ab	NT	-	-	-	-
	Cry1Ia	22	12	53	0.51 $\pm$ 0.07	4.29 $\pm$ 0.09

LC$_{50}$ values are expressed as ng/cm^2 for *L. botrana* and *O. nubilalis*, and in µg/mL for *L. decemlineata*. NT: non-toxic at 100 µg/mL.

2.2. BBMV of Susceptible Lepidopteran Insects Promoted Oligomerization of Cry1Ab But Not of Cry1Ia

The oligomer formation of Cry1Ab and Cry1Ia was studied by incubating the proteins with lepidopteran BBMV from *O. nubilalis* and *L. botrana*, susceptible hosts (Table 1). In order to favor oligomer detection, the milder SDS-PAGE denaturing conditions used in the bibliography to observe Cry1 oligomers were employed (see Section 4.8 in Materials and Methods). The results showed that Cry1Ab toxin was able to form oligomers after incubation with BBMV from *O. nubilalis*, but this incubation did not promote Cry1Ia oligomer formation (Figure 1). The oligomers observed (band of about 250 kDa) were associated to the *O. nubilalis* BBMV fraction (Figure 1a, lane P); these oligomeric structures could be inserted into the membranes or just bound to the surface of the BBMV due to interaction with specific membrane proteins. For Cry1Ab, the Western blot revealed another band with a molecular weight of approximately 60 kDa, corresponding to the monomeric form of Cry1Ab (Figure 1a, lane P). Nevertheless, the Cry1Ia protein associated with *O. nubilalis* BBMV was detected as a single band of about 50 kDa, corresponding to Cry1Ia monomers, and no band corresponding to Cry1Ia oligomers was detected (Figure 1b, lane P). The Cry1Ab and Cry1Ia monomeric proteins were also recovered in the supernatants, as bands of about 60 and 50 kDa, respectively (Figure 1, lanes S). Controls of Cry1Ab or Cry1Ia proteins without BBMV, but subjected to the same process as the rest of the samples, were conducted to assess the possible spontaneous formation of oligomers. The results (Figure 1, lanes C) showed the presence of Cry1Ab and Cry1Ia monomers mostly while the oligomers (tetramers) were not detected in these lanes, pointing to the fact that Cry1Ab and Cry1Ia tetramers did not form spontaneously in the solution in the absence of insect BBMV. In the case of Cry1Ia some minor bands of MW about 65 and 90 kDa were detected, which were most probably traces of Cry1Ia protoxin and partially trypsinized products; a third minor band of about 130 kDa (MW that does not match with the size of Cry1Ia dimers or trimers) was also observed.

The oligomer formation was also tested using BBMV from *L. botrana*. The results obtained were similar to the ones found with *O. nubilalis* BBMV. The incubation of Cry1Ia with BBMV from *L. botrana*, did not render bands with molecular weight consistent with Cry1Ia oligomeric structures (Figure 2b, lane P), whilst, after incubation of Cry1Ab with *L. botrana* BBMV, a clear band corresponding to a Cry1Ab oligomer (tetramer) was observed (Figure 2, lane P). On the other hand, monomers of both proteins were detected in the corresponding supernatants recovered after the BBMV-protein incubation assays (Figure 2, lanes S). In the Cry1Ia experiments, minor bands of MW of about 65 and 90 kDa were observed in the supernatant (S) and control (C) lanes that probably represent a minor fraction of partially trypsinized Cry1Ia.

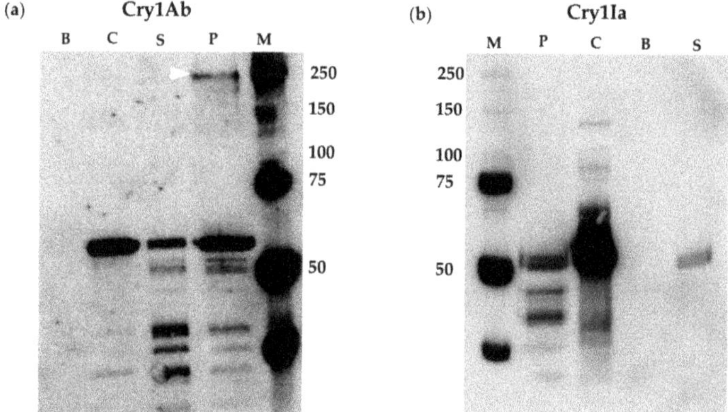

Figure 1. Cry1Ab and Cry1Ia oligomer formation promoted by *O. nubilalis* BBMV: (**a**) Cry1Ab; (**b**) Cry1Ia. Lanes B: *O. nubilalis* BBMV incubated without Cry1Ab or Cry1Ia proteins. Lanes C: Control of Cry1Ab or Cry1Ia proteins, incubated without BBMV. Lanes S: Supernatant obtained after incubation of Cry1Ab or Cry1Ia proteins, with the BBMV. Lanes P: Pellet obtained after incubation of Cry1Ab or Cry1Ia with the BBMV. Lanes M: Molecular weight marker. The arrowhead points to the Cry1Ab oligomer (about 250 kDa).

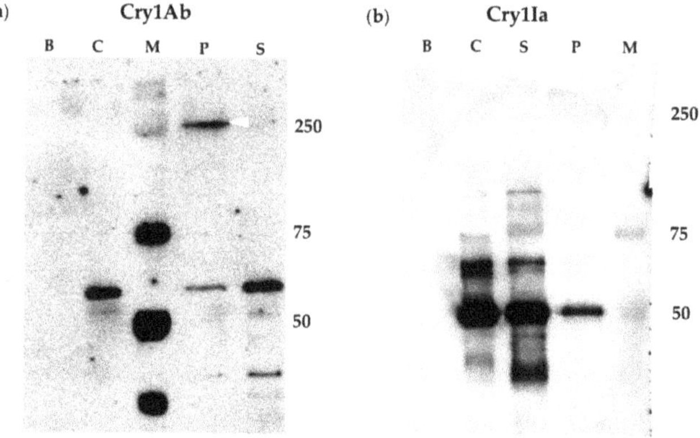

Figure 2. Cry1Ab and Cry1Ia oligomer formation promoted by *L. botrana* BBMV: (**a**) Cry1Ab; (**b**) Cry1Ia. Lanes B: *L. botrana* BBMV incubated without Cry1Ab or Cry1Ia proteins. Lanes C: Controls of Cry1Ab or Cry1Ia proteins incubated without BBMV. Lanes P: Pellet obtained after incubation of the respective protein with the *L. botrana* BBMV. Lanes S: Supernatant obtained after incubation of the respective protein with the *L. botrana* BBMV. Lanes M: molecular weight marker. The arrowhead points to the Cry1Ab oligomer (about 250 kDa).

2.3. Oligomerization of Cry1Ia Was Promoted by BBMV from L. decemlineata

Cry1Ia exhibits a dual toxic activity towards lepidopteran and coleopteran hosts, whereas Cry1Ab is only toxic to lepidopteran species. In this work, *L. decemlineata* BBMV was employed as a coleopteran Cry1Ia susceptible host tissue (Table 1) to study Cry1Ia oligomer promotion as well to study if these BBMV promoted the oligomerization of the Cry1Ab protein, non-toxic for this insect species (Table 1).

The results, summarized in Figure 3, showed that the incubation of either Cry1Ab or Cry1Ia toxins with the coleopteran BBMV provided bands of a molecular weight of about 250 kDa for both Cry proteins, corresponding to the respective oligomers. Interestingly, the oligomers were detected in both, pelleted (BBMV) and supernatant fractions (Figure 3, lanes P and S respectively). Bands corresponding to the monomers of both proteins were also observed in both fractions (pelleted BBMV and supernatant) for both Cry proteins. Incubation with Cry1Ab rendered also a band of about 150 kDa that could correspond to a dimer. It is worth highlighting that the band corresponding to the Cry1Ia oligomer had a molecular weight of about 250 kDa, which would indicate that Cry1Ia oligomer could be composed by more than four subunits, since the Cry1Ia monomer has a size of about 50 kDa. Moreover, a strong band of molecular weight higher than 250 kDa was also observed as being associated to the BBMV (Figure 3, lane P) that could correspond to Cry1Ia aggregates with high number of units.

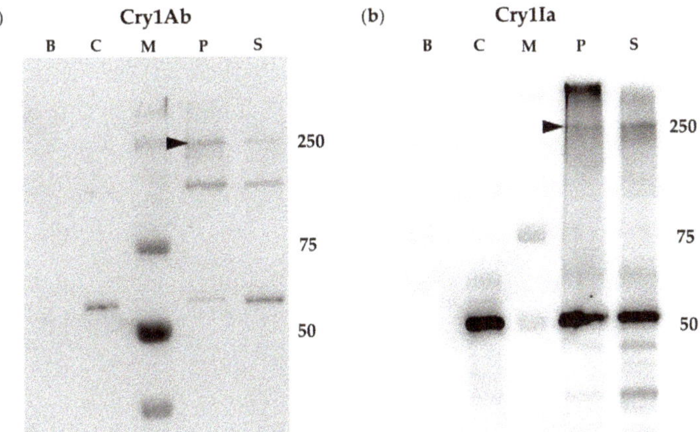

Figure 3. Cry1Ab and Cry1Ia oligomer formation promoted by *L. decemlineata* BBMV: (**a**) Cry1Ab; (**b**) Cry1Ia. Lanes B: *L. decemlineata* BBMV incubated without Cry1Ab or Cry1Ia proteins. Lanes C: Controls of Cry1Ab or Cry1Ia proteins incubated without BBMV. Lanes P: Pellet obtained after incubation of the respective protein with the *L. decemlineata* BBMV. Lanes S: Supernatant obtained after incubation of the respective protein with the *L. decemlineata* BBMV. Lanes M: molecular weight marker. The arrowheads in panels (**a,b**) point to the Cry1Ab (about 250 kDa) and the Cry1Ia (about 250 kDa) oligomer bands respectively.

2.4. Oligomerization Promoted by Sf21 Insect Cells

It has been described that Sf21 cells (insect cultured cells derived from *S. frugiperda* ovaries) are tolerant to both Cry1Ab and Cry1Ia proteins [40,41]. These insect cells were selected in order to check if oligomer formation could be promoted by tolerant insect tissues. Figure 4 shows the results of the incubation of Sf21 cells with each one of these proteins. In the case of Cry1Ab, the incubation of the protein with the insect cells resulted in the detection of two main bands of approximately 60 and 250 kDa associated with the cell fraction, corresponding to Cry1Ab monomers and oligomers, respectively (Figure 4a, lane P). Other minor bands detected were also present in the control lanes of Sf21 cells that had been incubated without Cry proteins, showing that these bands probably correspond to natural biotinylated proteins present in the cells (Figure 4, lanes B). In the case of Sf21 cells incubated with Cry1Ia, only a band with the molecular weight of the monomer (50 kDa) was found to be associated with the cells (Figure 4b, lane P). In summary, results pointed out the absence of Cry1Ia oligomer formation after incubation with Sf21 cells, in contrast to what is observed after incubation with Cry1Ab. It is worth noting that in the supernatants, only a main band with the molecular weight of the Cry1Ab or Cry1Ia monomers, was detected (Figure 4, lanes S).

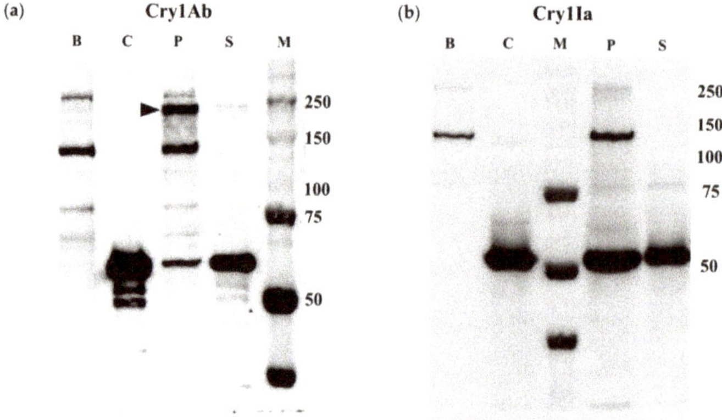

Figure 4. Cry1Ab and Cry1Ia oligomer formation promoted by Sf21 cells: (**a**) Cry1Ab; (**b**) Cry1Ia. Lanes B: Sf21 cells incubated without Cry1Ab or Cry1Ia proteins. Lanes C: Controls of Cry1Ab or Cry1Ia proteins incubated without Sf21 cells. Lanes P: Pellet obtained after incubation of the respective protein with the Sf21 cells. Lanes S: Supernatant obtained after incubation of the respective protein with the Sf21cells. Lanes M: molecular weight marker. The arrowhead in Cry1Ab panel points to the Cry1Ab oligomer.

3. Discussion

Cry1I proteins share sequence and structural similarities to the most-known three domain Cry proteins present in the parasporal crystal of Bt [10,29,42]. Therefore, so far, it has been assumed that their mode of action is similar to its crystal protein counterparts, despite their special features such as that the Cry1I proteins do not form crystals [5,43], their protoxin MW is smaller (about 81 kDa [29]), and they show dual toxic activity against lepidopteran and coleopteran insect pests [29,44].

The mode of action of the Cry proteins accumulated in the crystals is not completely understood [10,12]. It is commonly accepted that includes the solubilization of the crystals in the insect midgut to yield the protoxin form, and a proteolytic processing to produce the activated forms. Then, the "sequential binding model", suggests that the activated proteins bind to several midgut membrane receptors to finally form an oligomeric structure that inserts in the midgut membrane and forms pores, leading to cellular osmotic imbalance, cell lysis, septicaemia and eventually insect death [13,14]. In support of this model, several studies have reported that oligomerization plays a crucial role in the insecticidal activity of *B. thuringiensis* Cry toxins [18,38,45,46]. Moreover, Cry protein mutants that did not form oligomers, showed severely decreased toxicity [17,24,45]. Likewise, it has been shown that some Cry1A mutants that had lost their toxicity, were unable to oligomerize and to form pores [38,47]. However, despite these studies, the sequential binding to several insect membrane proteins as well as the oligomer formation and insertion events prior to pore formation are not clearly defined yet [12]. On the other hand, as an alternative to the "sequential binding model", the "signaling pathway model" has been proposed. This model claims that the Cry toxicity can be due to the activation of intracellular cell death pathways [9]. Indeed, some authors claim that both mechanisms could coexist [48].

In this context, the occurrence of an oligomerization step in the mode of action of Cry1Ia protein, and the study of Cry1Ia oligomerization promotion by membranes of susceptible and tolerant insect BBMV or insect cells, have been the main goals of this research.

In 2009, the occurrence of a spontaneous oligomerization of trypsinized Cry1Ie in solution was reported [28]. According to the mentioned research, the oligomer fraction contained a small amount of dimer and a large amount of aggregates larger than tetramers. The toxicity of the oligomers against *Plutella xylostella* was about 70 times lower than the toxicity of the monomer or the toxicity of the non-trypsinized Cry1Ie protein, which led the authors to claim that the Cry1Ie spontaneous aggregation most likely differ from the oligomers occur by the insect midgut membranes. In the present work, the oligomerization of Cry1Ia protein after incubation with susceptible insect BBMV or with non-susceptible cultured insect cells, has been determined for the first time.

To properly select the insect species for this study, the insecticidal activity of Cry1Ia and Cry1Ab protoxins towards two lepidopteran species (*O. nubilalis* and *L. botrana*) and a coleopteran insect (*L. decemlineta*) has been assessed. The toxicity data obtained were expected based on the published data [29,44,49,50]. Moreover, Sf21 cells were also used in this study, on the bases that Cry1Ia and Cry1Ab are not toxic for them [40,41].

So far, Cry oligomerization studies have focused on the incubation of the Cry proteins with BBMV of susceptible and resistant populations of the same insect, and, as a result, an association has been found between resistance and reduced oligomerization [51,52]. In the present work, we have used Cry1Ab as an oligomeric control protein [36–39], and have been able to clearly detect Cry1Ab oligomers in the expected tetramer form (about 250 kDa molecular weight size) after incubation with all tested insect tissues, regardless to their susceptibility. Thus, oligomers have been observed after incubation of the trypsinized Cry1Ab protein with *O. nubilalis* or *L. botrana* BBMV (susceptible insects), but also with *L. decemlineata* BBMV (non-susceptible insect) and Sf21 cells (non-susceptible cultured insect cells). The oligomeric structures were found associated to the insect BBMV or to the insect cells, indicating that they were either inserted or bound to the membrane proteins. The finding of Cry1Ab oligomers after incubation with tolerant insect BBMV or associated with tolerant insect cells could be explained by an improper insertion of oligomers into the membranes (and therefore inefficiency to produce damages), or by the inability to induce the post-pore subsequent events in the cells (e.g., no triggering of cell death mechanisms).

In this study, the Cry1Ab oligomer was mainly found to be associated to the membrane fractions in accordance with previous reports, but oligomers were also observed in the supernatants after incubation with BBMV from the non-susceptible insect *L. decemlineata*. This suggested that the Cry1Ab oligomers (tetramers and dimers), promoted by the coleopteran BBMV, are not only associated to the BBMV (whether inserted or not) but also free in the supernatant (Figure 3a, lane S). The results obtained resemble the ones shown by Rodríguez-Almazán et al. [47], with the Cry1Ab helix α-4 mutants which had a mutation in domain I, involved in membrane insertion and pore formation, and had lost drastically their insecticidal activity towards *M. sexta* larvae. Similarly, in our results, the incubation of Cry1Ab with BBMV from *L. decemlineata* rendered a relatively high amount of oligomeric structures (dimers and tetramers) in the supernatant, apparently indicating that these oligomers were not able to insert into the membranes, resulting in the absence of toxicity. Nevertheless, in this study, after the incubation of the Cry1Ia protein with *L. decemlineata* BBMV, a high proportion of Cry1Ia oligomers were also observed in the supernatants (Figure 3b, laneS), and in this case, the protein showed a high toxicity against this coleopteran pest (Table 1).

The Cry1Ia oligomers were observed after incubation of the protein with BBMV from the susceptible coleopteran *L. decemlineata*. The molecular weights of the bands that correspond to the Cry1Ia oligomers (about 250 kDa) point out that the Cry1Ia oligomer could be composed of more than four units.

After incubation of Cry1Ia with BBMV of susceptible lepidopterans (*O. nubilalis* and *L. botrana*) or with insect Sf21 cells, no oligomers were detected, suggesting that the Cry1Ia proteins associated to the lepidopteran membranes could be mainly in monomeric forms. In this case, oligomers would not be a limiting step in the toxicity of Cry1Ia proteins, and the toxicity could be mediated by intracellular signaling pathways [9]. Moreover, other hypotheses could be mentioned to explain the reason for the absence of Cry1Ia oligomers. Firstly, it can be suggested that the observed monomers come from disassembled oligomers produced due to the SDS-PAGE technique conditions, similarly to what was proposed by Ocelotl et al. [51] working with Cry1Ab oligomers. However, this reasoning would be in conflict with the observation of oligomers after the incubation of Cry1I with coleopteran

BBMV, which were detected using the same SDS-PAGE conditions. Secondly, it could be considered that the biotinlylation of Cry1I could interfere with oligomerization. This hypothesis was examined following the incubation of the biotin labelled Cry1Ab with *O. nubilalis* BBMV. The results showed no influence of biotin in oligomerization (Figure S1), similarly to what had been already claimed by other authors showing that biotinylation of Cry proteins does not prevent their oligomerization, binding and insecticidal activity [12,53,54]. Thirdly, it has to be noticed that, in the present study, the oligomer formation experiments were performed with trypsinized Cry1Ab and Cry1Ia proteins to mimic the in vivo conditions. It has been described that Cry1Ie protoxin and trypsinized protein have the same toxicity [28]. However, other Cry1I proteins such as Cry1Ia1 have shown some differences in toxicity amongst protoxins (more active) and trypsinized proteins [49]. We can speculate that in vitro trypsinization of Cry1Ia could alter the toxicity by impairing oligomer formation, maybe provoking a flawed ability to form oligomers when they are promoted by lepidopteran BBMV.

The spontaneous formation of Cry1Ab and Cry1Ia oligomers in solution has been questioned in this study. Cry1Ab submitted to the same experimental situation of treatments, but without being in contact with insect BBMV or cells, did not oligomerize (Figures 1, 2, 3 and 4a, lanes C). Regarding Cry1Ia, some bands of smaller sizes than the expected tetramer, and with MW sizes that were not multiples of 50 kDa (monomer size) were observed (Figures 1, 2, 3 and 4b, lanes C). Most probably, these bands are residual incomplete trypsinized Cry1Ia forms, dragged through the toxin purification process. In conclusion, although some studies have reported the spontaneous formation of Cry protein oligomers in solution without being in contact with the membrane-like environment (i.e., Cry4Ba [55] or Cry2Ab [24]), in our study neither Cry1Ab nor Cry1Ia formed oligomers without being exposed to insect BBMV or cultured insect cells.

In summary, our findings indicated that the oligomers of a classical 3D crystal forming Cry protein of Bt such as Cry1Ab were promoted and could be detected after incubation of activated Cry1Ab with susceptible or non-susceptible insect midgut BBMV and with non- susceptible Sf21 cells. In contrast, in the same assay conditions, Cry1Ia oligomers were detected only after incubation with BBMV of *L. decemlineata* (coleopteran susceptible host), but no Cry1Ia oligomeric structures were found following incubation of Cry1Ia with lepidopteran BBMV or with the Sf21 cells, regardless of its host susceptibility. Hence, our results, using trypsin processed Cry1Ab and Cry1Ia as an in vitro model of what might occur in vivo, suggest that: (1) The promotion of oligomers can occur by incubation of the Cry toxin with susceptible insect BBMV but also with non-susceptible insect tissues, and (2) The oligomerization may not be a determining step in the toxicity of Cry1I proteins.

4. Materials and Methods

4.1. Production and Purification of Cry Proteins

The Cry1Ab protein used for oligomer formation was obtained from a recombinant *Escherichia coli* strain GG094-208 (kindly supplied by Dr. R.A. de Maagd, Wageningen University, The Netherlands). Protein expression, inclusion bodies purification, solubilization and protoxin activation by trypsin, were performed as described previously [56]. The activated Cry1Ab was purified by anion-exchange chromatography using Äkta 100 explorer system (GE Healthcare, Amersham, UK) following Crava et al. [57]. The eluted fractions from the column were individually analysed by sodium dodecyl sulphate-12% polyacrylamide gel electrophoresis (SDS-PAGE).

The *cry1Ia7* gene was cloned and expressed in *E. coli*, BL21(DE3) cells [40]. The purification of the protein by affinity chromatography using a HisTrapTM FF crude column (GE Healthcare Bio-Sciences, Upsala, Sweden), protein dialysis and protoxin activation by trypsin, were performed as reported by Khorramnejad et al. [40]. The activated Cry1Ab and Cry1Ia were visualized after SDS-PAGE, and their concentration was estimated by densitometry using TotalLab Quant program version 12.3 (Newcastle, UK), employing bovine serum albumin as standard.

The Cry1Ab protein used in bioassays (protoxin) was obtained from a recombinant Bt strain that produced a crystal composed solely of the Cry1Ab protein, kindly supplied by Dr. Colin Berry, Cardiff University, Cardiff, UK. This Bt strain was grown in CCY medium [58] supplemented with erythromycin and the crystal formation was observed daily. When at least 95% of the cells were lysed, the fermentation process was stopped and then spores and crystals were collected by centrifugation (8600× g, 4 °C), washed with a saline solution (NaCl 1M, EDTA 10mM) and resuspended in KCl (10 mM). For Cry1Ia7 protoxin production, the recombinant *E. coli* BL21 (DE3) was grown and purified as described above. Both Cry1Ab and Cry1Ia7 protoxins were quantified by the method of Bradford [59] using BSA as standard, and kept at 4 °C until used.

4.2. Insect Rearing

Two different lepidopteran species, *L. botrana* (Lep: Tortricide) and *O. nubilalis* Hübner (Lep.: Crambidae), and one coleopteran species, *L. decemlineata* (Col.: Chrysomelidae) (the Colorado potato beetle, CPB) were used in this study. Both lepidopteran pests were maintained in the insectary of the Universidad Pública de Navarra (UPNA, Pamplona, Spain) at 25 °C ± 1 °C with 70% ± 5% RH and a photoperiod of 16/8 h (light/dark) on the artificial diet described by MacIntosh et al. [60]. The population of beetles was raised on potato plants (Desiré variety) grown throughout the year in a phytotron to provide a continuous supply of food substrate. This population was refreshed 1–2 times every year by the introduction of wild adults collected in the field during the spring-summer.

4.3. Insect Cell Line

The lepidopteran cell line Sf21, from ovaries of fall armyworm *S. frugiperda* (Lep.: Noctuidea), was obtained from Wageningen University (Wageningen, The Netherlands). The Sf21 cells were maintained at 25 °C on 1X Grace's medium (Gibco® Life technologies™, Carslbad, CA, USA) supplemented with 10% heat-inactivated fetal bovine serum (FBS) in 25 cm^2 cell culture flasks. Cells were passaged weekly. Cell concentrations were measured by using an automatic cell counter (Countess Automated Cell Counter from Invitrogen, Carlsbad, CA, USA).

4.4. Insect Bioassays

The bioassays for the three insect species were performed with first instar larvae. Five different protoxin concentrations, ranging from 0.39 to 100 µg/mL, were prepared to determine the concentration-mortality responses in order to calculate the mean lethal concentration (LC$_{50}$). For the lepidopteran insect species, *O. nubilalis* and *L. botrana*, the diet surface contamination assay was used [61], and for *L. decemlineata*, the leaf dip bioassay described by Iriarte et al. [62], was performed. For each bioassay, several protein concentrations were tested, using 28 larvae per concentration. Each bioassay was repeated at least three times. Total insect mortality was recorded after 7 days for lepidopteran insects, and after 4 days for *L. decemlineata*. The concentration-mortality data obtained for each insect species were analyzed after transformation of the concentration-response curve to fit a linear model using POLO-PC program (LeOra Software, Berkeley, CA, USA, 1987), based on the Probit analysis [63].

4.5. Midgut Isolation and BBMV Preparation

Midguts were dissected from fifth-instar larvae of *O. nubilalis* and forth-instar larvae of *L. decemlineata*. The dissected midguts were rinsed in ice-cold MET buffer (0.3 M mannitol, 5 mM EGTA, 17 mM Trsi-HCl, pH 7.5), snap frozen in liquid nitrogen and kept at −80 °C until use.

Brush border membrane vesicles (BBMV) were obtained from the dissected midguts of *O. nubilalis* and *L. decemlineata* and from the whole last instar larvae of *L. botrana*, following the differential magnesium precipitation method [64,65]. Proteins in the purified BBMV were quantified following Bradford protein assay [59] and stored at −80 °C.

4.6. Biotin Labelling

Trypsin activated Cry1Ia protein was biotinylated by using the protein biotinylation kit from GE Healthcare (GE Healthcare, Little Chalfont, UK), as described elsewhere [66]. The eluted fractions were quantified by NanoDrop 2000 spectrophotometer (ThermoFisher Scientific, Waltham, MA, USA), analyzed by 12% SDS-PAGE and verified by Western blot. The protein fractions were concentrated by using an Amicon Ultra-4 10K centrifugal filter device (Merck Millipore, Tullagreen, Ireland) and stored at 4 °C.

The interference of biotin in oligomerization was tested following incubation of the biotin labelled Cry1Ab with *O. nubilalis* BBMV. The detection of biotin labelled Cry1Ab oligomerization was performed following the same protocol than has been described for Cry1Ia. The results showed no influence of biotin in oligomerization (Figure S1), as had been already claimed by other authors that have used biotin labelled proteins to detect oligomers [12,20,53,54].

4.7. Oligomerization Assays with Sf21 Cells

The confluent monolayer growing Sf21 cells were suspended in fresh Grace's medium without FBS. The cell concentration was measured (by using the Countess Automated Cell Counter from Invitrogen, Carlsbad, California, USA), and 100 µL of cell suspension at a concentration of 2×10^6 cells/mL were seeded into 96-well plates. The oligomerization assays were performed as described by Portugal et al. [38] with slight modifications. In short, cells were incubated at 25 °C for at least 30 min. Later, 4 µg of activated toxins (biotinylated Cry1Ia or unlabeled Cry1Ab) were added to the cells (final concentration of 0.03 µg protein/µL) except in controls, which received 50 mM carbonate buffer pH 10.5. The plates were incubated for 3 h at 25 °C. After the incubation, the treated cells were collected and pelleted by centrifugation at 16,200× *g*, 4 °C for 15 min. The supernatants containing unbound proteins were kept for further analysis. The controls (proteins alone, without cells) were submitted to the same experimental conditions as treatments. After centrifugation of the controls, as there was no pellet due to the absence of cells, a dilution of supernatant (containing 200 ng of selected protein) was analyzed in the gel. The Sf21 cells in the pellet were washed once with 200 µL of 50 mM carbonate buffer pH 10.5, and recovered by centrifugation (45 min, 18,800× *g*). The final pellet was resuspended in 10 µL of buffer and heated at 50 °C for 3 min. The proteins present in the sample were separated by SDS-PAGE 10% and electrotransferred onto polyvinylidene difluoride (PVDF) Western Blotting membrane (Roche Diagnostics GmbH, Mannheim, Germany). The membrane was incubated overnight in blocking buffer (PBST; 0.1% Tween 20 in phosphate-buffered saline, supplemented with 5% skimmed milk) with gentle shaking, and washed three times with PBST, before incubation with the corresponding antibodies. Cry1Ab protein was detected with polyclonal rabbit anti-Bt Cry1Ab/1Ac (1:10,000; 60 min) from Abraxis (Warminster, PA, USA) followed by secondary antibody (1:20,000; 60 min) coupled with horseradish peroxidase (HRP) (Sigma-Aldrich, Saint Louis, MO, USA), whereas biotinylated Cry1Ia protein was detected by streptavidin-conjugated horseradish peroxidase (1:2000; 60 min) (GE Healthcare, Amersham, UK). Both Cry1Ab and Cry1Ia proteins were visualized by chemiluminescence using ECL™ prime western blotting detection reagent (GE Healthcare, Little Chalfont, UK) using an ImageQuant LAS400 image analyzer (GE Healthcare Bio-Sciences, Upsala, Sweden). The molecular weight marker used was Precision Plus Protein™ Dual Color Standard (Bio-Rad, Carlsbad, CA, USA). Each oligomerization assay was repeated at least three times.

4.8. Oligomerization Assays with BBMV

The BBMV were centrifuged for 10 min at 16,000× *g* and suspended in 50 mM carbonate buffer pH 10.5. The oligomerization protocol was set up after reviewing the procedures described in the literature for Cry1 proteins [17,18,37–39,45,47,51,52,54]. Finally, the oligomerization assays with BBMV were performed following Ocelotl et al. [51] who employed the milder SDS-PAGE denaturing conditions (heating the samples 3 min at 50 °C), with some modifications. Briefly, 2 µg of biotin labelled activated

Cry1Ia and activated Cry1Ab toxins were incubated for one hour with 5 µg of *L. botrana* or *L. decemlineata* BBMV, or with 20 µg of *O. nubilalis* BBMV, at 37 °C, in a final volume of 50 µL. Activated proteins incubated in the absence of BBMV and samples containing only BBMV were used as controls. Then, phenylmethylsulfoyl fluoride (PMSF) was added (final concentration 1 mM) and BBMV were recovered by centrifugation at 18,400× *g* for 45 min at 4 °C. The supernatant containing unbound protein was recovered and stored. The controls (samples with proteins and without BBMV) went through the same experimental conditions as treatments. After centrifugation of the controls, a dilution of the supernatant (containing 200 ng of selected protein) was analyzed in the gel to avoid the observation of the over saturated signal in the membrane. In the samples containing BBMV, the pellet was washed once with 100 µL ice-cold buffer. The final BBMV pellets were resuspended in 10 µl of the buffer. After incubating the samples for 3 min at 50 °C, the proteins were separated by 10% SDS-PAGE and blotted onto PVDF Western Blot membranes (Roche Diagnostics GmbH, Mannheim, Germany). After Western blot, Cry1Ab was detected with polyclonal rabbit anti-Bt Cry1Ab/1Ac, and Cry1Ia was detected by streptavidin-conjugated horseradish peroxidase as has been described in the previous section. The experiments were repeated, at least, three times.

Supplementary Materials: The following is available online at http://www.mdpi.com/2072-6651/12/2/133/s1, Figure S1: Biotin labelled Cry1Ab oligomer formation promoted by *O. nubilalis* BBMV.

Author Contributions: B.E. and Y.B. contributed to the design of the study. A.K. and M.D.-A. performed the experiments. A.K., Y.B., B.E., P.C. and M.D.-A. analyzed the data. A.K., Y.B. and M.D.-A. wrote the manuscript. All authors have read and agreed to the published version of the manuscript.

Funding: This work was supported by grants from the Spanish Ministry of Science, Innovation and Universities, the State Research Agency of Spain and the European FEDER founds (Refs. AGL2015-70584-C2 and RTI2018-095204-B-C21), and by the Generalitat Valenciana (GVPROMETEOII-2015-001). M. Domínguez received a predoctoral fellowship from the Universidad Pública de Navarra, Spain.

Acknowledgments: We are deeply grateful to Patricia Hernández-Martínez for her valuable comments. We thank Rosa Maria González-Martínez and Óscar Marin Vázquez for their support in laboratory assistance.

Conflicts of Interest: The authors declare no conflict of interest.

References

1. Van Frankenhuyzen, K. Insecticidal activity of *Bacillus thuringiensis* crystal proteins. *J. Invertebr. Pathol.* **2009**, *101*, 1–16. [CrossRef]
2. Ohba, M.; Mizuki, E.; Uemori, A. Parasporin, a new anticancer protein group from *Bacillus thuringiensis*. *Anticancer Res.* **2009**, *29*, 427–433. [PubMed]
3. Palma, L.; Muñoz, D.; Berry, C.; Murillo, J.; Caballero, P. *Bacillus thuringiensis* toxins: An overview of their biocidal activity. *Toxins* **2014**, *6*, 3296–3325. [CrossRef] [PubMed]
4. Chakroun, M.; Banyuls, N.; Bel, Y.; Escriche, B.; Ferré, J. Bacterial vegetative insecticidal proteins (Vip) from entomophatogenic bacteria. *Microbiol. Mol. Biol. Rev.* **2016**, *80*, 329–350. [CrossRef] [PubMed]
5. Kostichka, K.; Warren, G.W.; Mullins, M.; Mullins, A.D.; Craig, J.A.; Koziel, M.G.; Estruch, A.D. Cloning of a *cryV*-type insecticidal protein gene from *Bacillus thuringiensis*: The *cryV*-encoded protein is expressed early in stationary phase. *J. Bacteriol.* **1996**, *178*, 2141–2144. [CrossRef]
6. Jurat-Fuentes, J.L.; Jackson, T.A. Bacterial entomopathoges. In *Insect Pathology*, 2nd ed.; Vega, F.E., Kaya, H.K., Eds.; Academic Press: San Diego, CA, USA, 2012; pp. 265–349. [CrossRef]
7. Hoffman, C.; Vanderbruggen, H.; Höfte, H.; Van Rie, G.; Jansens, S.; Van Mellaert, H. Specificity of *Bacillus thuringiensis* delta-endotoxins is correlated with the presence of high-affinity binding sites in the brush border membrane of target insect midguts. *Proc. Natl. Acad. Sci. USA* **1988**, *85*, 7844–7848. [CrossRef]
8. Ferre, J.; VanRie, J. Biochemistry and genetics of insect resistance to *Bacillus thuringiensis*. *Annu. Rev. Entomol.* **2002**, *47*, 501–533. [CrossRef]
9. Zhang, X.; Candas, M.; Griko, N.B.; Taussig, R.; Bulla, L.A., Jr. A mechanism of cell death involving an adenylyl cyclase/PKA signaling pathway is induced by the Cry1Ab toxin of *Bacillus thuringiensis*. *Proc. Natl. Acad. Sci. USA* **2006**, *103*, 9897–9902. [CrossRef]

10. Adang, M.J.; Crickmore, N.; Jurat-Fuentes, J.L. Diversity of *Bacillus thuringiensis* crystal toxins and mechanism of action. *Adv. Insect Physiol.* **2014**, *47*, 39–87. [CrossRef]

11. Arenas, I.; Bravo, A.; Soberón, M.; Gómez, I. Role of alkaline phosphatase from *Manduca sexta* in the mechanism of action of *Bacillus thuringiensis* Cry1Ab toxin. *J. Biol. Chem.* **2010**, *285*, 12497–12503. [CrossRef]

12. Vachon, V.; Raynald, L.; Schwartz, J.L. Current models of the mode of action of *Bacillus thuringiensis* insecticidal crystal proteins: A critical review. *J. Invertebr. Pathol.* **2012**, *111*, 1–12. [CrossRef] [PubMed]

13. Pardo-López, L.; Soberón, M.; Bravo, A. *Bacillus thuringiensis* insecticidal three-domain Cry toxins: Mode of action, insect resistance and consequences for crop protection. *FEMS Microbiol. Rev.* **2013**, *37*, 3–22. [CrossRef] [PubMed]

14. Gomez, I.; Sanchez, J.; Miranda, R.; Bravo, A.; Soberon, M. Cadherin-like receptor binding facilitates proteolytic cleavage of helix alpha-1 in domain I and oligomer pre-pore formation of *Bacillus thuringiensis* Cry1Ab toxin. *FEBS Lett.* **2002**, *513*, 242–246. [CrossRef]

15. Rausell, C.; García-Robles, I.; Sánchez, J.; Muñoz-Garay, C.; Martínez-Ramírez, A.C.; Real, M.D.; Bravo, A. Role of toxin activation on binding and pore formation activity of the *Bacillus thuringiensis* Cry3 toxins in membranes of *Leptinotarsa decemlineata* (Say). *Biochim. Biophys. Acta* **2004**, *1660*, 99–105. [CrossRef]

16. Muñoz-Garay, C.; Portugal, L.; Pardo-López, L.; Jiménez-Juárez, N.; Arenas, I.; Gómez, I.; Sánchez-López, R.; Arroyo, R.; Holzenburg, A.; Savva, C.G.; et al. Characterization of the mechanism of action of the genetically modified Cry1AbMod toxin that is active against Cry1Ab-resistant insects. *Biochim. Biophys. Acta* **2009**, *1788*, 2229–2237. [CrossRef]

17. Tigue, N.; Jacoby, J.; Ellar, D.J. The α-helix residue, Asn[135], is involved in the oligomerization of Cry1Ac1 and Cry1Ab5 *Bacillus thuringiensis* toxins. *Appl. Environ. Microbiol.* **2001**, *67*, 5715–5720. [CrossRef]

18. Herrero, S.; González-Cabrera, J.; Ferré, J.; Bakker, P.L.; de Maagd, R.A. Mutations in the *Bacillus thuringiensis* Cry1Ca toxin demonstrate the role of domains II and III in specificity towards *Spodoptera exigua* larvae. *Biochem. J.* **2004**, *384*, 507–513. [CrossRef]

19. Pérez, C.; Muñoz-Garay, C.; Portugal, L.; Sánchez, J.; Gill, S.S.; Soberón, M.; Bravo, A. *Bacillus thuringiensis* spp. *israelensis* Cyt1Aa enhances activity of Cry11Aa toxin by facilitating the formation of a pre-pore oligomeric structure. *Cell Microbiol.* **2007**, *9*, 2931–2937. [CrossRef]

20. Fabrick, J.; Oppert, C.; Lorenzen, M.D.; Morris, K.; Oppert, B.; Jurat-Fuentes, J.L. A novel *Tenebrio molitor* cadherin is a functional receptor for *Bacillus thuringiensis* Cry3Aa toxin. *J. Biol. Chem.* **2009**, *284*, 18401–18410. [CrossRef]

21. Groulx, N.; McGuire, H.; Laprade, R.; Schwartz, J.L.; Blunck, R. Single molecule fluorescence study of the *Bacillus thuringiensis* toxin Cry1Aa reveals tetramerization. *J. Biol. Chem.* **2011**, *286*, 42274–42282. [CrossRef]

22. Carmona, D.; Rodríguez-Almazán, C.; Muñoz-Garay, C.; Portugal, L.; Pérez, C.; de Maagd, R.A.; Bakker, P.; Soberón, M.; Bravo, A. Dominant negative phenotype of *Bacillus thuringiensis* Cry1Ab, Cry11Aa and Cry4Ba mutants suggest hetero-oligomer formation among different Cry toxins. *PLoS ONE* **2011**, *6*, e19952. [CrossRef] [PubMed]

23. Khomkhum, N.; Leetachewa, S.; Angsuthanasombat, C.; Moonsom, S. Functional assembly of 260-kDa oligomers required for mosquito-larvicidal activity of the *Bacillus thuringiensis* Cry4Ba toxin. *Peptides* **2015**, *68*, 183–189. [CrossRef] [PubMed]

24. Xu, L.; Pan, Z.; Zhang, J.; Niu, L.; Li, J.; Chen, Z.; Liu, B.; Zhu, Y.; Chen, Q. Exposures of helices α4 and α5 is required for insecticidal activity of Cry2Ab by prompting assembly of prepore oligomeric structure. *Cell. Microbiol.* **2018**, e12827. [CrossRef] [PubMed]

25. Abe, Y.; Inoue, H.; Ashida, H.; Maeda, Y.; Kinoshita, T.; Kitada, S. Glycan region of GPI anchored-protein is required for cytocidal oligomerization of an anticancer parasporin-2, Cry46Aa1 protein, from *Bacillus thuringiensis* strain A1547. *J. Invertebr. Pathol.* **2017**, *142*, 71–81. [CrossRef]

26. López-Diaz, J.; Cantón, P.E.; Gill, S.S.; Soberón, M.; Bravo, A. Oligomerization is a key step in Cyt1Aa membrane insertion and toxicity but not necessary to synergize Cry11Aa toxicity in *Aedes aegypti* larvae. *Environ. Microbiol.* **2013**, *15*, 3030–3039. [CrossRef]

27. Palma, L.; Scott, D.J.; Harris, G.; Din, S.; Williams, T.L.; Reoberts, O.J.; Young, M.T.; Caballero, P.; Berry, C. The Vip3Ag4 insecticidal protoxins from *Bacillus thuringiensis* adopts a tetrameric configuration that is maintained on proteolysis. *Toxins* **2017**, *9*, 165. [CrossRef] [PubMed]

28. Guo, S.; Zhang, Y.; Song, F.; Zhang, J.; Huang, D. Protease-resistant core form of *Bacillus thuringiensis* Cry1Ie: Monomeric and oligomeric forms in solution. *Biotechnol. Lett.* **2009**, *31*, 1769–1774. [CrossRef]

29. Tailor, R.; Tippett, J.; Gibb, G.; Pells, S.; Pike, D.; Jordan, L.; Ely, S. Identification and characterization of a novel *Bacillus thuringiensis* delta-endotoxin entomocidal to coleopteran and lepidopteran larvae. *Mol. Microbiol.* **1992**, *6*, 1211–1217. [CrossRef]

30. Choi, S.K.; Shin, B.S.; Kong, E.M.; Rho, H.M.; Park, S.H. Cloning of a new *Bacillus thuringiensis cry1I*-type crystal protein gene. *Curr. Microbiol.* **2000**, *41*, 65–69. [CrossRef]

31. Xu, L.; Ferry, N.; Wang, Z.; Zhang, J.; Edwards, M.G.; Gatehouse, A.M.R.; He, K. A proteomic approach to study the mechanism of tolerance to Bt toxins in *Ostrinia furnacalis* larvae selected for resistance to Cry1Ab. *Transgenic Res.* **2013**, *22*, 1155–1166. [CrossRef]

32. Zhang, T.; He, M.; Gatehouse, A.M.R.; Wang, Z.; Edwards, M.G.; Li, Q.; He, K. Inheritance patterns, dominance and cross-resistance of Cry1Ab- and Cry1Ac-selected *Ostrinia furnacalis* (Guenée). *Toxins* **2014**, *6*, 2694–2707. [CrossRef] [PubMed]

33. Xu, L.; Wang, Z.; Zhang, J.; He, K.; Ferry, N.; Gatehouse, A.M.R. Cross-resistance of Cry1Ab-selected Asian corn borer to other Cry toxins. *J. Appl. Entomol.* **2010**, *134*, 429–438. [CrossRef]

34. Guo, J.; He, K.; Hellmich, R.L.; Bai, S.; Zhang, T.; Liu, Y.; Ahmed, T.; Wang, Z. Field trials to evaluate the effects of transgenic *cry1Ie* maize on the community characteristics of arthropod natural enemies. *Sci. Rep.* **2016**, *6*, 22102. [CrossRef] [PubMed]

35. Fan, C.; Wu, F.; Dong, J.; Wang, B.; Yin, J.; Song, X. No impact of transgenic *cry1Ie* maize on the diversity, abundance and composition of soil fauna in a 2-year field trial. *Sci. Rep.* **2019**, *9*, 10333. [CrossRef]

36. Rausell, C.; Muñoz-Garay, C.; Miranda-CasoLuengo, R.; Gómez, I.; Rudiño-Piñera, E.; Soberón, M.; Bravo, A. Tryptophan spectroscopy studies and black lipid bilayer analysis indicate that the oligomeric structure of Cry1Ab toxin from *Bacillus thuringiensis* is the membrane-insertion intermediate. *Biochem. J.* **2004**, *43*, 166–174. [CrossRef]

37. Gómez, I.; Sánchez, J.; Muñoz-Garay, C.; Matus, V.; Gill, S.S.; Soberón, M.; Bravo, A. *Bacillus thuringiensis* Cry1A toxins are versatile proteins with multiple modes of action: Two distinct pre-pores are involved in toxicity. *Biochem. J.* **2014**, *459*, 383–396. [CrossRef]

38. Portugal, L.; Gringorten, J.L.; Caputo, G.F.; Soberón, M.; Muñoz-Garay, C.; Bravo, A. Toxicity and mode of action of insecticidal Cry1A proteins from *Bacillus thuringiensis* in an insect cell line, CF-1. *Peptides* **2014**, *53*, 292–299. [CrossRef]

39. Nair, M.S.; Dean, D.H. Composition of the putative prepore complex of *Bacillus thuringiensis* Cry1Ab toxin. *Adv. Biol. Chem.* **2015**, *5*, 179–188. [CrossRef]

40. Khorramnejad, A.; Bel, Y.; Hernández-Martínez, P.; Talaei-Hassanloui, R.; Escriche, B. Insecticidal activity and cytotoxicity of *Bacillus thuringiensis* Cry1Ia7 protein. In *GMOs in Integrated Plant Protection*; Meissle, M., De Schrijver, A., Smagghe, G., Eds.; IOBC-WPRS Bulletin: Ghent, Belgium, 2018; Volume 131, pp. 56–63.

41. Martínez-Solis, M.; Pinos, D.; Endo, H.; Portugal, L.; Sato, R.; Ferré, F.; Herrero, S.; Hernández-Martínez, P. Role of *Bacillus thuringiensis* Cry1A toxins domains in the binding to the ABCC2 receptor from *Spodoptera exigua*. *Insect. Biochem. Mol. Biol.* **2018**, *101*, 47–56. [CrossRef]

42. Crickmore, N.; Zeigler, D.R.; Feitelson, J.; Schnepf, E.; Van Rie, J.; Lereclus, D.; Baum, J.; Dean, D.H. Revision of the nomenclature for the *Bacillus thuringiensis* pesticidal crystal proteins. *Microbiol. Mol. Biol. Rev.* **1998**, *62*, 807–813. [CrossRef]

43. Masson, L.; Erlandson, M.; Puzstai-Carey, M.; Brousseau, R.; Juárez-Pérez, V.; Frutos, R. A holistic approach for determining the entomopathogenic potential of *Bacillus thuringiensis* strains. *Appl. Environ. Microbiol.* **1998**, *46*, 4782–4788. [CrossRef]

44. Ruiz de Escudero, I.; Estela, A.; Porcar, M.; Martínez, C.; Oguiza, J.A.; Escriche, B.; Ferré, J.; Caballero, P. Molecular and insecticidal characterization of a Cry1I protein toxic to insects of the families Noctuidae, Tortricidae, Plutellidae and Chrysomelidae. *Appl. Environ. Microbiol.* **2006**, *72*, 4796–4804. [CrossRef] [PubMed]

45. Jiménez-Juárez, N.; Muñoz-Garay, C.; Gómez, I.; Saab-Rincon, G.; Damian-Almazo, J.Y.; Gill, S.S.; Soberón, M.; Bravo, A. *Bacillus thuringiensis* Cry1Ab mutants affecting oligomer formation are not-toxic to *Manduca sexta* larvae. *J. Biol. Chem.* **2007**, *282*, 21222–21229. [CrossRef]

46. Likitvivatanavong, S.; Katzenmeier, G.; Andsuthanasombat, C. Asn^{183} in α5 is essential for oligomerization and toxicity of the *Bacillus thuringiensis* Cry4Ba toxin. *Arch. Biochem. Biophys.* **2005**, *445*, 46–55. [CrossRef]

47. Rodríguez-Almazán, C.; Zavala, L.E.; Muñoz-Garay, C.; Jiménez-Juárez, N.; Pacheco, S.; Masson, L.; Soberón, M.; Bravo, A. Dominant negative mutants of *Bacillus thuringiensis* Cry1Ab toxin function as anti-toxins: Demonstration of the role of oligomerization in toxicity. *PLoS ONE* **2009**, *4*, e5545. [CrossRef] [PubMed]

48. Jurat Fuentes, J.L.; Adang, M.J. Cry toxin mode of action in susceptible and resistant *Heliothis virescens* larvae. *J. Invertebr. Pathol.* **2006**, *92*, 166–171. [CrossRef]

49. Sekar, V.; Held, B.; Tippet, J.; Amirhusin, B.; Robeff, P.; Wang, K.; Wilson, H.M. Biochemical and Molecular Characterization of the insecticidal fragment of CryV. *Appl. Environ. Microbiol.* **1997**, *63*, 2798–2801. [CrossRef]

50. Crava, C.M.; Bel, Y.; Escriche, B. Bacillus thuringiensis susceptibility variation among Ostrinia nubilalis populations. In *Future Research and Development in the Use of Microbial Agents and Nematodes for Biological Insect Control*; Ehlers, R.U., Crickmore, N., Enkerli, J., Glazer, I., Lopez-Ferber, M., Tkaczuk, C., Eds.; IOBC-WPRS Bulletin: Pamplona, Spain, 2009; Volume 45, pp. 171–174.

51. Ocelotl, J.; Sánchez, J.; Arroyo, R.; Gracía-Gómez, B.I.; Gómez, I.; Unnithan, G.C.; Tabashnik, B.E.; Bravo, A.; Soberón, M. Binding and oligomerization of modified and native Bt toxins in resistance and susceptible pink bollworm. *PLoS ONE* **2015**, *10*. [CrossRef]

52. Ocelotl, J.; Sánchez, J.; Gómez, I.; Tabashnik, B.E.; Bravo, A.; Soberón, M. ABCC2 is associated with *Bacillus thuringiensis* Cry1Ac toxin oligomerization and membrane insertion in diamondback moth. *Sci. Rep.* **2017**, *7*, 2386. [CrossRef]

53. Padilla, C.; Pardo-López, L.; de la Riva, G.; Gómez, I.; Sánchez, J.; Hernandez, G.; Nuñez, M.E.; Carey, M.P.; Dean, D.H.; Alzate, O.; et al. Role of tryptophan residues in toxicity of Cry1Ab toxin from *Bacillus thuringiensis*. *Appl. Environ. Microbiol.* **2006**, *72*, 901–907. [CrossRef]

54. Chakroun, M.; Sellami, S.; Ferré, J.; Tounsi, S.; Rouis, S. *Ephestia kuehniella* tolerance to *Bacillus thuringiensis* Cry1Aa associated with reduced oligomer formation. *Biochem. Biophys. Res. Commun.* **2017**, *482*, 808–813. [CrossRef] [PubMed]

55. Rodríguez-Almazán, C.; Reyes, E.Z.; Zúñiga-Navarrete, F.; Muñoz-Garay, C.; Gómez, I.; Evans, A.M.; Likitvivatanavong, S.; Bravo, A.; Gill, S.S.; Soberón, M. Cadherin binding is not a limiting step for *Bacillus thuringiensis* subsp. *israelensis* Cry4Ba toxicity to *Aedes aegypti* larvae. *Biochem. J.* **2012**, *443*, 711–717. [CrossRef] [PubMed]

56. Sayyed, A.H.; Gatsi, R.; Ibiza-Palacios, M.; Escriche, B.; Wright, D.J.; Crickmore, N. Common, but complex, mode of resistance of *Plutella xylostella* to *Bacillus thuringiensis* toxins Cry1Ab and Cry1Ac. *Appl. Environ. Microbiol.* **2005**, *11*, 6863–6868. [CrossRef] [PubMed]

57. Crava, C.M.; Bel, Y.; Jakubowska, A.K.; Ferré, J.; Escriche, B. Midgut aminopeptidase N isoforms from *Ostrinia nubilalis*: Activity characterization and differential binding to Cty1Ab and Cry1Fa proteins from *Bacillus thuringiensis*. *Insect. Biochem. Mol. Biol.* **2013**, *43*, 924–935. [CrossRef] [PubMed]

58. Stewart, G.S.; Johnstone, K.; Hagelberg, E.; Ellar, D.J. Commitment of bacterial spores to germinate. A measure of the trigger reaction. *Biochem. J.* **1981**, *196*, 101–106. [CrossRef] [PubMed]

59. Bradford, M.M. A rapid and sensitive method for the quantification of microgram quantities of protein using the principle of dye-binding. *Anal. Biochem.* **1976**, *72*, 248–254. [CrossRef]

60. MacIntosh, S.C.; Stone, T.B.; Sims, S.R.; Hunst, P.L.; Greenplate, J.T.; Marrone, P.G.; Perlak, F.J.; Fischhoff, D.A.; Fuchs, R.L. Specificity and efficacy of purified *Bacillus thuringiensis* proteins against agronomically important insects. *J. Invertebr. Pathol.* **1990**, *56*, 258–266. [CrossRef]

61. Beegle, C.C. Bioassay methods for quantification of *Bacillus thuringiensis* δ-endotoxin. In *Analytical Chemistry of Bacillus Thuringiensis*, 1st ed.; Hickle, L.A., Fitch, W.L., Eds.; American Chemical Society: Washington, DC, USA, 1990; pp. 14–21. [CrossRef]

62. Iriarte, J.; Bel, Y.; Ferrandis, M.D.; Andrew, R.; Murillo, J.; Ferré, J.; Caballero, P. Environmental distribution and diversity of *Bacillus thuringiensis* in Spain. *Syst. Appl. Microbiol.* **1998**, *21*, 97–106. [CrossRef]

63. Finney, D.J. *Probit Analysis*, 3rd ed.; Cambridge University Press: Cambridge, UK, 1971. [CrossRef]

64. Wolfersberger, M.G.; Luethy, P.; Maurer, A.; Parenti, P.; Sacchi, F.V.; Giordana, B.; Hanozet, G.M. Preparation and partial characterization of amino acid transporting brush border membrane vesicles from the larval midgut of the cabbage butterfly (*Pieris brassicae*). *Comp. Biochem. Physiol. A Physiol.* **1987**, *86*, 301–308. [CrossRef]

65. Escriche, B.; Silva, F.S.; Ferré, J. Testing suitability of brush border membrane vesicles prepared from whole larvae from small insects for binding studies with *Bacillus thuringiensis* CryIA(b) crystal protein. *J. Invertebr. Pathol.* **1995**, *65*, 318–320. [CrossRef]

66. Hernández-Martínez, P.; Vera-Velasco, N.M.; Martínez-Solís, M.; Ghislain, M.; Ferré, J.; Escriche, B. Shared binding sites for the *Bacillus thuringiensis* proteins Cry3Bb, Cry3Ca, and Cry7Aa in the African sweet potato pest *Cylas puncticollis* (Brentidae). *Appl. Environ. Microbiol.* **2014**, *80*, 7545–7550. [CrossRef] [PubMed]

Article

A *Bacillus thuringiensis* Chitin-Binding Protein is Involved in Insect Peritrophic Matrix Adhesion and Takes Part in the Infection Process

Jiaxin Qin [1,†], Zongxing Tong [1,†], Yiling Zhan [1,2], Christophe Buisson [3], Fuping Song [2], Kanglai He [2], Christina Nielsen-LeRoux [3,*] and Shuyuan Guo [1,*]

[1] School of Life Science, Beijing Institute of Technology, Beijing 100081, China
[2] State Key Laboratory for Biology of Plant Diseases and Insect Pests, Institute of Plant Protection, Chinese Academy of Agricultural Sciences, Beijing 100193, China
[3] Micalis Institute, INRAE, AgroParisTech, Université Paris-Saclay, 78350 Jouy-en-Josas, France
*   Correspondence: christina.nielsen-leroux@inrae.fr (C.N.-L.); guosy@bit.edu.cn (S.G.);
    Tel.: +33-01-3465-2101 (C.N.-L.); +86-10-6891-4495 (S.G.)
†   These authors contributed equally to this work.

Received: 10 March 2020; Accepted: 10 April 2020; Published: 13 April 2020

Abstract: *Bacillus thuringiensis* (Bt) is used for insect pest control, and its larvicidal activity is primarily attributed to Cry toxins. Other factors participate in infection, and limited information is available regarding factors acting on the peritrophic matrix (PM). This study aimed to investigate the role of a Bt chitin-binding protein (CBPA) that had been previously shown to be expressed at pH 9 *in vitro* and could therefore be expressed in the alkaline gut of lepidopteron larvae. A ΔcbpA mutant was generated that was 10-fold less virulent than wild-type Bt HD73 towards *Ostrinia furnacalis* neonate larvae, indicating its important role in infection. Purified recombinant *Escherichia coli* CBPA was shown to have a chitin affinity, thus indicating a possible interaction with the chitin-rich PM. A translational GFP–CBPA fusion elucidated the localization of CBPA on the bacterial surface, and the transcriptional activity of the promoter P*cbpA* was immediately induced and confirmed at pH 9. Next, in order to connect surface expression and possible *in vivo* gut activity, last instar *Galleria mellonella* (Gm) larvae (not susceptible to Bt HD-73) were used as a model to follow CBPA in gut expression, bacterial transit, and PM adhesion. CBPA-GFP was quickly expressed in the Gm gut lumen, and more Bt HD73 strain bacteria adhered to the PM than those of the ΔcbpA mutant strain. Therefore, CBPA may help to retain the bacteria, via the PM binding, close to the gut surface and thus takes part in the early steps of Bt gut interactions.

Keywords: *Bacillus thuringiensis*; chitin-binding protein; adhesion; peritrophic matrix

Key Contribution: Our findings add a new step to the understanding of *Bacillus thuringiensis* (Bt) infection and suggest that the chitin binding protein A (CBPA) plays an important role in the first steps of the infection for insects where the Bt bacteria itself plays a role, which is for instance the case for *Ostrinia*. Indeed, due to a better bacterial binding to the larval peritrophic matrix (PM), it may increase the efficacy of the gut cell and PM damaging virulence factors produced from the outgrown bacteria.

---

1. Introduction

Bacillus thuringiensis (Bt) is a prominent insect pathogen of the *Bacillus cereus* group. Several strains are used worldwide as microbial control agents against major agricultural and forest insect pests [1]. The primary insecticidal factors of Bt are the Cry toxins, which comprise parasporal protein crystals produced by Bt, and numerous studies have focused on the structural resolution of the crystals and on the mode of action of Cry toxins [2]. As an insect pathogen [3], the roles of Bt itself in pathogenesis has been much less investigated and may depend on the insect species, larval stage, and Bt strain.

For pathogenic bacteria, the successful establishment of infection generally requires adhesion, colonization, and host cell degradation or active invasion. The capacity for host cell and tissue adherence is a key feature of pathogenic bacteria [4,5]. Orally acting entomopathogenic bacteria including Bt face the peritrophic matrix (PM) just after ingestion. The PM is an important component of the insect digestive tract: It serves both as a physical barrier to separate food particles, digestive enzymes, and pathogens, and it serves as a biochemical barrier, sequestering or even inactivating ingested toxins [6]. Therefore, Bt must bypass the PM barrier to establish persistent infections [7] in order to develop and complete the process of infection and life cycle, ending with sporulation in the insect cadaver [8,9].

The PM is a chitin- and glycoprotein-rich matrix, separating intestinal cells from the gut content [10]. Chitin is a linear polymer of N-acetylglucosamine (GlcNAc) linked via β-1,4 linkage [11]. Chitinases can hydrolyze chitin, thus fragmenting the PM and suggesting that chitinases may be part of the enzymes involved in the degradation of the PM [12]. Chitinases can enhance the insecticidal activity of Bt, irrespective of chitinase activity derived from a chromosomal gene, the co-expression of chitinase with a Cry toxin gene, or even from the addition of commercial chitinases [13–16]. The most probable role of the endogenous chitinases of Bt is to weaken the integrity of the insect PM, facilitating the better access of the bacterial toxins and the bacteria to the gut epithelia [17].

Chitin-binding proteins (CBP) are present in numerous microorganisms. They belong to the 14, 18, or 33 families of the carbohydrate-binding domain proteins [18]. Various microorganisms simultaneously synthesize chitinases and CBPs [19]. The subcellular localization of CBP differs in accordance with the organism, most of them being secreted proteins [19,20]. Structural analyses have revealed the presence of the aromatic amino acid residues exposed on most CBPs, which play an important role in substrate binding [21–25]. From viruses to invertebrate organisms, CBP participates in various biological processes in different species, such as antifungal activity [26], synergistic effects with chitinase [19], and the detection of hydrophobic surfaces [27].

The alkaline pH of the midgut—in lepidopteron larvae, in particular—is needed for Bt to exert insecticidal activity, since an alkaline pH permits the solubilization of several Cry toxin crystals [1]. Hence, it is important for bacteria to adhere to host tissue and survive in this alkaline environment in order to pursue infection. Transcriptome gene microarray data previously indicated that *cbp3189* is up-regulated more than eight-fold after alkaline induction [28]. The protein encoded by *cbp3189* is referred to as chitin binding protein A (CBPA) in this study. CBPA is a conserved protein in Bt strains, as its amino acid sequence homology in 34 different Bt strains is greater than 97%, and, among them, nine strains have a 100% sequence identity. This gene encodes a protein containing a signal peptide and a transmembrane structure, and localization prediction has revealed that CBPA may be localized on the bacterial cell wall [29].

In this study, we addressed several questions related to the function, expression, and localization of CBPA. First, to determine whether CBPA plays a role during infection, a ΔcbpA mutant was constructed and assessed for virulence in *Ostrinia furnacalis* and *Galleria mellonella* larvae. Further, its subcellular localization in Bt and the activity of its promoter were assessed during in vitro growth. CBPA was expressed and purified from *Escherichia coli* through binding to chitin beads, and it was further analyzed for chitinase activity. Finally, the interaction of CBPA with the gut and the PM in vivo was assessed in *G. mellonella*, with a focus on the early stages of infection. Our results may provide functional insights into the role of CBPA in adhesion to the PM, thus improving the current understanding of the mode of action of Bt insecticidal strains, particularly for insects where the action of Cry toxins, spores, and out-grown bacteria are important for full virulence.

2. Results

2.1. ΔcbpA Mutant, ΔcbpA::cbpA-Complemented Strain Construction

To determine the role of CBPA in insect infection, an interruption mutant strain ΔcbpA mutant was constructed via homologous recombination. Cry1Ac protein expression levels were not changed in the ΔcbpA mutant in comparison with wild-type HD73 upon protein quantification (Figure 1A). Spore count results showed that the wild-type strain and ΔcbpA mutant contained equal CFU values (Figure 1), indicating that the absence of *cbpA* did not influence spore formation.

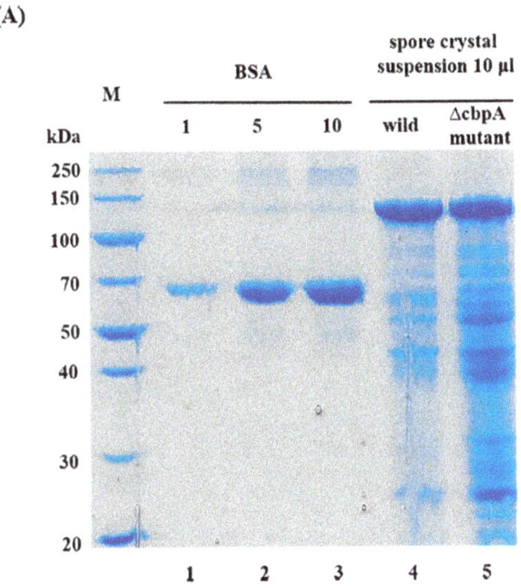

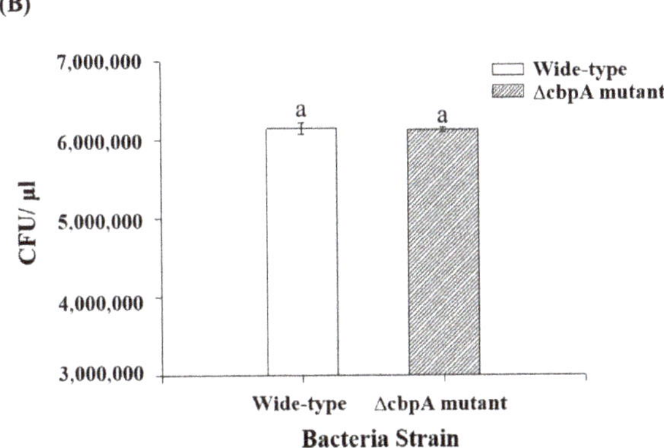

Figure 1. Comparison of Cry protein expression and spore formation in *Bacillus thuringiensis* (Bt) HD73 wild-type strain and ΔcbpA strain. (**A**) Cry protein production analyzed via SDS-PAGE. (**B**) Spore counts. 1. Bovine serum albumin (BSA) (1 µg); 2. BSA (5 µg); 3. BSA (10 µg); 4. Cry1Ac protein in wild-type (10-µL spore crystal suspension); 5. Cry1Ac protein in deletion mutant (10-µL spore crystal suspension). "a" indicates there was no significant difference (*p* > 0.05).

2.2. Role of CBPA in Ostrinia Furnacalis and Galleria Mellonella Mortality

As the Bt HD73 produces Cry1Ac that is toxic against the Asian corn borer (*Ostrina*), we selected this insect to evaluate the difference in mortality of larvae fed a diet supplemented with spore-crystal suspensions of wild-type or ΔcbpA mutant HD73 at various concentrations. Mortality induced by the ΔcbpA mutant was lower than that of the wild-type strain (Figure 2 and Table 1). Figure 2 shows the comparison of mortality rates between the wild-type HD73 strain and the ΔcbpA mutant strain for seven days post-feeding. For both strains, almost no change in mortality was observed after the fourth day of feeding. Mortality rates on the seventh day after feeding with seven different concentrations are listed in Table 1. The mortality between the wild-type and mutant strains was significantly different from the third concentration (Cry1Ac protein: 0.05 µg/g; spore: 5.4×10^7/g).

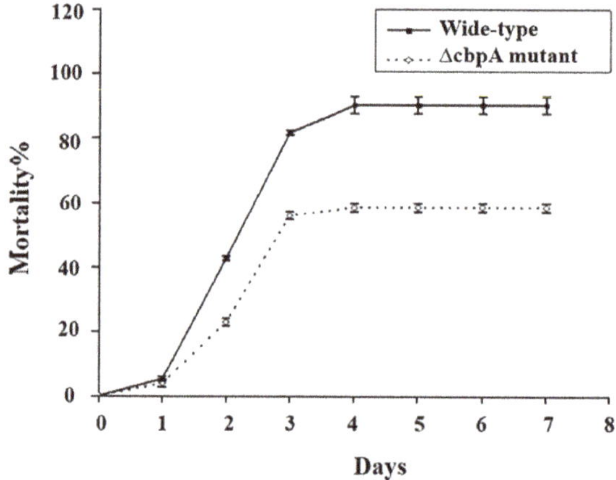

Figure 2. Comparison of mortality rates between the wild-type HD73 strain and the ΔcbpA mutant strain against the Asian corn borer (Cry protein: 2.5µg/g diet; spore: 2.7×10^9/g diet).

Table 1. Mortality of crystal and spore mixture against Asian corn borer larvae.

No.	Concentration of Cry Protein (µg/g)	Number of Spore (Numbers/g)	Mortality (%) Wild-Type	Mortality (%) ΔcbpA Mutant	Significance
1	0.005	5.4×10^6	0.0 ± 0.0	0.7 ± 0.7	0.374
2	0.025	2.7×10^7	4.7 ± 0.7	2.7 ± 1.3	0.251
3	0.050	5.4×10^7	17.0 ± 1.2	$8.7 \pm 0.7*$	0.003
4	0.250	2.7×10^8	35.3 ± 1.5	$19.7 \pm 0.7 *$	0.001
5	0.500	5.4×10^8	51.3 ± 1.8	$33.7 \pm 1.3*$	0.001
6	1.000	1.1×10^9	76.3 ± 1.3	$54.7 \pm 0.7*$	$<< 0.001**$
7	2.500	2.7×10^9	90.3 ± 2.7	$58.7 \pm 1.3 *$	$<<0.001***$

* Means within a line were significantly different ($p \leq 0.01$) via the t-test. ** significance: $p = 0.000130$. *** significance: $p = 0.000478$.

Further, the difference in LC_{50} was evaluated by administering larvae with a diet supplemented with spores of the HD73 wild-type strain, the ΔcbpA mutant, or Δ*cbpA*::*cbpA*-complemented strains at different doses (10^4–10^8 CFU/per gram diet) and with the concentration of Cry1Ac at 0.01 µg/g diet. Table 2 shows the LC_{50} values of the three strains. Deionized water was used as the negative control, and the mortality was 2% in this control. The inferred larval mortality of the ΔcbpA mutant was significantly lower than that of the wild-type strain, while that of the complemented strain reverted to levels of the wild-type strain. These data indicate that CBPA significantly contributes to Bt virulence in *Ostrinia*.

Table 2. LC$_{50}$ values of different strains against Asian corn borer.

	LC$_{50}$ (Spore CFU)	95% Confidence Interval
BtHD73	6.59×10^5	3.41×10^5–1.04×10^6
ΔcbpA mutant	4.85×10^6	2.25×10^6–7.61×10^6
ΔcbpA::*cbpA*	5.72×10^5	3.18×10^5–8.98×10^5

Each concentration of spore was mixed with Cry1Ac at a final concentration of 0.01 ug/g.

To further investigate the role of CBPA, we tested the mutant for virulence towards *G. mellonella*; this insect needs bacteria associated with Cry1Ca for complete virulence in the synergy model [30,31]. Therefore, *G. mellonella* is a suitable model to elucidate the role of Bt and *B. cereus* chromosomal carried factors, and large last instars are easy to manipulate for accurate feeding and for dissection. Infections were induced through controlled force-feeding at a dose of 5×10^6 mid log-phase vegetative bacteria (OD$_{600}$ = 1) or with spores associated with 3 µg of activated Cry1Ca toxin for each larva, as described previously [32]. The results (Figure S1) showed no differences in mortality between infection with wild-type HD73 and the ΔcbpA mutant strains under all tested conditions. Indeed, no mortality was observed with spores or log-phase bacteria alone, and 90–100% mortality was observed when associated with 3 µg of Cry1Ca for both strains. Therefore, under these infection conditions, no clear role of CBPA was elucidated in Bt HD-73 mortality towards *G. mellonella* last instars.

2.3. Localization of CBPA in Bt HD-73

GFP-conjugated CBPA was used to investigate the subcellular localization of CBPA in HD73 bacteria cells. Samples were harvested at different growth stages (T4, T7, T8, and T10 after the onset of the stationary phase). GFP expression was visualized via laser-scanning confocal microscopy. The cell membrane was stained with FM 4-64 dye solution. Red fluorescence indicated the cell membrane, while green fluorescence indicated the expression of the GFP–CBPA fusion protein. No green fluorescence was detected during early growth; however, it stabilized from T8 onwards (Figure 3A). Figure 3B shows the green fluorescence at T12. Fluorescence observed on the bacteria cell surfaces was indicated by a yellow arrow and on the prespore surface with a red arrow. Consistent with the in silico predictions, GFP fusion experiments revealed that CBPA was located on the cell surface.

(A)

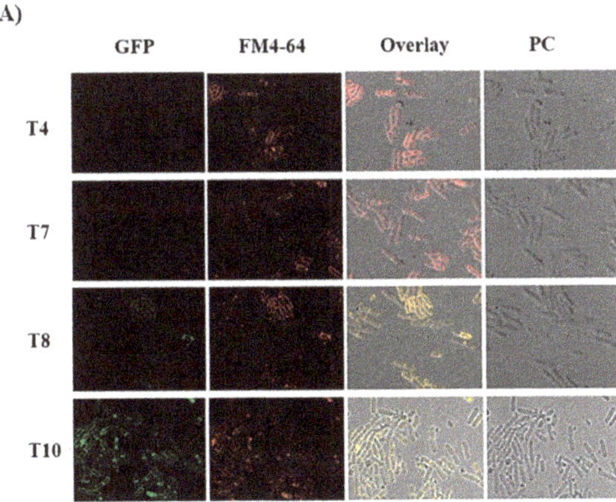

Figure 3. *Cont.*

(B)

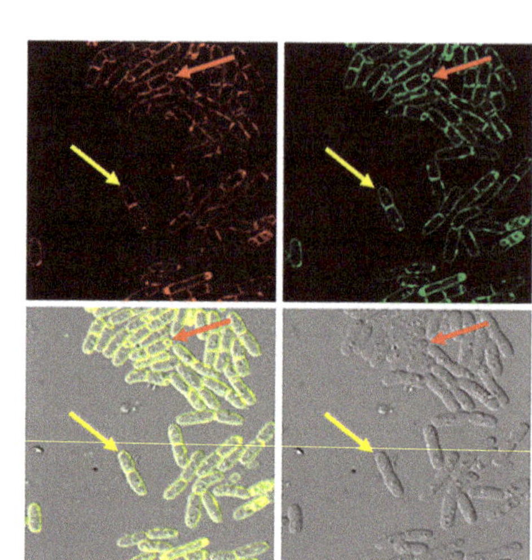

Figure 3. Green fluorescence detection of the chitin binding protein A (CBPA)–GFP fusion at different stages of culturing, observed via laser-scanning confocal microscopy. (**A**) Stages T4, T7, T8, and T10. (**B**) Localization of CBPA (T12). Yellow arrow denotes green fluorescence on the bacterial cell surface. Red arrow denotes green fluorescence on the spore surface. GFP (green fluorescent protein) signal in the bacterial cytosol. FM 4-64, (red fluorescent signal of bacterial membrane stain). The overlay shows green and red fluorescent signals. PC: phase-contrast microscopy.

2.4. Analysis of cbpA Promoter Activity under Alkaline Induction

To investigate the effect of alkali on CBPA expression, we selected an early culture stage wherein the CBPA protein was not expressed. Both the *cbpA-gfp* transduction fusion and the transcriptional activation of the *cbpA* promoter-*lacZ* fusion were analyzed under similar growth conditions. The bacteria were cultured to the late exponential growth stage up to an $OD_{600\,nm}$ between 1.5 and 2.0, and an NaOH solution was added to yield a final concentration of 24 mM (pH 9). Samples were maintained under this alkaline environment for 15 and 30 min. GFP expression was then visualized via laser-scanning confocal scanning microscopy (Figure 4A). Cells not treated with the NaOH solution were considered as the negative control. GFP was expressed after alkaline induction (pH 9) but not in the negative control, indicating that CBPA protein expression was rapidly induced under alkaline conditions. In the transcriptional *cbpA* promoter-*lacZ* fusion strain, β-galactosidase activity significantly increased after the addition of NaOH (15 or 30 min) in comparison with non-induced conditions (Figure 4B). Consequently, the cytological observation of the GFP fusion strain and the enzymatic activity analysis in P*cbpA-lacZ* promoter fusion strain yielded consistent results, indicating that CBPA expression was induced at an alkaline pH.

(A)

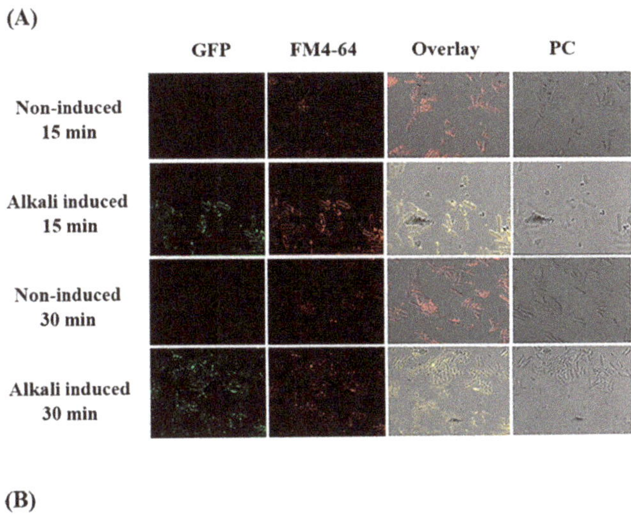

(B)

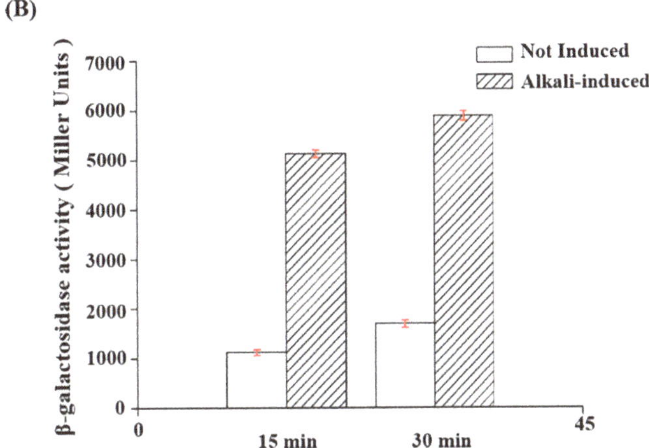

Figure 4. Analysis of transcriptional activity. (**A**) Observation of alkaline induction via laser-scanning confocal fluorescence microscopy. Row 1: non-induced for 15 min. Row 2: induced under alkaline conditions for 15 min. Row 3: non-induced for 30 min. Row 4: induced under alkaline conditions for 30 min. (**B**) Analysis of β-galactosidase activity of the P*cbpA-lacZ* fusion +/- alkaline induction.

2.5. Chitin Binding Ability and Chitinase Activity of CBPA

To follow-up on the in silico information indicating CBPA as a chitin binding protein, the next step was to analyze if CBPA really has chitin binding capacity. Therefore, we expressed CBPA as a heterologous recombinant protein. *cbpA* (gene 3189) from Bt HD73 was cloned and expressed in an *E. coli* BL21/DE3 strain. The expected size of the HD73-CBPA protein was 49.78 kDa. Protein expression was induced through an isopropyl-β-D-thiogalactopyranoside (IPTG) gradient, and expression was assessed via SDS-PAGE as a ~50-kDa protein at 0.4–1.0 mM IPTG (Figure 5A). Thereafter, CBPA was purified via chitin affinity chromatography and eluted at a gradient of 0.4 M NaCl, thus showing the chitin-binding capacity of CBPA (Figure 5B). The purified protein band was excised and analyzed via matrix-assisted laser desorption/ionization and time-of-flight peptide mass spectrometry analysis after in-gel digestion, confirming that the heterologous *E. coli*-cloned and -expressed CBPA protein contained the expected peptide composition. Having confirmed its chitin-binding capacity, we assessed for the eventual chitinase activity of the protein. CBPA displayed no chitobiosidase and endochitinase activity (Table 3). Indeed, the fluorescence intensity of chitinase degradation products from various substrates obtained with CBPA approached the same values as those of the negative control. Therefore, CBPA can probably not degrade chitin-rich structures, at least those analyzed herein.

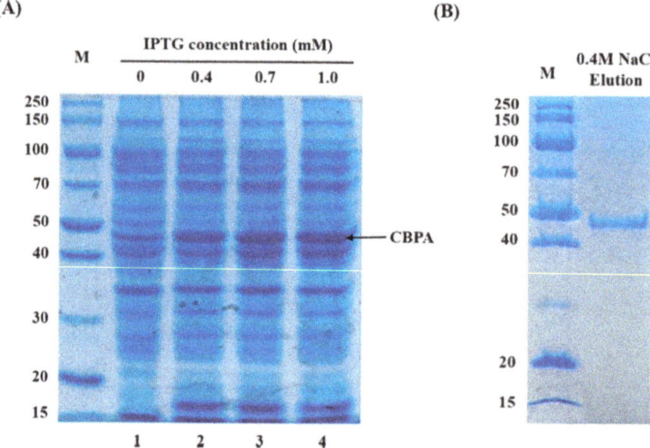

Figure 5. Expression and purification of CBPA proteins harvested for SDS-PAGE analysis. (A) Lane 1: the non-induced expression of CBPA in *E. coli* BL21/DE3. Lanes 2-4: induced expression of HD73-*cbpA* by the isopropyl-β-D-thiogalactopyranoside (IPTG) gradient of 0.4, 0.7 mM, and 1.0 mM. (B) purified CBPA eluted by a gradient of 0.4 M NaCl.

Table 3. Fluorescence value of chitinase activity.

Chitinase Substrate	Chitinase Activity (Substrate Degradation, μmol/min)		
	Negative Control	CPBA	Positive Control
4-Methylumbelliferyl N-acetyl-β-D-glucosaminide	4.79×10^4	4.88×10^4	3.47×10^6
4-Methylumbelliferyl B-D-N,N'-diacetylchitobioside hydrate	6.35×10^4	5.88×10^4	4.81×10^6
4-Methylumbelliferyl B-D-N,N,N''-triacetylchitotriose	5.01×10^4	4.79×10^4	2.21×10^6

2.6. Expression of CBPA-GFP Fusion in Vivo in G. mellonella

Despite the lack of an evident role of CBPA in virulence in the final instar of *Galleria* larvae, we aimed at determining the possibility of CPBA to bind the PM since chitin is a structural element of the PM in all insects. First, we investigated whether CBPA was expressed in the *Galleria* gut, since the aforementioned *in vitro* studies (Figures 3 and 4) indicated that CBPA was expressed on the surface of HD73 cells and under alkaline pH, which may occur in the *Galleria* larval gut. The pH of the *Galleria* midgut was measured via the injection of a liquid pH indicator into three sites of the gut; the pH was between 8.5 and 9 from the anterior to posterior midgut, as directly observed under binoculars with four times magnification. Thereafter, we assessed, via epi-fluorescence microscopy, the presence of fluorescent bacteria from the anterior and posterior midgut of *Galleria* larvae infected with mid log-phase Luria–Bertani (LB) grown vegetative HD73 bacteria carrying the CBPA–GFP plasmid fusion protein. Observations were recorded at 1 and 4 h post-ingestion. The CFU values and the fluorescence intensities were scored upon arbitrary visual observation (Figure 6). Greater CFU values were recovered at 1 than at 4 h post-ingestion (Figure 6A), indicating a relatively rapid intestinal transit and that fluorescent bacteria (Figure 6B) were more abundant at the early time point.

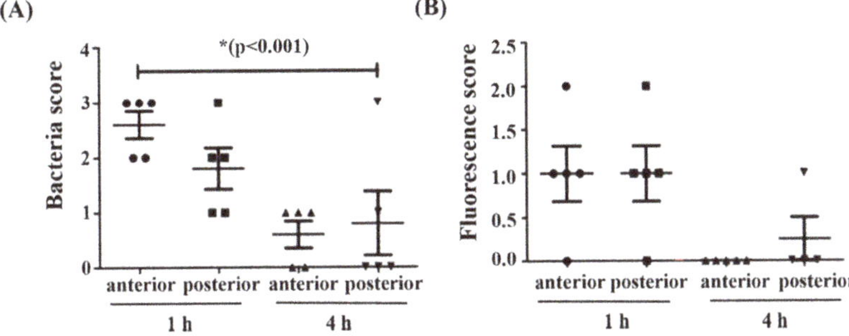

Figure 6. Arbitrary scores of bacteria (**A**) and the expression of the CBPA–GFP fusion protein (**B**) recovered in the *Galleria mellonella* intestine. At one hour and 4 h post-force-feeding with a wild-type HD73 (pHT*cbpA-gfp*) strain, samples from the anterior midgut and posterior midgut were analyzed via fluorescence microscopy at 1000× magnification from five chilled, dissected larvae. Scores are as follows: 0 = no bacteria and no fluorescence, 1 = few bacteria <10 per field, 2 = between 10 and 50 bacteria and 3 = more than 50 bacteria per observation field.

2.7. HD73 and HD73 ΔcbpA Intestinal Transit and Localization Assays

Since purified CBPA can bind to chitin, is expressed on the bacterial surface (Figures 3 and 5), and is activated in the gut of *Galleria*, we performed a tight analysis of the persistence of the HD73 wild-type and ΔcbpA mutant strains with vegetative bacteria, presuming that the PM binding capacity of CBPA *in vivo* would lead to a difference between the wild-type and the mutant strains during intestinal transit.

First, the presence of bacteria was estimated in whole larvae and dissected whole intestines (gut with the PM) (Figure 7A,B) at three time points. Immediately after ingestion (T0), no difference (≈5000 CFU) was observed between wild-type HD73 and ΔcbpA mutant strains, while at T3 h post-ingestion, a significant difference was observed between wild-type HD73 (≈100 CFU) and the ΔcbpA mutant (approximately 5000 CFU were still observed). At 24 h, no bacteria were observed in larvae fed with wild-type HD73, and approximately 100 CFU were recorded for larvae infected with the ΔcbpA mutant. A similar analysis was performed with the dissected whole intestines (gut and the PM) (Figure 7B). No difference was observed at T0; however, at T3 h post-ingestion, almost no bacteria were observed with the wild-type HD73 strain, while approximately 3000 CFU were still observed with the ΔcbpA mutant strain. Thus, bacteria not expressing CBPA persist longer in the gut than the wild-type bacteria, suggesting that the wild-type HD73 bacteria are more easily excreted with the PM during natural food bolus transit, resulting in feces production. The feces are surrounded by the PM [6].

Therefore, we further analyzed the speed of transit after force-feeding with spores in order to uncover the time where we would still find all bacteria in the insect before they would be excreted with the feces. Feces were collected and assessed for the presence of bacteria, and the mean numbers of feces per larvae were recorded at four time points. Feces were observed at 2 h (0.3 feces/larvae) and displayed an increase in the mean number of feces at 3 and 4 h to 1.5 feces per larva. The presence of the bacteria was found in feces from the 2 to early 3 h post ingestion, showing that the mean transit time under these conditions was approximately 2 h. Based on these observations, we thereafter tested for the presence of the bacteria adhering to the PM. Hence, we selected the time point of 1 h post-ingestion, since feces were excreted at 2 h per the aforementioned results and since at 3 h post-ingestion, only a few residual bacteria were observed in the wild-type HD73-treated larvae (Figure 7A,B). The CFU value recovered from the dissected PM alone (Figure 7C) was approximately 5000 for wild-type HD73 and three-fold lesser for the ΔcbpA mutant, which was significantly different from the wild-type. In addition, the *cbpA*-complemented ΔcbpA mutant recovered a better adhesion to the PM. One way ANOVA analysis and the Bonferroni's multiple comparison test showed significant differences between the PM from the ΔcbpA mutant and wild-type HD73, while the complemented strain PM had no significant differences with the others. The observed variations in CFU that were associated with the dissected intestine, separated intestine, and the PM may have been due to the difficulty of the dissection approach (see Materials and Methods). The results indicated that CBPA *in vivo* has an affinity for the PM (Figure 7C) and therefore may help retain the bacteria to this tissue, which could then increase the infection efficacy of Bt.

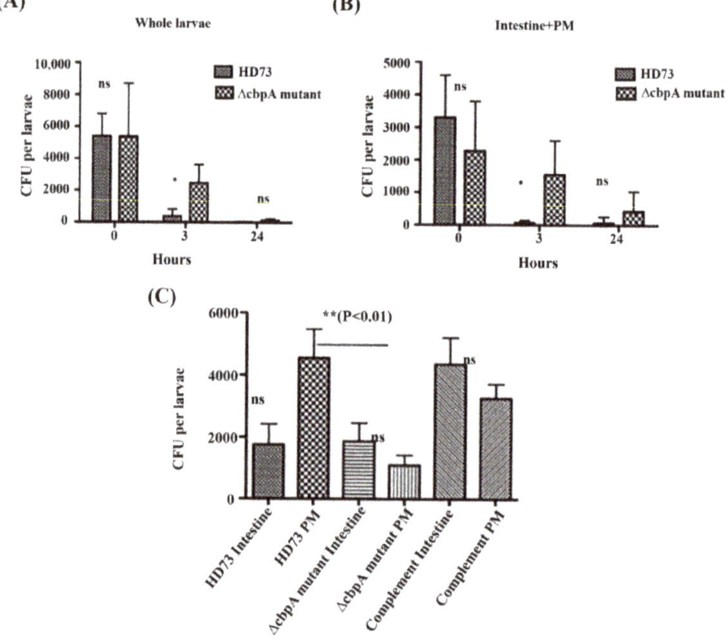

Figure 7. Presence of bacteria in *Galleria mellonella* whole larvae (**A**) dissected complete intestine (**B**), and dissected dissociated intestine and the PM (**C**) from chilled, fifth-instar larvae after force-feeding at stages T0, T3, and T24 h post-feeding for (**A,B**) and after 1 h for (**C**). Whole larvae, the larval intestine, and the PM were homogenized to determine the CFU counts for each sample. Assays were repeated at least three times with two replicates per sample time and sample type. CFU counts were analyzed with the PRISM software and one way ANOVA associated with the Bonferroni's multiple comparison test. ** <0.01 and * <0.05 level, ns (non-significant) different.

3. Discussion

During infection, a pathogen interacts with the host and circumvents the host's defense mechanisms. For many pathogens, the first step of colonization depends on the capacity to adhere to the host tissue via multiple factors [5]. Therefore, pathogenic bacteria produce surface molecules and appendages, such as flagella and pili. They sense host surfaces, thus facilitating their adhesion with host cells and thereby bringing secreted molecules close to the host cell targets. Pore-forming Cry endotoxins are the major *B. thuringiensis* insecticidal effectors. They bind to specific membrane receptors on the larval midgut epithelium [1,33]. Meanwhile, the spores and their outgrown vegetative form participate in infection [3]. Several proteases, lipases, chitinases, toxins, or adaptation factors are involved in pathogenesis [8,31] and in fulfilling the insect phase of the Bt life cycle. The germination and growth of *B. thuringiensis* in the gut of insect larvae have been previously photographically studied. It was reported that the spores of *B. thuringiensis* germinated at the surface of the epithelium 40–120 min after inoculation [34]. Additionally, histological studies have shown that the vegetative bacteria of *B. thuringiensis* are found in the gut lumen of *Chrysomela* [35]. However, thus far, limited information is available regarding the role of factors associated with Bt spores and newly outgrown vegetative bacteria in the very early stages of gut infection, particularly with respect to their interaction with the PM.

This study investigated the expression and functions of a yet unknown chitin-binding protein (CBPA) from the Bt HD73 strain and explored its role in insect infection. The gene encoding CBPA was previously identified among genes activated under alkaline conditions via an *in vitro* transcriptomic screening [28], and our previous *in silico* analysis indicated its putative chitin-binding function and its presence in several *B. cereus* genomes. Herein, we investigated the spatiotemporal aspects of its expression. The confocal microscopic imaging of Bt HD73 harboring a CBPA–GFP fusion protein revealed that CBPA was expressed in vitro at the late stationary growth stage and localized on both the bacterial and prespore surfaces (Figure 3). Furthermore, when the bacteria were exposed to an alkaline pH, the protein was expressed as soon as 15 min post-induction, as revealed through both CBPA–GFP fusion and a lacZ transcriptional promoter fusion (Figure 4). These observations indicate that expression is inducible at an alkaline pH, which is known for the Lepidopteran midgut environment. This has also been reported in the case of the Bt CBP-21 chitin-binding protein [36] and is concurrent with our former transcriptome findings [28] and with the present study, wherein the CBPA–GFP protein was observed in the *Galleria* midgut at 1 h post-ingestion with vegetative bacteria.

The widespread presence of CBP proteins in bacteria and other organisms implies their importance, with chitin binding being the most common function. CBP21 from the Bt HD1 strain binds chitin in insects [36], and CBP50 from the Bt *konkukian* serotype and CBP33A from *Lactococcus lactis* bind insoluble chitin (α and β), colloidal chitin, and cellulose [37]. Furthermore, *Streptomyces* can secrete small proteins that specifically bind α-chitin [38]. ChbB produced by *Bacillus amyloliquefaciens* preferentially binds β-chitin [39]. The present study showed the capacity of recombinant purified CBPA to bind chitin, which is concurrent with previous reports with similar CBP proteins.

Some bacteria simultaneously produce CBP and chitinase, thus improving the hydrolysis efficiency of chitinases. For example, CBP24 and CBP50 produced by Bt *serovar konkukian* act synergistically with bacterial chitinases for chitin degradation [40,41], CBP21 from *Serratia marcescens* exerts a synergistic effect with chitinase on its hydrolysis efficiency [19], and CBP33A from *Lactococcus lactis* can increase the hydrolysis efficiency of chitinase Chi18A [20]. As expected, chitinase activity was not observed for CBPA itself, but, based on our findings, we may suggest that CBPA on the bacterial surface can help target the bacteria to the chitin-rich PM, where the chitinases are of particular relevance. Indeed, the present results showed that the ΔcbpA mutant adhered less well to the *Galleria mellonella* PM, thus indicating that CBPA is involved in the adhesion of Bt HD73 out-grown vegetative bacteria to the PM (Figure 7C).

CBPA localized on the cell surfaces in a manner similar to CBP21 from the Bt HD1 strain, which was also reported to be present in the spore crystal preparation [36]. CBP21 and CBPA have low global sequence homology, and an *in silico* analysis indicated that CBPA comprises the chitin-binding-3 domain, two FN3 domains, and a CBM-5-12 domain, while CBP21 comprises a peptidase M73 domain. It was speculated that Bt-CBP21 interacts with Cry1Ac to potentiate its insecticidal activity [36], which may be concurrent with an earlier observation with strain Bt HD73, wherein Cry1Ac localized at the spore surface [42]. In the present study, Cry1Ac and spores from ΔcbpA mutant displayed higher LC_{50} (approximately 10-fold) values against *Ostrinia furnacalis* (Asian Corn borer) neonate larvae in comparison with the wild-type Bt HD73 strain, and the *cbpA*-complemented ΔcbpA mutant strain displayed a similar LC_{50} value to the wild-type strain, indicating the role of CBPA in virulence in the HD73 strain. As the first barrier in the digestive tube in most insects is the PM, orally acting pathogens require factors that can interfere with the PM. In the present study, CBPA increased the adherence to the PM, thus playing a role in the early stage of infection. In the susceptible insect, the Asian corn borer, the absence of CBPA strongly reduced mortality, which was not the case for *Galleria*, where no mortality was recorded with the wild-type HD73 strain or ΔcbpA mutant strain alone. Therefore, under the present conditions, no obvious function of CBPA in virulence was discerned in *Galleria*, probably because the strong synergism with Cry1Ca [31] concealed a rather subtle bacterial effect or because Cry1Ca is acting directly on the PM, consequently reducing the role of CBPA in that synergy model. However, the *Galleria* model was optimal to assess bacteria–PM interactions *in vivo*. Indeed, the transit studies in *Galleria* clearly indicated that CBPA plays a role *in vivo*, as its presence increases the capacity of vegetative bacteria to adhere to the PM.

Concurrent with previous reports, the present study proposes an additional step in the mode of action of *B. thuringiensis* (Figure 8). The present results indicated that CBPA can be induced in vegetative Bt cells in the alkaline midgut environment, thus facilitating the adhesion of Bt bacteria to the PM and thereby increasing the performance of various virulence factors. Accordingly, chitinases or enhancin-like proteins (Bel and MpbE) [43,44] may be produced and destabilize the chitin structure of the PM. This, along with the role of the active pore-forming Cry toxins known to damage the midgut cells resulting in reduced PM renewal and reduced intestinal transit time, further facilitates bacterial adhesion with intestinal cells and increases colonization. This might be followed up by bacterial translocation from the midgut to the insect hemocoel, through the action of non-specific adaptation and virulence factors, notably those from the PlcR regulon, which was earlier shown to being important for virulence toward *Galleria* [31]. Therefore, the present results further the current understanding of the complex pathogenesis and ecology of Bt, owing to studies on two insects: *Ostrina*, which is naturally sensitive to Bt HD73 and its Cry1Ac toxin, and the non-sensitive model insect *Galleria*, which allows for the easy manipulation of the PM. Further studies are required to analyze the importance of CBPA in other Bt strains and insects in order to understand the mode of action of CBPA and to validate our findings as a general feature in the early stages of Bt insect larva infection.

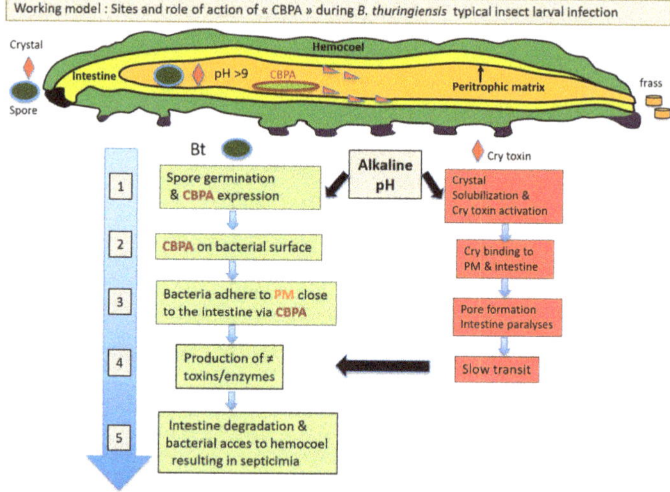

Figure 8. Proposed working model for the site of action of the chitin binding protein CBPA. The figure indicates where CBPA, during the oral infection with *B. thuringiensis* in a Cry toxin susceptible lepidopteron larva, plays a role. The green blocks of the steps refer to the spore/bacteria actions, and the red blocks refer to the role of the Cry toxins. The numbers, in the time scale arrow (in blue), indicates the order of events of which some occurs simultaneously. Our results showed that CBPA is expressed on the surface of vegetative bacteria (step 1) and is induced at alkaline pH. The proposed major role of CBPA is its adhesion to the peritrophic matrix (PM) (steps 1 and 3), which permits outgrown bacteria to bind to the PM and to be closer to the intestinal surface (4 and 5) in order to facilitate the tissue damaging action of secreted enzymes and toxins, e.g., from the PlcR regulon [31].

4. Materials and Methods

4.1. Bacterial Strains

The plasmids, primers, and sequences used herein are enlisted in Tables 4 and 5. Bt strains were cultured at 30 °C, and *Escherichia coli* was cultured at 37 °C with agitation at 220 rpm in an LB medium (1% NaCl, 1% tryptone, and 0.5% yeast extract) [45]. *B. thuringiensis* HD73 (the wild-type strain, producing crystals exclusively comprising the Cry1Ac toxin) was used to clone the target gene and monitor promoter activity as the recipient strain [46], as well as for bioassays and mutant construction. DNA sequences were obtained from the NCBI database (https://www.ncbi.nlm.nih.gov/) and compared using the Basic Local Alignment Search Tool (BLAST).

Table 4. Strains and plasmids used in this study.

Strain or Plasmid	Characeristics	Reference or Source
	E. coli strains	
JM110	$rpsL(str^r)$,*thr*,*leu*,*thi-1*,*lacY*,*galK*,*galT*,*ara*,*tonA*,*tsx*,*dam*,*dcm*,*supE44*,Δ(*lac-proAB*), [F',*traD36*,*proAB*,*laclqZ*Δ*M15*]	This laboratory
BL21/DE3	*E. coli* B, F-,*dcm*,*ompT*, *hsdS*(*rB-mB-*), *gal*, λ(*DE3*)	[47]
ET	Δ(*lac-proAB*) $RpsL(str^r)$, *thr*, *leu*, *endA*, *thi-1*, *lacY*, *galK*, *galT*, *ara*, *tonA*, *tsx*, *dam*, *dcm*, *supE44*, (F' *traD36proABlacI^qZ*Δ*M15*)	This laboratory
	B. thuringiensis subsp. *kurstaki* strains	
HD73	Contains *cry1Ac* gene	[46]
HD73(pHT-*gfp*)	HD73 strain containing plasmid pHT-*gfp*	[48]
HD73(pHT-*cbpA*-*gfp*)	HD73 strain containing the translational fusion plasmid pHT-*cbpA*-*gfp*	This study
HD (P_{cbpA}-*lacZ*)	HD73 strain containing plasmid pHTP$_{cbpA}$	This study
HD73(pRN5101Ω*cbpA*)	HD73 strain containing plasmid pRN5101Ω*cbpA*	This study
HD73(ΔcbpA)	HD73 mutant, Δ*cbpA*	This study
HD73(Δ*cbpA*::*cbpA*)	HD73(Δ*cbpA*) containing plasmid pHTC*cbpA*	This study
	Plasmids	
pET-21b	Expression vector, Amp^r, 5.4 kb	Novagen
BL21 (pET*cbpA*)	BL21(DE3) with pET*cbpA* plasmid	This study
pHT315	*B. thuringiensis-E. coli* shuttle vector, 6.5kb	[49]
pHT304-18Z	Promoterless *lacZ* vector, Ery^r Amp^r, 9.7 kb	[50]
pHTP$_{cbpA}$	pHT304-18Z carrying $P_{cbpA,}$ Amp^r Erm^r	This study
pET*cbpA*	pET-21b containing *cbpA* gene, Amp^r	This study
pETC*cbpA*	pHT304 carrying *cbpA* gene,Amp^r	This study
pHT P_{cbpA}-*3189*-*gfp*	pHT315 containing P_{cbpA}-*cbpA*-*gfp* gene	This study
pRN5101	Temperature-sensitive plasmid, 8.0 kb	[51]
pRN5101Ω*cbpA*	pRN5101 carrying partial *cbpA* deletion gene	This study
pDG780	Containing a kanamycin resistance gene	[52]

Table 5. Primers and sequences used in this study.

Primer	Sequence	Restriction Site
cbpA-a	CGCGGATCCGATGAACATGAATAATCGAT	*Bam*H I
cbpA-b	ACGCGTCGACTTACACTGTTTTCCATAAT	*Sal* I
cbpA-c	GCATGCCTGCAGGTCGACTCTAGAGGGCTGCTTTGAATTTGAAGGAAT	
cbpA-d	TGTAAAACGACGGCCAGTGAATTAAAGCCCATCATCTCTTAGTTCAT	
gfp-1	CGGGATCCAAGAGGCTGCTTTGAATTTGAAGG	*Bam*H I
gfp-2	CCGCCTCCACCTGACACTGTTTTCCATAATG	
gfp-3	CATTATGGAAAACAGTGTCAGGTGGAGGCGG	
gfp-4	CATGCATGCTTATTTGTATAGTTCATCCATGCC	*Sph*I(*Pae* I)
P_{cbpA}-F	TGCACTGCAGGGCTGCTTTGAATTTGAAGGAATC	*Pst* I
P_{cbpA}-R	CGGGATCCGTTCATGTCCCCTTCTTGTTATAC	*Bam*H I

Underline indicates the restriction enzyme site.

4.2. Insects

For *O. furnacalis* (Asian corn borer), larvae for bioassays and mortality tests (see below), were provided by the rearing at the Chinese Academy of Agriculture Sciences (Beijing, China). For mortality tests and other in vivo analyses, 5th instar larvae of the greater wax moth *G. mellonella* were used. Insects were reared at the INRAE-Micalis, Jouy en Josas, France, facilities at 27 °C and fed with pollen and bee wax (La ruche Roannaise, France). Prior to assays, the larvae weighting approximately 250 mg were selected and stored under starvation conditions for 24 h.

4.3. DNA Manipulation and Transformation

PCR amplifications were performed using Taq DNA polymerase and Pfu DNA polymerase (TIANGEN Biotechnologies Corporation, Beijing, China). PCR products were separated on agarose gels and recovered using the HiPure Gel Pure DNA Mini Kit (Magen Biotechnology Corporation, Guangzhou, China), and plasmid DNA was extracted from *E. coli* using the Plasmid Miniprep Kit (Axygen Biotechnology Corporation, Hangzhou, China) in accordance with the manufacturers' instructions. Restriction enzymes and T4 DNA ligase (Thermo Fisher Scientific, Beijing, China) were used in accordance with the manufacturer's instructions. Oligo-nucleotide primers were synthesized by Sangon Biotech (Shanghai, China). All constructs were confirmed by DNA sequencing (GENEWIZ, Beijing, China). *E. coli* cells were transformed via standard procedures [53], and Bt cells were transformed via electroporation, as described previously [54].

4.4. Cloning of the HD73-cbpA Gene

The *HD73-cbpA* gene (ID:14557228) was cloned from the wild-type HD73 genome via PCR, using specific primers *cbpA*-a and *cbpA*-b (Table 5) under the following cycling conditions: denaturation at 94 °C for 30 s, annealing at 50 °C for 1 min, and extension at 72 °C for 1 min for 34 cycles. The size of the PCR products was 1368 bp, which was digested with *Bam*H I and *Sal* I. Thereafter, the fragment of the *cbpA* gene was inserted into the expression vector pET21b (Novagen, Bloemfontein, South Africa) and digested with the aforementioned restriction enzymes. The recombinant plasmid was transformed into *E. coli* JM110 for amplification and preservation. The recombinant plasmid was selected using Amp^r on the vector to select transformants and to obtain potentially positive clones via PCR, followed by NCBI BLAST to verify the correct sequence. Finally, the recombinant plasmid was transformed into *E. coli* BL21/DE3 for expression.

4.5. Expression and Purification of CBPA

E. coli BL21-harboring pET*cbpA* were cultured in an LB medium up to the logarithmic phase (A_{600} = 0.8 to 1.0), and the culture was cooled to 20 °C and induced with IPTG at a step-down gradient of the final concentration from 0.4 to 1.0 mM at 150 rpm for 20 h. The cells were harvested via centrifugation (6000 rpm/min, 10 min, and 4 °C) and resuspended in a 20 mM Tris-HCl buffer (pH = 8.0). Thereafter, the supernatant (cytosol) and pellet of the crude protein extract were obtained via centrifugation (12,000 rpm/min, 20 min, and 4 °C), followed by bacterial cell lysis using an ultrasonic cell disruption system. Protein expression was analyzed via SDS-PAGE (10% resolving gel). The protein was incubated with chitin beads, and the bound protein was purified after elution with an NaCl solution containing a step-up gradient from 0 to 1.0 M.

4.6. Determination of Chitinase Activity

The chitinase activity of the CBPA protein was detected using the Sigma-Aldrich Chitinase Assay Fluorimetric kit (Sigma-Aldrich) in accordance with the manufacturer's instructions. First, the 4-MU standard solution was prepared, and fluorescence was measured. Thereafter, the CBPA protein (1.21 mg/mL) was added to three different substrates. Green Trichoderma chitinase was used as a positive control. The substrate reaction solution and the standard solution were equilibrated in a 37 °C water bath for 5–10 min. The standard sample and the reaction sample were prepared (10 μL) in accordance with the manufacturer's instructions before being subjected to agitation in parallel. The sample was incubated in a 37 °C warm bath for 30–60 min. Finally, 200 μL of a stop solution was added, and fluorescence was measured at an excitation wavelength of 360 nm and an emission wavelength of 450 nm. The concentration of the target solution was determined from a standard curve with chitinase as the positive control.

4.7. Construction and Expression of Recombinant gfp-conjugated cbpA

Both the 1379-bp fragment of the *cbpA* ORF and the 541-bp upstream sequence comprising the promoter were amplified via PCR with the specific primers *gfp*-1 and *gfp*-2 (Table 5), using the HD73 genome as the template. A 48-bp linker fragment (TCAGGTGGAGGCGGTTCAGGCGG AGGTGGCTCTGGCGGTGGCGGATCG) and a 717-bp GFP ORF were amplified via PCR with specific primers *gfp*-3 and *gfp*-4 (Table 5), using the Cry1Ac-GFP plasmid as the template [55]. The fusion fragment was amplified via overlapping PCR and inserted into the shuttle vector pHT315, as described previously [49], using the *Sph*I and *Bam*HI restriction sites. Thereafter, the recombinant plasmid was transformed into HD73 via electroporation.

4.8. Laser-Scanning Confocal Microscopy of CBPA-GFP Fusions

A single colony was inoculated in an LB medium, cultured overnight at 30 °C with agitation at 220 rpm, and 1% of the inoculum was seeded in 100 mL of the LB medium and incubated until the OD_{600} value approached 2.0–2.2, which is the end point of the exponential phage (T0) according to the previously established growth curve. One-milliliter bacterial aliquots were taken every 1 h for centrifugation to obtain the precipitation (30 °C, 12,000 rpm for 1 min) and washed twice with deionized water (200 µL). Thereafter, bacterial cells were resuspended in a specific amount of deionized water. Different samples were analyzed at time points T1, T2, etc., (Tn means n hours after T0 entrance into stationary phase). A red fluorescent membrane stain FM4-64 (Molecular Probes, Inc., Eugene, OR, USA) was suspended in dimethyl sulfoxide to a final concentration of 100 µM and incubated on ice for 1 min. Five-hundred nanoliters of the bacterial sample and an equal volume of FM4-64 were mixed and placed on a glass slide, covered with a coverslip, and sealed with a transparent nail polish. FM4-64-stained bacteria were observed using a 63× oil-immersion lens and scanned using a laser-scanning confocal microscope (Leica TCS SL; Leica Microsystems, Wetzlar, Germany). The FM4-64 was detected at an excitation wavelength of 514–543 nm; GFP was detected at 633 nm.

4.9. Construction of the Transcriptional Promoter P_{cbpA}-lacZ Fusion Gene

The sequence upstream from the *cbpA* gene, where the promoter fragment of P_{3189} is located, was cloned using the specific primers P_{cbpA}-F and P_{cbpA}-R (Table 5) from the wild-type HD73 genome, under the following cycling conditions: denaturation at 95 °C for 20 s, annealing at 50 °C for 20 s and extension at 72 °C for 1 min for 30 cycles. The fragment of the P_{cbpA} promoter was inserted into vector pHT304-18Z using the *Pst*I and *Bam*HI restriction sites. The vector pHT304-18Z harbored a promoter-less *lacZ* gene [50]. Thereafter, the recombinant plasmid pHTP$_{cbpA}$ was transformed into HD73 via electroporation, and positive strains were selected on the basis of the Erm^r phenotype and via PCR identification.

4.10. β-Galactosidase Assays

Bt strains containing *lacZ* fusion transcripts were cultured in LB at 30 °C and at 220 rpm with appropriate antibiotics up to an OD_{600} value of 1.5–2.0. Thereafter, NaOH was added to a final concentration of 24 mM (pH 9). No NaOH was added to the control culture. Two-milliliter aliquots were harvested from the experimental and control cultures at 15 and 30 min, and β-galactosidase activity (Miller units per milligram of protein) of the cell pellets was measured as described previously [56]. Final values were determined using the data processing software Original 8.0 and SPSS.

4.11. Construction of the HD-73 ΔcbpA Mutant

The interruption mutant was obtained via homologous recombination and insertion-replacement with a kanamycin resistance-encoding DNA cassette. The upstream gene fragment *cbpA*-u (317 bp) was obtained with the use of specific primers *cbpA*-A and *cbpA*-B and the downstream gene fragment 3189-d (504 bp) using primers *cbpA*-C and *cbpA*-D, with the genomic DNA of wild-type strain HD73 as the template. The kanamycin resistance gene cassette was a 1495-bp fragment. Primers *cbpA*-A and *cbpA*-D were used to ligate the aforementioned three fragments via overlapping PCR. Thereafter, the cassette fragment (2243 bp) was inserted into the temperature-sensitive suicide mutant erythromycin-resistant plasmid pRN5101 [51] at the *Bam*HI and *Sal*I restriction sites, finally yielding the pRN5101Ω*cbpA* plasmid. Positive transformant mutants were selected as reported previously [48].

4.12. Complementation of the HD-73 ΔcbpA Mutant

Oligonucleotide primers *cbpA*-c (with a pHT 304 upstream homologous arm) and *cbpA*-d (with a pHT 304 downstream homologous arm) (Table 5) were used to amplify the *cbpA* gene and its own promoter by using HD73 genomic DNA as the template. The amplified fragment (1946 bp) was ligated into the shuttle vector pHT304 to generate pHTC*cbpA* using a recombinant enzyme. The resulting plasmid (pHTC*cbpA*) was amplified in *E. coli* and introduced into the HD-73 ΔcbpA mutant strain. Genetically-complemented mutant HD73 strains (Δ*cbpA::cbpA*) were obtained by transforming pHTC*cbpA* into HD73ΔcbpA cells.

4.13. Crystal Spore Mixture Preparation for Asian Corn Borer Bioassays

A single colony of wild-type and mutant strains of Bt HD73 was inoculated in 20 mL of an LB liquid medium and cultured overnight at 30 °C with agitation. The activated bacteria were then transferred to 300 mL of an LB broth at a ratio of 1% and cultured at 180 rpm for 4–5 h to the logarithmic growth phase. Thereafter, the bacterial culture supernatant was transferred to 300 mL of an LP beef extract peptone medium at a ratio of 1% for approximately 40 h at 30°C, and the crystal cleavage rate was observed to be greater than 50%. The crystals, spores, and debris were harvested via centrifugation at 6000 rpm for 20 min (20 °C). The pellet was washed with water and centrifuged at 6000 rpm for 20 min (20 °C), and the supernatant was discarded. The pellet was resuspended in deionized water to obtain the final crystal and spore suspension.

4.14. Spore Preparation for Asian Corn Borer Bioassays

A single colony of the HD73, ΔcbpA mutant, and the complemented strains was inoculated in 20 mL of the LB broth and cultured overnight at 30 °C with agitation. Activated bacteria were transferred to 300 mL of a CCY liquid medium (MgCl$_2$·6H$_2$O: 0.5 mmol/L, MnCl$_2$·4H$_2$O: 0.01 mmol/L, FeCl$_3$·6H$_2$O: 0.05 mmol/L, ZnCl$_2$: 0.05 mmol/L, CaCl$_2$.6H$_2$O: 0.2 mmol/L, KH$_2$PO$_4$: 13 mmol/L, K$_2$HPO$_4$: 26 mmol/L, L-Glutamine: 20 mg/L, Casamino acids hydrolysate: 1 g/L, Yeast extract: 0.4 g/L, glycerol: 0.6g/L) based on a ratio of 1% and cultured at 220 rpm for 72 h up to a spore percentage of >99%, as reported previously [57]. The spores were harvested via centrifugation at 6000 rpm for 10 min (4 °C). The pellet was washed with water and centrifuged at 6000 rpm for 10 min, and the supernatant was discarded. The pellet was resuspended in deionized water to obtain a spore suspension.

4.15. Spore Counts

Spore suspensions of the wild-type HD73, ΔcbpA mutant, and the complemented strains were administered a heat treatment (65 °C for 30 min) to eliminate all vegetative cells. Thereafter, 100 μL of the 10^{-6} dilutions of each sample were plated on the LB agar medium and incubated at 30 °C for 12 h. The colony characteristics of the Bt culture were assessed as described previously [58].

4.16. Dose-Mortality Response Bioassays Against Asian Corn Borer

SDS-PAGE was performed to analyze protein profiles and concentrations, using bovine serum albumin (BSA) as the standard. Bioassays were performed as described previously [59]. Cry1Ac was prepared as described previously [60]. Insecticidal activity against the Asian corn borer, *O. furnacalis* (Guenée) was assayed by administering neonates with an artificial diet supplemented with a 0.01 μg Cry1Ac/g diet and different concentrations of spores (10^4–10^8 CFU/g diet) prepared from the HD73 wild-type, ΔcbpA mutant, and the complemented strains. The spores were heat-treated (65 °C for 30 min) to eliminate all vegetative cells and crystals. The feed was uniformly distributed into 48-well trays, each well containing approximately 200 mg of feed and infested with one neonate larva. Assays were carried out at 27 ± 1 °C with a L16/D 8-h photoperiod and 70–80% relative humidity. Survivors were recorded after 7 d. Deionized water was used as the negative control. Each assay was performed in triplicate.

4.17. In Vivo G. mellonella Virulence Assays with HD73 and ΔcbpA Mutant Strains

Oral force-feeding assays were performed as described previously [32,61]. A dose of 3–5 × 10⁶ spores or mid-log LB grown bacteria from HD-73 wild-type and ΔcbpA mutants was suspended in 10 µL of 0.3 mg/mL Cry1C activated toxin or in 1% NaCl water (negative control) using a needle and syringe for the accurate distribution to each larvae. At least 20 larvae per condition were incubated at 37 °C under starvation conditions, and mortality was scored at 24 h post-infection. Experiments were performed in triplicate. The inocula were evaluated via plating after serial dilution.

4.18. Expression of the CBPA-GFP Fusion Protein in Vivo in G. mellonella

CBPA expression during intestinal transit was scored via the fluorescence microscopic examination of intestinal samples from the anterior and posterior midgut of larvae infected with HD73 (pHT-*cbpA-gfp*). At least 5 larvae were dissected, and gut samples were observed using the Fluorescence Zeiss-Observer microscope with a 100× oil-immersion objective lens with a GFP filter at 1 and 3 h post-ingestion. CFU counts and fluorescence levels were scored through arbitrary visual observation, since only few bacteria were present, and the fluorescence intensity was too low for imaging or fluorescence-activated cell sorting analysis.

4.19. Intestinal Particle Transit Time for Final-Instar G. mellonella

Final-instar larvae starved for 24 h, similar to those in the virulence assays, were used. The transit time was recorded by observing the first appearance of feces and the presence of bacteria in the feces. Thirty larvae were force-fed with the same dose of spores as in the virulence assay. Larvae were placed individually in a 24-well micro-titer plate. Feces were harvested and enumerated at 1, 3, 4, and 24 h after force-feeding. The CFU value was evaluated in the feces at 1 and 2 h post-ingestion via the plating of 200 µL suspension in 1% NaCl water.

4.20. HD73 and HD-73 ΔcbpA Intestinal Transit and Localization Assays

To analyze the effect of *cbpA* deletion on bacterial transit following oral infection, a condition devoid of the Cry1C toxin was used. In total, 3–5 × 10⁶ mid-log-phase bacteria were force-fed to 5ᵗʰ instar *G. mellonella* larvae. CFU values were recorded after plating at different time points and with four different larval samples: whole larvae, dissected whole intestine, intestine without the PM, and the PM alone. First, alcohol (70%, 1 min)-cleaned larvae were homogenized in a PBS (phosphate buffed saline pH 7.4) buffer using an Ultraturax mixer at the rate of 2 larvae per 4 mL. Two completely dissected intestines were homogenized with Ultraturax in 1.5 mL 1% NaCl water. CFU values per larva were recorded at 0, 3, and 24 h post-feeding in at least 2 larvae and repeated 3–4 times. To record bacteria adhering to the PM, larvae were incubated on ice 1 h post-ingestion and gently dissected with the help of chirurgical scissors and tweezers to obtain the PM and intestine alone. The cooled larvae were placed on the dorsal under the binoculars and gently opened with the scissors from the rectum to the head, and the skin is maintained with needles. Fat body, silk glands, and other tissue were gently moved to only expose the digestive tube (DT) (the whole intestine from foregut to hindgut). The DT was cut at two sites, one just above the foregut and one above the rectum, and moved to a clean glass slide. Then, a small cut with the scissors was performed just above the hindgut, and the PM was gently pooled out and separated from the intestine with the fine tweezers; the PM and the intestine were placed directly in the respective tubes prior to homogenization. Two PMs were crushed with a small pestle and 5–6 glass beads in 200 µL 1% NaCl water, and an additional 200 µL of NaCl were added before serial dilution and plating. This experiment was performed in triplicate for each strain. The intestines alone were processed as for the above whole intestines method.

Supplementary Materials: The following are available online at http://www.mdpi.com/2072-6651/12/4/252/s1, Figure S1: Analysis of mortality of *Galleria mellonella*.

Author Contributions: Conceptualization, S.G., F.S. and C.N.-L.; validation, J.Q., Z.T., Y.Z. and C.B.; formal analysis, S.G., F.S. and C.N.-L.; investigation, J.Q., Z.T., C.B.; data Curation, K.H.; writing—original draft preparation, J.Q. and Z.T.; writing—review and editing, F.S., S.G. and C.N.-L.; visualization, Z.T.; funding acquisition, F.S. and C.N.-L. All authors have read and agreed to the published version of the manuscript.

Funding: This study was supported by the grants from The National Key Research and Development Program of China (2017YFD0200400) and from the National Natural Science Foundation (31530095). The project was also financially supported from France, INRAE, MICA department and the French National Research Program RISKOGM (11-MERES-RISKOGM-3-CVS-59).

Conflicts of Interest: We have no conflicts of interests.

References

1. Bravo, A.; Likitvivatanavong, S.; Gill, S.S.; Soberón, M. *Bacillus thuringiensis*: A story of a successful bioinsecticide. *Insect Biochem. Mol. Biol.* **2011**, *41*, 423. [CrossRef] [PubMed]

2. Adang, M.J.; Crickmore, N.; Jurat-Fuentes, J.L. Chapter Two–diversity of *Bacillus thuringiensis* crystal toxins and mechanism of action. *Adv. Insect Physiol.* **2014**, *47*, 39–87.

3. Ben, R.; Johnston, P.R.; Lereclus, D.; Nielsen-Leroux, C.; Crickmore, N. *Bacillus thuringiensis*: An impotent pathogen? *Trends Microbiol.* **2010**, *18*, 189.

4. Stones, D.H.; Krachler, A.M. Against the tide: The role of bacterial adhesion in host colonization. *Biochem. Soc. Trans.* **2016**, *44*, 1571–1580. [CrossRef]

5. Pizarrocerdá, J.; Cossart, P. Bacterial adhesion and entry into host cells. *Cell* **2006**, *124*, 715. [CrossRef]

6. Hegedus, D.; Erlandson, M.; Gillott, C.; Toprak, U. New insights into peritrophic matrix synthesis, architecture, and function. *Annu. Rev. Entomol.* **2009**, *54*, 285. [CrossRef]

7. Weiss, B.L.; Savage, A.F.; Griffith, B.C.; Wu, Y.; Aksoy, S. The peritrophic matrix mediates differential infection outcomes in the tsetse fly gut following challenge with commensal, pathogenic, and parasitic microbes. *J. Immunol.* **2014**, *193*, 773–782. [CrossRef]

8. Nielsen-Leroux, C.; Gaudriault, S.; Ramarao, N.; Lereclus, D.; Givaudan, A. How the insect pathogen bacteria *Bacillus thuringiensis* and *Xenorhabdus/Photorhabdus* occupy their hosts. *Curr. Opin. Microbiol.* **2012**, *15*, 220–231. [CrossRef]

9. Dubois, T.; Faegri, K.; Perchat, S.; Lemy, C.; Buisson, C.; Nielsen-LeRoux, C.; Lereclus, D. Necrotrophism is a quorum-sensing-regulated lifestyle in *Bacillus thuringiensis*. *PLoS Pathog.* **2012**, *8*, e1002629. [CrossRef]

10. Kuraishi, T.; Binggeli, O.; Opota, O.; Buchon, N.; Lemaitre, B. Genetic evidence for a protective role of the peritrophic matrix against intestinal bacterial infection in *Drosophila melanogaster*. *Proc. Natl. Acad. Sci. USA* **2011**, *108*, 15966–15971. [CrossRef]

11. Lehane, M.J. Peritrophic matrix structure and function. *Annu. Rev. Entomol.* **1997**, *42*, 525–550. [CrossRef]

12. Huber, M.; Cabib, E.; Miller, L.H. Malaria Parasite Chitinase and Penetration of the Mosquito Peritrophic Membrane. *Proc. Natl. Acad. Sci. USA* **1991**, *88*, 2807. [CrossRef]

13. Sirichotpakorn, N.; Rongnoparut, P.; Choosang, K.; Panbangred, W. Coexpression of Chitinase and the cry11Aa1 toxin genes in *Bacillus thuringiensis* serovar israelensis. *J. Invertebr. Pathol.* **2001**, *78*, 160. [CrossRef] [PubMed]

14. Smirnoff, W.A. Three years of aerial field experiments with *Bacillus thuringiensis* plus chitinase formulation against the spruce budworm. *J. Invertebr. Pathol.* **1974**, *24*, 344–348. [CrossRef]

15. Tantimavanich, S.; Pantuwatana, S.; Bhumiratana, A.; Panbangred, W. Cloning of a chitinase gene into *Bacillus thuringiensis subsp. aizawai* for enhanced insecticidal activity. *J. Gen. Appl. Microbiol.* **1997**, *43*, 341–347. [CrossRef]

16. Thamthiankul, S.; Moar, W.J.; Miller, M.E.; Panbangred, W. Improving the insecticidal activity of *Bacillus thuringiensis subsp. aizawai* against *Spodoptera exigua* by chromosomal expression of a chitinase gene. *Appl. Microbiol. Biotechnol.* **2004**, *65*, 183–192. [CrossRef] [PubMed]

17. Sampson, M.N.; Gooday, G.W. Involvement of chitinases of *Bacillus thuringiensis* during pathogenesis in insects. *Microbiology* **1998**, *144 Pt 8*, 2189. [CrossRef]

18. Hashimoto, M.; Ikegami, T.; Seino, S.; Ohuchi, N.; Fukada, H.; Sugiyama, J.; Watanabe, T. Expression and characterization of the chitin-binding domain of chitinase A1 from *Bacillus circulans WL-12*. *J. Bacteriol.* **2000**, *182*, 3045–3054. [CrossRef]

19. Vaajekolstad-Kolstad, G.; Horn, S.J.; van Aalten, D.M.; Synstad, B.; Eijsink, V.G. The non-catalytic chitin-binding protein CBP21 from *Serratia marcescens* is essential for chitin degradation. *J. Biol. Chem.* **2005**, *280*, 28492–28497. [CrossRef]

20. Vaaje-Kolstad, G.; Anne, A.C.; Mathiesen, G.; Eijsink, V.G.H. The chitinolytic system of *Lactococcus lactis ssp. lactis* comprises a nonprocessive chitinase and a chitin-binding protein that promotes the degradation of alpha-and beta-chitin. *FEBS J.* **2009**, *276*, 2402–2415. [CrossRef]

21. Bayer, E.A.; Shoham, Y.; Lamed, R. The cellulosome. *Glycomicrobiology* **2002**, 387–439.

22. Carrard, G.; Koivula, A.; Soderlund, H.; Beguin, P. Cellulose-binding domains promote hydrolysis of different sites on crystalline cellulose. *Proc. Natl. Acad. Sci. USA* **2000**, *97*, 10342–10347. [CrossRef] [PubMed]

23. Lehtiö, J.; Sugiyama, J.; Gustavsson, M.; Fransson, L.; Linder, M.; Teeri, T.T. The binding specificity and affinity determinants of family 1 and family 3 cellulose binding modules. *Proc. Natl. Acad. Sci. USA* **2003**, *100*, 484–489. [CrossRef] [PubMed]

24. Uchiyama, T.; Katouno, F.; Nikaidou, N.; Nonaka, T.; Sugiyama, J.; Watanabe, T. Roles of the exposed aromatic residues in crystalline chitin hydrolysis by chitinase A from *Serratia marcescens* 2170. *J. Biol. Chem.* **2001**, *276*, 41343. [CrossRef]

25. Van Aalten, D.M.; Synstad, B.; Brurberg, M.B.; Hough, E.; Riise, B.W.; Eijsink, V.G.; Wierenga, R.K. Structure of a two-domain chitotriosidase from *Serratia marcescens* at 1.9-A resolution. *Proc. Natl. Acad. Sci. USA* **2000**, *97*, 5842–5847. [CrossRef]

26. Xu, H.; Xie, W.; Gong, Z. Characteristics and antifungal activity of a chitin binding protein from *Ginkgo biloba*. *FEBS Lett.* **2000**, *478*, 123.

27. Kamakura, T.; Yamaguchi, S.; Saitoh, K.; Teraoka, T.; Yamaguchi, I. A novel gene, CBP1, encoding a putative extracellular chitin-binding protein, may play an important role in the hydrophobic surface sensing of *Magnaporthe grisea* during appressorium differentiation. *Mol. Plant Microbe Interact. MPMI* **2002**, *15*, 437. [CrossRef]

28. Kao, G.W.; Qiu, L.L.; Peng, Q.; Zhang, J.; Li, J.; Song, F.P. The analysis of major metabolic pathways in *Bacillus thuringiensis* under alkaline stress. *Acta Microbiol. Sin.* **2016**, *56*, 485–495.

29. Zhan, Y.L.; Guo, S.Y. Three-dimensional (3D) structure prediction and function analysis of the chitin-binding domain 3 protein HD73_3189 from *Bacillus thuringiensis HD73*. *Bio Med Mater. Eng.* **2015**, *26*, S2019–S2024. [CrossRef]

30. Li, R.S.; Jarrett, P.; Burges, H.D. Importance of spores, crystals, and δ-endotoxins in the pathogenicity of different varieties of *Bacillus thuringiensis* in *Galleria mellonella* and *Pieris brassicae*. *J. Invertebr. Pathol.* **1987**, *50*, 277–284. [CrossRef]

31. Aamer Mehmood, M.; Hussain, K.; Latif, F.; Rizwan Tabassum, M.; Gull, M.; Shahid Gill, S.; Iqbal, Z. The plcR regulon is involved in the opportunistic properties of *Bacillus thuringiensis* and *Bacillus cereus* in mice and insects. *Microbiology* **2000**, *146 Pt 11*, 2825.

32. Bouillaut, L.; Ramarao, N.C.; Gilois, N.; Gohar, M.; Lereclus, D.; Nielsen-Leroux, C. FlhA influences *Bacillus thuringiensis* PlcR-regulated gene transcription, protein production, and virulence. *Appl. Environ. Microbiol.* **2005**, *71*, 8903. [CrossRef] [PubMed]

33. Vachon, V.; Laprade, R.; Schwartz, J.L. Current models of the mode of action of *Bacillus thuringiensis* insecticidal crystal proteins: A critical review. *J. Invertebr. Pathol.* **2012**, *111*, 1–12. [CrossRef] [PubMed]

34. Chiang, A.S.; Yen, D.F.; Peng, W.K. Germination and proliferation of *Bacillus thuringiensis* in the gut of rice moth larva, *Corcyra cephalonica*. *J. Invertebr. Pathol.* **1986**, *48*, 96–99. [CrossRef]

35. Bauer, L.S.; Pankratz, H.S. Ultrastructural effects of *Bacillus thuringiensis var. san diego* on midgut cells of the cottonwood leaf beetle. *Jinvertebrpathol* **1991**, *60*, 15–25. [CrossRef]

36. Arora, N.; Sachdev, B.; Gupta, R.; Vimala, Y.; Bhatnagar, R.K. Characterization of a Chitin-Binding Protein from *Bacillus thuringiensis HD-1*. *PLoS ONE* **2013**, *8*, e66603. [CrossRef]

37. Mehmood, M.A.; Xiao, X.; Hafeez, F.Y.; Gai, Y.B.; Wang, F.P. Molecular characterization of the modular chitin binding protein Cbp50 from *Bacillus thuringiensis serovar konkukian*. *Antonie Van Leeuwenhoek* **2011**, *100*, 445–453. [CrossRef] [PubMed]

38. Schrempf, H. Characteristics of chitin-binding proteins from *Streptomycetes*. *Exs* **1999**, *87*, 99–108.

39. Chu, H.H.; Hoang, V.; Hofemeister, J.; Schrempf, H. A *Bacillus amyloliquefaciens* ChbB protein binds beta- and alpha-chitin and has homologues in related strains. *Microbiology* **2001**, *147*, 1793–1803. [CrossRef]

40. Mehmood, M.A.; Latif, M.; Hussain, K.; Gull, M.; Latif, F.; Rajoka, M.I. Heterologous expression of the antifungal β-chitin binding protein CBP24 from *Bacillus thuringiensis* and its synergistic action with bacterial chitinases. *Protein Pept. Lett.* **2015**, *22*, 39–44. [CrossRef]

41. Aamer Mehmood, M.; Hussain, K.; Latif, F.; Rizwan Tabassum, M.; Gull, M.; Shahid Gill, S.; Iqbal, Z. Synergistic Action of the Antifungal β-chitin Binding Protein CBP50 from *Bacillus thuringiensis* with Bacterial Chitinases. *Curr. Proteom.* **2014**, *11*, 23–26. [CrossRef]

42. Du, C.; Nickerson, K.W. *Bacillus thuringiensis HD-73* Spores have surface-localized Cry1Ac toxin: Physiological and pathogenic consequences. *Appl. Environ. Microbiol.* **1996**, *62*, 3722–3726. [CrossRef] [PubMed]

43. Arora, N.; Ahmad, T.; Rajagopal, R.; Bhatnagar, R.K. A constitutively expressed 36 kDa exochitinase from *Bacillus thuringiensis HD-1*. *Biochem. Biophys. Res. Commun.* **2003**, *307*, 620–625. [CrossRef]

44. Fang, S.; Wang, L.; Guo, W.; Zhang, X.; Peng, D.; Luo, C.; Sun, M. *Bacillus thuringiensis* Bel Protein Enhances the Toxicity of Cry1Ac Protein to *Helicoverpa armigera* Larvae by DegradingInsect Intestinal Mucin. *Appl. Environ. Microbiol.* **2009**, *75*, 5237–5243. [CrossRef] [PubMed]

45. Bertani, G. Lysogeny at mid-twentieth century: P1, P2, and other experimental systems. *J. Bacteriol.* **2004**, *186*, 595–600. [CrossRef] [PubMed]

46. Liu, G.; Song, L.; Shu, C.; Wang, P.; Deng, C.; Peng, Q.; Lereclus, D.; Wang, X.; Huang, D.; Zhang, J.; et al. Complete Genome Sequence of *Bacillus thuringiensis subsp. kurstaki* Strain HD73. *Genome Announcements.* **2013**, *1*, e0008013. [CrossRef] [PubMed]

47. Jr, R.S.M.; Sasaki, K. Protein D, a putative immunoglobulin D-binding protein produced by *Haemophilus influenzae*, is glycerophosphodiester phosphodiesterase. *J. Bacteriol.* **1993**, *175*, 4569–4571.

48. Yang, J.; Peng, Q.; Chen, Z.; Deng, C.; Shu, C.; Zhang, J.; Song, F. Transcriptional regulation and characteristics of a novel N-acetylmuramoyl-l-alanine amidase gene involved in *Bacillus thuringiensis* mother cell lysis. *J. Bacteriol.* **2013**, *195*, 2887. [CrossRef]

49. Arantes, O.; Lereclus, D. Construction of cloning vectors for *Bacillus thuringiensis*. *Gene* **1991**, *108*, 115–119. [CrossRef]

50. Agaisse, H.; Lereclus, D. Structural and functional analysis of the promoter region involved in full expression of the cryIIIA toxin gene of *Bacillus thuringiensis*. *Mol. Microbiol.* **1994**, *13*, 97–107. [CrossRef]

51. Gennaro, M.L.; Iordanescu, S.; Novick, R.P.; Murray, R.W.; Steck, T.R.; Khan, S.A. Functional organization of the plasmid pT181 replication origin. *J. Mol. Biology* **1989**, *205*, 355–362. [CrossRef]

52. Wang, G.J.; Zhang, J.; Song, F.P.; Wu, J.; Feng, S.L.; Huang, D.F. Engineered *Bacillus thuringiensis GO33A* with broad insecticidal activity against lepidopteran and coleopteran pests. *Appl. Microbiol. Biotechnol.* **2006**, *72*, 924. [CrossRef] [PubMed]

53. Sambrook, J.; Fritsch, E.F.; Maniatis, T. *Molecular Cloning: A Laboratory Manual, Volume 1*; Cold Spring Harbor Laboratory Press: New York, NY, USA, 2012.

54. Lereclus, D.; Arantès, O.; Chaufaux, J.; Lecadet, M. Transformation and expression of a cloned δ-endotoxin gene in *Bacillus thuringiensis*. *FEMS Microbiol. Lett.* **1989**, *51*, 211–217. [CrossRef]

55. Yang, H.; Wang, P.; Peng, Q.; Rong, R.; Liu, C.; Lereclus, D.; Huang, D. Weak transcription of the cry1Ac gene in nonsporulating *Bacillus thuringiensis* Cells. *Appl. Environ. Microbiol.* **2012**, *78*, 6466–6474. [CrossRef] [PubMed]

56. Miller, B.J.H. Assay of β galactosidase. In *Experiments in Molecular Genetics*; Cold Spring Harbor Laboratory Press: New York, NY, USA, 2015; pp. 352–359.

57. Clements, M.O.; Moir, A. Role of the gerI operon of *Bacillus cereus 569* in the response of spores to germinants. *J. Bacteriology* **1998**, *180*, 6729. [CrossRef]

58. Van Cuyk, S.; Deshpande, A.; Hollander, A.; Duval, N.; Ticknor, L.; Layshock, J.; Omberg, K.M. Persistence of *Bacillus thuringiensis subsp. kurstaki* in urban environments following spraying. *Appl. Environ. Microbiol.* **2011**, *77*, 7954–7961. [CrossRef]

59. Zhang, T.; He, M.; Gatehouse, A.M.; Wang, Z.; Edwards, M.G.; Li, Q.; He, K. Inheritance patterns, dominance and cross-resistance of Cry1Ab-and Cry1Ac-selected (Guenée). *Bautechnik* **2014**, *74*, 395–400.
60. Zhou, Z.; Wang, Z.; Liu, Y.; Liang, G.; Shu, C.; Song, F.; Zhang, J. Identification of ABCC2 as a binding protein of Cry1Ac on brush border membrane vesicles from *Helicoverpa armigera* by an improved pull-down assay. *Microbiologyopen* **2016**, *5*, 659–669. [CrossRef]
61. Ramarao, N.; Nielsen-Leroux, C.; Lereclus, D. The insect *Galleria mellonella* as a powerful infection model to investigate bacterial pathogenesis. *J. Vis. Exp. Jov.* **2012**, *70*, e4392. [CrossRef]

Article

Potential of Cry10Aa and Cyt2Ba, Two Minority δ-endotoxins Produced by *Bacillus thuringiensis* ser. *israelensis*, for the Control of *Aedes aegypti* Larvae

Daniel Valtierra-de-Luis [1], Maite Villanueva [1,2], Liliana Lai [1], Trevor Williams [3] and Primitivo Caballero [1,2,4,*]

[1] Departamento de Agronomía, Biotecnología y Alimentación, Universidad Pública de Navarra, 31006 Pamplona, Spain; daniel.valtierra@unavarra.es (D.V.-d.-L.); maite.villanueva@unavarra.es (M.V.); liliana.lai@unavarra.es (L.L.)
[2] Bioinsectis SL, Avda Pamplona 123, 31192 Mutilva, Spain
[3] Instituto de Ecología AC, 91073 Xalapa, Mexico; trevor.inecol@gmail.com
[4] Institute for Multidisciplinary Applied Biology Research (IMAB), Universidad Pública de Navarra, 31006 Mutilva, Spain
* Correspondence: pcm92@unavarra.es

Received: 17 April 2020; Accepted: 26 May 2020; Published: 29 May 2020

Abstract: *Bacillus thuringiensis* ser. *israelensis* (Bti) has been widely used as microbial larvicide for the control of many species of mosquitoes and blackflies. The larvicidal activity of Bti resides in Cry and Cyt δ-endotoxins present in the parasporal crystal of this pathogen. The insecticidal activity of the crystal is higher than the activities of the individual toxins, which is likely due to synergistic interactions among the crystal component proteins, particularly those involving Cyt1Aa. In the present study, Cry10Aa and Cyt2Ba were cloned from the commercial larvicide VectoBac-12AS® and expressed in the acrystalliferous Bt strain BMB171 under the *cyt1Aa* strong promoter of the pSTAB vector. The LC$_{50}$ values for *Aedes aegypti* second instar larvae estimated at 24 hpi for these two recombinant proteins (Cry10Aa and Cyt2Ba) were 299.62 and 279.37 ng/mL, respectively. Remarkable synergistic mosquitocidal activity was observed between Cry10Aa and Cyt2Ba (synergistic potentiation of 68.6-fold) when spore + crystal preparations, comprising a mixture of both recombinant strains in equal relative concentrations, were ingested by *A. aegypti* larvae. This synergistic activity is among the most powerful described so far with Bt toxins and is comparable to that reported for Cyt1A when interacting with Cry4Aa, Cry4Ba or Cry11Aa. Synergistic mosquitocidal activity was also observed between the recombinant proteins Cyt2Ba and Cry4Aa, but in this case, the synergistic potentiation was 4.6-fold. In conclusion, although Cry10Aa and Cyt2Ba are rarely detectable or appear as minor components in the crystals of Bti strains, they represent toxicity factors with a high potential for the control of mosquito populations.

Keywords: *Bacillus thuringiensis*; *Aedes aegypti*; minor proteins; synergy; mosquito control; Bti

Key Contribution: Cry10Aa and Cyt2Ba are found as minor components in the crystals of some strains of *Bacillus thuringienis* ser. *israelensis*. Both proteins have a high insecticidal activity against insects (e.g., *Aedes aegypti* larvae) and when ingested together they exhibit one of the strongest synergistic activities that have been described so far.

---

1. Introduction

Bacillus thuringiensis ser. *israelensis* (Bti) was the first Bt serotype found to be toxic for dipteran species [1]. Bti forms parasporal inclusion bodies composed of insecticidal proteins (δ-endotoxins) that are widely used as the basis for microbial larvicides against dipteran species of medical importance, including mosquitoes, blackflies and chironomids [2,3]. Bti based products are considered to be powerful and highly selective larvicides for the control of disease vectors [4–6]. Indeed, Bti has been used to control mosquitoes for more than 35 years with almost no resistance report in vector populations [7,8]. The absence of resistance is likely due to the different modes of action and the synergistic effects of the multiple crystal proteins present in Bti-based products [9–11].

The parasporal crystal of Bti contains large amounts of four toxins: Cry4A, Cry4B, Cry11A and Cyt1A [12]. In addition, Cry10Aa and Cyt2Ba have also been described in some Bti strains, although these are expressed and accumulate in the crystal in much smaller quantities that the four main components [13,14]. All six of these proteins are encoded in the fully sequenced Bti plasmid pBtoxis [15]. The Cry10Aa protein was cloned and named CryIVC, according to the existing classification at that time, but was described as a protein with a low larvicidal potency against *A. aegypti* (Diptera; Culicidae) [16]. Later, the *cry10Aa* gene was identified as part of an operon that comprises two open reading frames (*orf1* and *orf2*) separated by a 66 bp gap [15]. Cloning of the complete operon, linked to the strong promoter of the *cyt1A* gene, revealed that Cry10Aa was expressed at high levels and exhibited high larvicidal activity, both alone and in combination with Cyt1A [17]. In contrast, although present at relatively low abundance in the Bti crystal [18], Cyt2Ba exhibited activity against *A. aegypti* larvae, but lower than the better-studied Cyt1Aa protein [19].

The interactions among the Cry and Cyt proteins of Bti have received more attention than any of the other Bt serovars [20–24]. Interactions involving the Cyt1A protein have attracted particular attention given the capacity of this protein to enhance the insecticidal activity of Cry proteins in strains of Bti [9,17,20,24], and those of Bt strains belonging to other subspecies [25]. Conversely, studies on the interactions of Cyt2Ba with other components of the Bti crystal are restricted to a single report of low synergistic activity of Cyt2Ba with the Cry4Aa protein [21].

The objective of this study was to quantify the larvicidal activity of the δ-endotoxins Cry10A and Cyt2Ba, which are minor components of the parasporal crystal of some Bti strains, against *A. aegypti*. To produce high amounts of these minority proteins, two recombinant Bt strains were constructed. One of these recombinants produced a crystal whose only component were the two Cry10Aa proteins, while the other only produced Cyt2Ba. We provide evidence that these proteins interacted synergistically to a remarkable degree when simultaneously ingested by *A. aegypti* larvae.

2. Results

2.1. Insecticidal Cry and Cyt Genes Identified in Bti Strain from VectoBac-12AS®

A bioinformatic analysis of the genome of the Bti strain isolated from the commercial product VectoBac-12AS®, revealed that this strain contains a complex of insecticidal genes, including *cry* genes (*cry4Aa*, *cry4Ba*, *cry10Aa*, *cry11Aa*, and *cry60Aa/cry60Ba*) and *cyt* genes (*cyt1Aa*, *cyt2Ba*, and *cyt1Ca*). Unfortunately, it was not possible to obtain the complete sequence of the *cry4Aa* and *cry4Ba* genes, because they appeared distributed in various contigs. The rest of the *cry* genes shared 100% identity and similarity with some of the gene variants that have been previously described. Thus, *cry10Aa* was completely identical to *cry10Aa3* [15], *cry11Aa* was identical to *cry11Aa1* [26], and *cry60Aa/cry60Ba* were identical to *cry60Aa2/cry60Ba2* [27]. The three *cyt* genes (*cyt1Aa*, *cyt2Ba*, and *cyt1Ca*) identified in the VectoBac-12AS® strain were also fully identical to *cyt1Aa1* [28], *cyt2Ba1* [18], and *cyt1Ca1* [15], respectively.

2.2. Cloning of Cyt2Ba, Cry10Aa and Cry11Aa

The pairs of primers designed for *cyt2Ba*, *cry10Aa* and *cry11Aa* amplified fragments of 1536, 3813 and 2634 bp, respectively. The *cry10A* cloned fragment contained two open reading frames (*orf1* and *orf2*) in the nucleotide sequence, codifying for proteins of 680 and 489 amino acids, respectively. The *cry11Aa* amplicon encoded a protein of 646 amino acids but it also contained the *p19* gene located before *cry11Aa*, in line with the usual order of genes in this operon. The amplicon was cloned in a pSTAB plasmid containing the *p20* chaperon, which improves Cry11Aa synthesis and crystal formation [29]. The amplified fragment of *cyt2Ba* contained a single ORF which codified for a protein of 263 residues. The *cry4A* and *cry4Ba* genes, previously described and cloned by Delecluse et al. [30], were used in this study.

2.3. Characterization of Bt Recombinant Strains Expressing cyt2Ba, cry10Aa, cry4Aa, cry4Ba and cry11Aa

Daily microscopical observation of the growth of the recombinant Bt strains in CCY medium confirmed that all of them produced spores and crystals between 36 and 48 h after the medium was inoculated. As expected, vegetative cells of BMB171 strain transformed with an empty plasmid produced endospores but no crystals.

SDS-PAGE showed that the recombinant BMB171-Cry10Aa expressed two proteins with molecular masses of approximately 68 and 56 kDa, which corresponded to the predicted sizes of the proteins encoded by *orf1* and *orf2*, respectively, of the *cry10Aa* operon (Figure 1, lane 3). Samples of spores and crystals from the rest of the recombinant strains (BMB171-Cyt2Ba, 4Q2-81-Cry4Aa, 4Q2-81-Cry4Ba and BMB171-Cry11Aa) generated characteristic major bands of approximately 29, 134, 128, and 73 kDa, respectively (Figure 1). The electrophoretic mobility of all these bands correlated well with the molecular mass of the proteins Cyt2Ba (lane 2), Cry4Aa (lane 4), Cry4Ba (lane 5) and Cry11Aa (lane 6).

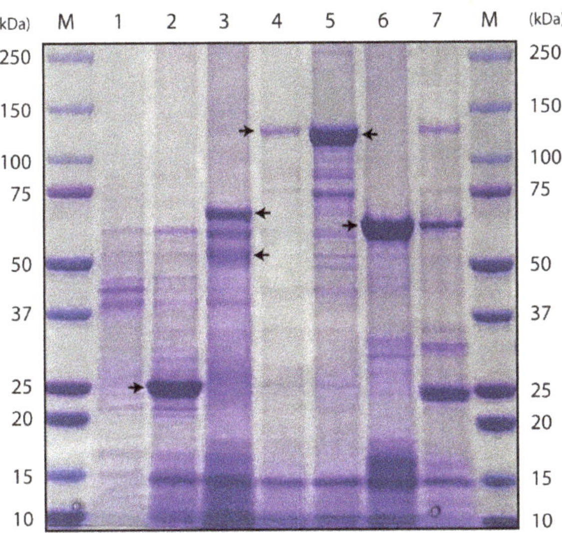

Figure 1. SDS-PAGE gel showing the protein profiles of the recombinant Bt strains and the strain present in VectoBac-12AS®. Lane M, molecular mass marker; lane 1, BMB171 acrystalliferous strain with an empty plasmid; lane 2, BMB171-Cyt2Ba; lane 3, BMB171-Cry10Aa; lane 4, 4Q2-81-Cry4Aa; lane5, 4Q2-81-Cry4Ba; lane 6, BMB171-Cry11Aa; lane 7, wild-type Bti strain from VectoBac-12AS®. Arrows indicate major protein bands.

2.4. Mosquitocidal Activity of the δ-Endotoxins Produced by Bti

Single-concentration bioassays involving an estimated LC_{30} concentration of inoculum in all cases were performed on mixtures of Cyt2Ba with each of the Cry4Aa, Cry4Ba, Cry10Aa and Cry11Aa proteins. The results of these assays indicated that *A. aegypti* second instar larvae treated with combinations of Cry10Aa+Cyt2Ba and Cry4Aa+Cyt2Ba experienced high mortality compared to the mortality values observed in insects treated with each of the toxins separately (Table 1). In contrast, no evidence of potentiation of larval mortality was observed for mixtures of Cry4Ba+Cyt2Ba or Cry11Aa+Cyt2Ba. For this reason, the 1:1 mixtures of Cry10Aa+Cyt2Ba and of Cry4Aa+Cyt2Ba were selected for subsequent concentration-mortality studies.

Table 1. Mortality of *A. aegypti* second instar larvae at 24 h after inoculation with individual Bti δ-endotoxins and the binary combinations Cyt2Ba/Cry10Aa, Cyt2Ba/Cry4Aa, Cyt2Ba/Cry4Ba and Cyt2Ba/Cry11Aa.

Treatment [1]	Concentration (ng/mL)	Mortality (% ± SD)
Cry10Aa	40	26 ± 5
Cyt2Ba	40	31 ± 8
Cry10Aa+Cyt2Ba (1:1)	80	93 ± 6
Cry4Aa	10	31 ± 10
Cyt2Ba	10	28 ± 23
Cry4Aa+Cyt2Ba (1:1)	20	100 ± 0
Cry4Ba	0.02	32 ± 18
Cyt2Ba	0.02	10 ± 9
Cry4Ba+Cyt2Ba (1:1)	0.04	43 ± 19
Cry11Aa	1.5	40 ± 16
Cyt2Ba	1.5	21 ± 14
Cry11Aa+Cyt2Ba (1:1)	3	46 ± 21

[1] Control insects experienced no mortality in all cases.

Table 2 shows the raw mortality data of a series of concentrations for Cry10Aa, Cyt2Ba and the combination of both. Analogously, Table 3 shows the raw mortality data of a series of concentrations for Cry4Aa, Cyt2Ba and the combination of both.

Table 2. Mortality of *A. aegypti* second instar larvae at 24 h after inoculation with Cry10Aa, Cyt2Ba and combination of both.

Cry10Aa			Cyt2Ba			Cry10Aa+Cyt2Ba		
ng/mL	Dead/Total	Mortality (% ± SD)	ng/mL	Dead/Total	Mortality (% ± SD)	ng/mL	Dead/Total	Mortality (% ± SD)
2000	39/50	78 ± 2%	4000	84/92	91.3 ± 10%	300	115/124	92.7 ± 9%
666	30/50	60 ± 10%	1333	64/85	75.3 ± 17%	60	86/114	75.4 ± 9%
222	23/41	56.1 ± 2%	444	39/94	41.5 ± 9%	12	75/135	55.6 ± 11%
74	12/44	27.3 ± 12%	148	28/78	35.9 ± 7%	2.4	53/131	40.5 ± 6%
24.7	6/46	13 ± 17%	49.4	22/86	25.6 ± 6%	0.48	34/116	29.3 ± 6%
8.2	1/42	2.4 ± 3%	16.4	21/84	25 ± 13%	0.096	23/120	19.2 ± 14%

Control insects experienced no mortality in all cases.

Table 3. Mortality of *A. aegypti* second instar larvae at 24 h after inoculation with Cry4Aa, Cyt2Ba and combination of both.

Cry4Aa			Cyt2Ba			Cry4Aa+Cyt2Ba		
ng/mL	Dead/Total	Mortality (% ± SD)	ng/mL	Dead/Total	Mortality (% ± SD)	ng/mL	Dead/Total	Mortality (% ± SD)
486	63/70	90 ± 11%	4000	84/92	91.3 ± 10%	54	139/163	85.3 ± 13%
162	53/71	74.6 ± 17%	1333	64/85	75.3 ± 17%	27	111/155	71.6 ± 17%
54	42/70	60 ± 13%	444	39/94	41.5 ± 9%	13.5	87/200	43.5 ± 14%
18	27/73	37 ± 11%	148	28/78	35.9 ± 7%	6.74	58/142	40.8 ± 24%
6	13/64	20.3 ± 10%	49.4	22/86	25.6 ± 6%	3.36	11/102	10.8 ± 16%
2	7/71	9.9 ± 7%	16.4	21/84	25 ± 13%	1.68	5/73	6.8 ± 8%

Control insects experienced no mortality in all cases.

Regression lines were performed for the individual toxins and the mixture of toxins (Figure 2) which were then used to estimate median lethal concentrations (LC$_{50}$) (Table 4).

Table 4. Logit regression of concentration-mortality results of wild-type (VectoBac-12AS®) and recombinant proteins and their mixtures in *A. aegypti* second instar larvae at 24 h.

Treatment [a]	Regression		LC_{50} Observed	FL (95%) [b]		LC_{50} Expected [c]			FL (95%) [b]	
	Slope ± SE	Intercept ± SE	(ng/mL)	Lower	Upper	(ng/mL)	Synergistic Factor [d]	Potency	Lower	Upper
Cyt2Ba	0.59 ± 0.13	−3.33 ± 0.79	279.37	190.20	410.38	-	-	1	-	-
Cry10Aa	0.74 ± 0.09	−4.26 ± 0.53	299.62	245.06	366.34	-	-	0.93	0.78	1.12
Cry4Aa	0.78 ± 0.02	−2.77 ± 0.09	34.63	29.73	40.34	-	-	8.07	6.40	10.17
Cry10Aa+Cyt2Ba [e]	0.45 ± 0.05	−0.64 ± 0.14	4.22	3.25	5.50	289.27	68.55	66.20	58.52	74.61
Cry4Aa+Cyt2Ba [e]	1.22 ± 0.17	−3.17 ± 0.46	13.41	12.55	14.33	61.62	4.60	20.83	15.15	28.63
VectoBac-12AS®	1.51 ± 0.29	3.44 ± 0.73	1.02×10^{-1}	9.34×10^{-2}	1.11×10^{-1}	-	-	2.73×10^3	2.03×10^3	3.69×10^3

[a] Inocula comprised spore+crystal mixtures. Control insects experienced no mortality in all cases. [b] FL: Fiducial limits (95%). [c] Expected LC_{50} calculated by the method of Tabashnik (1992). [d] Synergism factor defined as the ratio of the expected LC_{50} and the observed LC_{50}. [e] Toxins were present in equal amounts in the experimental inocula.

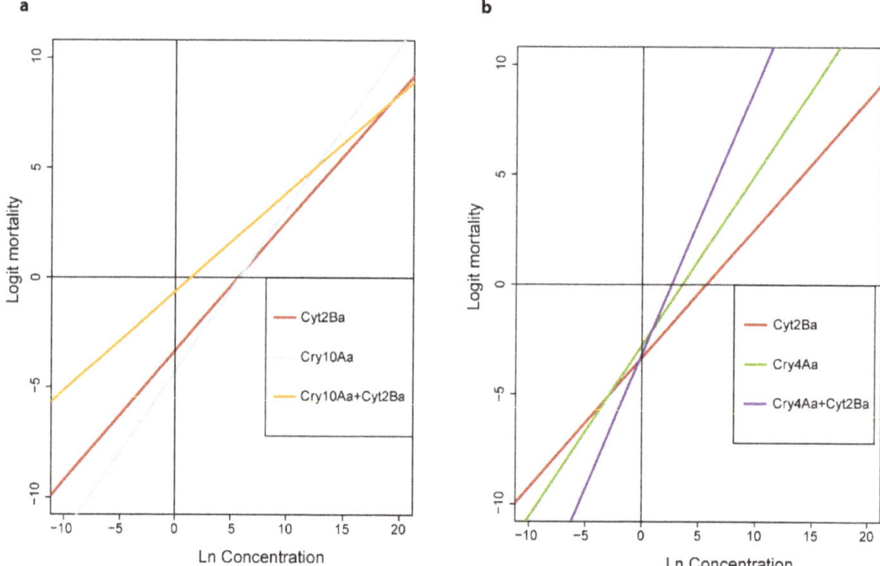

Figure 2. Graphical representation of the logit regression lines for the individual toxins and the toxin combinations. (**a**) Regression lines for Cyt2Ba, Cry10Aa and Cry10Aa+Cyt2Ba. (**b**) Regression lines for Cyt2Ba, Cry4Aa and Cry4Aa+Cyt2Ba.

Recombinant Cry10Aa and Cyt2Ba proteins exhibited a high insecticidal activity against *A. aegypti* second instar larvae when inoculated individually. The LC_{50} values estimated for Cry10Aa and Cyt2Ba were 299.62 ng/mL and 279.37 ng/mL, respectively. The VectoBac-12AS® wild-type strain, incorporated into the bioassays as a positive control, had an LC_{50} value of 1.02×10^{-1} ng/mL and the BMB171 strain with the empty plasmid resulted in no mortality (Table 4). The slopes of the regression lines corresponding to Cry10Aa and Cyt2Ba did not differ significantly ($F_{1,8} = 0.620$, $p = 0.454$), whereas the slope of the mixture of Cry10Aa+Cyt2Ba was significantly lower than that of the individual toxins ($F_{2,12} = 7.359$, $p = 0.008$). The observed LC_{50} value for Cry10Aa+Cyt2Ba was 4.22 ng/mL whereas the expected LC_{50} value was 289.27 ng/mL, assuming additive action of each of the toxins [31]. The estimated potentiation of Cyt2Ba and Cry10Aa proteins when ingested together and in the same relative proportions, was 68.6-fold (Table 4).

In contrast, the slopes of the regressions of the individual Cyt2Ba and Cry4Aa toxins differed significantly ($F_{1,8} = 11.405$, $p = 0.010$). The LC_{50} value for Cry4Aa was estimated at 34.63 ng/mL. The observed LC_{50} value for the binary combination of Cry4Aa+Cyt2Ba was 13.41 ng/mL, whereas the expected LC_{50} value was 61.62 ng/mL, assuming additive action of each of the toxins [31]. These results indicate potentiation in the Cry4Aa+Cyt2Ba protein mixture by a factor of 4.6 (Table 4).

3. Discussion

The δ-endotoxins that constitute the major parasporal crystal components of Bti strains (Cry4Aa, Cry4Ba, Cry11Aa and Cyt1Aa) are the best studied of the Bt crystal proteins, both in terms of the insecticidal properties of individual proteins and the interactions among them in the digestive tract of susceptible mosquito species. The present study provides evidence that additional proteins, such as Cry10Aa and Cyt2Ba, which are usually present as minor components in Bti strains, are also important toxicity factors that act in a highly synergistic manner when these proteins are inoculated simultaneously in *A. aegypti* larvae. Cyt2Ba was previously described as a synergy factor for Cry4Aa and in this study we quantified the effects of the interaction on larval mosquito mortality [21].

The insecticidal activity of Cry10Aa in fourth instar larvae of *A. aegypti* was previously estimated at LC_{50} = 2061 ng/mL [17], which is about 7-fold higher than the value that we estimated in second instar larvae of the same species. A decrease in the susceptibility of larvae to infection by pathogens with increasing growth stage is common in insects [32], including their susceptibility to Bt toxins [22,33,34].

Several previous studies have described the larvicidal activity of Cyt2Ba protein in mosquito species belonging to the genera *Culex*, *Aedes* and *Anopheles* [19,35,36], although for a given toxin concentration the mortality that was recorded in Cyt2Ba-treated *A. aegypti* larvae was lower than that produced by the Cyt1Aa protein [19]. The estimated 24 h LC_{50} value of Cyt2Ba in second instar larvae of *A. aegypti* obtained in this study was approximately 27-fold lower than the value estimated by others [36]. This may be due to differences in the origin of the mosquito population and history of exposure to Bt toxins, and the fact that Wirth et al. [36] used lyophilized powdered inoculum rather than the freshly-prepared spore + crystal preparations that we employed.

The high potential of Bti proteins against mosquito larvae is mainly attributed to the interactions that occur among the component toxins. Although present at low abundance, Cry10Aa and Cyt2Ba contribute to the insecticidal activity of Bti by potentiation of toxin interactions [5]. Cry10Aa shows synergistic activity with Cyt1Aa [17] and Cry4Aa [37], whereas Cyt2Ba shows synergistic interaction with Cry4Aa [21] and *L. sphaericus* [36] in *A. aegypti*. When large amounts of these proteins were produced in an acrystalliferous Bt strain in the present study, very high levels of toxicity against *A. aegypti* larvae were obtained.

The LC_{50} value obtained here for the Cry10Aa/Cyt2Ba mixture was about 19 times lower than the value described for the Cry10Aa/Cyt1Aa mixture. This degree of potentiation is one of the highest observed so far for Bti crystal proteins, only comparable to that described for Cyt1A with Cry4Aa and Cry11Aa [38], or Cry4Ba in mixtures with Cyt2Aa2 from Bt *darmstadiensis* against *A. aegypti* larvae [39]. It seems, therefore, that Cyt2 proteins have a greater involvement in toxin synergy than has been attributed to date.

The molecular mechanisms underlying synergistic or other combinatorial effects between Bt insecticidal proteins have been the subject of several studies, although none have focused specifically on Cyt2Ba. The synergistic interaction of Cyt1Aa and Cyt2Aa with Cry4Ba appears to involve binding to the Cry protein through the domain II loops [9,40]. For the interaction between Cyt1Aa and Cry11Aa, specific charged residues have been identified on the Cyt1Aa protein that are involved in binding to Cry11Aa prior to insertion in the midgut epithelial cell membrane [41], although others have proposed that Cyt1Aa is a membrane-bound receptor that uses the exposed charged residues to bind Cry11Aa, thereby facilitating the interaction of the Cry protein with the target cell membrane [42,43]. Oligomerization of the Cyt1Aa toxin is essential for its toxicity in *A. aegypti* [44].

Bti has high larvicidal activity against mosquitoes, although its repeated use can lead to the appearance of resistance. The major components of the crystal, such as Cyt1A, Cry11A, or Cry4 are likely to be the main targets of such resistance. In this study, we demonstrated that Cry10Aa and Cyt2Ba, minor components of the parasporal crystal, have a high mosquitocidal potency and marked synergistic activity when present in a mixture. The optimization of culture conditions that result in improved production of Cry10A and Cyt2Ba may offer a rapid means to produce more effective Bti-based mosquitocidal products.

4. Conclusions

The toxicities of Cry10Aa and Cyt2Ba against *A. aegypti* are comparable to the major toxins of Bti and show one of the strongest potentiation effects observed for Bti crystal components to date. This potentiation was much stronger than occurred between Cyt2Ba and Cry4Aa. Further study of the minor crystal components of Bti is likely to provide additional opportunities for the development of safe and effective tools for the biological control of mosquito vectors of medical importance.

5. Materials and Methods

5.1. Bacterial Strains and Plasmids

B. thuringiensis ser. *israelensis* (Bti) was isolated from the commercial insecticide VectoBac-12AS® (Kenogard, Barcelona, Spain). *Escherichia coli* XL1 blue was used for transformation. The recombinant vector pSTAB [45] was used as the protein expression vector, engineered with the gene of interest. The acrystalliferous Bt strain BMB171 was used as the host strain for protein expression [46]. Bt recombinant strains 4Q2-81 pHT606:*cry4Aa* and 4Q2-81 pHT611:*cry4Ba* were kindly provided by Dr. Colin Berry (Cardiff University, Cardiff, UK) [30]. The Bt strains were grown in CCY medium containing 13 mM KH_2PO_4, 26 mM K_2HPO_4, 10 mL/L Nutrient stock solution (comprising L-glutamine, casein hydrolysate, casitone, yeast extract and glycerol), 1 ml/L metal salts solution [47] at 28 °C with continuous shaking at 200 rpm. All *E. coli* strains were cultured at 37 °C with continuous shaking (200 rpm) in Luria-Bertani (LB) broth (1% tryptone, 0.5% yeast extract, and 1% NaCl, pH 7.0). When required for selective growth, LB medium was supplemented with 20 µg/mL erythromycin (Em) and 100 µg/mL ampicillin (Amp).

5.2. Insect Culture

A laboratory colony of *A. aegypti* was started using eggs obtained from Dr. Susana Vilchez, (Universidad de Granada, Granada, Spain). The colony was maintained, under controlled environmental conditions (25 ± 1 °C and 85% RH, and a 16 h:8 h light: dark photoperiod), in the insectary facilities of the Instituto Multidisciplinario de Biología Aplicada (IMBA), Universidad Pública de Navarra, Spain. Adults of both sexes were maintained in BugDorm-1 insect rearing cages (MegaView Science, Taichung, Taiwan) and had continuous access to 20% sucrose solution and intermittent access (3 h/day) to defibrinated horse blood (Thermo Scientific, Waltham, MA, USA) to complete their gonotrophic cycle. Larvae were reared in 250 mL glass beakers (40–50 larvae/beaker) with 100 mL distilled water and brewer's yeast (1 mg/mL) as food.

5.3. Total DNA Extraction and Genomic Sequencing

Genome sequencing was performed to ensure that our Bti clone contained all the expected plasmids and genes, some of which may be lost during laboratory culture. Total genomic DNA (chromosomal + plasmid) was extracted from VectoBac-12AS® strain, following the protocol for DNA isolation from Gram-positive bacteria using the Wizard® Genomic DNA Purification Kit (Promega, Madison, WI, USA). A DNA library was prepared from total DNA and was subsequently sequenced in an Illumina NextSeq500 Sequencer (Genomics Research Hub Laboratory, School of Biosciences, Cardiff University, Cardiff, UK).

5.4. Identification of Cry and Cyt Insecticidal Genes in VectoBac-12AS®

Genomic raw sequence data were processed and assembled using CLC Genomics Workbench 10.1.1. Reads were trimmed, filtered by low quality and reads of less than 50 bp were eliminated. Processed reads were assembled de novo using stringent criteria of at least 95 bp overlap and 95% identity. Reads were then mapped back to the contigs for assembly. Genes were predicted using GeneMark [48].

To assist in the identification of potential insecticidal proteins, local BLASTP [49] was deployed against a database built in our laboratory comprising the amino acid sequences of known Bt toxins available at http://www.lifesci.sussex.ac.uk/home/Neil_Crickmore/Bt [50], as well as other protein toxins of interest.

5.5. Amplification, Cloning and Sequencing of Cyt2Ba, Cry11Aa and Cry10Aa

Primers were designed to amplify the full-length coding sequence of *cyt2Ba*, *p19-cry11Aa* (including *p19* and *cry11Aa* genes) and the *cry10Aa* operon including *orf1* and *orf2* (Table 5). Primer sequences included XbaI and PstI restriction sites for *cyt2Ba*, as well as SalI and PstI restriction sites for *p19-cry11Aa* and SalI and PaeI restriction sites for *cry10Aa*. PCR reactions were performed, from total genomic DNAs, using Phusion DNA polymerase (NEB, Ipswich, UK) and amplicons were gel-purified using NucleoSpin Extract II kit (Macherey-Nagel, Düren, Germany). Purified products were then ligated into pJET1.2/blunt plasmid (CloneJET PCR Cloning Kit, Waltham, MA, USA) following the manufacturer's instructions. Ligation mixtures were transformed into *E. coli* XL1-Blue using standard procedures [51]. Colony-PCR was applied in order to check positive clones from which plasmid DNA was purified, using the NucleoSpin^R plasmid kit (Macherey-Nagel Inc., Bethlehem, PA, USA) following the manufacturer's instructions. Subsequently, pJET plasmids were verified by sequencing (STABVida, Caparica, Portugal), digested with the appropriate combination of restriction enzymes, electrophoresed in 1% agarose gel and ligated into pre-digested pSTAB vector using the Rapid DNA ligation kit (Thermo Scientific) to obtain the recombinant plasmids pSTAB-*cyt2Ba* and pSTAB-*cry10Aa*. To clone *cry11Aa* the amplicon was ligated in a pSTAB in which *p20* gene was previously introduced, to obtain the recombinant plasmid pSTAB-*p19-cry11Aa-p20*. Ligation products were then electroporated into *E. coli* XL1 blue cells following standard protocols [51]. Positive clones were verified by colony-PCR and plasmids were purified and verified by restriction endonuclease digestion and electrophoresis. Once pSTAB-*cyt2Ba*, pSTAB-*cry10Aa* and pSTAB-*p19-cry11Aa-p20* were obtained, they were introduced into the acrystalliferous Bt strain BMB171.

Table 5. Sequences of PCR and sequencing primers.

Primer Name	Primer Sequence	Reference
Cyt2B-Fw-XbaI	5′-TT<u>CTAGAG</u>ATAATGAAGGAGGGGAGTC-3′	This study
Cyt2B-Rv-PstI	5′-C<u>CTGCAG</u>CAAAATTAAATTGCTGAGTTACTATAATAAC-3′	This study
Cry10A-Fw-SalI	5′-AT<u>GTCGAC</u>TTGCAACAGAAAAGAGTTGTGTC-3′	[17]
Cry10A-Rv-PaeI	5′-GAG<u>CATGC</u>ACATTTCCCCACAATTTTCA-3′	[17]
Cry10A-test-Fw	5′-CGAAATTGTCAGACATAGAGAG-3′	This study
Cry10A-test-Rv	5′-GAATTACCAAGTCTCCACCTG-3′	This study
p20-Fw-PstI	5′-C<u>CTGCAG</u>GGATAAAATTGGAGGATAATTGATG-3′	This study
p20-Rv-PaeI	5′-G<u>GCATGC</u>GTTTCCAGTGCATTCAATTTAC-3′	This study
p19-Fw-SalI	5′-GT<u>GTCGAC</u>GTTTTTTAAAATTGCATAGAAGGG-3′	This study
Cry11A-Rv-PstI	5′-CT<u>CTGCAG</u>GTGCTAACATGACTTCTACTTTAG-3′	This study
Cry11A-test	5′-GGTCATAATTTATGAATAAAAATATGAC-3′	This study

Restriction enzyme sites are underlined.

Bacillus electrocompetent cells were generated as described previously [52]. Briefly, bacteria were grown in 300 mL of Brain heart infusion broth (Pronadisa) at 28 °C under shaking conditions (200 rpm) until the culture reached an OD600 nm of 0.4. Glycine was then added to the culture at 2% and bacterial cells were incubated for another hour, at 28 °C under shaking conditions (200 rpm). Bacterial cells were kept on ice for 5 min, centrifuged at 9000× *g* (4 °C) for 10 min and the pellet was washed three times with F buffer (272 mM sucrose, 0.5 mM MgCl$_2$, 0.5 mM K$_2$HPO$_4$, 0.5 mM KH$_2$PO$_4$, pH 7.2). Cells were then resuspended in 600 µL of ice-cold F buffer and stored in aliquots of 50 µL at −80 °C. Plasmids were transformed into the BMB171 strain by electroporation, as described previously [53]. Positive clones were selected by colony-PCR. BMB171 was also transformed with an empty plasmid as a negative control.

5.6. Expression of Cyt2Ba, Cry10Aa, Cry4Aa, Cry4Ba and Cry11Aa Recombinant Proteins and SDS-PAGE Analysis

Wild-type Bti and recombinant Bt strains were grown at 28 °C, under shaking conditions (200 rpm), in CCY medium supplemented with 20 µg/mL erythromycin, if required. Crystal formation was observed daily under the optical microscope. After 2–3 days, when ~95% of the cells had lysed, the mixture of spores and crystals was collected by centrifugation at 10,000× *g*, for 10 min at 4 °C. The pellet was washed once with saline solution (1 M NaCl, 10 mM EDTA) and three times with 10 mM KCl. The spore + crystal mixture was finally resuspended in 10 mM KCl and kept at 4 °C until used. Samples of spores and crystals were mixed with 2x sample buffer (Bio-Rad, Hercules, CA, USA), boiled at 100 °C for 5 min, and then subjected to electrophoresis as previously described [54], using Criterion TGX™ 4–20% Precast Gel (Bio-Rad). Gels were stained with Coomassie brilliant blue R-250 (Bio-Rad) and then distained in 30% ethanol and 10% acetic acid. For protein quantification, a 10 µL volume of spore and crystal suspension was solubilized in vitro in 1 mL of alkaline solution (50 mM Na_2CO_3, 10 mM DTT, pH 11.3) for 2 h at 37 °C. The protein concentration of each preparation was measured by the Bradford assay (Bio-Rad), using bovine serum albumin (BSA) as a standard.

5.7. Mosquitocidal Activity of the δ-Endotoxins Produced by Bti

The toxicity of Cry10Aa and Cyt2Ba was determined by bioassay against *A. aegypti* second instar larvae. Concentration-mortality bioassays were performed following a modified method described previously [33]. Groups of 10–15 second instar larvae were placed in one well of a 6-well cell culture plate (Costar) and they were exposed to one concentration of Bt (spores+crystals). Each well contained 5 mL of Bt suspension with the corresponding toxin concentration and 0.5 mg of brewer's yeast as food. Toxin concentrations were 2000, 666, 222, 74, 24.7 and 8.2 ng/mL for Cry10Aa; 4000, 1333, 444, 148, 49.4 and 16.4 ng/mL for Cyt2Ba and $4 \times 10^{-1}, 2 \times 10^{-1}, 1 \times 10^{-1}, 5 \times 10^{-2}, 2.5 \times 10^{-2}, 1.2 \times 10^{-2}$ ng/mL for Bti (VectoBac-12AS®) as the positive control. Each bioassay was performed at least three times, depending on the toxin. Control insects were mock-infected. Insects were incubated at 25 °C and 16 h:8 h L:D photoperiod. Mortality was recorded at 24 h post-treatment. The concentration-mortality raw data are represented in Tables 2 and 3. Graphical representation of logit regressions for the individual toxins are summarized in Figure 2. These regressions were used to estimate the median lethal concentration (LC_{50}) for the toxins.

To study the synergistic larvicidal activity of Cyt2Ba in binary mixtures with other components of the Bti crystal, a series of preliminary bioassays were made, using a single protein concentration (below 30% mortality). The binary combinations studied, as well as the concentration of proteins used in each case, are shown in Table 1. For those binary combinations that resulted in the highest mortality of inoculated insects, quantitative bioassays were performed in order to determine the potentiation between Cyt2Ba and other toxins. Concentration-mortality bioassays were performed for Cry4Aa at concentrations of 486, 162, 54, 18, 6, 2 ng/mL. Mixtures of Cyt2Ba with either Cry10Aa or Cry4Aa in equal proportions were tested at concentrations of 300, 60, 12, 2.4, 4.8×10^{-1} and 9.6×10^{-2} ng/mL, and 54, 27, 13.5, 6.74, 3.36 and 1.68 ng/mL, respectively. Each bioassay was performed between five and ten times. In all other aspects, the bioassay procedure and data curation was as described above. Graphical representation of logit regressions for all toxin mixtures are summarized in Figure 2. These regressions were used to estimate the median lethal concentration (LC_{50}) for the mixture of the toxins.

5.8. Statistical Analysis

Concentration-mortality data were subjected to logit regression to estimate the median lethal concentration (LC_{50}) for individual toxins and the mixture of toxins. The significance of treatment and interaction terms was determined by sequential removal of terms from the complete regression model. The observed and expected LC_{50} values for the individual toxins and the toxin mixture in *A. aegypti* were used to evaluate the interaction of Cyt2Ba with Cry10Aa and Cry4Aa. To calculate the expected LC_{50} values for the toxin mixture under the null hypothesis of no interaction the "simple similar action" model was used [31]. This model assumes that concentration-response regression lines for different components of a mixture are parallel and is suitable for testing synergism in chemically similar compounds such as Bt toxins. Because Cyt2Ba and Cry4Aa regression lines are not parallel the synergism factor calculated is only correct for the LC_{50} single point.

The expected LC_{50} was calculated as follows:

$$LC_{50(m)} = \left[\frac{r_A}{LC_{50(A)}} + \frac{r_B}{LC_{50(B)}} \right]^{-1}$$

where $LC_{50(m)}$ is the expected LC_{50} of the mixture of toxin A and toxin B, $LC_{50(A)}$ is the observed LC_{50} for toxin A alone, $LC_{50(B)}$ is the observed LC_{50} for toxin B alone and r_A and r_B represent the relative proportions of toxin A and toxin B in the mixture, respectively. All statistical procedures were performed using R software (v.3.5.1).

Author Contributions: Conceptualization, D.V.-d.-L., M.V. and P.C.; Data curation, D.V.-d.-L.; Formal analysis, D.V.-d.-L. and T.W.; Funding acquisition, P.C.; Investigation, D.V.-d.-L., M.V. and L.L.; Methodology, D.V.-d.-L. and M.V.; Project administration, P.C.; Resources, P.C.; Software, D.V.-d.-L.; Supervision, M.V., T.W. and P.C; Validation, D.V.-d.-L. and M.V.; Visualization, D.V.-d.-L. and M.V.; Writing—original draft, D.V.-d.-L. and M.V.; Writing—review & editing, D.V.-d.-L., M.V., T.W. and P.C. All authors have read and agreed to the published version of the manuscript.

Funding: This research was funded by Spanish Ministry of Science and Innovation (RTI2018-095204-B-C22). D.V.-d.-L. received a doctoral grant from Universidad Pública de Navarra, Pamplona, Spain. L.L. received a doctoral grant from the European Union's H2020 research and innovation programme under Marie Sklodowska-Curie grant agreement N° 801586.

Acknowledgments: The authors thank Colin Berry (Cardiff University, Cardiff, UK) for providing Bt recombinant strains 4Q2-81 pHT606:*cry4Aa* and 4Q2-81 pHT611:*cry4Ba* used in this study.

References

1. Goldberg, L.J.; Margalith, J. A bacterial spore demonstrating rapid larvicidal activity against *Anopheles sergentii, Uranotaenia unguiculata, Culex univitattus, Aedes aegypti* and *Culex pipiens. Mosq. News* **1977**, *1*, 355–362.

2. Federici, B.A.; Park, H.; Bideshi, D.K. Overview of the basic biology of *Bacillus thuringiensis* with emphasis on genetic engineering of bacterial larvicides for mosquito control. *Open Toxinol. J.* **2010**, *3*, 154–171. [CrossRef]

3. Margalith, Y.; Ben-Dov, E. Biological control by Bacillus thuringiensis subsp. israelensis. In *Insect Pest Management: Techniques for Environmental Protection*; Recheigl, J.E., Recheigl, N.A., Eds.; CRC Press: Boca Raton, FL, USA, 2000.

4. Lacey, L.A. Bacillus thuringiensis serovariety israelensis and Bacillus sphaericus for mosquito control. *J. Am. Mosq. Control Assoc.* **2007**, *23*, 133–163. [PubMed]

5. Ben-Dov, E. *Bacillus thuringiensis* subsp. *israelensis* and its dipteran-specific toxins. *Toxins* **2014**, *6*, 1222–1243. [PubMed]

6. Boyce, R.; Lenhart, A.; Kroeger, A.; Velayudhan, R.; Roberts, B.; Horstick, O. *Bacillus thuringiensis israelensis* (Bti) for the control of dengue vectors: Systematic literature review. *Trop. Med. Int. Health* **2013**, *18*, 564–577.

7. Paris, M.; Tetreau, G.; Laurent, F.; Lelu, M.; Despres, L.; David, J.P. Persistence of *Bacillus thuringiensis israelensis* (Bti) in the environment induces resistance to multiple Bti toxins in mosquitoes. *Pest Manag. Sci.* **2011**, *67*, 122–128.

8. Vasquez, M.I.; Violaris, M.; Hadjivassilis, A.; Wirth, M.C. Susceptibility of *Culex pipiens* (Diptera: Culicidae) field populations in cyprus to conventional organic insecticides, *Bacillus thuringiensis* subsp. *israelensis*, and methoprene. *J. Med. Entomol.* **2009**, *46*, 881–887. [CrossRef]

9. Cantón, P.E.; Reyes, E.Z.; De Escudero, I.R.; Bravo, A.; Soberón, M. Binding of *Bacillus thuringiensis* subsp. *israelensis* Cry4Ba to Cyt1Aa has an important role in synergism. *Peptides* **2011**, *32*, 595–600.

10. Wirth, M.C.; Park, H.W.; Walton, W.E.; Federici, B.A. Cyt1A of *Bacillus thuringiensis* delays evolution of resistance to Cry11A in the mosquito *Culex quinquefasciatus*. *Appl. Environ. Microbiol.* **2005**, *71*, 185–189. [CrossRef]

11. Wirth, M.C.; Georghiou, G.P.; Federici, B.A. CytA enables CryIV endotoxins of *Bacillus thuringiensis* to overcome high levels of CryIV resistance in the mosquito, *Culex quinquefasciatus*. *Proc. Natl. Acad. Sci. USA* **1997**, *94*, 10536–10540. [CrossRef]

12. Ibarra, J.E.; Federici, B.A. Isolation of a relatively nontoxic 65-kilodalton protein inclusion from the parasporal body of *Bacillus thuringiensis* subsp. *israelensis*. *J. Bacteriol.* **1986**, *165*, 527–533. [CrossRef] [PubMed]

13. Lee, S.G.; Eckblad, W.; Lee, A.B. Diversity of protein inclusion bodies and identification of mosquitocidal protein in *Bacillus thuringiensis* subsp. *israelensis*. *Biochem. Biophys. Res. Commun.* **1985**, *126*, 953–960. [CrossRef]

14. Garduno, F.; Thorne, L.; Walfield, A.M.; Pollock, T.J. Structural relatedness between mosquitocidal endotoxins of *Bacillus thuringiensis* subsp. *israelensis*. *Appl. Environ. Microbiol.* **1988**, *54*, 277–279. [CrossRef] [PubMed]

15. Berry, C.; Ben-dov, E.; Jones, A.F.; Murphy, L.; Quail, M.A.; Holden, M.T.G.; Harris, D.; Zaritsky, A.; Parkhill, J. Complete sequence and organization of pBtoxis, the toxin-coding plasmid of *Bacillus thuringiensis* subsp. *israelensis*. *Appl. Environ. Microbiol.* **2002**, *68*, 5082–5095. [CrossRef] [PubMed]

16. Thorne, L.; Garduno, F.; Thompson, T.; Decker, D.; Zounes, M.; Wild, M.; Waldfield, A.; Pollock, T.J. Structural similarity between the Lepidoptera- and Diptera-specific insecticidal endotoxin genes of *Bacillus thuringiensis* subsp. *"kurstaki"* and *"israelensis."*. *J. Bacteriol.* **1986**, *166*, 801–811. [CrossRef] [PubMed]

17. Hernández-Soto, A.; Del Rincón-Castro, M.C.; Espinoza, A.M.; Ibarra, J.E. Parasporal body formation via overexpression of the Cry10Aa toxin of *Bacillus thuringiensis* subsp. *israelensis*, and Cry10Aa-Cyt1Aa synergism. *Appl. Environ. Microbiol.* **2009**, *75*, 4661–4667.

18. Guerchicoff, A.; Ugalde, R.A.; Rubinstein, C.P. Identification and characterization of a previously undescribed cyt gene in *Bacillus thuringiensis* subsp. *israelensis*. *Appl. Environ. Microbiol.* **1997**, *63*, 2716–2721. [CrossRef]

19. Juárez-Pérez, V.; Guerchicoff, A.; Rubinstein, C.; Delecluse, A. Characterization of Cyt2Bc toxin from *Bacillus thuringiensis* subsp. *medellin*. *Appl. Environ. Microbiol.* **2002**, *68*, 1228–1231.

20. Crickmore, N.; Bone, E.J.; Williams, J.A.; Ellar, D.J. Contribution of the individual components of the δ-endotoxin crystal to the mosquitocidal activity of *Bacillus thuringiensis* subsp. *israelensis*. *FEMS Microbiol. Lett.* **1995**, *131*, 249–254. [CrossRef]

21. Manasherob, R.; Itsko, M.; Sela-Baranes, N.; Ben-Dov, E.; Berry, C.; Cohen, S.; Zaritsky, A. Cyt1 Ca from *Bacillus thuringiensis* subsp. *israelensis*: Production in *Escherichia coli* and comparison of its biological activities with those of other Cyt-like proteins. *Microbiology* **2006**, *152*, 2651–2659. [CrossRef]

22. Otieno-Ayayo, Z.N.; Zaritsky, A.; Wirth, M.C.; Manasherob, R.; Khasdan, V.; Cahan, R.; Ben-Dov, E. Variations in the mosquito larvicidal activities of toxins from *Bacillus thuringiensis* ssp. *israelensis*. *Environ. Microbiol.* **2008**, *10*, 2191–2199. [CrossRef] [PubMed]

23. Poncet, S.; Delecluse, A.; Klier, A.; Rapoport, G. Evaluation of synergistic interactions among CryIVA, CryIVB, and CryIVD toxic components of *B. thuringiensis* subsp. *israelensis* crystals. *J. Invertebr. Pathol.* **1995**, *66*, 131–135. [CrossRef]

24. Wu, D.; Johnson, J.J.; Federici, B.A. Synergism of mosquitocidal toxicity between CytA and CryIVD proteins using inclusions produced from cloned genes of *Bacillus thuringiensis*. *Mol. Microbiol.* **1994**, *13*, 965–972. [CrossRef] [PubMed]

25. Wirth, M.C.; Georghiou, G.P.; Malik, J.I.; Abro, G.H. Laboratory selection for resistance to *Bacillus sphaericus* in *Culex quinquefasciatus* (Diptera: Culicidae) from California, USA. *J. Med. Entomol.* **2000**, *37*, 534–540. [CrossRef] [PubMed]

26. Donovan, W.P.; Dankocsik, C.; Gilbert, M.P. Molecular characterization of a gene encoding a 72-kilodalton mosquito-toxic crystal protein from *Bacillus thuringiensis* subsp. *israelensis*. *J. Bacteriol.* **1988**, *170*, 4732–4738. [CrossRef] [PubMed]

27. Anderson, I.; Sorokin, A.; Kapatral, V.; Reznik, G.; Bhattacharya, A.; Mikhailova, N.; Burd, H.; Joukov, V.; Kaznadzey, D.; Walunas, T.; et al. Comparative genome analysis of *Bacillus cereus* group genomes with *Bacillus subtilis*. *FEMS Microbiol. Lett.* **2005**, *250*, 175–184. [CrossRef]

28. Waalwijlc, C.; Dullemans, A.M.; Van Workum, M.E.S.; Visser, B. Molecular cloning and the nucleotide sequence of the Mr 28 000 crystal protein gene of *Bacillus thuringiensis* subsp. *israelensis*. *Nucleic Acids Res.* **1985**, *13*, 5753–5763.

29. Wu, D.; Federici, B.A. Improved production of the insecticidal CryIVD protein in *Bacillus thuringiensis* using cryIA(c) promoters to express the gene for an associated 20 kDa protein. *Appl. Microbiol. Biotechnol.* **1995**, *42*, 697–702. [CrossRef]

30. Delecluse, A.; Poncet, S.; Klier, A.; Rapoport, G. Expression of cryIVA and cryIVB genes, independently or in combination, in a crystal-negative strain of *Bacillus thuningiensis* subsp. *israelensis*. *Appl. Environ. Microbiol.* **1993**, *59*, 3922–3927. [CrossRef]

31. Tabashnik, B.E. Evaluation of synergism among *Bacillus thuringiensis* toxins. *Appl. Environ. Microbiol.* **1992**, *58*, 3343–3346. [CrossRef]

32. Stiles, B.; Paschke, J.D. Midgut pH in different instars of three *Aedes* mosquito species and the relation between pH and susceptibility of larvae to a nuclear polyhedrosis virus. *J. Invertebr. Pathol.* **1980**, *35*, 58–64. [CrossRef]

33. McLaughlin, R.E.; Dulmage, H.T.; Alls, R.; Couch, T.L.; Dame, D.A.; Hall, I.M.; Rose, R.I.; Versoi, P.L. U.S. standard bioassay for the potency assessment of *Bacillus thuringiensis* serotype H-14 against mosquito larvae. *Bull. Entomol. Soc. Am.* **1984**, *30*, 26–29. [CrossRef]

34. Zehnder, G.W.; Gelernter, W.D. Activity of the M-ONE formulation of a new strain of *Bacillus thuringiensis* against the colorado potato beetle (Coleoptera: Chrysomelidae): Relationship between susceptibility and insect life stage. *J. Econ. Entomol.* **1989**, *82*, 756–761. [CrossRef]

35. Nisnevitch, M.; Cohen, S.; Ben-Dov, E.; Zaritsky, A.; Sofer, Y.; Cahan, R. Cyt2Ba of *Bacillus thuringiensis israelensis*: Activation by putative endogenous protease. *Biochem. Biophys. Res. Commun.* **2006**, *344*, 99–105. [CrossRef] [PubMed]

36. Wirth, M.C.; Delecluse, A.; Walton, W.E. Cyt1Ab1 and Cyt2Ba1 from *Bacillus thuringiensis* subsp. *medellin* and *B. thuringiensis* subsp. *israelensis* synergize *Bacillus sphaericus* against *Aedes aegypti* and resistant *Culex quinquefasciatus* (Diptera: Culicidae). *Appl. Environ. Microbiol.* **2001**, *67*, 3280–3284. [CrossRef]

37. Delécluse, A.; Bourgouin, C.; Klier, A.; Rapoport, G. Specificity of action on mosquito larvae of *Bacillus thuringiensis israelensis* toxins encoded by two different genes. *MGG Mol. Gen. Genet.* **1988**, *214*, 42–47. [CrossRef]

38. Khasdan, V.; Ben-Dov, E.; Manasherob, R.; Boussiba, S.; Zaritsky, A. Toxicity and synergism in transgenic *Escherichia coli* expressing four genes from *Bacillus thuringiensis* subsp. *israelensis*. *Environ. Microbiol.* **2001**, *3*, 798–806. [CrossRef]

39. Promdonkoy, B.; Promdonkoy, P.; Panyim, S. Co-expression of *Bacillus thuringiensis* Cry4Ba and Cyt2Aa2 in *Escherichia coli* revealed high synergism against *Aedes aegypti* and *Culex quinquefasciatus* larvae. *FEMS Microbiol. Lett.* **2005**, *252*, 121–126. [CrossRef]

40. Lailak, C.; Khaokhiew, T.; Promptmas, C.; Promdonkoy, B.; Pootanakit, K.; Angsuthanasombat, C. *Bacillus thuringiensis* Cry4Ba toxin employs two receptor-binding loops for synergistic interactions with Cyt2Aa2. *Biochem. Biophys. Res. Commun.* **2013**, *435*, 216–221. [CrossRef]

41. López-Diaz, J.A.; Cantón, P.E.; Gill, S.S.; Soberón, M.; Bravo, A. Oligomerization is a key step in Cyt1Aa membrane insertion and toxicity but not necessary to synergize Cry11Aa toxicity in *Aedes aegypti* larvae. *Environ. Microbiol.* **2013**, *15*, 3030–3039. [CrossRef]

42. Perez, C.; Fernandez, L.E.; Sun, J.; Folch, J.L.; Gill, S.S.; Soberon, M.; Bravo, A. *Bacillus thuringiensis* subsp. *israelensis* Cyt1Aa synergizes Cry11Aa toxin by functioning as a membrane-bound receptor. *Proc. Natl. Acad. Sci. USA* **2005**, *102*, 18303–18308. [PubMed]

43. Pérez, C.; Muñoz-Garay, C.; Portugal, L.C.; Sánchez, J.; Gill, S.S.; Soberón, M.; Bravo, A. *Bacillus thuringiensis* ssp. *israelensis* Cyt1Aa enhances activity of Cry11Aa toxin by facilitating the formation of a pre-pore oligomeric structure. *Cell. Microbiol.* **2007**, *9*, 2931–2937.

44. Anaya, P.; Onofre, J.; Torres-Quintero, M.C.; Sánchez, J.; Gill, S.S.; Bravo, A.; Soberón, M. Oligomerization is a key step for *Bacillus thuringiensis* Cyt1Aa insecticidal activity but not for toxicity against red blood cells. *Insect Biochem. Mol. Biol.* **2020**, *119*, 103317. [CrossRef] [PubMed]

45. Park, H.W.; Ge, B.; Bauer, L.S.; Federici, B.A. Optimization of Cry3A yields in *Bacillus thuringiensis* by use of sporulation-dependent promoters in combination with the STAB-SD mRNA sequence. *Appl. Environ. Microbiol.* **1998**, *64*, 3932–3938. [PubMed]

46. Li, L.; Yang, C.; Liu, Z.; Li, F.; Yu, Z. Screening of acrystalliferous mutants from *Bacillus thuringiensis* and their transformation properties. *Wei Sheng Wu Xue Bao.* **2000**, *40*, 85–90.

47. Stewart, S.D.; Adamczyk, J.J.; Knighten, K.S.; Davis, F.M. Impact of Bt cottons expressing one or two insecticidal proteins of *Bacillus thuringiensis* Berliner on growth and survival of Noctuid (Lepidoptera) larvae. *J. Econ. Entomol.* **2001**, *94*, 752–760. [CrossRef]

48. Borodovsky, M.; McIninch, J. GENMARK: Parallel gene recognition for both DNA strands. *Comput. Chem.* **1993**, *17*, 123–133.

49. Altschul, S.F.; Gish, W.; Miller, W.; Myers, E.W.; Lipman, D.J. Basic local alignment search tool. *J. Mol. Biol.* **1990**, *215*, 403–410. [CrossRef]

50. Crickmore, N.; Zeigler, D.R.; Feitelson, J.; Schnepf, E.; Van Rie, J.; Lereclus, D.; Baum, J.; Dean, D.H. Revision of the nomenclature for the *Bacillus thuringiensis* pesticidal crystal proteins. *Microbiol. Mol. Biol. Rev.* **1998**, *62*, 807–813. [CrossRef]

51. Sambrook, J.; Russell, D. *Molecular Cloning: A laboratory Manual*, 3rd ed.; Harbor, C.S., Ed.; Cold Spring Harbor Laboratory: New York, NY, USA, 2001.

52. Dominguez-Arrizabalaga, M.; Villanueva, M.; Fernandez, A.B.; Caballero, P. A strain of *Bacillus thuringiensis* containing a novel cry7Aa2 gene that is toxic to *Leptinotarsa desemlineata* (Say) (Coleoptera: Chrysomelidae). *Insects* **2019**, *10*, 259. [CrossRef]

53. Lee, J.C. Electrotransformation of Staphylococci. *Electroporation Protoc. Microorg.* **1995**, *47*, 209–216.

54. Laemmli, U.K. Cleavage of structural proteins during the assembly of the head of Bacteriophage T4. *Nature* **1970**, *225*, 680–685. [CrossRef] [PubMed]

Article

Protein-Lipid Interaction of Cytolytic Toxin Cyt2Aa2 on Model Lipid Bilayers of Erythrocyte Cell Membrane

Sudarat Tharad [1,*], **Boonhiang Promdonkoy** [2] **and José L. Toca-Herrera** [1,*]

[1] Institute for Biophysics, Department of Nanobiotechnology, University of Natural Resources and Life Sciences (BOKU), 1190 Vienna, Austria

[2] National Center for Genetic Engineering and Biotechnology, National Science and Technology Development Agency, Pathumthani 12120, Thailand; boonhiang@biotec.or.th

* Correspondence: sudarattharad@gmail.com (S.T.); jose.toca-herrera@boku.ac.at (J.L.T.-H.)

Received: 3 March 2020; Accepted: 1 April 2020; Published: 3 April 2020

Abstract: Cytolytic toxin (Cyt) is a toxin among *Bacillus thuringiensis* insecticidal proteins. Cyt toxin directly interacts with membrane lipids for cytolytic action. However, low hemolytic activity is desired to avoid non-specific effects in mammals. In this work, the interaction between Cyt2Aa2 toxin and model lipid bilayers mimicking the erythrocyte membrane was investigated for Cyt2Aa2 wild type (WT) and the T144A mutant, a variant with lower hemolytic activity. Quartz crystal microbalance with dissipation (QCM-D) results revealed a smaller lipid binding capacity for the T144A mutant than for the WT. In particular, the T144A mutant was unable to bind to the phosphatidylcholine lipid (POPC) bilayer. However, the addition of cholesterol (Chol) or sphingomyelin (SM) to the POPC bilayer promoted binding of the T144 mutant. Moreover, atomic force microscopy (AFM) images unveiled small aggregates of the T144A mutant on the 1:1 sphingomyelin/POPC bilayers. In contrast, the lipid binding trend for WT and T144A mutant was comparable for the 1:0.4 POPC/cholesterol and the 1:1:1 sphingomyelin/POPC/cholesterol bilayers. Furthermore, the binding of WT and T144A mutant onto erythrocyte cells was investigated. The experiments showed that the T144A mutant and the WT bind onto different areas of the erythrocyte membrane. Overall the results suggest that the T144 residue plays an important role for lipid binding.

Keywords: Cyt2Aa2 toxin; protein-lipid binding; erythrocyte membrane; AFM; QCM-D

Key Contribution: The alanine replacement of the threonine 144 residue reduces the lipid binding ability of the Cyt2Aa2 toxin onto model lipid bilayers (and erythrocyte cells). In particular, it was found that the Cyt2Aa2 T144A mutant did not bind onto POPC bilayers.

---

1. Introduction

The most widely known bacteria as a bioinsecticidal agent is *Bacillus thuringiensis* (Bt). It is a Gram-positive rod shape bacterium originally hosted in soil. In the last few decades Bt has been used to control insect larvae especially for pest insects in the form of a bioactive agent or a transgenic plant. The active proteins, Crystal (Cry) and Cytolytic (Cyt) toxins are produced as crystalline proteins during the sporulation phase of the Bt growth cycle [1]. After toxin ingestion by insect larvae, the protein crystals are solubilized and concomitantly activated by proteases in alkaline condition of the mid gut [2–4]. Consequently, the toxins interrupt a cell membrane permeability of the gut cells leading to cell burst because of the osmotic pressure imbalance [5,6]. However, both toxins disrupt the cell membrane with different mechanisms. Cry toxin requires a protein receptor for the cell membrane binding whereas Cyt toxin interacts directly with the membrane lipids [7–9], in particular with the unsaturated phospholipids [10]. Cry toxin has been used more in crop fields to control insect larvae than Cyt toxin because of its efficiency and specificity [11]. Nevertheless, the long-term application of Cry toxin has led to insect larvae resistance [12]. Accordingly, Cyt toxin has been taken into the strategy to overcome such resistance. Thus, the Cyt toxin is able to be a receptor for the Cry toxin and these toxins can synergize their activities together [13].

Cytolytic toxin Cyt2Aa2 is produced from *Bacillus thuringiensis* subsp. *darmstadiensis* [14]. The Cyt toxin shows the cytolytic activity against a broad range of cell types, e.g., insect cells, mammalian cells [7], and bacterial cells [15]. Previous experiments suggest that the protein-lipid binding mechanism of the Cyt toxin is driven by (i) pore formation [16,17], (ii) detergent-like action [18,19], and (iii) carpet action (protein aggregate) [20]. However, the precise mechanism is still unclear and devotes further investigation. In particular, hemolytic activity has been tested to determine the cytolytic activity of Cyt2Aa2 toxin against erythrocyte cells (in relation to mammalian cells). Therefore, we have tried to obtain a variant with lower hemolytic activity by performing amino acid mutation (in order to reduce the non-specific target to mammalian cells). The effect of the amino acid point mutation of the Cyt2Aa2 molecule on the toxin activity has been reported. Previous studies have shown that the amino acids located in the helix A, helix C [21], and helixD-beta4 loop [22,23] alter the activity of the Cyt2Aa2 toxin. Particularly, we have investigated the amino acid mutation of T144A (alanine replacement of T144 residue) placed in the helixD-beta4 loop (Figure S1) because it keeps its larvicidal activity, although its hemolytic activity is reduced [24].

To elucidate the influence of the point mutation T144A on the interaction of Cyt2Aa2 protein with model lipid bilayers (which mimic the erythrocyte membrane), we have carried out binding studies with the Cyt2Aa2 wild type (WT) and the mutant Cyt2Aa2-T144A. The prepared lipid bilayers containing phospholipid (POPC), sphingomyelin (SM), and cholesterol (Chol) were mixed in various molar ratios to build the different membranes, which can be found in erythrocyte cells [25,26]. The combination of quartz crystal microbalance with dissipation (QCM-D) and atomic force microscopy (AFM) indicated that the T144A mutant had a lower binding capability than the WT, especially for the POPC bilayer. Moreover, the T144A mutant formed small aggregates on the 1:1 SM/POPC bilayer (showing a different binding trend from the Cyt2Aa2 WT). Finally, both toxins were also exposed to lysed erythrocytes cells. It was found that WT and T144A bound to different parts of the erythrocyte membrane.

2. Results

2.1. Determination of the Cyt2Aa2-Lipid Interaction with Different Model Lipid Bilayers that Mimic the Erythrocyte Membrane by QCM-D

The lipid components of the cell membrane, phospholipid (POPC), sphingomyelin (SM) and cholesterol (Chol), were mixed in different molar ratios in order to form different lipid bilayers that could mimic the erythrocyte cell membrane. The interaction of both Cyt2Aa2 wild type (WT) and T144A mutant with pure POPC (Figure 1A), 1:0.4 POPC/Chol (Figure 1B), 1:1 SM/POPC (Figure 1C), and 1:1:1 SM/POPC/Chol bilayers was investigated with QCM-D (Figure 1D). The results showed that Cyt2Aa2 WT (black plot) interacted with all kind of lipid bilayers, whereas no interaction between the T144A mutant and the POPC bilayer could be detected. In this case, the frequency (ΔF) and the dissipation (ΔD) signals did not change with time (blue plot). The measurements indicated that the lipid binding of both the Cyt2Aa2 WT and T144A mutant was saturated for ΔF values between -25 to -40 Hz. The difference in dissipation (ΔD) achieved values between 2×10^{-6} to 5×10^{-6} (Table 1). Here it is worth remembering that ΔF relates to changes in the mass adsorption (on the sensor surface), and that ΔD refers to the viscoelastic properties of the formed hybrid protein-lipid layer.

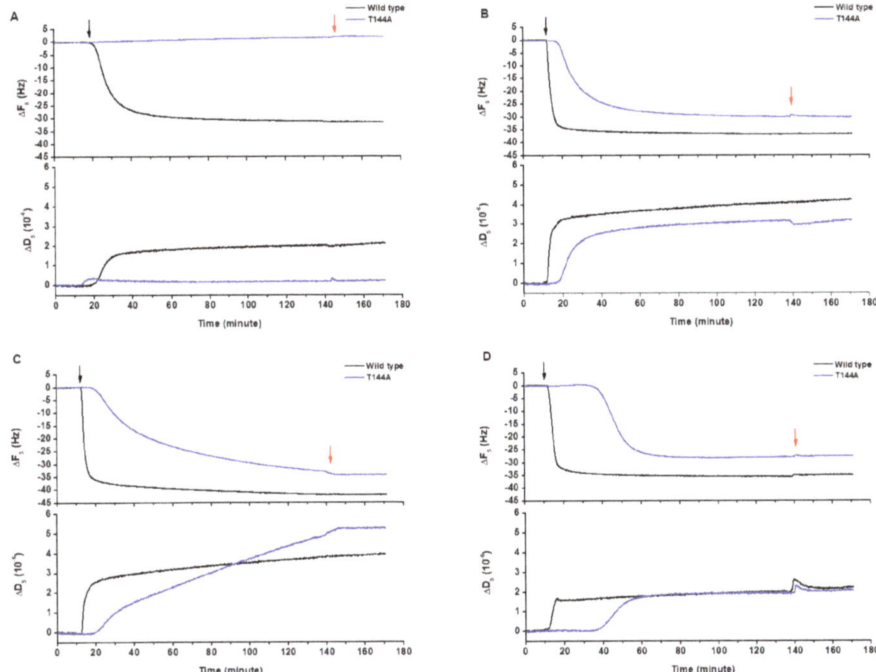

Figure 1. Protein-lipid binding of Cyt2Aa2 wild type and the T144A mutant on different lipid bilayers. The lipid bilayers were formed on the surface of silica sensors. Once the bilayer was built, the value of the frequency was set to zero. Thus, the reported difference in frequency relates to the adsorption of the protein toxin on the lipid bilayers. The protein solution (25 µg/mL) was filled into the quartz crystal microbalance with dissipation (QCM-D) chamber, and then the flow was paused in order to evaluate the Cyt2Aa2-lipid binding for 2 h. The black arrow and red arrow indicate protein exposure and buffer rinsing, respectively. (**A**) phospholipid (POPC), (**B**) 1:0.4 POPC/cholesterol (Chol), (**C**) 1:1 sphingomyelin (SM)/POPC and (**D**) 1:1:1 SM/POPC/Chol.

Table 1. ΔF, ΔD, and lipid binding rate values for wild type (WT) and T144A on different lipid bilayers.

Lipid Composition (Mole Ratio)	ΔF_5 (Hz)		ΔD_5 (10^{-6})		Lipid Binding Rate, Γ (min)	
	WT	**T144A**	**WT**	**T144A**	**WT**	**T144A**
POPC	-33.0 ± 3.8	1.3 ± 0.4	2.9 ± 1.1	0.0 ± 0.2	9.5 ± 0.3	No binding
1:0.4 POPC/Chol	-38.0 ± 1.7	-29.1 ± 0.6	3.1 ± 1.0	2.3 ± 0.7	2.1 ± 0.2	11.2 ± 2.0
1:1 SM/POPC	-40.1 ± 3.4	-24.4 ± 10.6	5.0 ± 0.2	3.0 ± 2.2	2.2 ± 0.6	52.1 ± 27.3
1:1:1 SM/POPC/Chol	-30.2 ± 4.4	-25.4 ± 2.4	2.3 ± 2.0	2.2 ± 0.2	2.5 ± 0.4	10.9 ± 1.9

In addition, ΔD-ΔF plots can be used to compare the binding behavior between the WT and the mutant T144A. For binding onto POPC bilayers (Figure 2A), the ΔD-ΔF signal for the T144A remained mostly constant with increasing time, while the WT showed a proportional increasing of ΔD and ΔF. However, the WT and the T144A seemed to bind in a similar way onto 1:0.4 POPC/Chol and 1:1:1 SM/POPC/Chol bilayers (Figure 2B,D), suggesting similar viscoelastic properties. On the contrary, a different trend occurred for the binding onto 1:1 SM/POPC bilayers (Figure 2C). It can be observed that for the same frequency change (ΔF) the mutant induced a final less rigid protein-lipid layer.

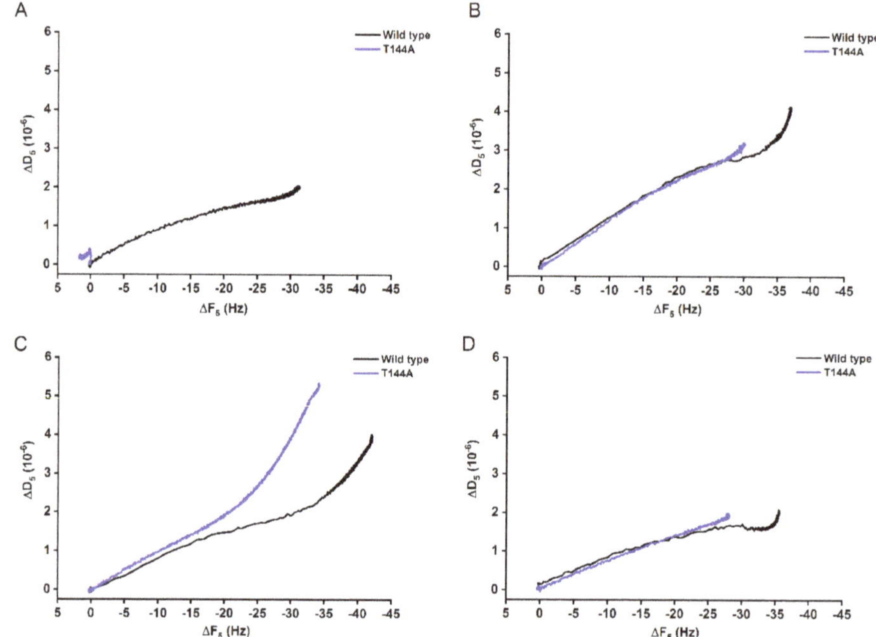

Figure 2. ΔD-ΔF plots of the binding of Cyt2Aa2 WT (black) and the T144A mutant (blue) on different model lipid bilayers. The dissipation value (ΔD) was plotted against the frequency value (ΔF) to elucidate the interplay between the protein binding and the viscoelasticity of the hybrid protein-lipid layer. The similarity of the slopes indicates an analogous qualitative behavior. (**A**) POPC, (**B**) 1:0.4 POPC/Chol, (**C**) 1:1 SM/POPC, and (**D**) 1:1:1 SM/POPC/Chol bilayers.

Furthermore, the binding kinetics were determined by fitting the experimental data to a single exponential decay equation:

$$F_t = F_0 + Ae^{-t/\Gamma} \tag{1}$$

(see Figure S2). In Table 1, the constant decay (Γ) indicates the lipid binding rate. Thus, a lower value means a faster binding rate, and vice versa. It can be observed that the WT showed lower Γ values than the T144A mutant for all types of model lipid bilayers. Hence, the binding rate of the WT was faster than the T144A mutant. Remarkably, the Γ values of the WT corresponding to the 1:0.4 POPC/Chol, 1:1 SM/POPC and 1:1:1 SM/POPC/Chol bilayers (ca. 2.0 min) were approximately five times smaller than the rate for the POPC bilayers (ca. 10.0 min). It seemed that the lipid bilayers containing either cholesterol or sphingomyelin favored the binding of Cyt2Aa2 WT. Similarly, the binding of the T144A mutant could be detected when cholesterol or sphingomyelin were present in the lipid bilayers. However, the Γ values indicate that the lipid binding ability of T144A mutant onto 1:1 SM/POPC bilayers (Γ = 52.1 min) was lower than 1:0.4 POPC/Chol (Γ = 11.2 min) and 1:1:1 SM/POPC/Chol (Γ = 10.9 min) bilayers, respectively. Unlike the WT case, sphingomyelin promoted a lower binding capability of the T144A mutant than cholesterol. These findings indicate that the replacement of threonine 144 with alanine results in a reduction of the lipid binding ability of Cyt2Aa2 toxin, especially for POPC bilayers.

2.2. AFM Imaging of the Cyt2Aa2 (WT and Mutant) Interaction with Different Model Lipid Bilayers

AFM experiments were carried out in order to investigate the topographic structure of the different Cyt2Aa2-lipid layers. The model lipid bilayers were successfully formed on the silica surface via lipid vesicle fusion and revealed a smooth surface (Figure S3). Subsequently, the protein solutions with the Cyt2Aa2 WT and the T144A mutant were incubated with the lipid bilayers. The surface topography of the hybrid protein-lipid layers was visualized after 30 min of incubation. Figure 3 shows no binding of the T144A mutant on POPC bilayers, which agrees with the QCM-D results. In contrast, the WT toxin almost covered the whole lipid bilayer surface (black areas refer to protein-free lipid bilayer). A longer incubation time of 120 min did not promote the binding of the T144A mutant on the POPC bilayer (Figure S4A). Subsequently, cholesterol and sphingomyelin were included into the lipid mixtures. Cyt2Aa2 WT bound onto the lipid surfaces reaching saturation. Thus, the lipid surfaces were fully covered with Cyt2Aa2 WT (note that the black areas disappeared). In addition, the binding between the T144A mutant and the lipid bilayers could be observed. The binding behavior of the T144A mutant onto 1:0.4 POPC/Chol and 1:1:1 SM/POPC/Chol bilayers showed a similar trend than the trend depicted by the WT; the protein-free membrane (black area) was observed prior to reaching a saturation after 120 min (see Figure S4). Remarkably, the T144A mutant formed small protein aggregates onto 1:1 SM/POPC bilayers. These aggregates seemed to be different from the ones observed for the WT and the T144A mutant on other lipid membranes (Figure 3).

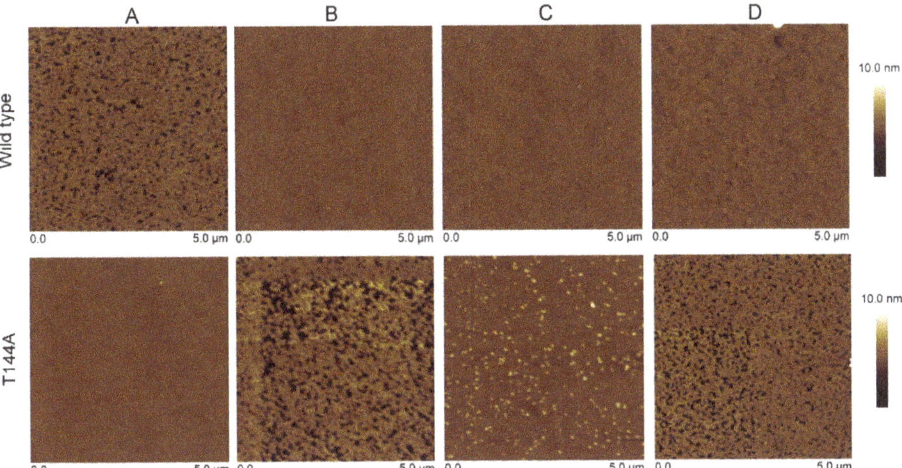

Figure 3. Atomic force microscopy (AFM) height images showing the interaction of the Cyt2Aa2 WT and the T144A mutant with the different lipid bilayers. First, the lipid bilayers were formed on silica surfaces. After, both protein solutions (WT and mutant) were exposed to the lipid bilayers for 30 min. The AFM images were collected in tapping mode with a scan rate of 1–2 Hz. Note that the scan size of every image is 5 μm × 5 μm. The vertical scale (until 10 nm) is indicated on the right. Image processing was carried out with the Nanoscope program. (**A**) POPC, (**B**) 1:0.4 POPC/Chol, (**C**) 1:1 SM/POPC, and (**D**) 1:1:1 SM/POPC/Chol bilayers.

Furthermore, the influence of sphingomyelin (SM) on the binding capability of the Cyt2Aa2 toxins was determined. For this purpose, 1:1 SM/DOPC (1,2-dioleoyl-*sn*-glycero-3-phosphocholine) bilayers were exposed to the toxins. For the SM/DOPC bilayers, a phase separation was observed where the sphingomyelin domains appeared as a liquid disordered-solid phase (l_d-S_o). In particular, the S_o domains of SM were distributed over the lipid bilayer surface being about 1 nm thicker than the DOPC-enriched domains (Figure S5). The 1:1 SM/DOPC bilayers were firstly exposed to the T144A mutant. The observed protein aggregates looked similar to the aggregates found on the SM/POPC bilayers. Subsequently, the WT protein was introduced into the system. The AFM micrographs indicate that the WT fully occupied the remaining areas (DOPC-enriched domains). Furthermore, no protein could be observed on SM domains (Figure 4). This suggests unfavorable binding of the Cyt2Aa2 toxins onto SM bilayers. Concordantly, AFM and QCM-D results support each other (i.e., no binding of the T144A mutant on the POPC bilayer). Moreover, AFM topography studies provided additional information about the T144A-lipid complex formation onto SM/POPC bilayers and its binding inability onto sphingomyelin bilayers.

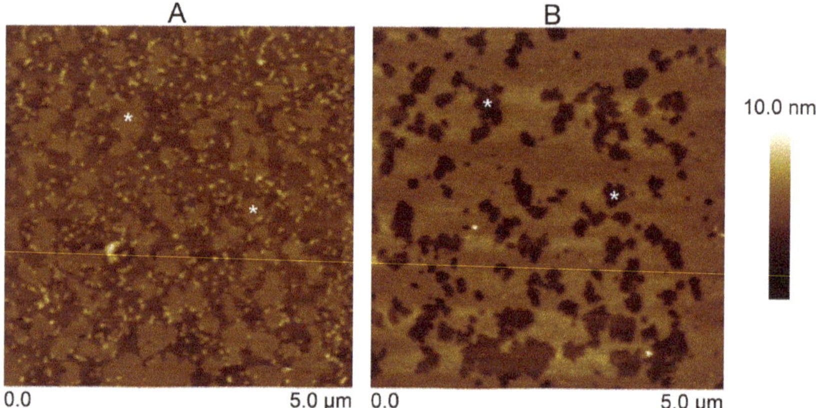

Figure 4. AFM height images of the Cyt2Aa2 wild type and the T144A mutant on 1:1 SM/DOPC (1,2-dioleoyl-*sn*-glycero-3-phosphocholine) bilayers. The SM domains are indicated as white asterisks. The lipid bilayers were initially exposed to the T144A mutant (25 μg/mL) for 2 h (**A**). After buffer rinsing, the Cyt2Aa2 wild type solution (25 μg/mL) was exposed to the lipid bilayers for 1 h (**B**). The topographic images were collected in tapping mode at a scan rate of 1–2 Hz. Note that the scan size of both images is 5 μm × 5 μm. The vertical scale (until 10 nm) is indicated on the right. Image analysis was performed with the Nanoscope program.

2.3. Cyt2Aa2 (WT and Mutant) Toxin Interaction with Erythrocyte Cell Membranes

Sheep erythrocytes presented a round and concave shape with diameter of ~3.0 μm under the light microscope. To prepare the erythrocyte membrane layers, a low salt solution (1/3 dilution PBS) was used to break the cell attached on the supporter surface. The ghost erythrocytes appeared as flat cells because of the releasing of cytoplasmic fluid (Figure S6). AFM images revealed a size of ca. 3.0–4.0 μm for the erythrocytes, which agreed with light microscopy observations.

After the erythrocytes were lysed, it was assumed that two types of erythrocyte membrane could be observed: (i) a single layer (inner cytoplasmic membrane) formed by cell opening (inside membrane facing up), and (ii) a double layer presenting the outer surface of the membrane (inner and outer cytoplasmic membranes) (Figure 5). Unlike the model lipid bilayers, the erythrocyte cytoplasmic membrane had a rougher surface. The membranes were firstly exposed to the T144A mutant. The binding of the T144A revealed a change in the topography of the height surface (area surrounding the asterisks). However, some areas of the membrane remained free of the T144A protein (no binding). In a second step, Cyt2Aa2 WT was introduced into the system. Cyt2Aa2 WT bound to the remaining areas leading to a smoother surface (compared to the erythrocyte-T144A ones) (Figure 5). The experiments with model lipid bilayers showed that the T144A mutant could not bind to the POPC bilayer bilayers. Therefore, it was not expected that the mutant would bind on the POPC areas of the erythrocyte membrane. On the contrary, the results of Cyt2Aa2 WT could indicate that WT bind POPC-enriched domains of the erythrocyte membrane.

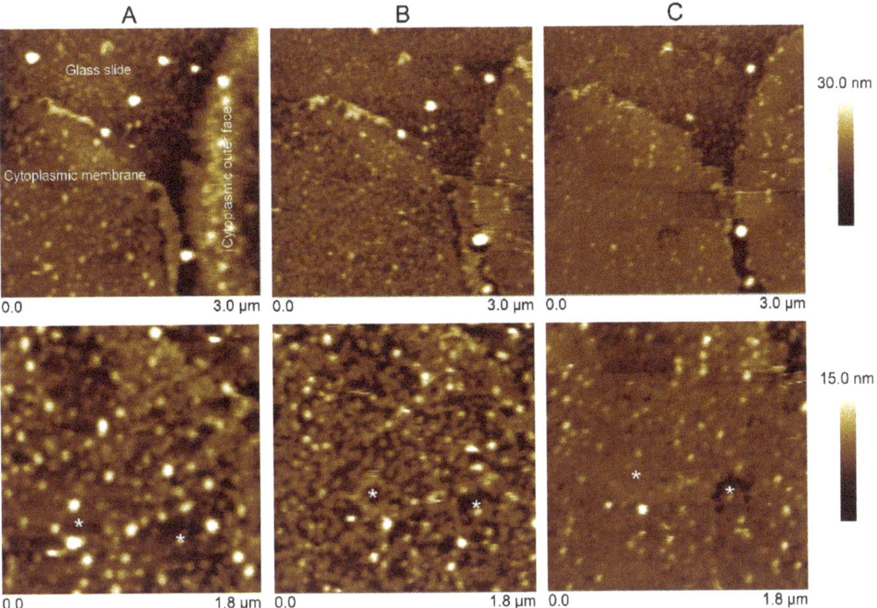

Figure 5. AFM images of Cyt2Aa2 toxin binding on the erythrocyte membrane. The erythrocyte membrane was prepared on a lysine-coated glass (**A**). The cell membranes were initially exposed to the T144A mutant (25 µg/mL) for 2 h (**B**). After buffer rinsing, the Cyt2Aa2 wild type solution (25 µg/mL) was exposed to the same membrane for 1 h (**C**). The topographic images were collected in tapping mode at a scan rate of 1–2 Hz. The images were analyzed with the Nanoscope program. The images have a scan size of 3 µm × 3 µm (upper panel) and 1.8 µm × 1.8 µm (lower panel). Note that the vertical scale differs: from 0 to 30 nm (upper panel), and from 0 to 15 nm (lower panel). The white asterisks mark the same areas on the erythrocyte membrane.

3. Discussion

In this work, we have studied the interaction of two Cyt2Aa2 proteins, the WT and the less hemolytic T144A mutant [24], with model lipid bilayers and erythrocyte membranes. QCM-D results showed that the T144A mutant could not bind to the POPC bilayer. In turn, the mutant protein retained its binding capability when exposed to 1:0.4 POPC/Chol, 1:1 SM/POPC, and 1:1:1 SM/POPC/Chol bilayers (Figure 1). Moreover, the ΔD–ΔF plots suggest that the binding behavior of the WT and T144A mutant followed similar adsorption trends except for 1:1 SM/POPC bilayers (Figure 2). The final values of ΔF and ΔD were not significantly different for the Cyt2Aa2 WT and the T144A mutant (Table 1). Our values are similar to the reported values for the binding of perforin on lipid membranes [27]. However, the mutant presented a lipid binding rate of at least five times lower than the WT, suggesting a smaller binding ability of the T144A mutant. The presence of either cholesterol or sphingomyelin in the lipid bilayer increased the binding rate of the WT and promoted the binding of the T144A mutant on the model lipid bilayers. At room temperature (25 °C), the POPC bilayer exists in a liquid disordered phase (l_d). Addition of cholesterol or sphingomyelin into lipid membranes reduces the fluidity of the lipid bilayer (less lateral diffusion) [28]. Less dynamic membranes seem to increase the possibility for the Cyt2Aa2-lipid interaction. Hence, Cyt2Aa2 WT bound faster on the 1:0.4 POPC/Chol, 1:1 SM/POPC and 1:1:1 SM/POPC/Chol bilayers than on the single POPC bilayer. In addition, it was observed that the T144A mutant could also bind onto these heterogeneous membranes, but not onto POPC bilayers. It seems that although the membranes of insect cells and mammalian cells are different in composition, both support the larvicidal activity and the low hemolytic activity of the T144A mutant.

In comparison to mammalian cells, insect cell membranes present a lower amount of cholesterol and sphingomyelin [29]. This might suggest that the binding of the Cyt2Aa2 toxin on insect cells is possibly promoted by other components of the cell membrane (e.g., membrane proteins) besides cholesterol and sphingomyelin. This different binding mechanism might play a role in the in vivo larvicial activity of the Cyt2Aa2 toxin.

Atomic force microscopy (AFM) was carried out to investigate the surface structure of the hybrid Cyt2Aa2-lipid bilayers. The AFM measurements confirmed that the T144A mutant does not bind onto POPC bilayers, while the WT toxin adsorbs on such lipid bilayers (Figure 3). With the addition of cholesterol into this lipid bilayer, the 1:0.4 POPC/Chol bilayer led to the binding of the T144A mutant. On the contrary, sphingomyelin (1:1 SM/POPC bilayer) also promoted the T144A mutant binding but a dissimilar binding was observed as well as small protein aggregates. Correspondingly, the ΔD–ΔF plots of QCM-D results suggest dissimilar binding behavior between the WT and the T144A mutant on the 1:1 SM/POPC bilayers (Figure 2). Furthermore, the binding behavior of Cyt2Aa2 WT and the T144A mutant were very much alike once cholesterol was included into the lipid membranes, as found for the 1:0.4 POPC/Chol and 1:1:1 SM/POPC/Chol bilayers (Figure 3). Moreover, both Cyt2Aa2 WT and the T144A mutant did not bind onto sphingomyelin (S_o) domains (Figure 4). These results suggest that although the lipid head group plays a role in the interaction between Cyt2Aa2 toxin and lipid bilayers, the lipid phase might also be taken into account [30,31]. This could be feasible since sphingomyelin contains the same choline head group as POPC. Besides, the small aggregates of the T144A mutant might imply a coexistence of a different fluid membrane in the 1:1 SM/POPC bilayer. This particular result indicates that cholesterol is more important for the binding behavior of the T144A mutant than for the binding of Cyt2Aa2 WT.

Furthermore, the interaction of Cyt2Aa2-lipid (WT and mutant) with sheep erythrocyte membranes was investigated. Unlike model lipid bilayers, a rougher surface was detected for the sheep erythrocyte membrane, which is comparable to the chicken erythrocyte membrane [32], and the human erythrocyte membrane [33]. The T144A mutant revealed a limited binding capability on the erythrocyte membrane, leaving part of the membrane uncovered. Further experiments indicated that the Cyt2Aa2 WT could bind on the remaining free areas (Figure 5). This binding study provides insight about the different properties of the erythrocyte membranes, e.g., lipid composition and lipid fluidity, when compared with model lipid membranes. Nowadays, a detection of the lipid phase coexistence in biological cell membranes is still a challenge for a biologist. Although lipid phase separation could not be observed in this experiment, the two Cyt2Aa2 proteins enabled us to distinguish the different components of the lipid membrane of sheep erythrocytes.

In conclusion, the combination of QCM-D and AFM permitted us to monitor the interaction between CytAa2 toxin with model lipid bilayers and supported erythrocyte membranes. The lower protein-lipid binding capability of the T144A mutant (in comparison with the WT) could lead to its small hemolytic activity. In particular, the alanine replacement of threonine 144 residue disables the binding properties of the Cyt2Aa2 toxin onto the POPC bilayer. Although certain hemolytic activity still remains for the T144A mutant, it can be said that the T144 residue located in the αD-$\beta 4$ loop plays an important role in the Cyt2Aa2-lipid binding. Furthermore, the modification of amino acid residues in the αD-$\beta 4$ loop of the Cyt2Aa2 toxin will be investigated for specific cell targeting. In future work, the effect of different amino acid properties (e.g., polar charge and positive charge) on the Cyt2Aa2-lipid interaction will be investigated. In addition, protein concentration, lipid phase and lipid charge will be taken into account for further investigations.

4. Materials and Methods

4.1. Reagents and Buffer

1-palmitoyl,2-oleoyl-sn-glycero-3-phosphocholine (POPC), 1,2-dioleoyl-*sn*-glycero-3-phospho choline (DOPC), chicken egg yolk sphingomyelin (SM), and cholesterol (Chol) were purchased from

Sigma-Aldrich (Darmstadt, Germany). The lipids were dissolved in chloroform and divided into 1 mg aliquots. Then, the organic solvent was evaporated under nitrogen stream and kept at −20 °C.

Phosphate buffered saline (PBS) pH 7.4 (137 mM NaCl, 2.7 mM KCl and 10 mM phosphate) was prepared from PBS tablet (Sigma-Aldrich, Darmstadt, Germany). The buffer tablet was dissolved in ultrapure water (Milli-Q, Merck, Darmstadt, Germany) and filtrated through a 0.22 μm filter (Whatman, GE Health care life science, Chicago, IL, USA).

4.2. Protein Preparation

The Cyt2Aa2 toxin from *Bacillus thuringiensis* subs. *darmstadiensis* was expressed in *Escherichia coli* as previously described by B. Promdonkoy [14]. The amino acid replacement at the threonine 144 residue with alanine was carried out by means of site-directed mutagenesis as described in a previous publication [24]. To obtain activated Cyt2Aa2 toxin (25 kDa), the Cyt2Aa2 inclusion was solubilized in 50 mM carbonate buffer, pH 10.0 at 30 °C for 1 h. The soluble Cyt2Aa2 toxin (29 kDa-protoxin) was collected by centrifugation at 10,000× g for 10 min. Then, the Cyt2Aa2 toxin was activated by 2% (w/w) chymotrypsin (Sigma-Aldrich, Darmstadt, Germany) at 30 °C for 2 h. The purity of the protein was determined by SDS-PAGE (Invitrogen, Waltham, MA, USA). Protein concentration was determined by UV adsorption (Hitachi, Tokyo, Japan). Stock protein solution was prepared to 2.0 mg/mL (80 μM) and kept at −20 °C.

4.3. Lipid Vesicle Preparation

The lipids were mixed in chloroform with the desired lipid ratios. After that, the organic solvent was evaporated under a gentle nitrogen stream to form lipid films. The residual solvent was removed by further keeping the lipid films under nitrogen stream for 1 h. Furthermore, the lipid films were hydrated with PBS solution to a concentration of 1 mg/mL and incubated above the melting transition temperature (T_m) for 2 h. The hydrated films were intermittently vortexed during incubation until complete suspension. The vesicles were homogenized by extrusion method for low T_m lipid mixtures (POPC/Chol system). The vesicles were pressed through a 50 nm Øpolycarbonate membrane for 21 times at room temperature by using a mini-extruder (Avanti, Alabaster, AL, USA). For the lipid mixtures with higher T_m (SM system) tip sonication with a 50% duty cycle of 10 min was used (Branson sonifier, Emerson, Ferguson, MO, USA). Then, the residual material was removed by centrifugation at 10,000× g for 10 min. After that, the vesicles size, in a range of 100–130 nm, was determined by Zetasizer Nano ZS (Malvern Instrument, Worcestershire, UK). The vesicle solutions were stored at a temperature higher than T_m and were used within a week.

4.4. Supported Erythrocyte Cell Membrane Preparation

Sheep blood (Oxoid, Thermo scientific, Waltham, MA, USA) was removed by washing with PBS pH 7.4 three times. The sheep blood was gentle mixed with PBS in a ratio of 1:7. Then, the erythrocytes were collected by centrifugation at 3000× g and 4 °C for 5 min. The erythrocyte pellet was kept and resuspended in PBS pH 7.4, 2% (V/V), as a working solution.

The erythrocyte membrane was prepared on a poly-lysine coated glass. The round-glass cover slips were cleaned as follows: soaking in 1.0 M hydrochloric acid for 2 h, rinsing thoroughly with ultrapure water (MilliQ, Merck, Darmstadt, Germany), sonication in 70% (v/v) ethanol for 10 min, and final treatment with plasma cleaner (Diener electronic, Ebhausen, Germany). Prior cell attachment, the glass cover slips were coated with 30–70 kDa poly L-lysine (Sigma, Darmstadt, Germany). The glass slips were immersed in a 0.1 mg/mL lysine solution (in PBS) for 30 min at room temperature. The excess of lysine was removed by buffer rinsing. After that, the erythrocytes were attached on the glass surface by incubation over the surface for 30 min at room temperature. The unbound erythrocytes were removed and the attached cells were opened under shear flow by using a low content salt solution (1/3 dilution PBS; 45.7 mM NaCl, 0.9 mM KCl and 3.3 mM phosphate). Finally, the cell membrane was rinsed with PBS pH 7.4.

4.5. Quartz Crystal Microbalance with Dissipation (QCM-D) Measurement

The protein-lipid bilayer interaction was evaluated with quartz crystal microbalance with dissipation from Q-Sense E4 (Biolin Scientific, Gothenburg, Sweden) using silica-coated sensors (QSX 303, Biolin Scientific, Sweden). Before use, the sensors were subsequently cleaned as follows: sonication in 2% (*w/w*) SDS solution for 15 min, rinsing with ultra-pure water, drying under nitrogen stream, and organic residues-eliminating with UV/Ozone cleaner (Bioforce Nanosciences, Salt Lake City, UT, USA) for 30 min. The frequencies of the sensors were evaluated prior to running the experiments. The outcome of the experiments delivers changes in frequency and dissipation. The change in frequency (ΔF) is proportional to changes in the adsorbed mass (Δm) on the crystal surface through the Sauerbrey equation:

$$\Delta m = -\frac{C}{n}\Delta F \tag{2}$$

where (C) is the sensitivity constant (-17.7 ng cm^{-2} Hz^{-1}) and (n) is the overtone number. Simultaneously, the change in dissipation (ΔD) indicates the viscoelastic properties of the new forming layer on the crystal surface (in our case, of the hybrid protein-bilayer system). Low dissipation values are typical for a rigid (elastic) layer whereas high values relate to softer (viscoelastic) layers. The changes in frequency (ΔF) and dissipation (ΔD) values are presented for the 5th overtone unless otherwise stated.

The lipid bilayers were formed by the lipid vesicle fusion method. After a stable baseline with PBS solution was achieved, 0.1 mg/mL lipid vesicle solutions were slowly flowed in the QCM-D chamber with a flow rate of 50 µl/min. Once the characteristic patterns (for the frequency and dissipation) of lipid bilayer formation were observed, the excess of vesicles was removed by buffer rinsing. Some of the lipid bilayers were completely formed by additional water rinsing (through osmotic stress). Finally, all lipid bilayers were incubated under PBS flow until reaching a stable baseline.

To study the interaction of the toxin with the model lipid bilayers, the different Cyt2Aa2 toxin solutions were introduced into the system at a flow rate of 50 µL/min. After that, the flow was stopped in order to evaluate when the protein-lipid binding could reach a saturated state. Furthermore, the unbound protein was flushed from the chamber with PBS solution at a flow rate of 50 µL/min for 30 min. The experiments were carried out with at least three replications at 25 °C (298 K). The protein-lipid binding kinetics was determined by curve fitting. The frequency (ΔF) vs. time plots were fitted with a single exponential decay equation (Equation (1)). All the plots and data fitting were carried out with Origin 8.0 (OriginLab Corporation, Northampton, MA, USA).

4.6. Atomic Force Microscopy (AFM) Imaging

The AFM cantilevers with a nominal spring constant of 0.24 N/m (DNP-S10, Bruker, Billerica, MA, USA) and the silica substrate were mounted inside a closed fluid cell with an O-ring. The 1 cm × 1 cm silica wafers (IMEC, Leuven, Belgium) were cleaned before using with the following procedure: sonication in 2% (*w/w*) SDS solution for 15 min, rinsing with ultrapure water, and drying under nitrogen stream. Finally, the substrates were treated by plasma cleaner (Diener electronic, Ebhausen, Germany). The lipid bilayers were formed by means of lipid vesicle fusion. 0.1 mg/mL of lipid vesicle solutions were incubated over the silica surface for at least 10 min and then the vesicle excess was rinsed from the chamber. Afterwards, the two Cyt2Aa2 proteins, wild type (WT) and the T144A mutant (25 µg/mL or 1.0 µM), were incubated with the lipid bilayers or with supported erythrocyte membrane for the desired experimental time. The surface topography was imaged in tapping mode with a JV-scanner controlled by a NanoScope V controller (Bruker, Billerica, MA, USA) at a scan rate of 1–2 Hz. The images were processed and analyzed with the Nanoscope program. The experiments were carried out at room temperature (298 K).

Supplementary Materials: The following are available online at http://www.mdpi.com/2072-6651/12/4/226/s1, Figure S1: Three dimensional structure of the Cyt2A toxin (PDB 1CBY). Figure S2: Curve fitting of the frequency vs. time plots of the binding between the Cyt2Aa2 toxins and the model lipid bilayers. Figure S3: AFM images of the lipid bilayers. Figure S4: Time sequence AFM imaging of the binding of the Cyt2Aa2 toxins on lipid bilayers. Figure S5: AFM height images and profile analysis of hybrid Cyt2Aa2-1:1 SM/DOPC bilayers. Figure S6: Cell shape of sheep erythrocytes under light microscopy.

Author Contributions: Conceptualization, S.T. and J.L.T.-H.; methodology, S.T.; validation, J.L.T.-H.; formal analysis, S.T.; investigation, S.T. and J.L.T.-H.; resources, J.L.T.-H. and B.P.; data curation, S.T. and J.L.T.-H.; writing—original draft preparation, S.T.; writing—review & editing, J.L.T.-H. and B.P.; visualization, S.T.; supervision, J.L.T.-H.; project administration, J.L.T.-H.; funding acquisition, J.L.T.-H. All authors have read and agreed to the published version of the manuscript.

Funding: This research was funded by The Austrian Science Fund (FWF), grant number P29562-N28 and the APC was funded by BOKU Vienna Open Access Publishing Fund.

Acknowledgments: The authors thank Jacqueline Friedmann for technical support.

Conflicts of Interest: There are no conflicts to declare.

References

1. Hofte, H.; Whiteley, H.R. Insecticidal crystal proteins of *Bacillus thuringiensis*. *Microbiol. Rev.* **1989**, *53*, 242–255. [CrossRef] [PubMed]

2. Angsuthanasombat, C.; Crickmore, N.; Ellar, D.J. Effects on toxicity of eliminating a cleavage site in a predicted interhelical loop in *Bacillus thuringiensis* CryIVB delta-endotoxin. *FEMS Microbiol. Lett.* **1993**, *111*, 255–261. [PubMed]

3. Crickmore, N.; Bone, E.J.; Williams, J.A.; Ellar, D.J. Contribution of the individual components of the [delta]-endotoxin crystal to the mosquitocidal activity of *Bacillus thuringiensis* subsp. *israelensis*. *FEMS Microbiol. Lett.* **1995**, *131*, 249–254. [CrossRef]

4. Al-yahyaee, S.A.S.; Ellar, D.J. Maximal toxicity of cloned CytA δ-endotoxin from *Bacillus thuringiensis* subsp. *israelensis* requires proteolytic processing from both the N- and C-termini. *Microbiology* **1995**, *141*, 3141–3148. [CrossRef]

5. Butko, P. Cytolytic toxin Cyt1A and its mechanism of membrane damage: Data and hypotheses. *Appl. Environ. Microbiol.* **2003**, *69*, 2415–2422. [CrossRef]

6. Pardo-Lopez, L.; Soberon, M.; Bravo, A. *Bacillus thuringiensis* insecticidal three-domain Cry toxins: Mode of action, insect resistance and consequences for crop protection. *FEMS Microbiol. Rev.* **2013**, *37*, 3–22. [CrossRef]

7. Thomas, W.E.; Ellar, D.J. *Bacillus thuringiensis* var *israelensis* crystal delta-endotoxin: Effects on insect and mammalian cells in vitro and in vivo. *J. Cell Sci.* **1983**, *60*, 181–197.

8. Gill, S.S.; Cowles, E.A.; Pietrantonio, P.V. The mode of action of *Bacillus thuringiensis* endotoxins. *Annu. Rev. Entomol.* **1992**, *37*, 615–636. [CrossRef]

9. Pigott, C.R.; Ellar, D.J. Role of receptors in *Bacillus thuringiensis* crystal toxin activity. *Microbiol. Mol. Biol. Rev.* **2007**, *71*, 255–281. [CrossRef]

10. Thomas, W.E.; Ellar, D.J. Mechanism of action of *Bacillus thuringiensis* var *israelensis* insecticidal delta-endotoxin. *FEBS Lett.* **1983**, *154*, 362–368. [CrossRef]

11. Ben-Dov, E. *Bacillus thuringiensis* subsp. *israelensis* and its dipteran-specific toxins. *Toxins (Basel)* **2014**, *6*, 1222–1243. [CrossRef] [PubMed]

12. Melo, A.L.; Soccol, V.T.; Soccol, C.R. *Bacillus thuringiensis*: Mechanism of action, resistance, and new applications: A review. *Crit. Rev. Biotechnol.* **2016**, *36*, 317–326. [CrossRef] [PubMed]

13. Gomez, I.; Pardo-Lopez, L.; Munoz-Garay, C.; Fernandez, L.E.; Perez, C.; Sanchez, J.; Soberon, M.; Bravo, A. Role of receptor interaction in the mode of action of insecticidal Cry and Cyt toxins produced by *Bacillus thuringiensis*. *Peptides* **2007**, *28*, 169–173. [CrossRef] [PubMed]

14. Promdonkoy, B.; Chewawiwat, N.; Tanapongpipat, S.; Luxananil, P.; Panyim, S. Cloning and characterization of a cytolytic and mosquito larvicidal delta-endotoxin from *Bacillus thuringiensis* subsp. *darmstadiensis*. *Curr. Microbiol.* **2003**, *46*, 94–98. [CrossRef]

15. Cahan, R.; Friman, H.; Nitzan, Y. Antibacterial activity of Cyt1Aa from *Bacillus thuringiensis* subsp. *israelensis*. *Microbiology* **2008**, *154*, 3529–3536. [CrossRef]

16. Knowles, B.H.; Blatt, M.R.; Tester, M.; Horsnell, J.M.; Carroll, J.; Menestrina, G.; Ellar, D.J. A cytolytic delta-endotoxin from *Bacillus thuringiensis* var. *israelensis* forms cation-selective channels in planar lipid bilayers. *FEBS Lett.* **1989**, *244*, 259–262. [CrossRef]

17. Promdonkoy, B.; Ellar, D.J. Investigation of the pore-forming mechanism of a cytolytic delta-endotoxin from *Bacillus thuringiensis*. *Biochem. J.* **2003**, *374*, 255–259. [CrossRef]

18. Butko, P.; Huang, F.; Pusztai-Carey, M.; Surewicz, W.K. Membrane permeabilization induced by cytolytic delta-endotoxin CytA from *Bacillus thuringiensis* var. *israelensis*. *Biochemistry* **1996**, *35*, 11355–11360. [CrossRef]

19. Manceva, S.D.; Pusztai-Carey, M.; Russo, P.S.; Butko, P. A detergent-like mechanism of action of the cytolytic toxin Cyt1A from *Bacillus thuringiensis* var. *israelensis*. *Biochemistry* **2005**, *44*, 589–597. [CrossRef]

20. Tharad, S.; Toca-Herrera, J.L.; Promdonkoy, B.; Krittanai, C. *Bacillus thuringiensis* Cyt2Aa2 toxin disrupts cell membranes by forming large protein aggregates. *Biosci. Rep.* **2016**, *36*. [CrossRef]

21. Promdonkoy, B.; Rungrod, A.; Promdonkoy, P.; Pathaichindachote, W.; Krittanai, C.; Panyim, S. Amino acid substitutions in alphaA and alphaC of Cyt2Aa2 alter hemolytic activity and mosquito-larvicidal specificity. *J. Biotechnol.* **2008**, *133*, 287–293. [CrossRef] [PubMed]

22. Promdonkoy, B.; Pathaichindachote, W.; Krittanai, C.; Audtho, M.; Chewawiwat, N.; Panyim, S. Trp132, Trp154, and Trp157 are essential for folding and activity of a Cyt toxin from *Bacillus thuringiensis*. *Biochem. Biophys. Res. Commun.* **2004**, *317*, 744–748. [CrossRef] [PubMed]

23. Pathaichindachote, W.; Rungrod, A.; Audtho, M.; Soonsanga, S.; Krittanai, C.; Promdonkoy, B. Isoleucine at position 150 of Cyt2Aa toxin from *Bacillus thuringiensis* plays an important role during membrane binding and oligomerization. *BMB Rep.* **2013**, *46*, 175–180. [CrossRef] [PubMed]

24. Suktham, K.; Pathaichindachote, W.; Promdonkoy, B.; Krittanai, C. Essential role of amino acids in alphaD-beta4 loop of a *Bacillus thuringiensis* Cyt2Aa2 toxin in binding and complex formation on lipid membrane. *Toxicon* **2013**, *74*, 130–137. [CrossRef] [PubMed]

25. van Meer, G.; Voelker, D.R.; Feigenson, G.W. Membrane lipids: Where they are and how they behave. *Nat. Rev. Mol. Cell Biol.* **2008**, *9*, 112–124. [CrossRef] [PubMed]

26. Leidl, K.; Liebisch, G.; Richter, D.; Schmitz, G. Mass spectrometric analysis of lipid species of human circulating blood cells. *Biochim. Biophys. Acta* **2008**, *1781*, 655–664. [CrossRef]

27. Stewart, S.E.; Bird, C.H.; Tabor, R.F.; D'Angelo, M.E.; Piantavigna, S.; Whisstock, J.C.; Trapani, J.A.; Martin, L.L.; Bird, P.I. Analysis of Perforin Assembly by Quartz Crystal Microbalance Reveals a Role for Cholesterol and Calcium-independent Membrane Binding. *J. Biol. Chem.* **2015**, *290*, 31101–31112. [CrossRef]

28. Rabinovich, A.L.; Kornilov, V.V.; Balabaev, N.K.; Leermakers, F.A.M.; Filippov, A.V. Properties of unsaturated phospholipid bilayers: Effect of cholesterol. *Biochem. (Moscow) Suppl. Ser. A Membr. Cell Biol.* **2007**, *1*, 343–357. [CrossRef]

29. Marheineke, K.; Grunewald, S.; Christie, W.; Reilander, H. Lipid composition of *Spodoptera frugiperda* (Sf9) and *Trichoplusia ni* (Tn) insect cells used for baculovirus infection. *FEBS Lett.* **1998**, *441*, 49–52. [CrossRef]

30. Giocondi, M.-C.; Seantier, B.; Dosset, P.; Milhiet, P.-E.; Le Grimellec, C. Characterizing the interactions between GPI-anchored alkaline phosphatases and membrane domains by AFM. *Pflügers Arch. -Eur. J. Physiol.* **2008**, *456*, 179–188. [CrossRef]

31. Crespo-Villanueva, A.; Gumi-Audenis, B.; Sanz, F.; Artzner, F.; Meriadec, C.; Rousseau, F.; Lopez, C.; Giannotti, M.I.; Guyomarc'h, F. Casein interaction with lipid membranes: Are the phase state or charge density of the phospholipids affecting protein adsorption? *Biochim. Biophys. Acta (BBA)-Biomembr.* **2018**, *1860*, 2588–2598. [CrossRef] [PubMed]

32. Tian, Y.; Cai, M.; Xu, H.; Wang, H. Studying the membrane structure of chicken erythrocytes by in situ atomic force microscopy. *Anal. Methods* **2014**, *6*, 8115–8119. [CrossRef]

33. Picas, L.; Rico, F.; Deforet, M.; Scheuring, S. Structural and mechanical heterogeneity of the erythrocyte membrane reveals hallmarks of membrane stability. *ACS Nano* **2013**, *7*, 1054–1063. [CrossRef] [PubMed]

Article

Structural and Functional Insights into the C-terminal Fragment of Insecticidal Vip3A Toxin of *Bacillus thuringiensis*

Kun Jiang [1,†], Yan Zhang [1,†], Zhe Chen [1,†], Dalei Wu [1,2], Jun Cai [3] and Xiang Gao [1,*]

[1] State Key Laboratory of Microbial Technology, Shandong University, Qingdao 266237, China; jiangkun@sdu.edu.cn (K.J.); yanzhang1991@sdu.edu.cn (Y.Z.); zhechen@mail.sdu.edu.cn (Z.C.); dlwu@sdu.edu.cn (D.W.)

[2] Helmholtz International Lab, Shandong University, Qingdao 266237, China

[3] Department of Microbiology, College of Life Sciences, Nankai University, Tianjin 300071, China; caijun@nankai.edu.cn

* Correspondence: xgao@email.sdu.edu.cn

† These authors contributed equally to this work.

Received: 10 June 2020; Accepted: 3 July 2020; Published: 5 July 2020

Abstract: The vegetative insecticidal proteins (Vips) secreted by *Bacillus thuringiensis* are regarded as the new generation of insecticidal toxins because they have different insecticidal properties compared with commonly applied insecticidal crystal proteins (Cry toxins). Vip3A toxin, representing the vast majority of Vips, has been used commercially in transgenic crops and bio-insecticides. However, the lack of both structural information on Vip3A and a clear understanding of its insecticidal mechanism at the molecular level limits its further development and broader application. Here we present the first crystal structure of the C-terminal fragment of Vip3A toxin (Vip3Aa11$_{200–789}$). Since all members of this insecticidal protein family are highly conserved, the structure of Vip3A provides unique insight into the general domain architecture and protein fold of the Vip3A family of insecticidal toxins. Our structural analysis reveals a four-domain organization, featuring a potential membrane insertion region, a receptor binding domain, and two potential glycan binding domains of Vip3A. In addition, cytotoxicity assays and insect bioassays show that the purified C-terminal fragment of Vip3Aa toxin alone have no insecticidal activity. Taken together, these findings provide insights into the mode of action of the Vip3A family of insecticidal toxins and will boost the development of Vip3A into more efficient bio-insecticides.

Keywords: *Bacillus thuringiensis*; Vip3A; 3D-structure; mode of action; biological control

Key Contribution: Here we showed the first atomic structure of the C-terminal fragment of Vip3A toxin. Our study revealed the general domain organization and the potential function of each domain of C-terminal Vip3A family toxin. It also showed the interesting convergent evolution between Vip3A toxin and Cry toxin.

1. Introduction

The entomopathogenic bacteria *Bacillus thuringiensis* (Bt), is the most widely used microbial insecticide in the world [1,2]. It is renowned for its ability to produce insecticidal crystal proteins (Cry toxins) during its sporulation phase, which have been widely used in the prevention and control of agricultural pests through the development of transgenic plants or Bt-based biopesticides [3–5]. However, many pests are not sensitive to Cry toxins, and the development of insect resistance has also been reported [1,6,7]. The successful application of Cry proteins, coupled with their limitations, has spurred on intensive research seeking to identify and characterize novel classes of insecticidal toxins that can be developed for agricultural purposes.

Vegetative insecticidal proteins (Vips), which are produced by Bt during its vegetative stages, have a wide spectrum of insecticidal activity, especially against lepidopteran pests [8]. To date, ~150 distinct Vip toxins have been identified, which have been classified into four families (Vip1, Vip2, Vip3 and Vip4) based on their sequence similarity [9]. Among the Vip toxin family, Vip3A toxins are the most abundant and most studied [8]. Compared with known Cry toxins, Vip3A toxins share no sequence homology, bind to different receptors [10–13], and lack cross-resistance [14–17], therefore they are considered as ideal options to complement and expand the use of Bt in crop protection and resistance management. At present, the Vip3Aa toxin is the only family member that has been used in commercial transgenic crops together with Cry toxins, and no field-evolved resistance has yet been reported [1,8,18]. However, the lack of structural information and incomplete understanding of its mechanisms of action have severely limited the further development of Vip3A as a tool in pest control.

Vip3A toxins are large proteins (~789 amino acids) consisting of a conserved N-terminus and a variable C-terminal region. The ~88kDa Vip3A protoxin could be digested by insect midgut juices into two fragments: a ~20 kDa fragment corresponding to the N-terminal 198 amino acids, and a ~65 kDa fragment corresponding to the C-terminal fragment of Vip3A protein, which is regarded as an essential step for its activation and toxicity [12,19–23]. Since their discovery in 1996 [24], Vip3A proteins have been the subject of intensive research. It has been reported that Vip3A stimulates membrane pore formation and apoptosis upon binding to target cells, which is proposed to be responsible for its cytotoxic effects [12,25–27]. The scavenger receptor class C like protein (Sf-SR-C) and the fibroblast growth factor receptor (Fgfr) have been reported as potential receptors for Vip3A [10,11]. Vip3Aa16 and Vip3Af1 have been subjected to in silico modelling, and three domains and five domains were proposed respectively [28,29]. Quan et al. propose a map of Vip3Af protein with five domains based on the altered protease digestion patterns through the Vip3Af alanine mutants [23]. In addition, Vip3Ag protoxin and the trypsin-activated toxin were found to be a potential tetrameric complex according to the surface topology obtained by transmission electron microscopy [20]. Recently, Zheng et al. reported the crystal structure of a Vip3B protoxin like protein: Vip3B2160 [30], which shares around 60% sequence identity to Vip3A. The overall structure of Vip3B2160 shows a five-domain organization and forms a novel tetramer structure assembly. However, the atomic structure of Vip3A is still not available, which makes it difficult to reveal the relationship between its structure and function accurately.

Here, we report the crystal structure of the C-terminal fragment of Vip3A toxin (Vip3Aa11$_{200-789}$). The structure shows a four-domain organization which is likely to be conserved for all insecticidal Vip3A family toxins. We identify conserved hydrophobic α-helices in domain II, which we predict to be involved in the membrane insertion process. Structure-guided cell binding assays reveal that domain III may have a central role in host cell targeting and binding of Vip3A toxins. Structural analysis indicates that Vip3A toxins have potential for glycan binding through domains IV and V. Together, our structural and functional studies provide new insights into the molecular mechanisms underlying the mode of action of insecticidal Vip3A toxins.

2. Results

2.1. Overall Structure of Vip3Aa11$_{200-end}$

We used a Vip3A toxin from Bt strain C9, which has been named Vip3Aa11 (GenBank accession No. AY489126.1) in this study. Full-length Vip3Aa11 consists of 789 amino acids, which have been demonstrated to be digested between residues K198 and D199 by insect midgut juice [19–21]. We initiated our crystallization trial with both Vip3Aa11 protoxin and Vip3Aa11$_{199-end}$. Using spare matrix crystallization screening, we only identified one condition that yielded needle-shaped crystals of Vip3Aa11 protoxin. However, the crystals diffracted to only ~15 Å and could not be improved despite extensive effort. No crystals were observed for the Vip3Aa$_{199-end}$ construct despite screening more than

1000 crystallization conditions. However, when we deleted the N-terminal amino acid (Asp199) from the Vip3Aa11$_{199-end}$, we obtained the crystal of Vip3Aa11$_{200-end}$, which diffracted to ~6 Å. Through the addition of an N-terminal MBP (Maltose Bind Protein) tag, we were able to isolate crystals with improved diffraction. The structure was solved using a combination of anomalous phasing with a selenomethionine derivative crystal of Vip3Aa$_{200-end}$ and molecular replacement using MBP as a model in native crystals. The final structure of Vip3Aa11$_{200-end}$ was refined to a 3.2 Å resolution with R and R$_{free}$ values of 0.1980 and 0.2389, respectively (Table S1).

The structure shows that the Vip3Aa11 C-terminal fragment is comprised of four domains (Figure 1A,B). Vip3Aa11 could be digested between residues K198 and D199, and residues 1–198 are lacking in our structure. For a better description based on the full-length Vip3A, we assume that Vip3Aa11$_{1-198}$ is a separate domain. Then the protoxin can be divided into five domains, starting from N-terminus: domain I, 1–198; domain II, 199–327; domain III, 328–518; domain IV, 537–667; and domain V, 679–789 in Vip3Aa11 (Figure 1A and Figure S1). The overall structure of the Vip3Aa11$_{200-end}$ resembles a lobster, wherein domains II and III form the body, and domains IV and V are the claws (Figure 1B,C). The connection between domain II and domain III is compact. However, domains III/IV and IV/V are connected by long and flexible loops, which indicates that the relative locations and orientations of these two domains could change under different biological circumstances. There are over 100 known proteins of the Vip3A family. Based on their high degree of sequence conservation and previous studies [8], they are very likely to share similar overall structures and domain compositions.

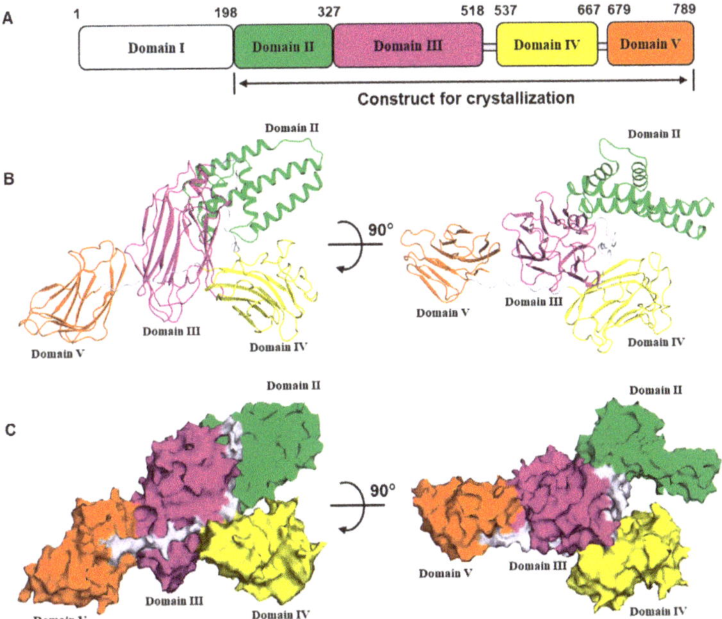

Figure 1. Overall structure of vegetative insecticidal protein Vip3Aa11$_{200-end}$. (**A**) Domain organization of Vip3A. (**B**) Two views of the overall structure of Vip3Aa11$_{200-end}$ monomer colored as in (**A**). (**C**) Two views of the surface model of Vip3Aa11$_{200-end}$ monomer colored as in (**A**). The black arrow indicates the angle of rotation around the central axis.

The crystal belongs to the P2$_1$ space group and four MBP-Vip3Aa11 molecules were found in one asymmetric unit (Figure S2). These four molecules form two dimers in different orientations. PISA [31] determined that there was limited interaction between the two dimers, indicating that their association was caused by crystal packing. Notably, the two Vip3Aa11 molecules in the "dimer" showed moderate conformational variations, with a core root mean square deviation (r.m.s.d) of 1.234 Å among 468 Cα atoms (Figure S3). Superimposition of separate domains between the two molecules revealed better alignment for domains III, IV and V, but not for domain II (Figure S3), suggesting that domain II might potentially be involved in the conformational changes during the activation of Vip3A toxins. Due to their high similarity, we used the monomeric structure of Vip3Aa11$_{200\text{–end}}$ for subsequent analysis.

2.2. Domain II Contains a Conserved Hydrophobic Architecture

Domain II of Vip3Aa11 (residues199–327) consists of five helices, which form two layers (Figure 2A). The outer layer facing the solution contains two short helices, α2 and α3, while the inner layer that contacts with domain III consists of three anti-parallel helices α1, α4 and α5. The outer layer contacts with the upper portion of the inner layer and is almost perpendicular to the inner layer. Among these five helices, helix α4 is the longest. It spans around 45 Å and contains 30 amino acid residues, starting from E267 at the N-terminus to L296 at the C-terminus. Electrostatic surface potentials analysis shows that the majority of charged and polar amino acid residues locate at the N-terminal and C-terminal ends of helix α4 (Figure 2B). For the middle portion of helix α4, from F274 to L289, 75% amino acid residues are hydrophobic residues. Sequence alignment through Vip3 family shows that the hydrophobic region of helix α4 is very much conserved and it is also the most agminated hydrophobic region of Vip3 family proteins (Figure S1). Close to helix α4, helix α1 also shows several conserved hydrophobic amino acid residues facing helix α4 (Figures S1 and S4).

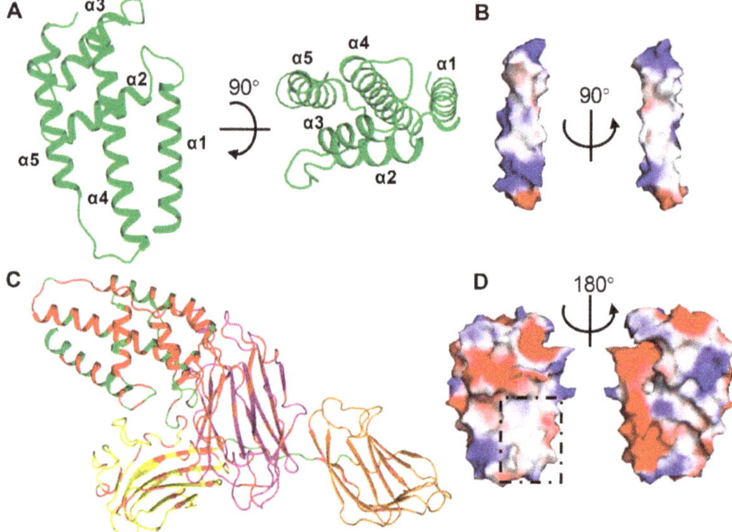

Figure 2. Domain II of Vip3Aa11 shows a conserved hydrophobic surface. (**A**) Two views of structure of Vip3Aa11 domain II shown as a ribbon cartoon. (**B**) Two views of the surface model of helix α4 from domain II show its surface charge distribution. (**C**) The highly conserved amino acid residues from Vip3 family sequence alignment (Figure S1) are highlighted in the Vip3Aa structure with red color. (**D**) Two views of the surface model of Vip3Aa11 domain II show its surface charge distribution. The conserved hydrophobic surface is highlighted by black square. (**B,D**) The surface is colored as the basis of electrostatic potential with positive charged surface in blue and negatively charged area in red. The black arrow indicates the angle of rotation around the central axis.

Based on the sequence alignment (Figure S1), all the conserved amino acid residues were highlighted on the Vip3Aa11$_{200-end}$ structure (Figure 2C). This shows that the sequence of domain II is highly conserved (about 62%), only slightly lower than that of domain I (about 68%). Electrostatic surface potential analysis shows that there is an obvious hydrophobic surface, which is mainly contributed to by the conserved helix α1 and α4 (Figure 2D).

2.3. Domain III Is Involved in Cell Binding of Vip3A Toxin

Domain III of Vip3Aa11 (residues 328–518) consists of twelve β strands and one short α-helix at the C-terminal end (Figure 3A). Twelve β strands comprise three antiparallel β sheets sharing a similar "Greek-key" topology (Figure 3A) with a hydrophobic center featuring highly conserved residues V349, F360, I362 and L370 from β sheet I, I425 and F427 from β sheet II, and I481, F492 and L505 from β sheet III (Figure 3B). The results from the DALI server [32] showed that the fold of domain III is similar to that of domain II of the three domain Cry (3d-Cry) family of insecticidal toxins, which has been shown to be involved in host cell receptor recognition and binding [4]. We therefore sought to explore whether domain III serves as a receptor binding domain for Vip3A toxins.

To explore this hypothesis, we used *Spodoptera frugiperda* cells (Sf9 cells), which have previously been shown to be specifically targeted by Vip3 toxins [11,33]. To determine which domain(s) of Vip3Aa11 interact with Sf9 cells, we carried out fluorescence-based cell binding assays using different C-terminal RFP-tagged Vip3Aa truncation derivatives (shown schematically in Figure 3C). As shown in Figure 3D,E and Figure S5, while domain IV-V does not show detectable binding to Sf9 cells, the binding of a construct featuring only domain I-III or II-III to Sf9 cells is indistinguishable from that of full-length Vip3Aa. The interaction of domain III alone with Sf9 cells is significantly stronger than that of the domain I-II construct, indicating that domain III may have a central role in Vip3A receptor binding to Sf9 cells. In addition, Domain II-III shows higher binding than Domain III alone to Sf9 cells, and structural analysis shows that domain II and domain III have close interaction, suggesting that the presence of domain II is also important for cell binding.

2.4. Domains IV and V Are Glycan Binding Motifs

Both domains IV and V are all β-sheets folds (Figure 4A,B). Unlike domains II and III, which have compact organization, domains III/IV and IV/V are connected by long and flexible loops (Figure 4A). In addition to these loops, there are several polar interactions between domains IV/V and domain III, that reduce the flexibility and fix domains IV and V at the observed positions and orientations (Figure 4A).

Domains IV and V are both built from two anti-parallel sheets of β sandwich, forming the "jelly-roll" topology. Despite showing only 17% sequence identity (Figure 4C), domains IV and V align very well structurally, with a root-mean-squared deviation (r.m.s.d) of 1.299 Å over 61 Cα atoms (Figure 4D). To examine the potential function of these two domains, we searched for their structural homologues using the DALI server [32]. The results for both domains show a very high similarity (Z score > 10) to family 16 carbohydrate binding module (CBM16) of S-Layer associated multidomain endoglucanase (RCSB ID 2ZEY). Superimposition of domains IV, V and CBM16 demonstrates that these three motifs share a similar fold (Figure 4E), suggesting that they are likely to share a related function as well. CBM16 is a carbohydrate-binding domain of the highly active mannanase from the thermophile *Thermoanaerobacterium polysaccharolyticum* with high specificity toward β-1,4-glucose or β-1,4-mannose polymers [34]. Analysis of the electrostatic surface potential shows that both domains IV and V have a surface pocket at a similar position to a sugar-binding pocket of the CBM16 domain (RCSB ID 2ZEY), although all three pockets have different shapes and charge distributions (Figure 4F). Taken together, our structural analysis indicates that domains IV and V of Vip3A both contain a conserved glycan binding motif and that these motifs may target different sugars.

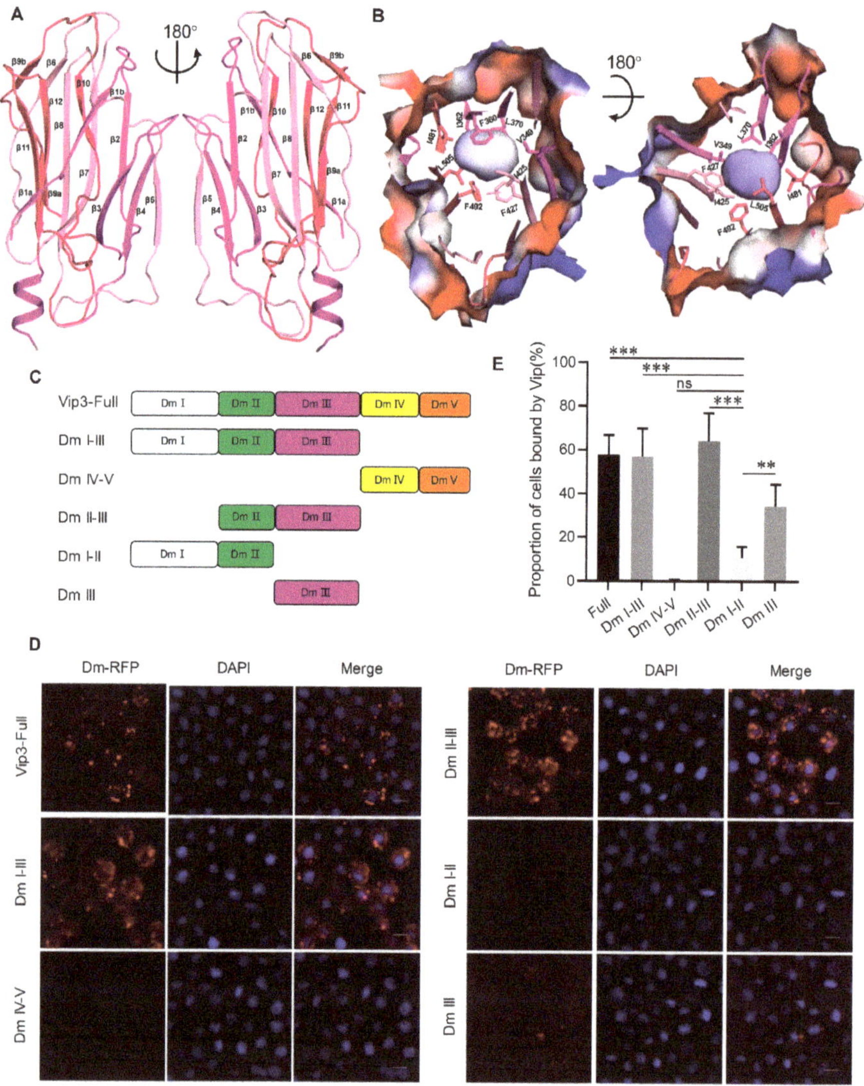

Figure 3. Domain III is a potential receptor binding domain. (**A**) Overall structure of Vip3Aa11 domain III shown as a ribbon cartoon. Two views of three antiparallel β sheets from domain III are shown in three different colors, the black arrow indicates the angle of rotation around the central axis. (**B**) Two views of the surface model of domain III of Vip3Aa11. Inside the domain III, there is a conserved hydrophobic core, and the conserved hydrophobic amino acid residues from three antiparallel β sheets are shown as sticks, the black arrow indicates the angle of rotation around the central axis. (**C**) The schematics of C-terminal RFP (red fluorescent protein)-tagged Vip3Aa and its truncation derivatives. (**D**) Fluorescence microscope images of Sf9 cells treated with Vip3Aa-RFP or its truncations, which were labeled with C-terminal RFP tag, for 6 h. Nuclei are stained with DAPI (blue). (**E**) Quantification of the number of Sf9 cells that can be bound by RFP-tagged Vip3Aa and its truncations of Figure 3D in a blind fashion (*n* = 100 cells per sample). Data are expressed as the mean ± SD from three independent experiments. ns, nonsignificant; **, *p* < 0.01; ***, *p* < 0.001. Scale bar: 10 μm.

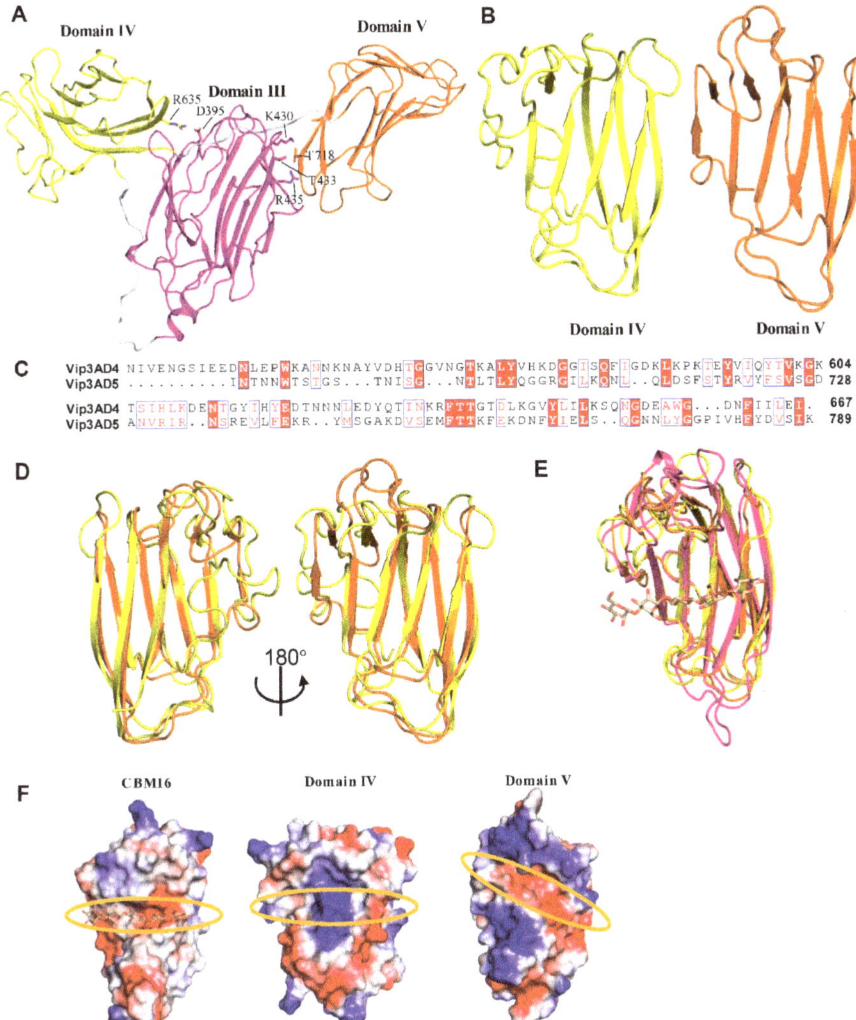

Figure 4. Domains IV and V of Vip3Aa11 have glycan binding motifs. (**A**) Domain architectures of domains III, IV and V of Vip3Aa11. The polar interactions between domain IV, V and domain III are shown as sticks. (**B**) Overall structure of Vip3Aa11 domain IV and V shown as a ribbon cartoon. (**C**) Amino acid sequence alignment between domain IV and domain V of Vip3Aa11. The identical residues are denoted in white characters and red background, and the similar residues are denoted in red. ClustalX2 was used to perform the sequence alignment [35]. ESPript-3.0 was used to generate the figure [36]. (**D**) Two views of structure superimposition between domain IV and domain V of Vip3Aa11 shown as a ribbon cartoon. Color of each domain is consistent with Figure 4B. (**E**) Structure superimposition between domains IV, V of Vip3Aa11 and glycan bound CBM16 (RCSB ID 2ZEY) shown as a ribbon cartoon. Domains IV and V are colored as Figure 4B, and CBM16 is shown in magenta color. The glycan in CBM16 is shown as stick in light brown color. (**F**) Surface charge distribution of the sugar-binding pocket of CBM16 (RCSB ID 2ZEY) and potential sugar-binding pocket of domains IV, V of Vip3Aa11, highlighted with the orange circle.

2.5. Purified Vip3Aa11$_{200-end}$ Has no Insecticidal Activity

The C-terminal fragment of Vip3A has been considered to be the toxic core [8]. To verify whether the purified Vip3Aa$_{200-end}$ still have insecticidal activity, cytotoxicity assays and insect bioassays were carried out. As shown in Figure 5A, the purified full length Vip3Aa toxin has significant toxicity to Sf9 cells, while Vip3Aa$_{199-end}$, Vip3Aa$_{200-end}$ and MBP-Vip3Aa$_{200-end}$ have no toxicity to Sf9 cells. In addition, bioassay results showed that wild-type Vip3Aa was highly toxic against *S. exigua* at the concentration of 200 ng/cm^2. However, the purified Vip3Aa$_{199-end}$, Vip3Aa$_{200-end}$ and MBP-Vip3Aa$_{200-end}$ have no obvious insecticidal activity to *S. exigua* larvae even at the concentration of 2000 ng/cm^2 (Figure 5B). These results indicate that the purified C-terminal fragment of Vip3A alone has no insecticidal activity.

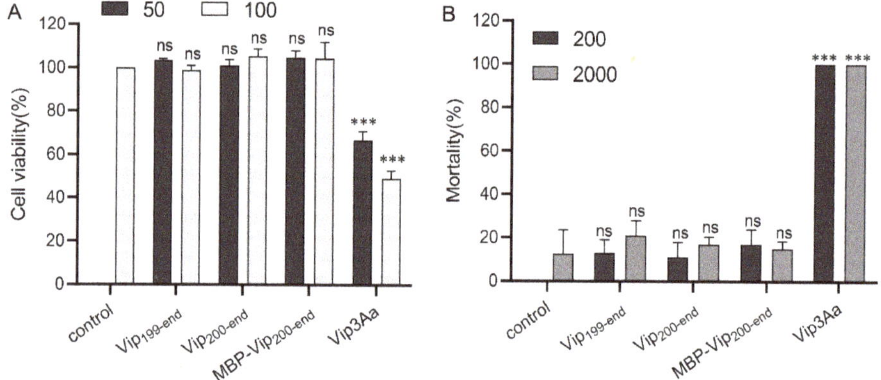

Figure 5. Cytotoxicity assays and insect bioassays of different Vip3A constructs. (**A**) Cell viability of Sf9 treated with different Vip3A constructs (50 and 100 µg/mL). (**B**) Mortality analysis of *S. exigua* caused by different Vip3A constructs (200 and 2000 ng/cm^2). Data are expressed as the mean ± SD from three independent experiments; ns, nonsignificant; *** $p < 0.001$; one-way ANOVA with Dunnett's method.

3. Discussion

Vip3A toxins show a wide spectrum of specific insecticidal activities and are functionally distinct compared to the Cry toxins. These features make them good candidates for combined application with Cry toxins in transgenic crops to broaden the insecticidal spectrum and to prevent or delay resistance [1,8,11]. The structural features and insecticidal mechanisms of Cry toxins have been studied in detail, which has been crucial to their widespread application [2–4,6]. However, despite the fact that Vip3A toxins were identified almost 25 years ago [24], their mode of action remains poorly understood. One of the main reasons for this is the lack of a high-resolution three-dimensional structure, which significantly impedes detailed molecular-level functional and mechanistic studies, and thus limits the development of their insecticidal potential. In this study, we report the first crystal structure of the C-terminal fragment of Vip3A toxin, which provides a badly-needed framework to explore the molecular-level functional details of Vip3A-family toxins.

Although the amino acid sequence similarity between the Vip3A family toxin and the 3d-Cry toxin is very low, our three-dimensional structural analysis showed interesting convergent evolution between these two families. Domain II of Vip3A has an all α-helix fold, including two conserved hydrophobic α-helices. Similarly, domain I of 3d-Cry also has an all α-helix fold and two hydrophobic α-helices, although it has additional α helices surrounding the conserved hydrophobic helices [4,37]. Several studies have reported that domain I of 3d-Cry toxin is involved in its membrane insertion and pore formation processes through its conserved hydrophobic α-helices [4]. This therefore suggests that domain II of Vip3A may also take part in these processes through its conserved hydrophobic α-helices.

Both domain III of Vip3A and domain II of 3d-Cry are comprised of three β sheets with a conserved hydrophobic core. Extensive studies of domain II of 3d-Cry toxins showed that it plays a key role in the recognition of midgut receptors [4]. The results of our cell binding assay indicate that Vip3A domain III is also central to cell binding. Furthermore, the binding ability of domains II-III to Sf9 cells is similar to that of full-length Vip3Aa and stronger than domain III alone, and the Vip3Aa11$_{200-end}$ structure shows that domain II and domain III have very compact interaction, which revealed that domain II is also involved in the binding of Vip3A to sensitive cells.

Domain III of 3d-Cry toxins was predicted to bind glycans with a classic glycan binding motif [38–40]. Based on amino acid sequence analysis, previous studies also predicted that all Vip3A proteins contain a carbohydrate-binding motif (CBM_4_9 superfamily; pfam02018) in the C-terminus (amino acids 536 to 652 in Vip3Aa) [8]. In the present structure, we found that, instead of the single CBM found in Cry toxins, there were two potential different CBM domains in the C-terminus of the Vip3A toxin, forming domains IV and V, respectively. Our structural analysis indicates that the putative glycan-binding pockets of these two domains differ significantly, suggesting that they are likely to have different glycan binding specificities. This multiplicity of CBMs in Vip3A toxins may increase the diversity of their target polysaccharides. However, in our cell binding assay, domains IV-V did not show binding ability to the Sf9 cells (Figure 3D,E), which may be due to the lack of the specific glycans recognized by domain IV-V on the Sf9 cells' surface. The effect of domains IV and V on the toxicity of Vip3A toxins in insect midgut needs further study.

Taken together, we find here that although the overall structure and domain organization are very different between Vip3 toxin and 3d-Cry toxin, these two families are comprised of functionally and structurally related modules that are assembled in different ways, which may expand the insecticidal spectrum of Bt and make Bt more powerful and efficient to target and kill its hosts.

In addition, the ~65 kDa C-terminal fragment of Vip3A used to be considered as the toxic core [8]. However, recent studies indicated that the ~20 kDa N-terminal fragment and the ~65 kDa C-terminal fragment of Vip3A still bind together after digestion, and the N-terminus is required for the stability and toxicity of Vip3A [20,21,41]. Moreover, several studies further demonstrated that Vip3A remains tetrameric after being activated by trypsin or midgut fluid [20,22]. In our work, the C-terminal fragment of Vip3Aa alone has shown no toxicity through cytotoxicity assays and insect bioassays, and it forms a dimer in the crystal structure, which is consistent with the fact that the C-terminal fragment of Vip3B2160 will form a dimer instead of a tetramer without the N-terminal 21-kDa segment [30]. It is possible that, without N-terminal assistance, the C-terminal fragment cannot correctly assemble into an active tetramer; or, maybe without the protection of N-terminal, the C-terminal fragment loses stability and is degraded by protease.

Vip3A and Vip3B share about 65% sequence similarity and have different insecticidal specificity [42], and recently the C-terminal fragment was found to be related to insecticidal specificity of Vip3 [42,43]. Our structure provides a good opportunity to further study the mechanism of insecticidal specificity between Vip3A and Vip3B. The recently reported structure of Vip3B2160 showed a five-domain organization [30] (Figure S6). When these two Vip3 protein structures are superposed, the C-terminal fragment of Vip3B2160 shows similar folds and organization to Vip3Aa11 (Figure 6A). According to our division, the domain I of Vip3B2160, which is lacked in Vip3Aa11$_{200-end}$ structure, formed a unique fold containing five α-helices wrapping around domain II. Domains III, IV and V of Vip3B2160 have similar folds as their counterparts from Vip3Aa11, respectively (Figure 6C–E). Although domain V in the two structures share similar folds, their positions in their respective structures are obviously different, which suggests that the location of domain V is flexible, and this flexibility of domain V may be related to the insecticidal specificity of Vip3 toxins. However, there are dramatic conformational differences in their domain II (Figure 6B); in the Vip3B2160 structure, the highly conserved hydrophobic α-helix (corresponding to the helix α4 in Vip3Aa11$_{200-end}$ domain II) is surrounded by other helices from domains I and II (Figure 6A, Figure S6). In Vip3Aa11$_{200-end}$ structure, the helices α1 and α2 of domain II have significant conformational changes and expose the hydrophobic region in domain II

(Figure 6B). Hence, we hypothesize that the structural difference in domain II between the full-length and cleaved Vip3 proteins may represent the conformational change after the proteolysis of Vip3A toxins inside the insect midgut. In this scenario, once the cleavage site between domain I and II is processed by insect midgut juice, the α-helices of domain II may undergo a dramatic structural shift that enables helix α1 to rotate and form a hairpin-like structure with helix α4. However, a complex structure of ~20 kDa N-terminal fragment and the ~65 kDa C-terminal fragment of Vip3 after protease digestion will be needed to further prove this hypothesis and to further understand the function of the N-terminal fragment for Vip3 insecticidal activity.

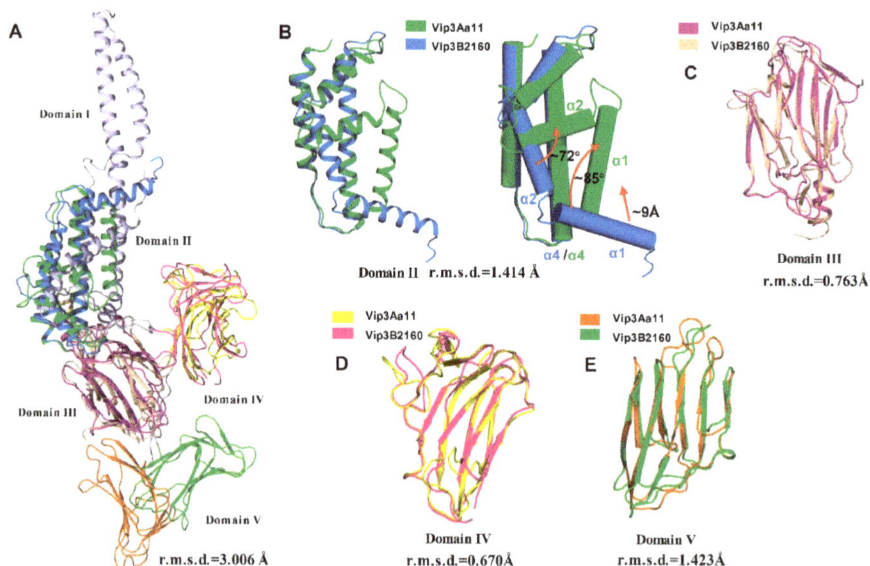

Figure 6. Structural comparation between corresponding domains of Vip3Aa11$_{200-end}$ and Vip3B2160. (**A**) Structural comparation of Vip3Aa11$_{200-end}$ and Vip3B2160. The domain I of Vip3B2160 which is lacking in Vip3Aa11$_{200-end}$ structure is colored in light purple. (**B**) Structural overlay of domain II between Vip3A (green) and Vip3B2160 (blue). The cylindrical cartoon shows a detailed view of the conformational changes. (**C–E**) Structure superimposition for domain III, domain IV and domain V between Vip3Aa11$_{200-end}$ and Vip3B2160. Each domain is color-coded as the indication.

Collectively, these data provide important structural and functional insights into Vip3A family toxins as well as a valuable resource to guide future studies and to re-evaluate the previous genetic and functional studies that are crucial for the development of Vip3A as a new generation of bio-insecticides.

4. Materials and Methods

4.1. Bacterial Strains, Cell Lines and Plasmids

E. coli BL21(DE3) for plasmid constructions and protein purification were cultured at 37 °C in lysogeny broth (LB) or agar. Methionine auxotrophic *E. coli* strain B834 (DE3) (Novagen, Madison, WI, USA) were used for selenomethionine-substituted (SeMet) Vip3Aa$_{200-end}$ expressing. The S. frugiperda Sf9 cells (Thermo fisher Scientific, Grand Island, NY, USA) were maintained and propagated in Sf-900 II SFM (Gibco, Grand Island, NY, USA) culture medium at 27 °C.

The DNA of Vip3Aa$_{200-end}$ was amplified from the Vip3Aa11 gene (GenBank accession No. AY489126.1) using oligonucleotide primer Vip200-F and Vip200-R and cloned into the pET28a vector with an N-terminal 6×His-MBP tag. Plasmids used for RFP (red fluorescent protein) and C-terminal RFP tagged Vip3Aa (Vip3Aa-RFP) expression were constructed as described by Jiang et al. [11]. The different Vip3Aa DNA truncations were amplified from the Vip3Aa11 gene using oligonucleotide primer pairs, DmI-III-F and DmI-III-R, DmIV-V-F and DmIV-V-R, DmI-II-F and DmI-II-R, DmII-III-F and DmII-III-R, and DmIII-F and DmIII-R, and cloned into the pET28a vector with a C-terminal RFP-6×His tag, respectively. All plasmids were generated by the Gibson assembly strategy [44]. The nucleotide sequences of recombinant plasmid were verified by DNA sequencing. All the primers used in this study are shown in Table S2.

4.2. Protein Expression and Purification

Native His-MBP-Vip3Aa$_{200-end}$ (Vip3Aa$_{200-end}$) protein was expressed in *E. coli* B21(DE3) at 25 °C for 48 h in autoinduction Terrific broth (TB) medium. The cells were harvested by centrifugation at 5000× *g* at 4 °C for 15 min and the pellet was resuspended in lysis buffer (20 mM Tris–HCl pH 8 and 150 mM NaCl). After the cells were lysed by high pressure cell crusher (Union-Biotech co., LTD, shanghai, China), the supernatant was collected after centrifuged at 12000× *g* at 4 °C for 60 min. The proteins were purified using Ni-NTA agarose resin, washed with 20 mM Tris-HCl, 150 mM NaCl, 20 mM imidazole, pH 8.0, and then eluted with 300 mM imidazole. The Vip3Aa$_{200-end}$ proteins were further purified by HiTrap Q HP ion-exchange chromatography and Superdex 200 gel filtration chromatography (GE Healthcare Life Sciences, Marlborough, MA, USA). Fractions containing the Vip3Aa$_{200-end}$ protein were concentrated to ~7 mg/mL for crystallization. The expression and purification steps of other Vip3Aa truncations were the same as those of Vip3Aa$_{200-end}$.

SeMet-substituted Vip3Aa$_{200-end}$ was expressed in *E. coli* B834(DE3) strain. Briefly, the cells were cultured in the LB medium at 37 °C along with shaking until the OD600 of the bacterial culture reached 1.0. The cells were harvested by centrifugation at 4000× *g* at 4 °C for 15 min and the pellet was washed once with PBS. The pellet was resuspended in 1 L Medium A (M9 medium plus 0.4% glucose, 1 mM MgCl$_2$, 1 mM CaCl$_2$, 1 mg Biotin, 1 mg thiamin, 50 mg EDTA, 8.3 mg FeCl$_3$, 0.84 mg ZnCl$_2$, 0.13 mg CuCl$_2$, 0.1 mg CoCl$_2$, 0.1 mg H$_3$BO$_3$ and 0.016 mg MgCl$_2$) and incubated for 3 h at 37 °C. We added 50 mg seleno-methionine in the medium and incubated for a further 30 min. The protein was incubated to express for a further 10 h by adding 200 mM IPTG (isopropyl-β-D-thiogalactopyranoside). The SeMet-Vip3Aa$_{200-end}$ was purified by the same procedure as for the native Vip3Aa$_{200-end}$ protein.

4.3. Crystallization, Data Collection and Structural Determination

The purification of His$_6$-tagged MBP-Vip3Aa$_{200-end}$ used for crystallization is described above. MBP-Vip3Aa$_{200-end}$ (5 mg/mL) was used to perform initial spare matrix crystal screening with a crystallization robot. After crystal optimization trials, MBP-Vip3Aa$_{200-end}$ (7 mg/mL) crystals grew in 3 days at 18 °C using the hanging-drop vapor-diffusion method in a mix of 1 μL of protein with 1 μL of reservoir solution consisting of 0.1 M sodium acetate pH 4.2, 0.5 M potassium formate, 0.1 M ammonium sulfate and 11% PEG4000. SeMet MBP-Vip3Aa$_{200-end}$ crystals grew in a similar condition.

A native data set with the space group of P2$_1$2$_1$2$_1$ was collected at 3.62 Å (native I). A weak selenomethionine (SeMet) derivative data set was collected at 3.9 Å with the same symmetry as the native I crystal for the amino-acid assignment using the difference Fourier map of the SeMet derivative. After further crystallization optimization, another native crystal (native II) was obtained with the space group of P2$_1$ that could diffract to around 3.2 Å. Diffraction data were collected on BL17U1 and BL18U beamlines at Shanghai Synchrotron Radiation Facility (Shanghai, China) and processed by HKL2000 [45].

Molecular replacement was carried out to identity the MBP positions in the native crystals [46] by PHASER [47]. The initial phases were further improved with multi-crystal averaging [48]. Model building was performed manually in COOT [49], and the sequence assignment was helped with the SeMet anomalous difference map. Figures were prepared using PyMol (v.2.3.2, https://pymol.org/, Schrödinger, New York, NY, USA). Structure refinement was done by PHENIX [50]. The data collection and refinement statistics are summarized in Table S1.

4.4. Immunofluorescence

Sf9 cells with a density of 5×10^4 cells per ml were seeded into 6-well culture plates separately. After overnight culture, the cells were respectively treated with RFP tagged Vip3Aa or its truncations (0.15 μM) for 6 h. After treatment, the cells were washed three times with PBS to remove unbound proteins, and fixed with 4% paraformaldehyde at 37 °C for 15 min. The cell nuclei were labeled with DAPI (0.2 μg/mL) for 30 min. Cell images were captured using a Nikon TI-E inverted fluorescence Microscope (Nikon, NIKON TI-E, Tokyo Metropolis, Japan).

4.5. Cytotoxicity Assays

Cell viability assays were performed as described by Jiang et al. [25]. Briefly, cells with a density of 5×10^4 cells per ml were seeded into 96-well culture plates separately. After overnight incubation, the cells were treated with different Vip3Aa toxins for 72 h. WST-8 reagent was then added to each well. After incubating at 27 °C for 2 h, the absorbance was measured in microplate reader at 450 nm. Treatment with protein buffer was used as a control. All tests were performed in triplicate and were repeated at least three times. Cell viability (%) = average absorbance of treated group/average absorbance of control group × 100%.

4.6. Bioassay

Bioassays were assessed using surface contamination method with *S. exigua* first instar larvae and maintained in a rearing chamber at 27 °C, with 50% relative humidity, and 16:8 h light:dark photoperiod. The artificial diet was poured in a 1.8-cm^2 24 well plate (about 5 mm thick per hole). 200 and 2000 ng/cm^2 concentrations of Vip3Aa proteins (full length Vip3Aa, $Vip3Aa_{199-end}$ $Vip3Aa_{200-end}$ and MBP-$Vip3Aa_{200-end}$) were spread on the diet. A tris buffer (20 mM Tris-HCl, 300mM NaCl, pH 8.0) was used as a blank control. Three independent replicates and 16 first instar larvae of *S. exigua* were used for each concentration. Mortality was recorded after 5 days.

4.7. Statistical Analysis

All functional assays were performed at least three times independently. Data were shown as means ± SD. All statistical data were calculated using GraphPad Prism version 8.0 (GraphPad Software, San Diego, CA, USA). One-way ANOVA followed by Dunnett's test were used to identify statistically significant differences between treatments. Significance of mean comparison is annotated as follow: ns, nonsignificant; **, $p < 0.01$; ***, $p < 0.001$. A p value of < 0.05 was considered to be statistically significant.

4.8. Data Availability

Coordinate for the atomic structure has been deposited in the RCSB Protein Data Bank under RCSB ID: 6VLS. The data that support the findings of this study are available from the corresponding author upon reasonable request.

Supplementary Materials: The following are available online at http://www.mdpi.com/2072-6651/12/7/438/s1, Figure S1. Sequence alignment of selected Vip3 family members. Each domain is indicated by the lines bellow sequences, colored as in Figure 1A. Secondary structural elements of Vip3Aa11 are shown above the sequences. The conserved hydrophobic amino acid residues discussed in domain II and domain III are marked with green and magenta triangles, respectively. The potential cleavage site between domain III and domain IV is highlighted with blue triangle. ClustalX2 was used for the sequence alignment. ESPript-3.0 was used to generate the figure. Figure S2. Structure of MBP-$Vip3Aa11_{200-end}$ in the $P2_1$ space group. Two views of MBP-$Vip3Aa11_{200-end}$ structure in one asymmetric unit. There are four molecules of MBP-$Vip3Aa11_{200-end}$ in one asymmetric unit and they are arranged into two copies of dimer in the different orientations. The molecule A, B, C and D are shown in green, cyan, magenta and yellow, respectively. The MBP (Maltose Bind Protein) tags are shown in silver color in all four molecules. The interaction area between molecule B and C is less than 500 $Å^2$, as calculated by PISA server. Figure S3. Structural alignment between Molecule A and B from the $Vip3Aa11_{200-end}$ dimer. Structure superimposition for the $Vip3Aa11_{200-end}$ and each domain between molecule A and B from the $Vip3Aa11_{200-end}$ dimer structure. Molecule A is colored as Figure 1A, and Molecule B is shown in cyan color. The root means square deviation (r.m.s.d) of each alignment is listed. Figure S4. Two hydrophobic helices from domain II. The hydrophobic amino acid residues are shown as sticks and labelled with residue numbers. The amino acid residues involved in the hydrophobic (red) and polar (yellow) interactions between α1 and α4 helices are shown as sticks, and the polar interactions are shown in black dashes. Figure S5. Images of Sf9 cells treated with RFP. Fluorescence microscope images of Sf9 cells treated with RFP protein only for 6 h as control. The images are representative of three independent experiments. Nuclei are stained with DAPI (blue), Scale bar, 10 μm. Figure S6. Overall structure of Vip3B2160. Domains I, II, III, IV and V are colored in light purple, blue, light brown, magenta and green, respectively. Table S1. X-ray and refinement statistics. Table S2: Primers used in this study

Author Contributions: Conceptualization, K.J. and X.G.; methodology, K.J., Y.Z., Z.C., D.W., J.C. and X.G.; software, Z.C. and X.G.; validation, K.J., Y.Z., Z.C. and X.G.; formal analysis, K.J., D.W., J.C. and X.G.; investigation, K.J., Y.Z., Z.C. and X.G.; resources, J.C. and X.G.; data curation, X.G.; writing—original draft preparation, K.J., D.W. and X.G.; writing—review and editing, K.J. and X.G.; visualization, K.J., Y.Z., Z.C., D.W. and X.G.; supervision, X.G.; project administration, K.J. and X.G.; funding acquisition, K.J. and X.G. All authors have read and agreed to the published version of the manuscript.

Funding: This work was supported by the National Natural Science Foundation of China (31901943 and 31770143), the Major Basic Program of Natural Science Foundation of Shandong Province (ZR2019ZD21), China Postdoctoral Science Foundation funded project (2019T120585 and 2019M652370), Youth Interdiscipline Innovative Research Group of Shandong University (2020QNQT009) and Taishan Young Scholars (tsqn20161005).

Acknowledgments: We thank Jiawei Wang for providing the suggestion for structure determination, Casey Flower and Jorge Galan for constructive proofreading of this manuscript, the staffs from BL17U1/BL18/BL19U1 beamlines of National Facility for Protein Science Shanghai (NFPS) at Shanghai Synchrotron Radiation Facility (SSRF) for assistance during data collection, and Xiaoju Li from Shandong University Core facilities for life and environmental sciences for her help with the XRD.

Conflicts of Interest: The authors declare no conflict of interest.

References

1. Tabashnik, B.E.; Carrière, Y. Surge in insect resistance to transgenic crops and prospects for sustainability. *Nat. Biotechnol.* **2017**, *35*, 926–935. [CrossRef] [PubMed]

2. Jouzani, G.S.; Valijanian, E.; Sharafi, R. *Bacillus thuringiensis*: A successful insecticide with new environmental features and tidings. *Appl. Microbiol. Biot.* **2017**, *101*, 2691–2711. [CrossRef]

3. Pardo-López, L.; Soberón, M.; Bravo, A. *Bacillus thuringiensis* insecticidal three-domain Cry toxins: Mode of action, insect resistance and consequences for crop protection. *FEMS Microbiol. Rev.* **2013**, *37*, 3–22. [CrossRef] [PubMed]

4. Adang, M.N.; Crickmore, N.; Jurat-Fuentes, J.L. Diversity of *Bacillus thuringiensis* crystal toxins and mechanism of action. In *Advances in Insect Physiology 2014, Volume 47: Insect Midgut and Insecticidal Proteins*; Dhadialla, T.S.S., Ed.; Academic Press: San Diego, CA, USA, 2014.

5. Jurat-Fuentes, J.L.; Crickmore, N. Specificity determinants for Cry insecticidal proteins: Insights from their mode of action. *J. Invertebr. Pathol.* **2017**, *142*, 5–10. [CrossRef] [PubMed]

6. De Almeida Melo, A.L.; Soccol, V.T.; Soccol, C.R. *Bacillus thuringiensis*: Mechanism of action, resistance, and new applications: A review. *Crit. Rev. Biotechnol.* **2016**, *36*, 317–326. [CrossRef]

7. Tabashnik, B.E.; Carrière, Y. Global patterns of resistance to Bt crops highlighting *Pink Bollworm* in the United States, China, and India. *J. Econ. Entomol.* **2019**, *112*, 2513–2523. [CrossRef]

8. Chakroun, M.; Banyuls, N.; Bel, Y.; Escriche, B.; Ferré, J. Bacterial vegetative insecticidal proteins (Vip) from entomopathogenic bacteria. *Microbiol. Mol. Biol. Rev.* **2016**, *80*, 329–350. [CrossRef]

9. Crickmore, N.; Baum, J.; Bravo, A.; Lereclus, D.; Narva, K.; Sampson, K.; Schnepf, E.; Sun, M.; Zeigler, D.R. Bacillus Thuringiensis Toxin Nomenclature. 2018. Available online: http://www.btnomenclature.info/ (accessed on 8 June 2020).

10. Jiang, K.; Hou, X.; Han, L.; Tan, T.; Cao, Z.; Cai, J. Fibroblast growth factor receptor, a novel receptor for vegetative insecticidal protein Vip3Aa. *Toxins* **2018**, *10*, 546. [CrossRef]

11. Jiang, K.; Hou, X.-Y.; Tan, T.-T.; Cao, Z.-L.; Mei, S.-Q.; Yan, B.; Chang, J.; Han, L.; Zhao, D.; Cai, J. Scavenger receptor-C acts as a receptor for *Bacillus thuringiensis* vegetative insecticidal protein Vip3Aa and mediates the internalization of Vip3Aa via endocytosis. *PLoS Pathog.* **2018**, *14*. [CrossRef]

12. Lee, M.K.; Walters, F.S.; Hart, H.; Palekar, N.; Chen, J.S. Mode of action of the *Bacillus thuringiensis* vegetative insecticidal protein Vip3A differs from that of Cry1Ab delta-endotoxin. *Appl. Environ. Microbiol.* **2003**, *69*, 4648–4657. [CrossRef]

13. Sena, J.A.D.; Sara Hernández-Rodríguez, C.; Ferré, J. Interaction of *Bacillus thuringiensis* Cry1 and Vip3A proteins with *Spodoptera frugiperda* midgut binding sites. *Appl. Environ. Microbiol.* **2009**, *75*, 2236–2237. [CrossRef] [PubMed]

14. Kurtz, R.W.; McCaffery, A.; O'Reilly, D. Insect resistance management for Syngenta's VipCot (TM) transgenic cotton. *J. Invertebr. Pathol.* **2007**, *95*, 227–230. [CrossRef] [PubMed]

15. Jackson, R.E.; Marcus, M.A.; Gould, F.; Bradley, J.R., Jr.; Van Duyn, J.W. Cross-resistance responses of Cry1Ac-selected *Heliothis virescens* (Lepidoptera: Noctuidae) to the *Bacillus thuringiensis* protein Vip3A. *J. Econ. Entomol.* **2007**, *100*, 180–186. [CrossRef] [PubMed]

16. Chen, W.-B.; Lu, G.-Q.; Cheng, H.-M.; Liu, C.-X.; Xiao, Y.-T.; Xu, C.; Shen, Z.-C.; Wu, K.-M. Transgenic cotton coexpressing Vip3A and Cry1Ac has a broad insecticidal spectrum against lepidopteran pests. *J. Invertebr. Pathol.* **2017**, *149*, 59–65. [CrossRef]

17. Wang, Z.; Fang, L.; Zhou, Z.; Pacheco, S.; Gomez, I.; Song, F.; Soberón, M.; Zhang, J.; Bravo, A. Specific binding between *Bacillus thuringiensis* Cry9Aa and Vip3Aa toxins synergizes their toxicity against *Asiatic rice borer* (Chilo suppressalis). *J. Biol. Chem.* **2018**, *293*, 11447–11458. [CrossRef]

18. Xiao, Y.; Wu, K. Recent progress on the interaction between insects and *Bacillus thuringiensis* crops. *Philos. Trans. R. Soc. B.* **2019**, *374*. [CrossRef]

19. Zhang, L.; Pan, Z.-Z.; Xu, L.; Liu, B.; Chen, Z.; Li, J.; Niu, L.Y.; Zhu, Y.-J.; Chen, Q.-X. Proteolytic activation of *Bacillus thuringiensis* Vip3Aa protein by *Spodoptera exigua* midgut protease. *Int. J. Biol. Macromol.* **2018**, *107*, 1220–1226. [CrossRef]

20. Palma, L.; Scott, D.J.; Harris, G.; Din, S.-U.; Williams, T.L.; Roberts, O.J.; Young, M.T.; Caballero, P.; Berry, C. The Vip3Ag4 insecticidal protoxin from *Bacillus thuringiensis* adopts a tetrameric configuration that is maintained on proteolysis. *Toxins* **2017**, *9*, 165. [CrossRef]

21. Bel, Y.; Banyuls, N.; Chakroun, M.; Escriche, B.; Ferré, J. Insights into the structure of the Vip3Aa insecticidal protein by protease digestion analysis. *Toxins* **2017**, *9*, 131. [CrossRef]

22. Kunthic, T.; Surya, W.; Promdonkoy, B.; Torres, J.; Boonserm, P. Conditions for homogeneous preparation of stable monomeric and oligomeric forms of activated Vip3A toxin from *Bacillus thuringiensis*. *Eur. Biophys. J. Biophy.* **2017**, *46*, 257–264. [CrossRef]

23. Quan, Y.; Ferré, J. Structural domains of the *Bacillus thuringiensis* Vip3Af protein unraveled by tryptic digestion of alanine mutants. *Toxins* **2019**, *11*, 368. [CrossRef] [PubMed]

24. Estruch, J.J.; Warren, G.W.; Mullins, M.A.; Nye, G.J.; Craig, J.A.; Koziel, M.G. Vip3A, a novel *Bacillus thuringiensis* vegetative insecticidal protein with a wide spectrum of activities against lepidopteran insects. *Proc. Natl. Acad. Sci. USA* **1996**, *93*, 5389–5394. [CrossRef] [PubMed]

25. Jiang, K.; Mei, S.-Q.; Wang, T.-T.; Pan, J.-H.; Chen, Y.-H.; Cai, J. Vip3Aa induces apoptosis in cultured *Spodoptera frugiperda* (Sf9) cells. *Toxicon* **2016**, *120*, 49–56. [CrossRef] [PubMed]

26. Hernández-Martínez, P.; Gomis-Cebolla, J.; Ferré, J.; Escriche, B. Changes in gene expression and apoptotic response in *Spodoptera exigua* larvae exposed to sublethal concentrations of Vip3 insecticidal proteins. *Sci. Rep.* **2017**, *7*. [CrossRef]

27. Hou, X.; Han, L.; An, B.; Zhang, Y.; Cao, Z.; Zhan, Y.; Cai, X.; Yan, B.; Cai, J. Mitochondria and lysosomes participate in Vip3Aa-induced *Spodoptera frugiperda* Sf9 cell apoptosis. *Toxins* **2020**, *12*, 116. [CrossRef]

28. Banyuls, N.; Hernández-Martínez, C.S.; Van Rie, J.; Ferré, J. Critical amino acids for the insecticidal activity of Vip3Af from *Bacillus thuringiensis*: Inference on structural aspects. *Sci. Rep.* **2018**, *8*. [CrossRef]

29. Sellami, S.; Jemli, S.; Abdelmalek, N.; Cherif, M.; Abdelkefi-Mesrati, L.; Tounsi, S.; Jamoussi, K. A novel Vip3Aa16-Cry1Ac chimera toxin: Enhancement of toxicity against *Ephestia kuehniella*, structural study and molecular docking. *Int. J. Biol. Macromol.* **2018**, *117*, 752–761. [CrossRef]

30. Zheng, M.; Evdokimov, A.G.; Moshiri, F.; Lowder, C.; Haas, J. Crystal structure of a Vip3B family insecticidal protein reveals a new fold and a unique tetrameric assembly. *Protein. Sci.* **2019**. [CrossRef]

31. Krissinel, E.; Henrick, K. Inference of macromolecular assemblies from crystalline state. *J. Mol. Biol.* **2007**, *372*, 774–797. [CrossRef]

32. Holm, L. Benchmarking Fold Detection by DaliLite v.5. In *Bioinformatics*; Oxford University Press: Oxford, UK, 2019; Volume 35, pp. 5326–5327. [CrossRef]

33. Singh, G.; Sachdev, B.; Sharma, N.; Seth, R.; Bhatnagar, R.K. Interaction of *Bacillus thuringiensis* Vegetative insecticidal protein with ribosomal S2 protein triggers larvicidal activity in *Spodoptera frugiperda*. *Appl. Environ. Microbiol.* **2010**, *76*, 7202–7209. [CrossRef] [PubMed]

34. Bae, B.; Ohene-Adjei, S.; Kocherginskaya, S.; Mackie, R.I.; Spies, M.A.; Cann, I.K.O.; Nair, S.K. Molecular basis for the selectivity and specificity of ligand recognition by the family 16 carbohydrate-binding modules from *Thermoanaerobacterium polysaccharolyticum* ManA. *J. Biol. Chem.* **2008**, *283*, 12415–12425. [CrossRef] [PubMed]

35. Larkin, M.A.; Blackshields, G.; Brown, N.P.; Chenna, R.; McGettigan, P.A.; McWilliam, H.; Valentin, F.; Wallace, I.M.; Wilm, A.; Lopez, R.; et al. Clustal W and clustal X version 2.0. *Bioinformatics* **2007**, *23*, 2947–2948. [CrossRef] [PubMed]

36. Robert, X.; Gouet, P. Deciphering key features in protein structures with the new ENDscript server. *Nucleic. Acids. Res.* **2014**, *42*, W320–W324. [CrossRef] [PubMed]

37. Grochulski, P.; Masson, L.; Borisova, S.; Pusztaicarey, M.; Schwartz, J.L.; Brousseau, R.; Cygler, M. *Bacillus thuringiensis* cryla(a) insecticidal toxin-crystal-structure and channel formation. *J. Mol. Biol.* **1995**, *254*, 447–464. [CrossRef]

38. De Maagd, R.A.; Bakker, P.L.; Masson, L.; Adang, M.J.; Sangadala, S.; Stiekema, W.; Bosch, D. Domain III of the *Bacillus thuringiensis* delta-endotoxin Cry1Ac is involved in binding to *Manduca sexta* brush border membranes and to its purified aminopeptidase N. *Mol. Microbiol.* **1999**, *31*, 463–471. [CrossRef]

39. Jenkins, J.L.; Lee, M.K.; Valaitis, A.P.; Curtiss, A.; Dean, D.H. Bivalent sequential binding model of a *Bacillus thuringiensis* toxin to *gypsy moth* aminopeptidase N receptor. *J. Biol. Chem.* **2000**, *275*, 14423–14431. [CrossRef]

40. Jurat-Fuentes, J.L.; Adang, M.J. Characterization of a Cry1Ac-receptor alkaline phosphatase in susceptible and resistant *Heliothis virescens* larvae. *Eur. J. Biochem.* **2004**, *271*, 3127–3135. [CrossRef]

41. Shao, E.; Zhang, A.; Yan, Y.; Wang, Y.; Jia, X.; Sha, L.; Guan, X.; Wang, P.; Huang, Z. Oligomer formation and insecticidal activity of *Bacillus thuringiensis* Vip3Aa toxin. *Toxins* **2020**, *12*, 274. [CrossRef]

42. Zack, M.D.; Sopko, M.S.; Frey, M.L.; Wang, X.; Tan, S.Y.; Arruda, J.M.; Letherer, T.T.; Narva, K.E. Functional characterization of Vip3Ab1 and Vip3Bc1: Two novel insecticidal proteins with differential activity against lepidopteran pests. *Sci. Rep.* **2017**, *7*. [CrossRef]

43. Gomis-Cebolla, J.; Ferreira dos Santos, R.; Wang, Y.; Caballero, J.; Caballero, P.; He, K.; Jurat-Fuentes, J.L.; Ferré, J. Domain shuffling between Vip3Aa and Vip3Ca: Chimera stability and insecticidal activity against European, American, African, and Asian Pests. *Toxins* **2020**, *12*, 99. [CrossRef]

44. Gibson, D.G.; Young, L.; Chuang, R.-Y.; Venter, J.C.; Hutchison, C.A.; Smith, H.O. Enzymatic assembly of DNA molecules up to several hundred kilobases. *Nat. Methods* **2009**, *6*, 343–345. [CrossRef] [PubMed]

45. Otwinowski, Z.; Minor, W. Processing of X-ray diffraction data collected in oscillation mode. *Macromol. Crystallogr.* **1997**, *276*, 307–326. [CrossRef]

46. Waugh, D.S. Crystal structures of MBP fusion proteins. *Protein. Sci.* **2016**, *25*, 559–571. [CrossRef] [PubMed]

47. McCoy, A.J.; Grosse-Kunstleve, R.W.; Adams, P.D.; Winn, M.D.; Storoni, L.C.; Read, R.J. Phaser crystallographic software. *J. Appl. Crystallogr.* **2007**, *40*, 658–674. [CrossRef] [PubMed]

48. Aver, N.; Skeletonisation, S.K. Cowtan, Joint CCP4 and ESF-EACBM Newsletter on Protein. *Crystallography* **1994**, *31*, 34–38.

49. Emsley, P.; Cowtan, K. Coot: Model-building tools for molecular graphics. *Acta Crystallogr. D* **2004**, *60*, 2126–2132. [CrossRef] [PubMed]

50. Adams, P.D.; Afonine, P.V.; Bunkoczi, G.; Chen, V.B.; Davis, I.W.; Echols, N.; Headd, J.J.; Hung, L.-W.; Kapral, G.J.; Grosse-Kunstleve, R.W.; et al. PHENIX: A comprehensive Python-based system for macromolecular structure solution. *Acta. Crystallogr. D* **2010**, *66*, 213–221. [CrossRef]

Article

Domain Shuffling between Vip3Aa and Vip3Ca: Chimera Stability and Insecticidal Activity against European, American, African, and Asian Pests

Joaquín Gomis-Cebolla [1], Rafael Ferreira dos Santos [2], Yueqin Wang [3], Javier Caballero [4], Primitivo Caballero [4], Kanglai He [3], Juan Luis Jurat-Fuentes [2] and Juan Ferré [1,*]

[1] ERI de Biotecnología y Biomedicina (BIOTECMED), Department of Genetics, Universitat de València, 46100-Burjassot, Spain; joaquin.gomis@uv.es

[2] Department of Entomology and Plant Pathology, University of Tennessee, Knoxville, TN 37996, USA; rferrei1@utk.edu (R.F.d.S.); jurat@utk.edu (J.L.J.-F.)

[3] State Key Laboratory for Biology of Plant Diseases and Insect Pests, Institute of Plant Protection, Chinese Academy of Agricultural Sciences, Beijing 100193, China; yueqinqueen@126.com (Y.W.); hekanglai@caas.cn (K.H.)

[4] Institute for Multidisciplinary Applied Biology, Universidad Pública de Navarra, Campus Arrosadía, 31192 Mutilva, Navarra, Spain; javier.caballero@unavarra.es (J.C.); pcm92@unavarra.es (P.C.)

* Correspondence: Juan.Ferre@uv.es

Received: 20 December 2019; Accepted: 29 January 2020; Published: 4 February 2020

Abstract: The bacterium *Bacillus thuringiensis* produces insecticidal Vip3 proteins during the vegetative growth phase with activity against several lepidopteran pests. To date, three different Vip3 protein families have been identified based on sequence identity: Vip3A, Vip3B, and Vip3C. In this study, we report the construction of chimeras by exchanging domains between Vip3Aa and Vip3Ca, two proteins with marked specificity differences against lepidopteran pests. We found that some domain combinations made proteins insoluble or prone to degradation by trypsin as most abundant insect gut protease. The soluble and trypsin-stable chimeras, along with the parental proteins Vip3Aa and Vip3Ca, were tested against lepidopteran pests from different continents: *Spodoptera exigua, Spodoptera littoralis, Spodoptera frugiperda, Helicoverpa armigera, Mamestra brassicae, Anticarsia gemmatalis,* and *Ostrinia furnacalis*. The exchange of the Nt domain (188 N-terminal amino acids) had little effect on the stability and toxicity (equal or slightly lower) of the resulting chimeric protein against all insects except for *S. frugiperda*, for which the chimera with the Nt domain from Vip3Aa and the rest of the protein from Vip3Ca showed a significant increase in toxicity compared to the parental Vip3Ca. Chimeras with the C-terminal domain from Vip3Aa (from amino acid 510 of Vip3Aa to the Ct) with the central domain of Vip3Ca (amino acids 189–509 based on the Vip3Aa sequence) made proteins that could not be solubilized. Finally, the chimera including the Ct domain of Vip3Ca and the Nt and central domain from Vip3Aa was unstable. Importantly, an insect species tolerant to Vip3Aa but susceptible to Vip3Ca, such as *Ostrinia furnacalis*, was also susceptible to chimeras maintaining the Ct domain from Vip3Ca, in agreement with the hypothesis that the Ct region of the protein is the one conferring specificity to Vip3 proteins.

Keywords: Bacillus thuringiensis; Spodoptera spp., Helicoverpa armigera; Mamestra brassicae; Anticarsia gemmatalis; Ostrinia furnacalis

Key Contribution: Chimeric proteins between Vip3Aa and Vip3Ca were generated combining fragments of the Nt, the central part, and the Ct of the proteins. The exchange of the Nt domain had little effect on the stability and toxicity (equal or slightly lower), except for *S. frugiperda*, for which a gain of function was observed. Specificity to *O. furnacalis* followed the Ct domain from Vip3Ca.

---

1. Introduction

Bacillus thuringiensis (Bt) is an aerobic, spore-forming, Gram-positive, and entomopathogenic bacterium belonging to the *Bacillus cereus* group. The Bt bacterium produces a wide variety of insecticidal proteins [1] along with other virulence factors contributing to its pathogenicity [2]. Two major categories of insecticidal proteins produced by Bt are δ-endotoxins (Cry and Cyt toxins) that form crystals within the sporangium in the sporulation phase, and vegetative insecticidal proteins (Vip), which are secreted into the growth medium during vegetative growth [1,3,4]. The Vip proteins are classified into four groups (Vip1, Vip2, Vip3, and Vip4) based on their protein sequence similarity [4,5]. The Vip1 and Vip2 proteins act as binary toxins against coleopteran pests [1,4], while for the Vip4 protein no insecticidal activity has been reported yet.

The Vip3 proteins, mainly those of the Vip3A family, are active against a wide range of lepidopteran pests [1,4]. These proteins do not share structural homology with the Cry proteins, but the toxic action follows the same sequence of events: ingestion, activation by midgut proteases, binding to specific receptors in the midgut epithelium, and pore formation [1,4]. Recent studies indicate that Vip3 proteins (either as protoxins or in the activated form of toxin) spontaneously form tetramers in solution [6–10]. In addition, when the Vip3 proteins are activated by proteases, the oligomer remains stable and the cleaved Nt fragment (19 kDa) remains associated to the main fragment (65 kDa) of the protein [6–10]. In agreement with their diverse structure, Vip3 proteins do not share receptors with Cry proteins [11–15], but share receptors with other Vip3 proteins, either from the same (Vip3Aa, Vip3Af, Vip3Ae, and Vip3Ad) or different (Vip3Ca) protein families [14,16].

Five domains have been proposed for the structure of Vip3A proteins from in silico modelling [17,18]. Based on structural features and stability to trypsin, Quan and Ferré [19] identified five domains from Vip3Af: Domain I encompassing amino acids (aa) 12-198, domain II aa 199-313, domain III aa 314–526, domain IV aa 527–668, and domain V aa 669–788. As far as the structural role of the proposed Vip3Af domains, Quan and Ferré [19] found that domains I–III were required to form the tetrameric structure, the role for domain IV was unclear, and domain V was not necessary for oligomerization. Wang et al. [20] generated a disabled Vip3A protein with two site-engineered mutations (S175C and L177C) in domain I, which was not toxic but retained the ability to compete for the wild type binding sites. Taken together, these results suggest that domain I may be involved in post-binding events, such as membrane insertion, and domain V in binding recognition and specificity.

In this work, we capitalized on the high sequence similarity among Vip3 proteins to test, by domain shuffling, the compatibility of the proposed Vip3Af domains in protein stability and toxicity using representatives from two different Vip3 protein families (Vip3Aa45 and Vip3Ca2). Six chimeric Vip3 proteins (Vip3_ch1, Vip3_ch2, Vip3_ch3, Vip3_ch4, Vip3_ch5, and Vip3_ch6) were designed, where the amino acids (aa) phenylalanine and serine at positions 188 and 509 were chosen as the sites to generate the chimeric Vip3 proteins (Figure 1).

Sequence exchange at these sites coincided approximately with domains I, II+III, and IV+V in the proposed Vip3Af domain model. For the sake of simplicity, we named these domains as the Nt domain (domain I), the central domain (domains II+III), and the Ct domain (domains IV+V), respectively (Figure 1). The objectives of the current research were to determine which main regions of the Vip3 proteins are exchangeable while maintaining the stability and toxicity of the proteins, with the aim to evaluate if any of the new chimeric proteins might confer an increase in the toxicity compared to the parental proteins.

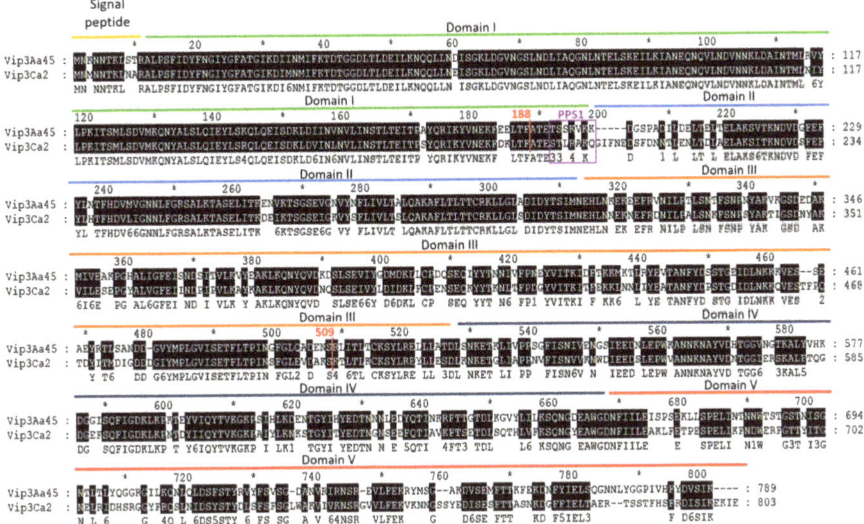

Figure 1. Protein sequence alignment of Vip3Aa45 and Vip3Ca2. Black background shading is used to highlight the conserved amino acid between proteins. The proposed structural domains (based on the Vip3Af proteolysis mutants) are indicated with colored lines above the sequences [19]. The purple box indicates the position of the cleavage site (PPS1), while the red vertical lines show the sites chosen to generate the chimeric proteins.

2. Results

2.1. Sequence Analysis of the Vip3Aa and Vip3Ca Proteins and Determination of the Vip3 Protein Fragment Combinations that Generate Stable Chimeric Proteins

The amino acid sequence alignment of the two Vip3 proteins indicate that most of the differences are located in their Ct domain (Figure 1). Chimeric proteins were constructed by exchanging the Nt domain (aa 1–188), the central domain (aa 189–509), and the Ct domain (aa 510–788), using as a reference the Vip3A sequence (Figure 1). The Nt domain is highly conserved, with only eight residue differences between the two proteins. The main difference in the protein sequence of the central domain between Vip3Ca and Vip3Aa was two insertions (in Vip3Ca), one located immediately after the main proteolytic processing site (PPS1) (^{188}GIFNE), and the other at aa position 464 (^{464}TF) [21]. To determine the combinations of the different Vip3 protein domains that generated soluble chimeric Vip3 proteins, all the possible combinations were expressed in *Escherichia coli* (Figure 2). The results indicated that the six chimeric proteins could be expressed, but only the Vip3_ch1, Vip3_ch2, Vip3_ch4, and Vip3_ch5 proteins could be solubilized from the respective inclusion bodies (Figure 2). The exchange of Nt domain did not affect the solubility of the generated chimeric proteins (Vip3_ch1 and Vip3_ch2) (Figure 2). However, the exchange of the Ct domain had, in most cases, a negative effect on the solubility of the chimeric protein. The Ct domain from Vip3Aa combined with the central domain from Vip3Ca produced insoluble proteins (Vip3_ch3 and Vip3_ch6), whereas the reciprocal combination produced a little soluble protein, with tendency to precipitate (Vip3_ch5), and a soluble one (Vip3_ch4) (Figure 2).

Figure 2. Summary of the combinations of the different Vip3 protein domains expressed in the heterologous *Escherichia coli* expression system. The "single" chimeric Vip3 proteins (Vip3_ch1, Vip3_ch2, Vip3_ch5, and Vip3_ch6) were obtained from the Vip3Aa and Vip3Ca as a template, while the "double" chimeric Vip3 proteins (Vip3_ch3 and Vip3_ch4) were amplified from the Vip3_ch5 and Vip3_ch6. The percentage of similarity of the different proteins vs. the parental proteins, Vip3Aa and Vip3Ca, was calculated with the NCBI Blast align tool.

2.2. Proteolytic and Thermal Stability of The Parental and Chimeric Proteins

To determine whether the chimeric proteins were stable to the activation by proteases, the proteins were digested with 1% trypsin (w:w). The results showed that Vip3Aa, Vip3Ca, and the chimeric proteins Vip3_ch1, Vip3_ch2, and Vip3_ch4 were processed into the two expected protein fragments of 65-67 kDa and 19-22 kDa (Figure 3). However, the proteolytic pattern of the Vip3_ch5 chimera differed from the rest of the Vip3 proteins; this phenomenon could be due to (i) instability to proteases, (ii) instability of the Vip3_ch5 protein in solution, or (iii) problems in the production and purification of the protein. (Figure 3B). Thermal stability of the more soluble and highly purified Vip3 proteins (Vip3Aa, Vip3Ca, Vip3_ch1, Vip3_ch2, and Vip3_ch4) resistant to the trypsin treatment was tested by the thermofluor method (Figure S1). The parental protein Vip3Aa showed two thermal transitions (melting temperature, Tm, of Vip3A-Peak (1): 59.4 ± 0.4 and Tm of Vip3A-Peak (2): 75.5 ± 0.0), while Vip3Ca only showed one thermal transition (Tm of Vip3C-Peak (2): 73.7 ± 0.0) (Figure S1). The chimeric proteins also showed two thermal transitions, but with the first negative peak less pronounced than in the parental Vip3Aa, indicating that the first denaturation involved a lesser part of the protein (Figure S1).

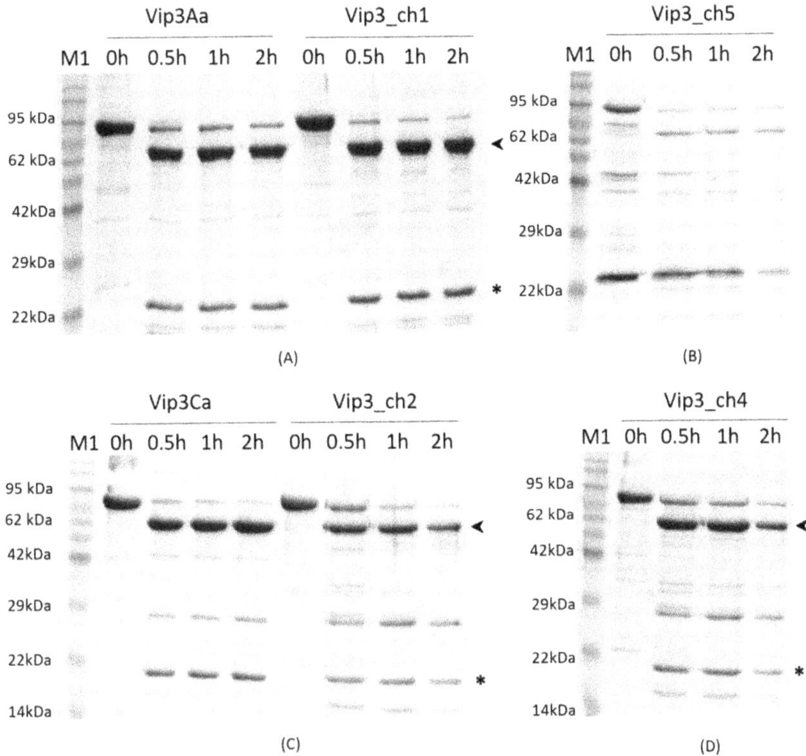

Figure 3. Time course of trypsin activation of Vip3 parental and chimeric protoxins. The reaction was carried out using 1% trypsin (w:w) at 37 °C for increasing incubation periods. (**A**) Vip3Aa protein and Vip3_ch1, (**B**) Vip3_ch5. (**C**) Vip3Ca and Vip3_ch2; (**D**) Vip3_ch4. The arrowheads indicate the protein bands corresponding to the 62–67 kDa fragment, while the asterisks indicate the protein bands corresponding to the 19–22 kDa fragment. M1: Molecular Mass Marker.

2.3. Insecticidal Activity of the Parental and Chimeric Vip3 Proteins

The insecticidal activity of the soluble chimeric proteins (Vip3_ch1, Vip3_ch2, Vip3_ch4, and Vip3_ch5) was compared to that of the parental proteins by testing eight insect species with different susceptibilities to Vip3Aa and Vip3Ca (Table 1). The Vip3Aa protein was toxic for all the insect species tested except for *Ostrinia furnacalis* (for this insect species the Vip3Aa protein is only toxic at very high concentration). The Vip3Ca protein showed high toxicity to *O. furnacalis* and moderate toxicity to *A. gemmatalis*; for the other insect species tested, this protein was slightly active at very high concentrations (Table 1).

Table 1. Susceptibility of lepidopteran insect pests to parental (Vip3Aa and Vip3Ca) and chimeric proteins.

Insect Family	Insect Genus	Insect Species (Instar Tested)	Concentration	% of Corrected Mortality (Mean ± SD *) [ß]					
				Vip3Aa	Vip3Aa Chimeras		Vip3Ca	Vip3Ca Chimeras	
					Vip3_ch1	Vip3_ch5		Vip3_ch2	Vip3_ch4
Noctuidae	Spodoptera	*S. exigua* (neonate)	µg/cm^2 4.1	85.9 ± 4.6	86.7 ± 3.3	6.5 ± 0.2	-	-	-
			0.5 [ζ]	70.8 ± 10.4 (a)	68.2 ± 0.5 (a)	6.3 ± 6.3 (b)	-	-	-
			7	-	-	-	51.8 ± 1.4	41.5 ± 10.3	9.4 ± 0.0
			0.7 [ζ]	-	-	-	29.7 ± 1.6 (c)	32.6 ± 4.5 (c)	3.4 ± 0.3 (d)
		S. littoralis (neonate)	4.1	98.4 ± 1.6	100.0 ± 1.6	7.8 ± 1.6	-	-	-
			0.5 [ζ]	98.4 ± 1.6 (b)	70.0 ± 17.0 (c)	3.1 ± 3.1 (d)	-	-	-
			7	-	-	-	32.3 ± 2.1	34.3 ± 12.6	4.6 ± 1.6
			0.7 [ζ]	-	-	-	6.3 ± 6.3 (e)	11.5 ± 0.8 (e)	4.7 ± 4.7 (f)
		S. frugiperda (neonate)	4.1	97.0 ± 3.1	98.4 ± 1.6	1.6 ± 1.6	-	-	-
			0.5 [ζ]	91.0 ± 0.0 (g)	59.4 ± 9.4 (h)	1.6 ± 1.6 (i)	-	-	-
			7	-	-	-	53.0 ± 6.6	98.4 ± 1.6	1.6 ± 1.6
			0.7 [ζ]	-	-	-	6.2 ± 6.2 (j)	92.2 ± 7.8 (k)	1.6 ± 1.6 (l)
	Helicoverpa	*H. armigera* ‡ (neonate)	2.5	100 ± 0.0	66.5 ± 11.5	17.0 ± 4.0	-	-	-
			0.3 [ζ]	83.5 ± 3.5 (m)	9.0 ± 1.0 (o)	6.5 ± 2.5 (p)	-	-	-
			4	-	-	-	61.5 ± 9.5	82.0 ± 4.0	2.0 ± 2.0
			0.4 [ζ]	-	-	-	15.5 ± 5.5 (q)	15.0 ± 2.0 (q)	1.0 ± 1.0 (r)
Noctuidae	Anticarsia	*A. gemmatalis* (neonate)	µg/cm^2 4.1	98.4 ± 1.5	6.3 ± 6.3	7.8 ± 4.7	-	-	-
			0.5 [ζ]	98.4 ± 1.5 (s)	0.0 ± 0.0 (t)	6.3 ± 6.3 (v)	-	-	-
			7	-	-	-	90.0 ± 3.3	54.7 ± 14.1	7.8 ± 7.8
			0.7 [ζ]	-	-	-	54.7 ± 11.8 (w)	21.9 ± 6.3 (y)	1.6 ± 1.6 (z)
	Mamestra	*M. brassicae* ‡ (L2)	µg/ml 2.5	79.3 ± 0.9	69.0 ± 6.0	22.0 ± 7.0	-	-	-
			0.3 [ζ]	56.5 ± 8.5 (aa)	31.0 ± 2.0 (ab)	14.0 ± 1.0 (ac)	-	-	-
			4	-	-	-	65.6 ± 12.3	26.0 ± ND [δ]	15.5 ± 4.0
			0.4 [ζ]	-	-	-	30.5 ± 1.5 (ad)	7.0 ± ND [δ] (ae)	10.0 ± 1.0 (af)
Crambidae	Ostrinia	*O. furnacalis* * (neonate)	µg/g 50	55.2 ± 1.0	14.6 ± 4.2	21.8 ± 3.1	96.8 ± 1.0	98.9 ± 1.0	85.4 ± 2.1
			5 [ζ]	16.6 ± 4.2 (ah)	8.3 ± 0.0 (ai)	16.6 ± 0.0 (ah)	91.6 ± 2.1 (aj)	79.2 ± 2.1 (ak)	65.6 ± 1.0 (al)

* Standard deviation of the mean. [δ] ND: not possible to calculate the standard deviation of the mean of the Vip3_ch2 in *M. brassicae* colony because the assay was done with one replicate. [ß] The percentage of mortality in the different treatments was corrected by subtracting the value of mortality observed for the buffer treatment (negative control). ‡ In the case of *H. armigera* and *M. brassicae*, the concentration used in the surface contamination and droplet feeding method assays were adapted from Ruiz de Escudero et al. 2014 [22] (2.5 and 0.3 µg/cm^2) and Palma et al. 2012 [21] (4 and 0.4 µg/cm^2). [ζ] For each insect species, the mortality values at lowest dose (discriminant concentration used) followed by the same letter were not statistically different based on the based on the overlap of standard deviation of the mean.

Regarding the chimeric proteins, the exchange of the Nt domain in Vip3Aa (Vip3_ch1 chimera) decreased the insecticidal activity (compared to Vip3Aa) against all the insect species tested (detected when testing at lower concentrations 0.5 and 0.3 ug/cm^2, 0.3 ug/mL and 5 ug/g) except for *S. exigua*. In the case of *A. gemmatalis* and *O. furnacalis*, this chimera completely lost toxicity (Table 1). In the case of the Vip3Ca protein, the exchange of the Nt domain (Vip3_ch2 chimera) led to different outcomes depending on the insect species considered. Insecticidal activity did not significantly differ from that of Vip3Ca in *S. exigua*, *S. littoralis*, *H. armigera*. The insecticidal activity of the Vip3_ch2 chimera decreased in *A. gemmatalis*, *M. brassicae*, and *O. furnacalis*. Most interestingly, in the case of *S. frugiperda* the Vip3_ch2 chimera showed a strong gain of function compared to the Vip3Ca parental protein with mortality values similar to the ones of the most active parental protein, Vip3Aa (Table 1). Chimera Vip3_ch4 (with the central domain from Vip3Aa and the flanking ones from Vip3Ca) was nontoxic for any of the insect species tested, except for *O. furnacalis* (Table 1). In the case of Vip3_ch5, the chimeric protein did not cause any damage to any of the insects tested, most likely due to the instability of this protein or problems in the production and purification.

The toxicity of the three proteins active against *O. furnacalis* was confirmed by determining the LC_{50} values (Table 2 and Table S1). The results indicated that, though similarly toxic, Vip3Ca was the most toxic protein (LC_{50} = 1.2 µg/g) followed by Vip3_ch2 (LC_{50} = 2.3 µg/g) and Vip3_ch4 (LC_{50} = 3.9 µg/g). In the case of the chimera with gain of function for *S. frugiperda*, the LC_{50} value was determined and compared to the most toxic parental protein, Vip3Aa. The results indicated that the toxicity of Vip3_ch2 (LC_{50} = 133 ng/cm^2) did not significantly differ from that of Vip3Aa (LC_{50} = 162 ng/cm^2), but was significantly increased compared to the Vip3Ca protein (LC_{50} > 7000 ng/cm^2) (Table 1, Table 2 and Table S1).

Table 2. Determination of the lethal concentration (LC_{50}) of the parental and selected chimeric Vip3 proteins in *Ostrinia furnacalis* and *Spodoptera frugiperda*.

Insect Species	Toxin	Number of Insects Tested	Slope Factor			Lethal Concentration				Goodness of Fit			
			Slope	SE *	CI$_{95}$ †	LC_{50} [c]	SE *	CI$_{95}$ †	R^2	Absolute Sum Squares	Sy.x ‡	Df [¥]
O. furnacalis						µg/g						
	Vip3Ca	768	1.4	0.06	1.3–1.6	1.2 (a)	1.0	1.1–1.3	0.99	129	2.7	18
	Vip3_ch2	768	1.0	0.05	0.9–1.1	2.3 (b)	1.0	2.0–2.5	0.99	195	3.3	18
	Vip3_ch4	768	1.2	0.07	1.0–1.3	3.9 (c)	1.0	3.5–4.5	0.98	327	4.3	18
S. frugiperda						ng/cm^2						
	Vip3Aa	512	1.6	0.25	1.1–2.2	162.0 (d)	1.1	130–202	0.95	1616	9.2	19
	Vip3_ch2	336	1.6	0.25	1.1–2.2	133.1 (d)	1.1	107–166	0.97	1086	8.2	16

* Standard error of the slope and lethal dose concentration, respectively † Confidence interval at 95% for the slope and lethal dose concentration, respectively. ‡ Quantification of the standard deviation of the residuals (vertical distance of the point from the fit line or curve) expressed as % of mortality. At higher value, the data shows a greater variance and lower goodness of fit (R^2). [¥] Degree of freedom. [c] For each insect species, the LC values followed by the same letter were not statistically different based on of the statistical analysis extra-sum-square F test analysis (α 0.05) (Table S1).

3. Discussion

Insecticidal proteins in the Vip3A family have been incorporated in commercial transgenic crop varieties [23] due their potent and broad spectrum activity against lepidopteran pests [4]. In contrast, members of the Vip3B and Vip3C protein families have a narrow insecticidal spectrum and a moderate activity [8,21,24–28]. In the case of Vip3Ca, among the 25 species of insects tested [21,25–28], only for *O. furnacalis* and *Mythimna separata* its toxicity was comparable to the most toxic Cry or Vip3 proteins (Cry1Ab for *O. furncacalis* or Vip3Aa for *M. separata*) [26,27]. The present work focused on testing the compatibility of domains exchanged between a member of the Vip3A family and one of the Vip3C family, with the possibility of generating novel proteins with increased insecticidal activity.

The results show that the exchange of the Nt domain does not affect the solubility and trypsin stability of the resulting chimeric Vip3 proteins (Vip3_ch1 and Vip3_ch2) (Figure 2). Similar results were obtained in another study testing the exchange of the Nt domain between Vip3Ab and Vip3Bc [8]. This is not a surprising result since the Nt domain is extremely conserved among Vip3 proteins, suggesting a structural role or a possible role in post binding events, such as pore formation. Wang et al. [20] generated a Vip3A protein mutated at the Nt domain (S175C and L177C) which was able to compete with binding of the wild type protein but not cause mortality, thus supporting the previous hypothesis. With regards to the Ct domain, of the four chimeras produced only one was soluble and stable to treatment with trypsin (Figure 2). These results suggest that the interaction of the Ct domain with the other domains in the 3D structure of Vip3 is more specific and critical to the physicochemical properties of the molecule. Furthermore, results from thermofluor assays showed that the chimeric proteins had a thermal stability intermediate between that of the two parental proteins and that the Tm values and the presence/intensity of one or two thermal transitions depended on the interaction between the respective Vip3 domains (Figure S1). Specifically, the denaturation curve profile for the Vip3Aa, Vip3_ch1, Vip3_ch2, and Vip3_ch4 proteins indicates that these proteins have two motifs with different thermal stability, while the Vip3Ca protein would be more stable because of its single denaturation curve. Further understanding of the structure in this family of proteins would shed light on this aspect.

The results from the insecticidal spectrum of the chimeric proteins indicated that, in general (and considering that we only tested one species in the Crambidae family), the activity follows taxonomical relationships at the family level. Thus, species in Noctuidae presented a closer susceptibility profile to both parental and chimeric proteins when compared to the tested species in Crambidae (Table 1). This observation is in agreement with the results of Zack et al. [8], where the Noctuid insects (*Helicoverpa zea*, *S. frugiperda*, and *Pseudoplusia includens*) showed more similar susceptibility profiles for the parental (Vip3Ab and Vip3Bc) proteins and their chimeric proteins (generated by exchange of the Nt domain), compared to the Crambidae insects (*Ostrinia nubilalis*). That study also showed that the chimeras were less toxic than the parental proteins to *H. zea*, *S. frugiperda*, *O. nubilalis*, and *P. includens* [8]. Similarly, our results with the chimeras indicate that, with the exception of Vip3_ch2, the chimera proteins are similarly or less toxic than the parental proteins (Table 1). The Vip3_ch2 chimera, a Vip3C protein with the Nt domain from Vip3Aa, displayed gain of function only with *S. frugiperda* but not with other closely related species of the same genus (Table 1). A similar result was recently reported in which a "modified Vip3C protein" (i.e., ARP150v2, 98% similarity to the Vip3_ch2) had higher toxicity against *S. frugiperda* than the parental Vip3Ca protein [29]. Sequence analysis indicates that ARP150v2 is a chimera in which the Nt domain of Vip3Ca has been replaced by that of Vip3Af1. We do not have a clear explanation for this unique increase in toxicity, but due to the specificity of the phenomenon, the reason has to be more efficient interaction with the receptors and/or facilitated post-binding events, such as membrane insertion or pore formation. Further research testing the mode of action of this family of proteins should clarify this particular phenomenon.

4. Conclusions

In summary, we present evidence for the relative importance of different Vip3 protein domains in stability and toxicity, and an example of how the design of chimeric Vip3 proteins may lead to novel proteins with improved and expanded insecticidal activity. Specifically, the Vip3_ch2 protein, a Vip3C protein with the Nt domain from Vip3Aa, showed a gain of function for *S. frugiperda*. In addition, the Vip3_ch4 protein showed that for the toxicity of the Vip3C protein in *O. furnacalis*, the specificity is provided by the Ct domain.

5. Materials and Methods

5.1. Design and Construction of Chimeric Vip3 Proteins

An overlap PCR method was used to generate the chimeric proteins from the parental Vip3Aa45 (JF710269.1) and Vip3Ca2 (JF916462.1) proteins [21,30]. To construct the Vip3A and Vip3C chimeric proteins, amino acids (aa) stretches at positions 188 (^{188}FATET) and 509 (^{509}SRLIT) of the Vip3Aa protein were used to define the protein fragments to exchange: fragment I (aa 0 - 188), fragment II (aa

189 - 508), fragment III (aa 509 - 788) (Figure 1). Six chimeric proteins were generated and classified as "single" (Vip3_ch1, 2, 5, and 6) or "double" (Vip3_ch3 and 4), depending on whether they were amplified from the parental or the Vip3_ch5 and 6 proteins, respectively (Table S2).

To generate the chimeric genes, first a PCR was performed to amplify the necessary fragments separately with the annealing primers (Tables S2 and S3). The PCR reaction contained, in a final volume of 50 μL, 50 ng of the DNA template, 0.25 U of Kapa Hifi DNA polymerase, 5 μL of five-fold reaction buffer, 0.6 mM of each dNTPs, and 0.3 μM of the respective primers. PCR amplifications were carried out as follows: 5 min denaturation at 95 °C, 35 cycles of amplification ((20 s of denaturation at 98 °C, 15 s of annealing at 60 °C, and 30 s of extension at 72 °C), and an extra extension step of 5 min at 72°C). The amplicons were purified form the agarose gel and a second PCR (overlap step + "amplification step") was performed with the respective DNA fragments (Table S2). First, the "overlap step" was conducted in a final volume of 50 μL with 100 ng (total amount) of the respective DNA fragments (Table S2) in an equimolecular ratio, 0.25 U of Kapa Hifi DNA polymerase, 5 μL of five-fold reaction buffer, 0.6 mM of each dNTPs. PCR amplifications were carried out as follows: 5 min denaturation at 95 °C, 15 cycles of amplification ((20 s of denaturation at 98 °C, 30 s (Vip3_ch1, 2, 4, 5, and 6)/1 min (Vip3_ch3) of annealing at 55 °C (Vip3_ch1, 2, 4, 5, and 6) or 50 °C (Vip3_ch3), 2 min of extension at 72 °C) and an extra extension step of 5 min at 72 °C. Second, the "amplification step" was performed with the respective end primers (Table S2), adding 0.3 μM of each to the PCR reactions. The PCR reactions were carried out in the conditions described for the "overlap step". In addition, for the Vip3_ch3 protein a nested-PCR with the DNA amplified in the second PCR was carried out (PCR reaction: final volume 50 μL, 7 ng of the Vip3 chimera 3, 0.25 U of Kapa Hifi DNA polymerase, 5 μL of five-fold reaction buffer, 0.6 mM of each dNTPs, and 0.3 μM of the respective primers (Table S2). Conditions for this nested-PCR amplification were 5 min denaturation at 95 °C, 35 cycles of amplification (20 s of denaturation at 98 °C, 60 s of annealing at 50 °C, and 2 min of extension at 72 °C), and an extra extension step of 5 min at 72 °C). Amplicons were purified from an agarose gel, ligated into the pGEM®-T Easy plasmid or pCR®2.1-TOPO®, cloned in *E. coli* DH10β, and sequenced with the sequencing primers (Table S3).

For expression, the full length genes were amplified from the pGEM®-T Easy or pCR®2.1-TOPO® with the end primers (Table S3). The PCR reactions contained, in a final volume of 50 μL, 50 ng of the respective Vip3 chimeric genes, 0.25 U of Kapa Hifi DNA polymerase, 5 μL of five-fold reaction buffer, 0.6 mM of each dNTPs, and 0.3 μM of the respective end primers (Table S2). Conditions for PCR amplifications were as follows: 5 min denaturation at 95 °C, 35 cycles of amplification ((20 s of denaturation at 98°C, 60 s of annealing at 55 °C (Vip3_ch1, 2, 4, 5, and 6) 50 °C (Vip3_ch3), 2 min of extension at 72 °C), and an extra extension step of 5 min at 72 °C). Amplicons were purified and together with the expression vector (pET30a (+)) were digested with BamHI and NotI for 2 h at 37 °C. The pET30a (+) plasmid was dephosphorylated for 2 h at 37 °C with alkaline phosphatase. The linearized/dephosphorylated pET30a (+) and the digested chimeric genes were purified prior to ligation using T4 DNA Ligase overnight at 4 °C. Ligation reactions were transformed in *E. coli* BL21 (D3) and transformants confirmed by sequencing with the sequencing primers (Table S3).

5.2. Expression and Purification of Vip3Aa, Vip3Ca, and Chimeric Proteins

5.2.1. Expression of the Parental and Chimeric Vip3 Proteins

The Vip3Ca protein was expressed following the conditions described by Gomis-Cebolla et al. (2017) [16]. For expression of Vip3Aa and the chimeric proteins, a single colony was inoculated in 7 mL of LB-K medium (LB medium containing 50 μg/mL kanamycin) and grown overnight at 37 °C and 180 rpm. A 1/100 dilution of the culture in 700 mL LB-K medium was further incubated at 37 °C and 180 rpm. When the OD was 0.7–0.8, 1 mM IPTG (final concentration) was added for induction. Induced cultures were grown overnight at 37 °C and 180 rpm, and the cells were collected by centrifugation at 8800× *g* for 30 min at 4 °C. Cell pellets for the Vip3Aa and Vip3Ca proteins were lysed by chemical lysis. Briefly, three milliliters of lysis buffer-I (50 mM sodium phosphate buffer, 0.5 M NaCl, pH 8.0, containing 3 mg/mL lysozyme, 10 µg/mL DNase, 10 mM DTT, and 100 µM PMSF) per gram of pellet were added to the samples. The pellets were resuspended with an Ultra Turrax T25 digital homogenizer (IKA, Janke & Kunkel-Str. 10 Staufen, DE) at 16,000× g and incubated at 37 °C for 60 min with strong shaking (200 rpm). Then, the lysate was sonicated on ice applying five cycles (1 min pulse at 70 W, 10 s off, 1 min pulse at 70 W). Insoluble materials were separated by centrifugation at 31,000× g for 15 min and 4 °C. The soluble cellular fractions were filtered through sterile 0.45 µm cellulose acetate filters. In the case of chimeric proteins, three milliliters of lysis buffer-II (50 mM sodium phosphate buffer, 0.5 M NaCl, pH 8.0, containing 3 mg/mL lysozyme, 10 mM DTT and 100 µM PMSF) per gram were added to the pellets and the samples were resuspended as described above, and then incubated at 37 °C for 60 min with strong shaking (200 rpm). After incubation, 8 mg of deoxycholic acid sodium salt per gram of pellet was added and incubated 30 min at 37 °C with gentle shaking (100 rpm), after which 40 µg/mL of DNase was added to eliminate the viscosity of the lysates and further incubated for 30 min at 37 °C with gentle shaking. The lysates of the chimeric proteins were then sonicated on ice applying five cycles (1 min pulse at 70 W, 10 s off, 1 min pulse at 70 W), centrifuged at 31,000× g for 15 min at 4 °C and the soluble cellular fraction was filtered through sterile 0.45 µm cellulose acetate filters. In the case of the Vip3 chimeras 3 and 6, they formed inclusion bodies that were not possible to dissolve in the conditions used in the present study (Figure S2).

5.2.2. Purification of Vip3Aa, Vip3Ca, and Chimeric Vip3 Proteins by Isoelectric Point Precipitation

For the insect toxicity assays, two independent batches of the Vip3Aa, Vip3Ca and the chimeric proteins (Vip3_ch1, Vip3_ch2, Vip3_ch4, and Vip3_ch5), were purified by isoelectric point precipitation (IPP) in three steps (Figure 4A) [31]. First, the soluble cellular fractions of the Vip3 proteins were diluted three-fold with 50 mM sodium phosphate buffer pH 8.0, dialyzed overnight against the dialysis buffer (20 mM sodium phosphate buffer, 150 mM NaCl, pH 8), centrifuged at 14,000× g for 15 min at 4 °C, and then filtered through 0.45 µm cellulose acetate filters. Second, the pH of the respective lysates was lowered with acetic acid to pH 5.5 for Vip3Aa, pH 5.9 for Vip3Ca, pH 5.0 for Vip3_ch1, pH 5.5 for the Vip3_ch2, pH 5.2 for the Vip3_ch4, pH 5.2 for the Vip3_ch5. The Vip3 proteins were recovered by centrifugation (14,000× g for 15 min at 4 °C). Third, the pellets were resuspended in storage buffer (20 mM Tris buffer, 150 mM NaCl pH 8.6) for 1 h with shaking at 4 °C, and then centrifuged (14,000× g for 15 min at 4 °C) and filtered through 0.45 µm cellulose acetate filters. The Vip3 proteins were quantified by densitometry and the ratio of Vip3 protein/total protein (w:w) was calculated. The samples were stored at −80 °C and lyophilized prior to their use or shipping at room temperature to other laboratories.

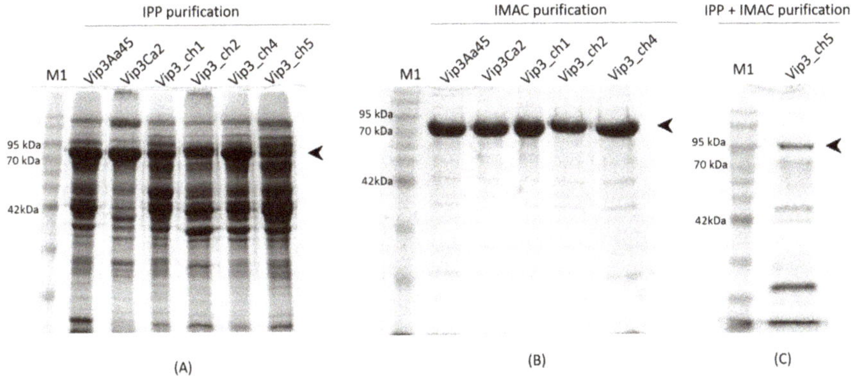

Figure 4. SDS-PAGE analysis of the purified parental (Vip3Aa and Vip3Ca) and chimeric (Vip3_ch1, Vip3_ch2, Vip3_ch4, and Vip3_ch5) proteins. (**A**) Parental proteins and chimeric Vip3 proteins (5 μg) purified by isoelectric point precipitation (IPP). (**B**) Parental proteins and chimeric Vip3 proteins (5 μg) purified by ion metal affinity chromatography (IMAC) on a Hi-Trap chelating HP column (GE Healthcare). (**C**) Vip3_ch5 protein (2 μg) purified by IPP and IMAC on a Hi-Trap chelating HP column (GE Healthcare). The arrowheads indicate the protein band corresponding to the chimeric Vip3 proteins. M1: Molecular Mass Marker.

5.2.3. Purification of Vip3Aa, Vip3Ca, and Chimeric Vip3 Proteins by Ion Metal Affinity Chromatography

For the proteolysis and thermal shift assays, the parental proteins (Vip3Aa and Vip3Ca) and the chimeric Vip3 proteins (Vip3_ch1, Vip3_ch2, Vip3_ch4, and Vip3_ch5) were purified by ion metal affinity chromatography (IMAC) on a His-Trap FF crude lysate column (GE Healthcare) (Figure 4B). The soluble cellular fractions of the Vip3 proteins (Vip3Aa, Vip3Ca, Vip3_ch1, 2, 4, and 5) were diluted three-fold with 50 mM sodium phosphate buffer pH 8.0, dialyzed against the dialysis buffer to eliminate the presence of DTT and deoxycholic acid. Samples were dialyzed for 10-16 h at 4 °C, and the dialysis buffer exchanged twice. The unclarified lysates were filtered through 0.45 μm cellulose acetate filters to discard protein aggregates. The soluble Vip3 protein fractions were loaded into a His-Trap FF crude lysate column equilibrated in binding buffer (20 mM phosphate buffer, 150 mM NaCl, 10 mM imidazole, pH 8). After washing the column with binding buffer to eliminate unbound molecules, Vip3 proteins were eluted using elution buffer (20 mM phosphate buffer, 150 mM NaCl, 150 mM imidazole, pH 8) into 2 mL tubes containing 0.1 mM EDTA (pH 8.0).

Since the Vip3_ch5 was expressed in low quantities (data not shown), first the protein was partially purified by IPP as described above, and then filtered through 0.45 μm cellulose acetate filter, prior to loading into the His-Trap FF crude lysate column (Figure 4C).

To avoid protein precipitation, buffer exchange was performed against storage buffer (20 mM Tris 500 mM NaCl, pH 8.6) by dialysis overnight at 4 °C. The concentration and quality of the purified proteins were estimated with the Bradford assay [32] using BSA as standard and by SDS-PAGE, respectively. The Vip3 proteins were snap frozen in liquid nitrogen and stored at −80 °C until used.

5.3. Thermal and Protease Stability of the Parental and the Chimeric Vip3 Proteins

The parental (5 μg) and chimeric (5 μg of Vip3_ch1, Vip3_ch2, Vip3_ch4, and 2 μg of Vip3_ch5) proteins were subjected to proteolysis with 1% (w/w) bovine trypsin (SIGMA T8003, Sigma-Aldrich, Madrid, Spain) for different time intervals (0, 0.5, 1, and 2 h) at 37 °C. The proteolytic reactions were stopped with 1 mM of AEBSF protease inhibitor for 10 min at room temperature, and then the samples were resolved by SDS-12%PAGE and stained with Coomassie brilliant blue R-250 (Sigma 1125530025,

Sigma-Aldrich, Madrid, Spain). The size of the protoxin and trypsin-activated fragments were analyzed using the TotalLab 1D v 13.01 software.

The Tm of the parental proteins and the chimeric proteins resistant to trypsin treatment were determined using the environmentally sensitive extrinsic dye SYPRO-Orange [33]. The thermal shift reactions were prepared at room temperature (RT) and contained, in a final volume of 180 μL, 1 μM of the respective Vip3 proteins (filtered through 0.45 μm cellulose acetate filter), 15× of SPYRO-Orange (diluted in storage buffer-I) and storage buffer-I up to 180 μL. Eight replicates (20 μL) of the parental and chimeric proteins plus a negative control (15× of SPYRO-Orange and storage buffer-I) were incubated for 5 min at RT and centrifuged at 141× *g* for 1 min prior to analysis with the StepOnePlus™ Real-Time PCR System (Thermo Fisher Scientific, Waltham, MA, USA). Thermal shift assays program was carried out as follows: Reporter: ROX; Passive Reference: None; Run Method: Mode Continuous, Program 2 min at 25 °C, Temperature Ramp 4% (4 °C/min), 2 min at 99 °C. The data were exported to an Excel file to determine the Tm of the respective Vip3 proteins by plotting the negative of the first derivative of the fluorescence as a function of temperature-dFv/dT, where Fv and T at (t+1)-t represent the increment of fluorescence and temperature between each measurement.

$$-dFv/dT = (Fv(t + 1) - Fv(t))/(T(t + 1) - T(t)) \tag{1}$$

The Tm values of the respective Vip3 proteins were compared with one-way ANOVA analysis and datasets statistically significant ($\alpha < 0.05$), were analyzed by the multiple comparison Tukey post hoc test ($\alpha < 0.05$).

5.4. Insect Colonies and Toxicity Assays

Insects were reared and bioassays performed at the insectaries of the University of Valencia (for *S. exigua* and *S. littoralis*, Spain), Public University of Navarra (for *H. armigera, M. brassicae*, Spain), University of Tennessee (for *S. frugiperda* and *A. gemmatalis*, Knoxville, TN, USA), and Chinese Academy of Agricultural Sciences (for *O. furnacalis*, Haidan district, Beijing, China) at 25 °C, 70% RH, 16:8 L/D photoperiod (*S. exigua, S. frugiperda, S. littoralis, M. brassicae, H. armigera*, and *A. gemmatalis*) and 27 °C, 80% RH, 16:8 h L/D photoperiod (*O. furnacalis*), respectively. The insect colonies of *S. exigua, S. littoralis, H. armigera, M. brassicae*, and *O. furnacalis* had been reared for several generations in laboratory conditions without exposure to any insecticide. In the case of *S. frugiperda* and *A. gemmatalis* the insects were purchased from Benzon Research Inc. (Carlisle, PA, USA). The laboratory insect colonies of *S. exigua, S. littoralis, M. brassicae*, and *H. armigera* were reared in a growth wheat germ-based semi-synthetic diet [34], while *O. furnacalis* had been reared using standard rearing techniques without exposure to any insecticide [35]. In the case of *S. frugiperda* and *A. gemmatalis*, they were reared with meridic diet (#F9772, Frontier Agricultural Sciences, Newark, DE, USA). The same diets and rearing conditions were used in the bioassays with the parental proteins and chimeric Vip3 proteins.

Different methodologies were used in the bioassays depending on the insect species tested. For *S. exigua, S. littoralis, S. frugiperda, H. armigera*, and *A. gemmatalis*, bioassays were performed on neonates using surface contamination. Briefly, two pairs of different concentrations were dispensed on the diet surface. Prior to the sample application, the surface of the diet was sterilized under UV light for 10 min. A volume of 50–75 μL of each concentration was applied on the surface of solidified diet (2 cm² multiwell plates, Bio-Cv-16, C-D International) and let dry in a flow hood. Once dried, one larva was transferred to each well using a brush. In the case of *O. furnacalis*, bioassays were performed on neonates using diet incorporation assays [36], while for *M. brassicae* bioassays were performed on L2 instar larvae using a droplet feeding method [37]. To determine the effect of the domain exchange on toxicity, bioassays with Vip3 proteins were performed with two different concentrations (chosen as to give a discriminant mortality, range of mortality for each insect species between 1% and 99%) in at least two different insect generations (Table 1). Thirty-two neonates were used for each protein concentration for *S. exigua, S. littoralis, S. frugiperda*, and *A. gemmatalis*; 28 in the case of *H. armigera* and

M. brassicae; for *O. furnacalis* 42 neonates were tested. Mortality (number of dead larvae) was scored after 7 days for *S. exigua*, *S. littoralis*, *S. frugiperda*, *H. armigera*, *M. brassicae*, and *O. furnacalis*; while for *A. gemmatalis* mortality was scored after 5 days. Only data from bioassays with <10% control mortality were considered.

Determination of the LC_{50} (concentration of protein killing 50% of tested individuals) for the toxic parental and chimeric proteins was done for *O. furnacalis* (concentration range 0.04–50 µg/g) and *S. frugiperda* (concentration range 0.01–3 µg/cm^2). For *S. frugiperda* a set of 16–32 neonates per concentration (7–8 concentrations of the respective Vip3 proteins) were tested under the same conditions as described above for bioassays, and bioassays replicated twice. The number of dead larvae was recorded after 7 days of exposure. In the case of *O. furnacalis*, neonates were introduced to individual wells of 48-well trays containing 9–11 concentrations of purified toxin, which were tested with a total of 96 larvae per concentration. Trays were incubated as per the rearing conditions above and mortality and survivor weight were recorded after 7 days of exposure. Bioassays were repeated with two insect generations. The storage buffer was used to dilute the parental and chimeric Vip3 proteins and as negative control. Bioassay data were subjected to nonlinear regression using the software GraphPad Prism7 to obtain the LC_{50} of the parental proteins and chimeric Vip3 proteins, which were compared the parental proteins vs. the chimeric proteins with the statistical analysis extra-sum-square *F* test (α 0.05) (Table S1).

Supplementary Materials: The following are available online at http://www.mdpi.com/2072-6651/12/2/99/s1, Figure S1. Thermal shift assays and multiple comparison of the thermal transitions of the parental and chimeric proteins. The dashed vertical lines in the thermal shift assays curves indicate the Tm (measured in °C) of respective thermal transitions. C- indicates the fluorescence intensity due to the SPYRO-Orange 15× in 20 mM Tris 500 mM NaCl pH 8.6. The thick line indicates the comparison of the Tm by one-way ANOVA (α 0.05). The dashed line indicates the multiple comparison analyzed by Tukey's range test (α 0.05). "****" indicates a p value less than 0.0001 and "ns" indicates a p value greater than 0.05. Figure S2. Expression of the chimeric Vip3 proteins (Vip3_ch3 and Vip3_ch6). (A) SDS-PAGE of different dilutions of the pellet and supernatant of the Vip3_ch3 and Vip3_ch6 proteins. (B) Western blot analysis of different dilutions of the pellet and supernatant of the respective chimeric Vip3 proteins. The dilutions of the lysates were made with 50 mM phosphate buffer, 500 mM NaCl (pH 8.0), while the pellets were dissolved in the same volume of the supernatant and the dilutions were made with 50 mM phosphate buffer, 500 mM NaCl (pH 8.0). The arrowhead indicates the protein band corresponding to the chimeric Vip3 proteins. M1: Molecular Mass Marker "PINK Plus Prestained Protein Ladder" (Genedirex). M2: Molecular Mass Marker "Precision Plus Protein™ Dual Color Standards" (Biorad) developed with "Precision Protein™ Strep Tactin-HRP conjugate. Availability of data and material: Sequences encoding the Vip3 chimeras have been deposited in GenBanK with the following accession numbers: Vip3 chimera 1 (MH363727), Vip3 chimera 2 (MH363728), Vip3 chimera 3 (MH363729), Vip3 chimera 4 (MH363730), Vip3 chimera 5 (MH363731), and Vip3 chimera 6 (MH363732). Table S1. Comparison analyses of the respective dose-response assays (LC50 values) of the parental and chimeric Vip3 proteins in S. frugiperda and O. furnacalis. Table S2. Construction of the chimeric Vip3 proteins from the Vip3Aa and Vip3Ca proteins. Table S3. Primers used in construction and sequencing of the genes encoding the chimeric Vip3 proteins.

Author Contributions: J.F. and J.G.-C. conceived and designed the experiments. J.G.-C., R.F.d.S., Y.W. and J.C. performed the experiments. J.G.-C., and J.F. analyzed the data. J.G.-C., J.L.J.-F., P.C., K.H. and J.F. wrote the paper. All authors have read and agreed to the published version of the manuscript.

Funding: This research was supported to JF by the Spanish Ministry of Economy and Competitivity (Grants Ref. AGL2015-70584-C02-1-R and RTI2018-095204-B-C21), by the Generalitat Valenciana (GVPROMETEOII-2015-001), and by European FEDER funds. JGC was recipient of a PhD grant from the Spanish Ministry of Economy and Competitivity (grant ref. BES-2013-065848 and EEBB-I-17-12367). Support was also provided to JLJ-F by an Agriculture and Food Research Initiative Foundational Program competitive grant (No. 2018-67013-27820) from the USDA National Institute of Food and Agriculture, and to KH by a grant "Key Project for Breeding Genetically Modified Organisms" from China (grant ref. 2016ZX08003-001).

Acknowledgments: We thank R. González-Martínez for their help in rearing the insect colonies.

Conflicts of Interest: The authors declare no conflict of interest.

References

1. Palma, L.; Muñoz, D.; Berry, C.; Murillo, J.; Caballero, P. *Bacillus thuringiensis* toxins: An overview of their biocidal activity. *Toxins* **2014**, *6*, 3296–3325. [CrossRef]

2. Jouzani, G.S.; Valijanian, E.; Sharafi, R. *Bacillus thuringiensis*: A successful insecticide with new environmental features and tidings. *Appl. Microbiol. Biotechnol.* **2017**, *101*, 2691–2711.

3. Adang, M.N.; Crickmore, N.; Jurat-Fuentes, J.L. Diversity of *Bacillus thuringiensis* crystal toxins and mechanism of action. In *Advances in Insect Physiology 2014, Volume 47: Insect Midgut and Insecticidal Proteins*; Dhadialla, T.S.S., Ed.; Academic Press: San Diego, CA, USA, 2014.

4. Chakroun, M.; Banyuls, N.; Bel, Y.; Escriche, B.; Ferré, J. Bacterial Vegetative Insecticidal Proteins (Vip) from Entomopathogenic Bacteria. *Microbiol. Mol. Biol. Rev.* **2016**, *80*, 329–350. [CrossRef] [PubMed]

5. Crickmore, N.; Zeigler, D.R.; Schnepf, E.; Van Rie, J.; Lereclus, D.; Baum, J.; Bravo, A.; Dean, D.H. Bacillus thuringiensis Toxin Nomenclature. Available online: http://www.lifesci.sussex.ac.uk/Home/Neil_Crickmore/Bt/ (accessed on 2 December 2019).

6. Kunthic, T.; Watanabe, H.; Kawano, R.; Tanaka, Y.; Promdonkoy, B.; Yao, M.; Boonserm, P. pH regulates pore formation of a protease activated Vip3Aa from *Bacillus thuringiensis*. *Biochim. Biophys. Acta Biomembr.* **2017**, *1859*, 2234–2241. [CrossRef] [PubMed]

7. Palma, L.; Scott, D.J.; Harris, G.; Din, S.U.; Williams, T.L.; Roberts, O.J.; Young, M.T.; Caballero, P.; Berry, C. The Vip3Ag4 insecticidal protoxin from *Bacillus thuringiensis* adopts a tetrameric configuration that is maintained on proteolysis. *Toxins* **2017**, *9*, 165. [CrossRef] [PubMed]

8. Zack, M.D.; Sopko, M.S.; Frey, M.L.; Wang, X.; Tan, S.Y.; Arruda, J.M.; Letherer, T.T.; Narva, K.E. Functional characterization of Vip3Ab1 and Vip3Bc1: Two novel insecticidal proteins with differential activity against lepidopteran pests. *Sci. Rep.* **2017**, *7*, 11112. [CrossRef] [PubMed]

9. Gomis-Cebolla, J. Mining of new insecticidal protein genes plus determination of the insecticidal spectrum and mode of action of Bacillus thuringiensis Vip3Ca protein. Ph.D. Thesis, University of Valencia, Valencia, Spain, 22 February 2019.

10. Kunthic, T.; Surya, W.; Promdonkoy, B.; Torres, J.; Boonserm, P. Conditions for homogeneous preparation of stable monomeric and oligomeric forms of activated Vip3A toxin from *Bacillus thuringiensis*. *Eur Biophys. J.* **2017**, *46*, 257–264. [CrossRef] [PubMed]

11. Lee, M.K.; Miles, P.; Chen, J.S. Brush border membrane binding properties of *Bacillus thuringiensis* Vip3A toxin to *Heliothis virescens* and *Helicoverpa zea* midguts. *Biochem. Biophys. Res. Commun.* **2006**, *339*, 1043–1047. [CrossRef]

12. Sena, J.A.; Hernández-Rodríguez, C.S.; Ferré, J. Interaction of *Bacillus thuringiensis* Cry1 and Vip3Aa proteins with *Spodoptera frugiperda* midgut binding sites. *Appl. Environ. Microbiol.* **2009**, *75*, 2236–2237. [CrossRef]

13. Gouffon, C.; Van Vliet, A.; Van Rie, J.; Jansens, S.; Jurat-Fuentes, J.L. Binding sites for *Bacillus thuringiensis* Cry2Ae toxin on heliothine brush border membrane vesicles are not shared with Cry1A, Cry1F, or Vip3A toxin. *Appl. Environ. Microbiol.* **2011**, *77*, 3182–3188. [CrossRef]

14. Chakroun, M.; Ferré, J. In vivo and in vitro binding of Vip3Aa to *Spodoptera frugiperda* midgut and characterization of binding sites by ^{125}I radiolabeling. *Appl. Environ. Microbiol.* **2014**, *80*, 6258–6265. [CrossRef] [PubMed]

15. Abdelkefi-Mesrati, L.; Rouis, S.; Sellami, S.; Jaoua, S. Prays oleae midgut putative receptor of *Bacillus thuringiensis* vegetative insecticidal protein Vip3LB differs from that of cry1Ac toxin. *Mol. Biotechnol.* **2009**, *43*, 15–19. [CrossRef]

16. Gomis-Cebolla, J.; Ruiz de Escudero, I.; Vera-Velasco, N.M.; Hernández-Martínez, P.; Hernández-Rodríguez, C.S.; Ceballos, T.; Palma, L.; Escriche, B.; Caballero, P.; Ferré, J. Insecticidal spectrum and mode of action of the *Bacillus thuringiensis* Vip3Ca insecticidal protein. *J. Invertebr. Pathol.* **2017**, *142*, 60–67. [CrossRef] [PubMed]

17. Banyuls, N.; Hernández-Rodríguez, C.S.; Van Rie, J.; Ferré, J. Critical amino acids for the insecticidal activity of Vip3Af from *Bacillus thuringiensis*: Inference on structural aspects. *Sci. Rep.* **2018**, *8*, 7539. [CrossRef] [PubMed]

18. Sellami, S.; Jemli, S.; Abdelmalek, N.; Cherif, M.; Abdelkefi-Mesrati, L.; Tonusi, S.; Jamoussi, K. A novel Vip3Aa16-Cry1Ac chimera toxin: Enhancement of toxicity against *Ephestia kuehniella* structural study and molecular docking. *Int. J. Biol. Macromol.* **2018**.

19. Quan, Y.; Ferré, J. Structural Domains of the *Bacillus thuringiensis* Vip3Af protein unraveled by tryptic digestion of alanine mutants. *Toxins* **2019**, *11*, 368. [CrossRef] [PubMed]

20. Wang, Y.; Wang, J.; Fu, X.; Nageotte, J.R.; Silverman, J.; Bretsnyder, E.C.; Chen, D.; Rydel, T.J.; Bean, G.J.; Li, K.S.; et al. *Bacillus thuringiensis* Cry1Da_7 and Cry1B.868 protein interactions with novel receptors allow control of resistant fall armyworms, *Spodoptera frugiperda* (J.E. Smith). *Appl. Environ. Microbiol* **2019**, *85*, e00579-19. [CrossRef]

21. Palma, L.; Hernández-Rodríguez, C.S.; Maeztu, M.; Hernández-Martínez, P.; Ruíz de Escudero, I.; Escriche, B.; Muñoz, D.; Van Rie, J.; Ferré, J.; Caballero, P. Vip3Ca, a novel class of vegetative insecticidal proteins from *Bacillus thuringiensis*. *Appl. Environ. Microbiol.* **2012**, *78*, 7163–7165. [CrossRef]

22. Ruíz de Escudero, I.; Banyuls, N.; Bel, Y.; Maeztu, M.; Escriche, B.; Muñoz, D.; Caballero, P.; Ferré, J. A screening of five *Bacillus thuringiensis* Vip3A proteins for their activity against lepidopteran pests. *J. Invertebr. Pathol.* **2014**, *117*, 51–55.

23. GM Approval Database. Available online: http://www.isaaa.org/gmapprovaldatabase/ (accessed on 2 December 2019).

24. Rang, C.; Gil, P.; Neisner, N.; Van Rie, J.; Frutos, R. Novel vip3-related protein from Bacillus thuringiensis. *Appl. Environ. Microbiol.* **2005**, *71*, 6276–6281. [CrossRef]

25. Lemes, A.R.N.; Figueiredo, C.S.; Sebastião, I.; Marques da Silva, L.; Da Costa Alves, R.; De Siqueira, H.Á.A.; Lemos, M.V.F.; Fernandes, O.A.; Desidério, J.A. Cry1Ac and Vip3Aa proteins from *Bacillus thuringiensis* targeting Cry toxin resistance in *Diatraea flavipennella* and *Elasmopalpus lignosellus* from sugarcane. *Peer J.* **2017**, *5*, e2866. [CrossRef]

26. Gomis-Cebolla, J.; Wang, Y.; Quan, Y.; He, K.; Walsh, T.; James, B.; Downes, S.; Kain, W.; Wang, P.; Leonard, K.; et al. Analysis of cross-resistance to Vip3 proteins in eight insect colonies, from four insect species, selected for resistance to *Bacillus thuringiensis* insecticidal proteins. *J. Invertebr. Pathol.* **2018**, *155*, 64–70. [CrossRef] [PubMed]

27. Yang, J.; Quan, Y.; Sivaprasath, P.; Shabbir, M.Z.; Wang, Z.; Ferré, J.; He, K. Insecticidal activity and synergistic combinations of ten different Bt toxins against *Mythimna separata* (Walker). *Toxins* **2018**, *10*, 454. [CrossRef]

28. Lemes, A.R.N.; Figueiredo, C.S.; Sebastião, I.; Desidério, J.A. Synergism of the *Bacillus thuringiensis* Cry1, Cry2, and Vip3 Proteins in *Spodoptera frugiperda* Control. *Appl. Biochem. Biotechnol.* **2019**, *188*, 798–809.

29. Kahn, T.W.; Chakroun, M.; Williams, J.; Walsh, T.; James, B.; Monserrate, J.; Ferré, J. Efficacy and resistance management potential of a modified Vip3C protein for control of *Spodoptera frugiperda* in maize. *Sci. Rep.* **2018**, *8*, 16204. [CrossRef] [PubMed]

30. Palma, L.; Ruiz de Escudero, I.; Maeztu, M.; Caballero, P.; Muñoz, D. Screening of vip genes from a Spanish *Bacillus thuringiensis* collection and characterization of two Vip3 proteins highly toxic to five lepidopteran crop pests. *Biol. Control* **2013**, *66*, 141–149. [CrossRef]

31. Chakroun, M.; Bel, Y.; Caccia, S.; Abdelkefi-Mesrati, L.; Escriche, B.; Ferré, J. Susceptibility of *Spodoptera frugiperda* and *S. exigua* to *Bacillus thuringiensis* Vip3A insecticidal protein. *J. Invertebr. Pathol.* **2012**, *110*, 334–339. [CrossRef]

32. Bradford, M.M. A rapid and sensitive method for the quantitation of microgram quantities of protein utilizing the principle of protein-dye binding. *Anal. Biochem.* **1976**, *72*, 248–254. [CrossRef]

33. Lavinder, J.J.; Hari, S.B.; Sullivan, B.J.; Magliery, T.J. High-throughput thermal scanning: A general rapid dye-binding thermal shift screen for proteinengineering. *J. Am. Chem. Soc* **2009**, *131*, 3794–3795. [CrossRef]

34. Greene, G.L.; Leppla, N.C.; Dickerson, W.A. Velvetbean caterpillar: A rearing procedure and artificial medium. *J. Econ. Entomol.* **1976**, *69*, 487–488. [CrossRef]

35. Song, Y.; Zhou, D.; He, K. Studies on mass rearing of Asian corn borer: Development of a satisfactory non-agar semi-artificial diet and its use. *Acta Phytophylacica Sin.* **1999**, *26*, 324–328.

36. He, K.; Wang, Z.; Wen, L.; Bai, S.; Ma, X.; Yao, Z. Determination of baseline susceptibility to Cry1Ab protein for Asian corn borer (Lep., Crambidae). *J. Appl. Entomol.* **2005**, *129*, 407–412. [CrossRef]

37. Hughes, P.R.; Wood, H.A. A synchronous peroral technique for the bioassay of insect viruses. *J. Invert. Pathol.* **1981**, *37*, 154–159. [CrossRef]

Article

Reduced Membrane-Bound Alkaline Phosphatase Does Not Affect Binding of Vip3Aa in a *Heliothis virescens* Resistant Colony

Daniel Pinos [1], Maissa Chakroun [1,†], Anabel Millán-Leiva [1], Juan Luis Jurat-Fuentes [2], Denis J. Wright [3], Patricia Hernández-Martínez [1] and Juan Ferré [1,*]

[1] Department of Genetics, Instituto de Biotecnología y Biomedicina (BIOTECMED), Universitat de València, 46100 Burjassot, Spain; daniel.pinos@uv.es (D.P.); chakrounmaissa7@gmail.com (M.C.); anabel.millan@uv.es (A.M.-L.); patricia.hernandez@uv.es (P.H.-M.)

[2] Department of Entomology and Plant Pathology, University of Tennessee, Knoxville, TN 37996, USA; jurat@utk.edu

[3] Department of Life Sciences, Imperial College London, Silwood Park Campus, Ascot, Berks SL5 7PY, UK; d.wright@imperial.ac.uk

* Correspondence: juan.ferre@uv.es

† Current address: Centre de Biotechnologie de Sfax, Sfax, Tunisia.

Received: 7 May 2020; Accepted: 17 June 2020; Published: 19 June 2020

Abstract: The Vip3Aa insecticidal protein from *Bacillus thuringiensis* (Bt) is produced by specific transgenic corn and cotton varieties for efficient control of target lepidopteran pests. The main threat to this technology is the evolution of resistance in targeted insect pests and understanding the mechanistic basis of resistance is crucial to deploy the most appropriate strategies for resistance management. In this work, we tested whether alteration of membrane receptors in the insect midgut might explain the >2000-fold Vip3Aa resistance phenotype in a laboratory-selected colony of *Heliothis virescens* (Vip-Sel). Binding of ^{125}I-labeled Vip3Aa to brush border membrane vesicles (BBMV) from 3rd instar larvae from Vip-Sel was not significantly different from binding in the reference susceptible colony. Interestingly, BBMV from Vip-Sel larvae showed dramatically reduced levels of membrane-bound alkaline phosphatase (mALP) activity, which was further confirmed by a strong downregulation of the membrane-bound alkaline phosphatase 1 (*HvmALP1*) gene. However, the involvement of HvmALP1 as a receptor for the Vip3Aa protein was not supported by results from ligand blotting and viability assays with insect cells expressing *HvmALP1*.

Keywords: *Bacillus thuringiensis*; insecticidal proteins; insect resistance; tobacco budworm

Key Contribution: The biochemical characterization of a Vip3Aa-resistant colony of *H. virescens* shows that binding to receptors in the midgut is not affected and discards the role of mALP as a Vip3Aa receptor. This study suggests that Vip3A resistance may occur through mechanisms other than those commonly found for Cry proteins.

---

1. Introduction

The polyphagous pest *Heliothis virescens* (L.) (Lepidoptera: Noctuidae) is well known for producing substantial economic losses, particularly in cotton production, due to its ability to evolve resistance to different synthetic control products such as methyl parathion or pyrethroids [1,2]. As an alternative approach, genetically modified crops expressing Cry and Vip3A insecticidal protein genes from *Bacillus thuringiensis* (Bt crops) were introduced in 1996 for the control of this and other pests. However, extensive use threatens their effectiveness and cases of field-evolved practical resistance have already been reported for some lepidopteran and coleopteran pests [3].

Gene pyramiding has been proposed as an effective strategy for insect resistance management in Bt crops [4]. This approach consists of combined production of distinct insecticidal Bt proteins in the same plant, and its success heavily relies on the expressed insecticidal proteins having distinct mode of action, commonly defined as not sharing binding sites in target tissues [5,6].

Although the mechanism of action and receptors for Cry proteins have been widely studied [7], little is known about the biochemical mechanisms that underlie the action of Vip3A proteins. Several studies have shown that Vip3A proteins do not share binding sites with Cry1 or Cry2 proteins, yet their damage to the midgut epithelium resembles Cry action [8–11]. Supported by the lack of shared binding sites, transgenic corn and cotton varieties pyramided with Cry1, Cry2, and Vip3A genes are currently commercialized in several countries.

Knowledge of the biochemical and genetic factors involved in resistance is crucial to design management practices that delay the appearance of resistance and allow its rapid detection and ways to overcome it. The genetic potential to evolve resistance to Vip3A has already been shown in some laboratory-selected insect species such as *H. virescens* [12], *Spodoptera litura* [13], *Helicoverpa armigera* and *Helicoverpa punctigera* [14], *Spodoptera frugiperda* [15,16], and *Helicoverpa zea* [17]. However, the biochemical basis of resistance to Vip3A has only been studied in a laboratory-selected colony of *H. armigera*, for which alteration of binding sites was not the cause of resistance [18].

In the present study, we aimed to determine the biochemical basis of >2000-fold resistance to Vip3A in a *H. virescens* colony (Vip-Sel). In a previous study with this colony, resistance was shown to be polygenic, conferring little cross-resistance to Cry1Ab and no cross-resistance to Cry1Ac [19]. A transcriptomic analysis detected significant differences in gene expression compared to a susceptible strain, with 420 over-expressed and 1569 under-expressed genes in Vip-Sel [20]. Results herein support that Vip3Aa binding is not significantly altered in Vip-Sel compared to susceptible *H. virescens* and that membrane bound alkaline phosphatase (mALP) is not involved in Vip3Aa binding.

2. Results

2.1. Vip3Aa Binding to Midgut Brush Border Membrane Vesicles (BBMV)

In testing whether binding of Vip3Aa was altered in larvae from the Vip3A-resistant (Vip-Sel) compared to the reference susceptible (Vip-Unsel) colony, we measured binding of radiolabeled Vip3Aa to BBMV from the two colonies. Binding analyses showed specific Vip3Aa binding for BBMV from both colonies, with similar homologous competition curves (Figure 1a). A high percentage (35–40% of the input labeled toxin) of non-specific binding, i.e., not blocked by high concentrations of unlabeled Vip3Aa competitor, was detected, as previously reported [11,18]. The K_d and R_t values estimated from the competition curves (Table 1) indicated that Vip3Aa binds with low affinity to a high number of binding sites in BBMV from *H. virescens*. No major differences were found for these equilibrium binding parameters between the two *H. virescens* colonies, suggesting that binding alteration is not mechanistically related to Vip3Aa resistance in Vip-Sel.

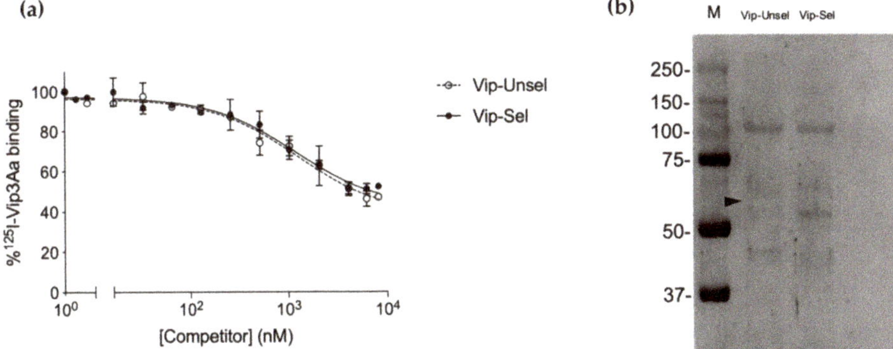

Figure 1. Analysis of ^{125}I-Vip3Aa binding to BBMV from susceptible (Vip-Unsel) and resistant (Vip-Sel) colonies of *H. virescens*. (**a**) Homologous competition binding assays of BBMV from the two colonies with ^{125}I-Vip3Aa, using increasing concentrations of unlabeled Vip3Aa as a competitor. Each data point represents the mean of two replicates performed in technical duplicates (±SEM). (**b**) Ligand blot of BBMV proteins from Vip-Unsel and Vip-Sel colonies probed with Vip3Aa. Lane M, protein molecular weight marker (in kDa) (Precision Plus Protein ™ Dual Color Standards, Bio-Rad, St. Louis, MO, USA). The black arrow indicates expected molecular weight of mALP (*ca.* 66 kDa).

Table 1. Equilibrium K_d (dissociation constant) and R_t (concentration of binding sites) binding parameters estimated from Vip3Aa homologous competition assays with BBMV from resistant (Vip-Sel) and susceptible (Vip-Unsel) *H. virescens* insects.

Strain	Mean ± SEM [1]	
	K_d (nM)	R_t (pmol/mg) [2]
Susceptible	138 ± 18	443 ± 66
Resistant	161 ± 34	443 ± 109 [1]

[1] Values are the mean of two replicates. [2] Values are expressed in picomoles per milligram of BBMV protein.

2.2. Reduced ALP Levels in the Vip3Aa-Resistant H. virescens Colony

During the evaluation of BBMV quality, we determined and compared the specific activities of alkaline phosphatase (ALP) and aminopeptidase-N (APN) as brush border membrane marker enzymes in midgut homogenates and BBMV preparations from Vip-Unsel and Vip-Sel colonies (Figure 2). The specific APN activity in midgut homogenates from both colonies was around 12 mU/mg, while in the BBMV preparations it was around 70 mU/mg, indicating an enrichment of APN activity of around 5.8 folds. Importantly, no significant differences (Student's *t*-test, $p > 0.05$) in APN activity were observed between the midgut homogenates or BBMV from Vip-Unsel and Vip-Sel colonies. In agreement with the 5.8-fold enrichment value from APN activity comparisons, specific ALP activity was 7.44 mU/mg in midgut homogenates and 42.5 mU/mg in the BBMV from the Vip-Unsel colony. In contrast, dramatically reduced ALP activity was detected in both midgut homogenate (1.15 mU/mg) and BBMV (1.88 mU/mg) samples from the Vip-Sel colony. While unexpected, this observation is in line with reports of reduced ALP levels in Cry1-resistant lepidopteran species, including *H. virescens* [21–24]. Consequently, we further explored the extremely reduced ALP activity in Vip-Sel to determine whether it was due to a loss of enzymatic function or reduced gene expression.

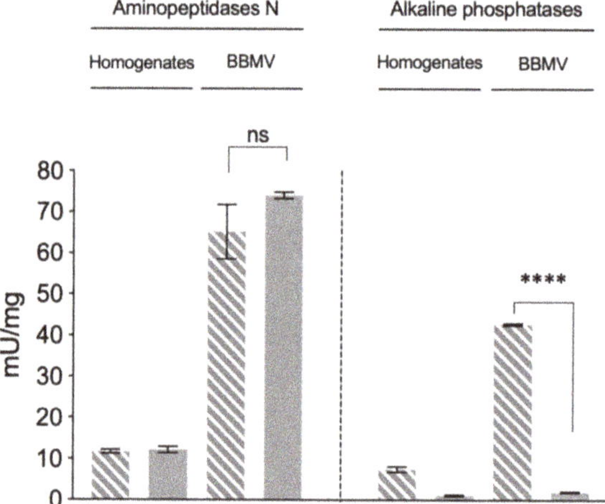

Figure 2. Enzymatic activities in homogenates and BBMV from the two colonies of *H. virescens* (dashed-grey bars: Vip-Unsel; grey bars: Vip-Sel). Each bar represents the mean of three replicates (±SEM). Asterisks represent significant difference (Student's *t*-test, **** $p < 0.0001$).

Electrophoretic comparison of BBMV proteins from the two *H. virescens* colonies showed a protein band of ~66 kDa for the Vip-Unsel colony that was almost imperceptible in the BBMV from the Vip-Sel colony (Figure 3a). Western blotting indicated the presence of ALP in the ~66-kDa protein band, and confirmed the highly reduced levels of this protein in the Vip-Sel colony (Figure 3b). The composition of the ~66-kDa protein band and its relative abundance in the two *H. virescens* colonies were determined by liquid chromatography coupled to mass spectrometry (LC-MS) analysis. The spectra for the most abundant protein detected and identified in the ~66-kDa band matched to membrane-bound alkaline phosphatase (mALP) from *H. virescens* (Genbank Accession No. ABR88230). According to the exponentially modified protein abundance index (emPAI) expressing the proportional protein content in a protein mixture, the abundance ratio of mALP between Vip-Unsel and Vip-Sel was 22.7 folds.

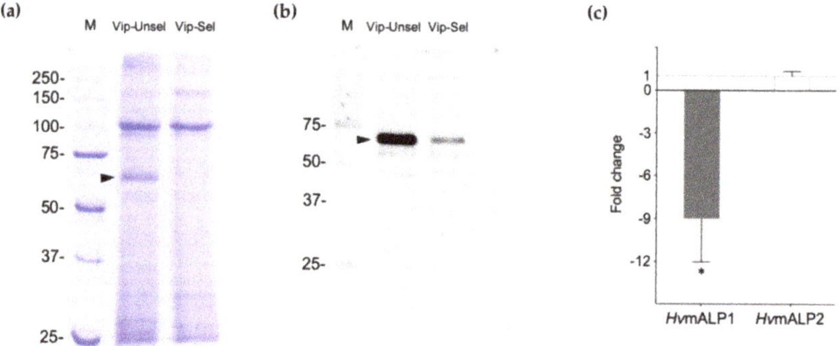

Figure 3. Analysis of membrane ALP levels in the susceptible (Vip-Unsel) and resistant (Vip-Sel) colonies of *H. virescens*. (**a**) Protein gel electrophoresis (SDS–PAGE) of BBMV from the two colonies. (**b**) Western blot performed with anti-ALP antibody against BBMV from the two colonies. The black arrow indicates mALP (ca. 66 kDa). Lane M, protein marker (molecular weight in kDa). (**c**) Membrane ALP expression levels in Vip-Sel colony using transcript levels in Vip-Unsel colony as a reference. Fold changes calculated by REST-MCS Software. Bars represent the mean of three independent experiments (±SD, * $p < 0.05$).

To test if the reduced mALP protein levels in Vip-Sel were controlled at the transcriptional level, we performed real-time quantitative PCR (RT-qPCR) with mRNA extracted from total RNA from the two colonies. Transcript levels for two *H. virescens* mALP genes, *HvmALP1* (Accession No. FJ416470.1) and *HvmALP2* (Accession No. FJ416471.1), were analyzed. Compared to insects from the Vip-Unsel colony, larvae from the Vip-Sel colony had significant (*p*-value < 0.05) nine-fold downregulation of the *HvmALP1* gene, while transcript levels for *HvmALP2* were not different between colonies (Figure 3c). These results support that reduced ALP enzyme activity in BBMV from Vip-Sel compared to Vip-Unsel is due to reduced expression of *HvmALP1* in the Vip-Sel colony.

2.3. Functional Role of HvmALP1 in Vip3Aa Binding

Since *H. virescens* ALP was proposed to play a role in binding of Cry1 proteins to the midgut membrane [25], we used ligand blotting to test whether mALP was involved in Vip3Aa binding. Binding of Vip3Aa to blots of resolved BBMV proteins was detected with anti-Vip3Aa antisera. No differences in the Vip3Aa-binding band pattern were detected between both colonies, in agreement with the binding results with radiolabeled Vip3Aa. However, no Vip3Aa binding was observed at the mALP position (~66 kDa) (Figure 1b).

To further discard mALP as a functional Vip3Aa receptor, we cloned and transiently expressed the *HvmALP1* gene in cultured (Sf21) insect cells and performed cell viability tests after challenge with Vip3Aa. Transfection was successful, as transfected cells showed ~5-fold increased specific ALP activity compared with non-transfected cells or cells transfected with the empty plasmid (Figure 4a). However, after a challenge with Vip3Aa, the viability of transfected cells was not significantly different (Student's *t*-test; *p* > 0.05) from that of the control cells (Figure 4b), confirming that mALP does not serve as a functional receptor for Vip3Aa during the toxicity process.

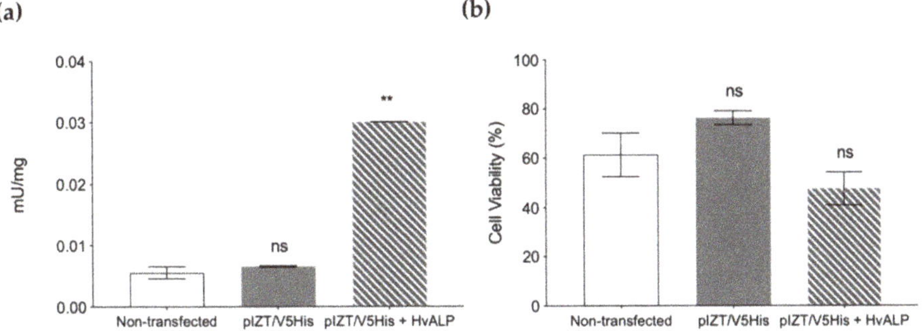

Figure 4. Specific ALP enzymatic activity and viability assays of Sf21 cells producing the HvmALP1 isoform. (**a**) Alkaline phosphatase enzymatic activity on non-transfected cells (empty bars), cells transfected with empty plasmid (grey bars, and plasmid with *HvmALP1* (dashed-grey bars). (**b**) Cell viability after 24 h of Vip3Aa intoxication (300 µg/mL final concentration) on the same three cell types. Each value represents the mean (±SEM). Means were compared by Student's *t*-test (** *p* < 0.01).

3. Discussion

The use of resistant insect strains isolated from the field or selected in the laboratory has been a powerful tool to understand the biochemical and genetic bases of resistance to Bt insecticidal proteins. Many studies have found that the alteration of membrane receptors is a common mechanism conferring high levels of resistance to Cry proteins [26–28]. In the case of Cry1 proteins, an important body of literature identifies aminopeptidase N, ABC transporters, cadherins and membrane alkaline phosphatases as main receptors, and identifies their alterations in association with resistance [29,30]. In contrast, three candidate receptors have been proposed for Vip3A proteins, including the *Spodoptera spp.* ribosomal protein S2 [31], the fibroblast growth factor receptor-like protein [32] and the scavenger receptor class C-like protein [33], yet their role in resistance has not been established.

In the present work, we aimed to determine whether alteration of membrane receptors was the basis for the observed 2040-fold resistance to Vip3Aa in the Vip-Sel colony of *H. virescens*. Results from binding assays with BBMV and radiolabeled Vip3Aa did not detect significant differences between the susceptible and resistant colonies, suggesting no involvement of binding site alteration in resistance. This conclusion was further supported by results from ligand blotting, where no differences between the binding patterns of Vip3Aa to BBMV proteins from the two colonies were observed. Similar results were reported for a laboratory-selected Vip3A-resistant colony of *H. armigera* [18], suggesting that high levels and narrow spectrum of Vip3A resistance may develop by mechanisms other than alteration of Vip3Aa binding sites.

Even though differences in binding were not found, a dramatic reduction in the ALP enzymatic activity was detected in midgut samples from the resistant compared to susceptible colony. Western blotting and RT-qPCR analyses showed that the decreased activity was due to a reduction in the amount of mALP protein, which was controlled at the transcriptional level, in agreement with a previous study [20]. Downregulation or reduced levels of mALP in the midgut membrane have been observed as a common phenomenon in resistance to Cry1Ac in *H. virescens* [25], *Helicoverpa zea* [21], *Plutella xylostella* [22], and *Helicoverpa armigera* [24]; to Cry1F in *S. frugiperda* [23]; to Cry1C in *Spodoptera litura* [34]; and even in *Aedes aegypti* resistant to Bt subsp. *israeliensis* (Bti) [35]. The fact that Cry1Ac and Cry1C do not share binding sites [36] suggests that the role of ALP downregulation in resistance may not be related to reduced Cry binding, but may represent a physiological response to resistance. In agreement with this hypothesis, susceptibility of Sf21 cells expressing HvmALP1 was not significantly different to Vip3Aa, supporting that ALP is not a functional receptor for Vip3Aa in *H. virescens*. In addition, in a Cry1Ac-resistant strain of *P. xylostella*, altered expression of different genes (including the *PxmALP*) was reported to be *trans*-regulated by upregulation of a mitogen-activated protein kinase, which was linked to resistance [22]. Similar *trans*-regulation of genes involved in resistance to Bt has also been observed for APN in *Trichoplusia ni* resistant to Cry1Ac [37] and *Ostrinia nubilalis* resistant to Cry1Ab [38], and for both APN and an ABCC transporter in *Bombyx mori* resistant to Cry1Ab [39]. Further research should test the involvement of this control mechanism in downregulation of *mALP* in Vip-Sel and other Bt-resistant colonies.

The two studies so far focused on the underlying mechanism of resistance to Vip3Aa proteins share a similar feature in that in vitro binding is not reduced [18] (and the present work). According to these results, mechanisms other than binding site alteration seem to be responsible for conferring specific and high-level resistance to Vip3A. This contrasts with the fact that the alteration of membrane receptors is a common mechanism conferring high levels of resistance to Cry proteins. Better knowledge of the mode of action of Vip3A proteins will help shed light on the biochemical basis of resistance to these proteins.

4. Conclusions

The results herein show lack of significant Vip3Aa binding alterations in a resistant colony of *H. virescens*. These observations are in contrast to most cases of high levels of resistance to Cry proteins for which decreased binding is commonly detected. In addition, this study provides evidence of downregulation of membrane bound alkaline phosphatase (mALP) in the Vip3Aa-resistant colony, although results do not support involvement of mALP as a receptor for the Vip3A protein.

5. Materials and Methods

5.1. Insects

Two colonies of *H. virescens* originating from the same field population collected in Arkansas (USA) were used in this study: Vip-Sel (Vip3Aa-resistant) and Vip-Unsel (Vip3Aa susceptible). The process of selection of the Vip-Sel colony with Vip3Aa has been previously described [12,19]. After 13 generations of selection, the LC_{50} of the Vip-Sel colony was 2300 µg/mL, representing a 2040-fold resistance ratio relative to the control Vip-Unsel colony. Both colonies were reared at the Imperial College London, Silwood Park campus (UK), and frozen larvae were sent for analysis to the Universitat de València (Spain).

5.2. BBMV Preparation and Enzyme Activity Assays

Brush border membrane vesicles (BBMV) from 3rd instar *H. virescens* larval midguts from Vip-Sel and Vip-Unsel colonies were prepared according to the differential magnesium precipitation method [40]. Isolated BBMV were flash frozen in liquid nitrogen and kept at −80 °C until used. The protein concentration of the BBMV preparations was determined by the method of Bradford using bovine serum albumin (BSA) as a standard [41].

Alkaline phosphatase (ALP) and leucine aminopeptidase (APN) activities were used as brush border membrane enzymatic markers to determine the quality of the BBMV preparations [42]. Specific ALP activity was determined by chromogenic detection of *p*-nitrophenyl phosphate (*p*NPP) substrate hydrolysis into *p*-nitrophenol, and specific APN activity was detected by hydrolysis of L-leu-*p*-nitroanilide substrate into *p*-nitroanilide. In both cases, chromogenic variation was measured on 1 µg of either BBMV or midgut homogenate at 405 nm (Infinite m200, Tecan, Mannedorf, Switzerland). Two different batches of BBMV were used and all enzymatic activity assays were performed in triplicate. Means values for enzyme activities from Vip-Unsel and Vip-Sel were compared by Student's *t*-test at a 5% level of significance.

For measuring specific ALP enzymatic activity in cultured Sf21 cells, a 1.6-mL suspension of each cell type (non-transfected, transfected with empty plasmid, and transfected with plasmid with *HvmALP1*) was used. Culture cells were centrifuged, washed twice with 300 µL of phosphate buffered saline (PBS) and then resuspended in 50 µL of PBS. Protein concentration was determined by the method of Bradford and specific ALP activity measured as above.

5.3. Vip3Aa Protein Expression and Purification

The Vip3Aa16 (Vip3Aa) protein (NCBI Accession No. AAW65132) was overexpressed in recombinant *Escherichia coli* BL21 carrying the *vip3Aa16* gene. Protein expression and lysis was carried out following the conditions described elsewhere [43]. Soluble Vip3Aa in the cell lysate was purified by two different methodologies. For binding and cell viability assays, Vip3Aa was partially purified by isoelectric point precipitation (IPP), activated with trypsin treatment and further purified by anion-exchange chromatography, as previously described [11]. For ligand assays, affinity chromatography purification was carried out using a HiTrap chelating HP column (GE Healthcare, Uppsala, Sweden) and then activated with trypsin, as described [11].

5.4. Vip3Aa Labeling and Binding Experiments

Purified Vip3Aa activated protein (25 µg) was labeled with 0.5 mCi of ^{125}I using the chloramine T method [11]. The labeled protein was separated from the excess of free ^{125}I in a PD10 desalting column (GE Healthcare, Uppsala, Sweden) and the purity of the ^{125}I-labeled Vip3Aa was checked by autoradiography. The specific activity of the labeled protein was 2.2 mCi/mg.

Binding assays were performed as described elsewhere [11]. Prior to being used, BBMV were centrifuged and resuspended in binding buffer (20 mM Tris, 150 mM NaCl, 1 mM $MnCl_2$, pH 7.4, 0,04% Blocking reagent from Sigma Aldrich, St. Louis, MO, USA). Competition binding experiments were conducted by incubating 1.4 µg of BBMV protein with 0.65 nM ^{125}I-Vip3Aa in a final volume of 0.1 mL of binding buffer for 90 min at 25 °C in the presence of increasing amounts of unlabeled Vip3Aa. After incubation, samples were centrifuged at 16,000× *g* for 10 min and the pellet was washed once with 500 µL of ice-cold binding buffer. Radioactivity retained in the pellet was measured in a model 2480 WIZARD[2] gamma counter. Data from the competition experiments were analyzed to determine equilibrium binding parameters, dissociation constant (K_d), and concentration of binding sites (R_t) using the LIGAND software [44].

5.5. Western and Ligand Blotting

For the detection of ALP proteins in BBMV by Western blotting, BBMV (20 µg) were suspended in ice-cold PBS and heat denatured before separation on a SDS–10% PAGE gel. The resolved BBMV proteins were transferred to a nitrocellulose filter (Protran 0.45 µm NC, GE Healthcare, Uppsala, Sweden) using a BioRad Mini Trans-Blot system (Bio-Rad, Hercules, CA, USA) at 4 °C in blotting buffer (39 mM Glycine, 48 mM Tris-HCl, 0.037% SDS, 10% methanol, pH 8.5) for 1 h at constant voltage (100 V). After transfer, the nitrocellulose filter was blocked in blocking buffer (PBS, 0.1% Tween 20, 5% skimmed milk powder) overnight at 4 °C. After blocking and washing with PBST (PBS, 0.1% Tween 20) three times (5 min each), incubation with primary antibody against the membrane-bound form of ALP from *Anopheles gambiae* (generously provided by M. Adang, University of Georgia, USA) was performed for 90 min at a 1:5000 dilution at room temperature (RT). The membrane was then washed with PBST three times for 5 min each and then incubated with secondary antibody (goat anti-rabbit conjugated to horseradish peroxidase (HRP) at a 1:10,000 dilution) for 1 h at RT. After being washed with PBST three times for 5 min each, the membrane was developed using enhanced chemiluminescence (ECL Prime Western Blotting detection reagent, GE Healthcare, Uppsala, Sweden) in an ImageQuant LAS 4000 (GE Healthcare, Uppsala, Sweden), according to the manufacturer's instructions.

Ligand blotting for the detection of BBMV proteins binding Vip3Aa protein was performed with BBMV proteins resolved and immobilized as described above for Western blotting. The nitrocellulose membrane was blocked for 1 h at 4 °C in blocking buffer (5% skimmed milk), and after three washes for 5 min each with PBST buffer, it was incubated overnight at 4 °C with blocking buffer (1% skimmed milk) supplemented with affinity chromatography-purified Vip3Aa at a final concentration of 4 µg/mL. After washing with PBST three times for 5 min each, the membrane was incubated with primary antibody against Vip3Aa at a 1:5000 dilution for 1 h at RT. After three washing steps with PBST (5 min each), membranes were incubated with secondary antibody (goat anti-rabbit conjugated to HRP) for 1 h at RT. To visualize the marker, Precision Protein™ Streptactin-HRP conjugate (Bio-Rad, St. Louis, MO, USA) was used following the manufacturer's instructions. Upon washing three times (5 min each) with PBST, the membrane was developed as described for Western blotting.

5.6. Proteomic Analysis

After resolving BBMV proteins from Vip-Sel and Vip-Unsel colonies by SDS–10% PAGE, the gel was stained with Coomassie blue (Thermo Scientific™, Waltham, MA, USA). The band corresponding to the expected molecular weight of ALP (~66 kDa) was cut out and subjected to analysis by nano-electron spray ionization (nano-ESI) followed by tandem mass spectrometry (qQTOF) in a 5600 TripleTOF (AB Sciex, Madrid, Spain) system. Results were analyzed with ProteinPilot v5.0 software and the relative amount of the proteins detected was estimated using the exponentially modified protein abundance index (emPAI) as described elsewhere [45].

5.7. RT-qPCR

Relative expression levels for *HvmALP1* and *HvmALP2* isoforms (accession numbers FJ416470 and FJ416471, respectively) were determined by reverse transcription quantitative polymerase-chain reaction (RT-qPCR). For this purpose, total RNA of dissected midguts from both colonies (Vip-Unsel and Vip-Sel) was isolated using RNAzol (MRC Inc., Cincinnati, OH, USA) according to the manufacturer's protocol. Each RNA (1 µg) was reverse-transcribed to cDNA using random hexamers and oligo (dT) by following the instructions provided in the Prime-Script RT Reagent Kit (Perfect Real Time from TaKaRa Bio Inc., Otsu Shiga, Japan). RT-qPCR was carried out in a StepOnePlus Real-Time PCR system (Applied Biosystems, Foster City, CA, USA). Reactions were performed using 5× HOT FIREpol EVAGreen qPCR Mix Plus (ROX) from Solis BioDyne (Tartu, Estonia) in a total reaction volume of 25 µL.

Specific primers for *HvmALP1*, *HvmALP2* and *Rps18* (endogenous control) genes were as described elsewhere [24]. The REST MCS software was used for gene expression analysis [46].

5.8. Expression Vector Construction

The full-length *HvmALP1* transcript was amplified from cDNA of *H. virescens* larvae and cloned into pET30a as described elsewhere [47]. Purified plasmid DNA was digested with *Eco*RI and *Not*I to excise the full-length sequence and ligate it in frame into *Eco*RI-*Not*I sites of the pIZT/V5His vector (Thermo Scientific™, Waltham, MA, USA), to generate the pIZT/V5His/*HvmALP1* construct. Ligation products were transformed into *E. coli* strain DH5α and transformants checked for correct insertion by sequencing (University of Tennessee Sequencing Facility, Knoxville, TN, USA). Purified plasmid was used to transform *E. coli* strain DH10β and liquid cultures of LB medium supplemented with Zeocin (25 µg/mL) were used to amplify the vector. To purify the plasmids for transfection, the NucleoSpin® Plasmid kit (Macherey-Nagel, Düren, Germany) was used. Double digestion with *Eco*RI and *Not*I (which cleaved the full-length *HvmALP1* insert) and 1% agarose gel electrophoresis were performed to check plasmid and/or insert integrity. The concentration of plasmid DNA was measured with a Thermo Scientific™ Nanodrop™ 2000 Spectrophotometer.

5.9. Transient Expression of HvmALP1 in Sf21 Cells

Cultured Sf21 insect cells, originally derived from *S. frugiperda*, were maintained in 25 cm^2 tissue culture flasks (Nunc T25 flasks, Thermo Scientific™, Waltham, MA, USA) at 25 °C with 4 mL of Gibco® Grace's Medium (1×) (Life Technologies™, Paisley, UK) supplemented with 10% heat-inactivated fetal bovine serum (FBS).

For transient expression, cells were seeded on 12-well plates with the same medium without FBS at ca. 70% confluency and transfected with 0.5 µg of the pIZT/V5His/*HvmALP1* or pIZT/V5His plasmid using Cellfectin® II Reagent (Thermo Scientific™, Waltham, MA, USA), following manufacturer's instructions. Five hours post-transfection, the medium was replaced with fresh medium containing 10% FBS. After 24 h, cells were examined using a confocal microscope (Olympus, FV1000MPE, Tokyo, Japan) equipped with the appropriate filter for green fluorescent protein (GFP) detection as transfection marker. The enzymatic activity of alkaline phosphatase was then measured as explained above.

5.10. Cell Viability Assays

Viability of transfected Sf21 cells exposed for 24 h to Vip3Aa was measured using the MTT (3-[4,5-dimethylthiazol-2-yl]-2,5-diphenyltetrazolium bromide) assay. Preliminary assays were performed to determine a final Vip3A concentration of 300 µg/mL as resulting in ~50% loss of viability in the control cell line (data not shown). Briefly, cells (100 µL per well) were transferred to 96-well ELISA plates (flat bottom) and incubated at 25 °C for at least 45 min. Then, 10 µL of trypsin-activated Vip3Aa toxin was added to each well (300 µg/mL final concentration). As negative and positive controls, 10 µL of Tris buffer (Tris 20 mM, NaCl 150 mM, pH 9) and 10 µL of 2% Triton X-100 were used, respectively. After 24 h of incubation at 25 °C, cell viability was assessed by applying 20 µL of CellTiter 96® AQueous One Solution Reagent (Promega, Madison WI, USA) to each well and incubating for 2 h at 25 °C. Absorbance was measured at 490 nm (Infinite m200, Tecan, Mannedorf, Switzerland). The percentage of viable cells was obtained as described elsewhere [48]. Mean values in the transfected cells against the non-transfected cells were compared by Student's *t*-test at 5% level of significance.

Author Contributions: Conceptualization, P.H.-M. and J.F.; Formal analysis, D.P., M.C., and A.M.-L.; Funding acquisition, J.L.J.-F. and J.F.; Methodology, D.P. and M.C.; Resources, J.L.J.-F. and D.J.W.; Supervision, P.H.-M. and J.F.; Writing—original draft, D.P.; and Writing—review and editing, J.L.J.-F., D.J.W., P.H.-M., and J.F. All authors have read and agreed to the published version of the manuscript.

Funding: This research was supported by the Spanish Ministry of Science, Innovation and Universities (grant no. RTI2018-095204-B-C21). D.P. is recipient of a PhD grant from the Spanish Ministry of Science, Innovation and Universities (grant ref. FPU15/05652). Support for JLJ-F was provided by the NC246 Multistate Hatch project from the USDA National Institute for Food and Agriculture (NIFA). The proteomic analysis was performed at the Proteomics Facility of the Servei Central de Suport a la Investigació Experimental (SCSIE) at the Universitat de València (Valencia, Spain) that belongs to ProteoRed, PRB2-ISCII, supported by grant PT13/0001, of the PE I+D+I 2013-2016, funded by ISCII and FEDERPT13/0001.

Conflicts of Interest: The authors declare no conflict of interest. The funders had no role in the design of the study; in the collection, analyses, or interpretation of data; in the writing of the manuscript, or in the decision to publish the results.

References

1. Wolfenbarger, D.A.; Bodegas, P.R.; Flores, R. Development of resistance in *Heliothis* spp. in the Americas, Australia, Africa, and Asia. *Bull. Entomol. Soc. Am.* **1981**, *27*, 181–185. [CrossRef]
2. Terán-Vargas, A.P. Insecticide resistance of tobacco budworm in Southern Tamaulipas Mexico. In Proceedings of the Beltwide Cotton Conference, Nashville, TN, USA, 9–12 January 1996; National Cotton Council of America: Memphis, TN, USA; pp. 784–786.
3. Tabashnik, B.E.; Carrière, Y. Global patterns of resistance to Bt crops highlighting Pink Bollworm in the United States, China, and India. *J. Econ. Entomol.* **2019**, *112*, 2513–2523. [CrossRef]
4. Roush, R.T. Bt-transgenic crops: Just another pretty insecticide or a chance for a new start in resistance management? *Pestic. Sci.* **1997**, *51*, 328–334. [CrossRef]
5. Moar, W.J.; Anilkumar, K.J. Plant science: The power of the pyramid. *Science* **2007**, *318*, 1561–1562. [CrossRef]
6. Carrière, Y.; Crickmore, N.; Tabashnik, B.E. Optimizing pyramided transgenic Bt crops for sustainable pest management. *Nat. Biotechnol.* **2015**, *33*, 161–168. [CrossRef]
7. Adang, M.; Crickmore, N.; Jurat-Fuentes, J.L. Diversity of *Bacillus thuringiensis* crystal toxins and mechanism of action. In *Advances in Insect Physiology*; Dhadialla, T.S., Gill, S., Eds.; Academic Press: San Diego, CA, USA, 2014; Volume 47, pp. 39–87. [CrossRef]
8. Lee, M.K.; Miles, P.; Chen, J. Brush border membrane binding properties of *Bacillus thuringiensis* Vip3A toxin to *Heliothis virescens* and *Helicoverpa zea* midguts. *Biochem. Biophys. Res. Commun.* **2006**, *339*, 1043–1047. [CrossRef]
9. Sena, J.A.D.; Hernández-Rodríguez, C.S.; Ferré, J. Interaction of *Bacillus thuringiensis* Cry1 and Vip3A proteins with *Spodoptera frugiperda* midgut binding sites. *Appl. Environ. Microbiol.* **2009**, *75*, 2236–2237. [CrossRef]
10. Gouffon, C.; Van Vliet, A.; Van Rie, J.; Jansens, S.; Jurat-Fuentes, J.L. Binding sites for *Bacillus thuringiensis* Cry2Ae toxin on heliothine brush border membrane vesicles are not shared with Cry1A, Cry1F, or Vip3A toxin. *Appl. Environ. Microbiol.* **2011**, *77*, 3182–3188. [CrossRef]
11. Chakroun, M.; Ferré, J. In vivo and in vitro binding of Vip3Aa to *Spodoptera frugiperda* midgut and characterization of binding sites by ^{125}I radiolabeling. *Appl. Environ. Microbiol.* **2014**, *80*, 6258–6265. [CrossRef]
12. Pickett, B. Studies on Resistance to Vegetative (Vip3A) and Crystal (Cry1A) Insecticide Toxin of *Bacillus thuringiensis* in *Heliothis virescens* (Fabricius). Ph.D. Thesis, Imperial College London, London, UK, 2009.
13. Barkhade, U.P.; Thakare, A.S. Protease mediated resistance mechanism to *Cry1C* and *Vip3A* in *Spodoptera litura*. *Egypt Acad. J. Biol. Sci.* **2010**, *3*, 43–50. [CrossRef]
14. Mahon, R.J.; Downes, S.J.; James, B. Vip3A resistance alleles exist at high levels in Australian targets before release of cotton expressing this toxin. *PLoS ONE* **2012**, *7*, e39192. [CrossRef]
15. Bernardi, O.; Bernardi, D.; Horikoshi, R.J.; Okuma, D.M.; Miraldo, L.L.; Fatoretto, J.; Medeiros, F.C.; Burd, T.; Omoto, C. Selection and characterization of resistance to the Vip3Aa20 protein from *Bacillus thuringiensis* in *Spodoptera frugiperda*. *Pest. Manag. Sci.* **2016**, *72*, 1794–1802. [CrossRef]
16. Yang, F.; Morsello, S.; Head, G.P.; Sansone, C.; Huang, F.; Gilreath, R.T.; Kerns, D.L. F_2 screen, inheritance and cross-resistance of field-derived Vip3A resistance in *Spodoptera frugiperda* (Lepidoptera: Noctuidae) collected from Louisiana, USA. *Pest. Manag. Sci.* **2018**, *74*, 1769–1778. [CrossRef]

17. Yang, F.; Santiago-González, J.C.; Little, N.; Reisig, D.; Payne, G.; Ferreira Dos Santos, R.; Jurat-Fuentes, J.L.; Kurtz, R.; Kerns, D.L. First documentation of major Vip3Aa resistance alleles in field populations of *Helicoverpa zea* (Boddie) (Lepidoptera: Noctuidae) in Texas, USA. *Sci. Rep.* **2020**, *10*, 5867. [CrossRef]

18. Chakroun, M.; Banyuls, N.; Walsh, T.; Downes, S.; James, B.; Ferré, J. Characterization of the resistance to Vip3Aa in *Helicoverpa armigera* from Australia and the role of midgut processing and receptor binding. *Sci. Rep.* **2016**, *6*, 24311. [CrossRef]

19. Pickett, B.R.; Gulzar, A.; Ferré, J.; Wright, D.J. *Bacillus thuringiensis* Vip3Aa toxin resistance in *Heliothis virescens* (Lepidoptera: Noctuidae). *Appl. Environ. Microbiol.* **2017**, *83*, e03506–e03516. [CrossRef]

20. Ayra-Pardo, C.; Ochagavía, M.E.; Raymond, B.; Gulzar, A.; Rodríguez-Cabrera, L.; Rodríguez de la Nova, C.; Morán-Bertot, I.; Terauchi, R.; Yoshida, K.; Matsumura, H.; et al. HT-SuperSAGE of the gut tissue of a Vip3Aa-resistant *Heliothis virescens* (Lepidoptera: Noctuidae) strain provides insights into the basis of resistance. *Insect Sci.* **2019**, *26*, 479–498. [CrossRef]

21. Caccia, S.; Moar, W.J.; Chandrashekhar, J.; Oppert, C.; Anilkumar, K.J.; Jurat-Fuentes, J.L.; Ferré, J. Association of Cry1Ac toxin resistance in *Helicoverpa zea* (Boddie) with increased alkaline phosphatase levels in the midgut lumen. *Appl. Environ. Microbiol.* **2012**, *78*, 5690–5698. [CrossRef]

22. Guo, Z.; Kang, S.; Chen, D.; Wu, Q.; Wang, S.; Xie, W.; Zhu, X.; Baxter, S.W.; Zhou, X.; Jurat-Fuentes, J.L.; et al. MAPK signaling pathway alters expression of midgut ALP and ABCC genes and causes resistance to *Bacillus thuringiensis* Cry1Ac toxin in diamondback moth. *PLoS Genet.* **2015**, *11*, e1005124. [CrossRef]

23. Jakka, S.R.K.; Gong, L.; Hasler, J.; Banerjee, R.; Sheets, J.J.; Narva, K.; Blanco, C.A.; Jurat-Fuentes, J.L. Field-evolved Mode 1 resistance of the fall armyworm to transgenic Cry1Fa-expressing corn associated with reduced Cry1Fa toxin binding and midgut alkaline phosphatase expression. *Appl. Environ. Microbiol.* **2016**, *82*, 1023–1034. [CrossRef]

24. Jurat-Fuentes, J.L.; Karumbaiah, L.; Jakka, S.R.K.; Ning, C.; Liu, C.; Wu, K.; Jackson, J.; Gould, F.; Blanco, C.A.; Portilla, M.; et al. Reduced levels of membrane-bound alkaline phosphatase are common to lepidopteran strains resistant to Cry toxins from *Bacillus thuringiensis*. *PLoS ONE* **2011**, *6*, e17606. [CrossRef]

25. Jurat-Fuentes, J.L.; Adang, M.J. Characterization of a Cry1Ac-receptor alkaline phosphatase in susceptible and resistant *Heliothis virescens* larvae. *Eur. J. Biochem.* **2004**, *271*, 3127–3135. [CrossRef]

26. Ferré, J.; Van Rie, J. Biochemistry and genetics of insect resistance to *Bacillus thuringiensis*. *Annu. Rev. Entomol.* **2002**, *47*, 501–533. [CrossRef]

27. Ferré, J.; Van Rie, J.; MacIntosh, S.C. Insecticidal genetically modified crops and insect resistance management (IRM). In *Integration of Insect-Resistant Genetically Modified Crops Within IPM Programs*; Romeis, J., Shelton, A.M., Kennedy, G.G., Eds.; Springer Science Business Media BV: Dordrecht, The Netherlands, 2008; pp. 41–86. [CrossRef]

28. Peterson, B.; Bezuidenhout, C.C.; Van den Berg, J. An overview of mechanisms of Cry toxin resistance in lepidopteran insects. *J. Econ. Entomol.* **2017**, *110*, 362–377. [CrossRef]

29. Pigott, C.R.; Ellar, D.J. Role of receptors in *Bacillus thuringiensis* crystal toxin activity. *Microbiol. Mol. Biol. Rev.* **2007**, *71*, 255–281. [CrossRef]

30. Sato, R.; Adegawa, S.; Li, X.; Tanaka, S.; Endo, H. Function and role of ATP-Binding Cassette Transporters as receptors for 3D-Cry toxins. *Toxins* **2019**, *11*, 124. [CrossRef]

31. Singh, G.; Sachdev, B.; Sharma, N.; Seth, R.; Bhatnagar, R.K. Interaction of *Bacillus thuringiensis* vegetative insecticidal protein with Ribosomal S2 protein triggers larvicidal activity in *Spodoptera frugiperda*. *Appl. Environ. Microbiol.* **2010**, *76*, 7202–7209. [CrossRef]

32. Jiang, K.; Hou, X.; Han, L.; Tan, T.; Cao, Z.; Cai, J. Fibroblast growth factor receptor, a novel receptor for vegetative insecticidal protein Vip3Aa. *Toxins* **2018**, *10*, 546. [CrossRef]

33. Jiang, K.; Hou, X.; Tan, T.; Cao, Z.; Mei, S.; Yan, B.; Chang, J.; Han, L.; Zhao, D.; Cai, J. Scavenger receptor-C acts as a receptor for *Bacillus thuringiensis* vegetative insecticidal protein Vip3Aa and mediates the internalization of Vip3Aa via endocytosis. *PLoS Pathog.* **2018**, *14*, e1007347. [CrossRef]

34. Gong, L.; Wang, H.; Qi, J.; Han, L.; Hu, M.; Jurat-Fuentes, J.L. Homologs to Cry toxin receptor genes in a *de novo* transcriptome and their altered expression in resistant *Spodoptera litura* larvae. *J. Invertebr. Pathol.* **2015**, *129*, 1–6. [CrossRef]

35. Tetreau, G.; Bayyareddy, K.; Jones, C.M.; Stalinski, R.; Riaz, M.A.; Paris, M.; David, J.; Adang, M.J.; Després, L. Larval midgut modifications associated with *Bti* resistance in the yellow fever mosquito using proteomic and transcriptomic apaproaches. *BMC Genom.* **2012**, *13*, 248. [CrossRef]

36. Jakka, S.R.K.; Ferré, J.; Jurat-Fuentes, J.L. Cry toxin binding sites and their use in strategies to delay resistance evolution. In *Bt Resistance—Characterization and Strategies for GM Crops Producing Bacillus thuringiensis Toxins*; Soberón, M., Gao, Y., Bravo, A., Eds.; CAB International: Boston, MA, USA, 2015; pp. 138–149. [CrossRef]

37. Tiewsiri, K.; Wang, P. Differential alteration of two aminopeptidases N associated with resistance to *Bacillus thuringiensis* toxin Cry1Ac in cabbage looper. *Proc. Natl. Acad. Sci. USA* **2011**, *108*, 14037–14042. [CrossRef]

38. Coates, B.S.; Sumerford, D.V.; Siegfried, B.D.; Hellmich, R.L.; Abel, C.A. Unlinked genetic loci control the reduced transcription of aminopeptidase N 1 and 3 in the European corn borer and determine tolerance to *Bacillus thuringiensis* Cry1Ab toxin. *Insect Biochem. Mol. Biol.* **2013**, *43*, 1152–1160. [CrossRef]

39. Chen, Y.; Li, M.; Islam, I.; You, L.; Wang, Y.; Li, Z.; Ling, L.; Zeng, B.; Xu, J.; Huang, Y.; et al. Allelic-specific expression in relation to *Bombyx mori* resistance to Bt toxin. *Insect Biochem. Mol. Biol.* **2014**, *54*, 53–60. [CrossRef]

40. Wolfersberger, M.; Luethy, P.; Maurer, A.; Parenti, P.; Sacchi, F.V.; Giordana, B.; Hanozet, G.M. Preparation and partial characterization of amino acid transporting brush border membrane vesicles from the larval midgut of the cabbage butterfly (*Pieris brassicae*). *Comp. Biochem. Physiol.* **1987**, *86*, 301–308. [CrossRef]

41. Bradford, M.M. A rapid and sensitive method for the quantitation of microgram quantities of protein utilizing the principle of protein-dye binding. *Anal. Biochem.* **1976**, *72*, 248–254. [CrossRef]

42. Hernández, C.S.; Rodrigo, A.; Ferré, J. Lyophilization of lepidopteran midguts: A preserving method for *Bacillus thuringiensis* toxin binding studies. *J. Invertebr. Pathol.* **2004**, *85*, 182–187. [CrossRef]

43. Abdelkefi-Mesrati, L.; Rouis, S.; Sellami, S.; Jaoua, S. *Prays oleae* midgut putative receptor of *Bacillus thuringiensis* vegetative insecticidal protein Vip3LB differs from that of Cry1Ac toxin. *Mol. Biol.* **2009**, *43*, 15–19. [CrossRef]

44. Munson, P.J.; Rodbard, D. Ligand: A versatile computerized approach for characterization of ligand-binding systems. *Anal. Biochem.* **1980**, *107*, 220–239. [CrossRef]

45. Ishihama, Y.; Oda, Y.; Tabata, T.; Sato, T.; Nagasu, T.; Rappsilber, J.; Mann, M. Exponentially modified protein abundance index (emPAI) for estimation of absolute protein amount in proteomics by the number of sequenced peptides per protein. *Mol. Cell. Proteom.* **2005**, *4*, 1265–1272. [CrossRef]

46. Pfaffl, M.W.; Horgan, G.W.; Dempfle, L. Relative expression software tool (REST) for group-wise comparison and statistical analysis of relative expression results in real-time PCR. *Nucleic Acids Res.* **2002**, *30*, e36. [CrossRef]

47. Perera, O.P.; Willis, J.D.; Adang, M.J.; Jurat-Fuentes, J.L. Cloning and characterization of the Cry1Ac-binding alkaline phosphatase (HvALP) from *Heliothis virescens*. *Insect Biochem. Mol. Biol.* **2009**, *39*, 294–302. [CrossRef]

48. Martínez-Solís, M.; Pinos, D.; Endo, H.; Portugal, L.; Sato, R.; Ferré, J.; Herrero, S.; Hernández-Martínez, P. Role of *Bacillus thuringiensis* Cry1A toxins domains in the binding to the ABCC2 receptor from *Spodoptera exigua*. *Insect Biochem. Mol. Biol.* **2018**, *101*, 47–56. [CrossRef]

Article

Oligomer Formation and Insecticidal Activity of *Bacillus thuringiensis* Vip3Aa Toxin

Ensi Shao [1,†], Aishan Zhang [1,†], Yaqi Yan [1], Yaomin Wang [1], Xinyi Jia [1], Li Sha [1], Xiong Guan [2], Ping Wang [3,*] and Zhipeng Huang [1,*]

[1] China National Engineering Research Center of JUNCAO Technology, College of Life Science, Fujian Agriculture and Forestry University, Fuzhou 350002, Fujian, China; es776@fafu.edu.cn (E.S.); aishanz712@126.com (A.Z.); yyq961202@163.com (Y.Y.); wangyaomin8409@163.com (Y.W.); jxy17735109183@163.com (X.J.); sha1@fafu.edu.cn (L.S.)

[2] Key Laboratory of Biopesticide and Chemical Biology (Ministry of Education), College of Plant Protection, Fujian Agriculture and Forestry University, Fuzhou 350002, Fujian, China; guanxfafu@126.com

[3] Department of Entomology, Cornell University, Geneva, NY 14456, USA

* Correspondence: pingwang@cornell.edu (P.W.); zhipengh6841@163.com (Z.H.); Tel.: +1-(315)-787-2348 (P.W.); +86-591-83789259 (Z.H.)

† These authors contributed equally to this work.

Received: 19 March 2020; Accepted: 21 April 2020; Published: 23 April 2020

Abstract: *Bacillus thuringiensis* (Bt) Vip3A proteins are important insecticidal proteins used for control of lepidopteran insects. However, the mode of action of Vip3A toxin is still unclear. In this study, the amino acid residue S164 in Vip3Aa was identified to be critical for the toxicity in *Spodoptera litura*. Results from substitution mutations of the S164 indicate that the insecticidal activity of Vip3Aa correlated with the formation of a >240 kDa complex of the toxin upon proteolytic activation. The >240 kDa complex was found to be composed of the 19 kDa and the 65 kDa fragments of Vip3Aa. Substitution of the S164 in Vip3Aa protein with Ala or Pro resulted in loss of the >240 kDa complex and loss of toxicity in *Spodoptera litura*. In contrast, substitution of S164 with Thr did not affect the >240 kDa complex formation, and the toxicity of the mutant was only reduced by 35%. Therefore, the results from this study indicated that formation of the >240 kDa complex correlates with the toxicity of Vip3Aa in insects and the residue S164 is important for the formation of the complex.

Keywords: *Bacillus thuringiensis*; Vip3A; *Spodoptera litura*; site-directed mutagenesis

Key Contribution: Our results correlated the formation of a >240 kDa protein complex with the insecticidal activity of Vip3Aa toxin. The residue S164 in Vip3Aa protein was identified to be important for the formation of the >240 kDa protein complex.

---

1. Introduction

The vegetative insecticidal proteins (VIPs) from *Bacillus thuringiensis* (Bt) have been used as important insecticidal proteins for control of insect pests [1–3]. Vip toxins are divided into four families, including Vip1, Vip2, Vip3 and Vip4 [3]. Vip1 and Vip2 proteins act as binary toxins against some species of coleopteran and hemipteran insect [4,5]. Only 1 Vip4 protein has so far been identified but shows no activity in insects [6]. Vip3 proteins contain approximately 787 amino acid residues, showing no sequence homology with Vip1, Vip2 and Vip4 proteins [3]. Vip3 proteins have a high insecticidal activity against a wide variety of lepidopteran pests [7]. Vip3A proteins do not share the binding sites with the Bt Cry proteins [8–11], so pyramiding Vip3A proteins and Cry proteins has been widely adopted in Bt-crops [12].

Although Vip3A toxins have already been applied in transgenic crops for the control of lepidopteran pests, current understanding of the mode of action of Vip3A proteins remains limited. It is commonly assumed that Vip3A toxins exert their insecticidal activity by going through a similar sequence of events as Bt Cry1A toxins [13]. So far, the structure of Vip3A toxin has not been solved. Its structural information has been derived only by *in-silico* modeling [14,15], though the structure of the Vip3B was recently reported [16]. Studies on proteolytic activation of Vip3A proteins have shown that by a proteolytical process Vip3A protoxins are cleaved to become several major fragments, generally including fragments of 62–66 kDa, 45 kDa, 33 kDa and 19–22 kDa [17–20]. The 62–66 kDa fragment from the C-terminus of Vip3A toxins has been determined to be the main product from proteolytic processing. The 45 kDa and 33 kDa fragments are products from further processing of the 62–66 kDa fragment [18]. The 19–22 kDa fragment contains the first 199 amino acids at the N-terminus of Vip3A [21]. It has been suggested that the 62–66 kDa fragment at the C-terminus in Vip3A toxin is the activated core of the toxin [22–24]. However, recent studies have indicated that both of the 19–22 kDa and the 62–66 kDa fragments are required for the stability and specificity of Vip3A toxins [20]. More recently, a ~340 kDa homo-tetramer, constituted by the 19–22 kDa and the 62–66 kDa fragments, has been identified from Vip3A after treatment with trypsin or insect midgut proteases [18]. However, whether the formation of this ~340 kDa homo-tetramer is essential for the insecticidal activity of Vip3A in insects remains unknown.

A recent study of Vip3Af by Ala scanning to cover 558 out of the 788 residues showed that the most Ala substitutions in Vip3Af significantly decreased the insecticidal activity, and the proteolytically processed fragments of the Vip3Af substitution mutants displayed six different patterns by SDS-PAGE analysis [14]. Further analysis indicated that Vip3Af mutants with different proteolytic patterns could form a variety of oligomeric products [21]. The substitution of the residue T167 or G168 by Ala in the predicted 19 kDa N-terminal fragment of Vip3Af did not change the proteolytic proccessing, but both substitutions significantly decreased the insecticidal activity [14]. Sequence alignments indicated that the amino acid residues from K152 to P171 are highly conserved among the Vip3A toxins [3].

Spodoptera litura is a polyphagous species and a major pest of many crops worldwide due to its vigorous defoliation [25]. *S. litura* is not susceptible to Bt Cry1A toxins but highly susceptible to Vip3A toxins [26,27]. In this study, we constructed Vip3Aa mutants by site directed mutagenesis and investigated the insecticidal activity of the mutants in *S. litura*. The amino acid residue S164 in Vip3Aa protein was identified to be critical for the toxicity of Vip3Aa. Investigation of the toxicity of Vip3Aa in *S. litura* by substitutions of S164 with different amino acid residues indicated that a protein oligomer formed with the 19 kDa and the 65 kDa fragments of Vip3Aa is the toxin core necessary for the insecticidal toxicity.

2. Results

2.1. Insecticidal Activity of Residue Substituted Vip3Aa Mutants Against Neonates of S. litura

The wild-type Vip3Aa protein and its mutants at K152, D154 and S164, respectively, were prepared through a glutathione S-transferase (GST) tagged protein purification system. Vip3Aa mutants with substitution of K152 or D154 with Ala were expressed as GST-Vip3Aa-K152A and GST-Vip3Aa-D154A fusion proteins. Vip3Aa mutants from substitution of S164 with Ala, Pro and Thr, respectively, were expressed as GST-Vip3Aa-S164A, GST-Vip3Aa-S164P and GST-Vip3Aa-S164T fusion proteins. The purified GST-Vip3Aa fusion proteins and the wild-type Vip3Aa (Vip3Aa-WT) were fed to neonates of *S. litura* to determine their insecticidal activity respectively. The bioassay results showed that the substitution mutations of K152A and D154A did not significantly change the toxicity of the toxin, in comparison with the wild-type Vip3Aa protein (Table 1). However, substitution of S164 with Ala or Pro completely abolished the toxicity of Vip3Aa. In contrast, substitution of S164 with the similar amino acid residue Thr only slightly reduced the insecticidal activity (35% reduction). Mortality of the control groups fed with 100 µg/mL or 250 µg/mL of purified GST tag protein were below 5% after 96 h feeding (results not shown).

Table 1. Insecticidal activity of Vip3Aa toxins in neonates of *S. litura*.

	LC$_{50}$ (95% CI, µg/mL)	Slope	X^2 (df, p)	Toxicity-Ratio
Vip3Aa-WT	1.69 (1.36–2.04)	3.50 ± 0.47	0.34 (4, 0.08)	1.0
GST-Vip3Aa-S164T	2.62 (2.24–3.17)	4.19 ± 0.47	8.88 (8, 1.11)	0.65
GST-Vip3Aa-S164A	>480	-	-	<0.0035
GST-Vip3Aa-S164P	>480	-	-	<0.0035
GST-Vip3Aa-K152A	2.23 (1.67–2.83)	3.16 ± 0.63	0.005 (3, 0.002)	0.76
GST-Vip3Aa-D154A	1.45 (1.36–1.55)	8.80 ± 0.99	2.74 (4, 0.69)	1.17

CI: confident interval.

2.2. Analysis Vip3Aa Fragments After Proteolytic Processing

To examine the difference in the proteolytic processing among the Vip3Aa-WT and three S164 mutants, each Vip3Aa protein was processed by trypsin or midgut proteases of *S. litura* and analyzed by SDS-PAGE after heat denaturation. The tryptic fragments from the three S164 mutants and Vip3Aa-WT contained major bands at 65 kDa, 35 kDa and 19 kDa and multiple weak bands from 29 kDa to 66.4 kDa (Figure 1). Vip3Aa proteins processed by midgut proteases of *S. litura* present a different pattern to the tryptic proteins. Besides the major fragments at 65 kDa, a band at 38 kDa and another at 30 kDa were observed in the midgut proteases processed Vip3Aa-WT and three S164 mutants. The band of 19 kDa was weak or invisible after in vitro proteolytic processing of Vip3Aa by the midgut proteases of *S. litura* (Figure 1). It could be observed that after proteolytic treatment with trypsin or midgut proteases, Vip3Aa-WT and three S164 mutants showed the same protein patterns. All trypsin digested Vip3Aa proteins contained a strong band at 19 kDa. Several other protein fragments were observed with the molecular weight from 14.3 kDa to 19 kDa from Vip3Aa-WT and three S164 mutants, although some bands were weak (Figure 1).

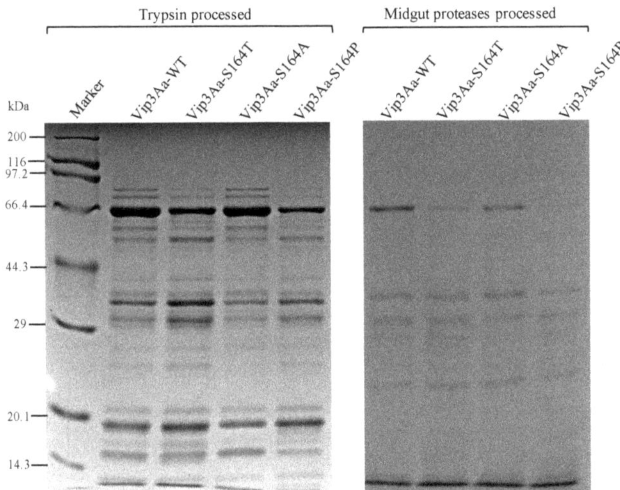

Figure 1. Analysis of Vip3Aa proteins after treatment by trypsin or midgut proteases of *S. litura*. Purified Vip3Aa-WT, Vip3Aa-S164T, Vip3Aa-S164A and Vip3Aa-S164P were *in vitro* digested by commercial trypsin or midgut proteases of *S. litura*. Processed proteins were mixed with 5 × SDS-PAGE sample buffer followed by heat denaturation and analyzed by the electrophoretic analysis in an SDS-PAGE gel.

2.3. Analysis of Vip3Aa Protein Complexes by Native PAGE After Proteolytic Processing

The same or similar protein digestion patterns were observed by SDS-PAGE from Vip3Aa-WT and three S164 mutants after proteolytic processing by either trypsin or midgut proteases of *S. litura*. Protein fragments from trypsin- or gut proteases-processed Vip3Aa proteins were then analyzed by native PAGE to identify the protein complexes. Two similar major bands, representing the protein complex 1 and 2, were observed from Vip3Aa-WT, Vip3Aa-S164T, Vip3Aa-S164A and Vip3Aa-S164P. However, a band, representing the protein complex 3, with a higher molecular weight than the two bands were observed from the trypsin-processed Vip3Aa-WT, Vip3Aa-S164T and Vip3Aa-S164P but not from Vip3Aa-S164A (Figure 2a). For the midgut proteases-processed Vip3Aa proteins, the band of protein complex 3 was observed from Vip3Aa-WT and Vip3Aa-S164T but not from Vip3Aa-S164A and Vip3Aa-S164P (Figure 2b). To estimate the molecular weight of the three protein complexes, the trypsin-processed Vip3Aa was analyzed by native SDS-PAGE. Two clear bands at ~240 kDa and ~200 kDa were observed (Figure 2c). The two bands in Figure 2c are assumed to the relatively dominant protein complex 1 and 2 in Figure 2a, and the molecular weight of complex 3 is predicted to be >240 kDa.

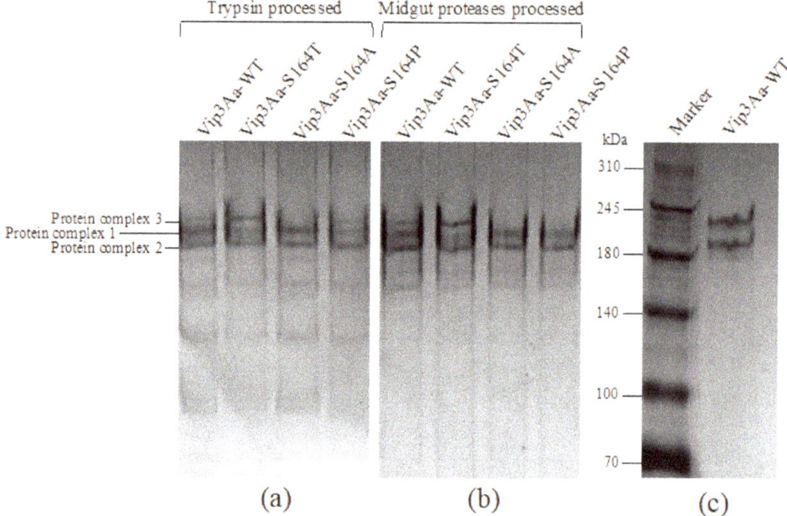

Figure 2. Analysis of native Vip3Aa proteins after proteolytic processing. Protease treated Vip3Aa-WT, Vip3Aa-S164T, Vip3Aa-S164A and Vip3Aa-S164P by either commercial trypsin or midgut proteases of *S. litura* larvae were analyzed by the electrophoretic analysis without heat denaturation. (**a**) Vip3Aa proteins after tryptic processing were analyzed in a native gel; (**b**): Vip3Aa proteins after processing by midgut proteases were analyzed in a native gel; (**c**): Vip3Aa proteins after tryptic processing were mixed with 5 × SDS-PAGE sample buffer without β-mercaptoethanol and analyzed in an SDS-PAGE gel. Protein complex 1, protein complex 2 and protein complex 3 in panel (**a**) indicate gel bands sliced from each lane in the native gel.

2.4. Composition of the Three Protein Complexes Formed from Vip3Aa Toxins after Tryptic Processing

To analyze the compositions of the three protein complexes, the bands corresponding to protein complexes 1, 2 and 3 (Figure 2a) were excised from the native PAGE gel (Figure 2a). The gel slices were mixed with SDS-PAGE sample buffer, heat denatured and loaded on SDS-PAGE gel to separate the proteins. All protein complexes contained the 65 kDa major fragment and multiple weak bands from 29 kDa to 66.4 kDa (Figure 3). Difference in composition was observed between the wild type and the mutants in the fragments below 20 kDa. A 19 kDa fragment was observed from protein complex 3 of trypsin-processed Vip3Aa-WT and Vip3Aa-S164T (Figure 3a). In comparison with the 19 kDa fragment, a slightly smaller fragment (17 kDa) was observed from the protein complex 3 from Vip3Aa-S164P (Figure 3a). An even smaller 15 kDa fragment was observed from protein complex 1 of Vip3Aa-WT and three mutants (Figure 3b). No protein bands below 20 kDa were observed from the protein complex 2 (Figure 3b). In addition, a peptide showing molecular weight around 95 kDa was observed from the protein complex 1 and 3 but not from the protein complex 2 (Figure 3).

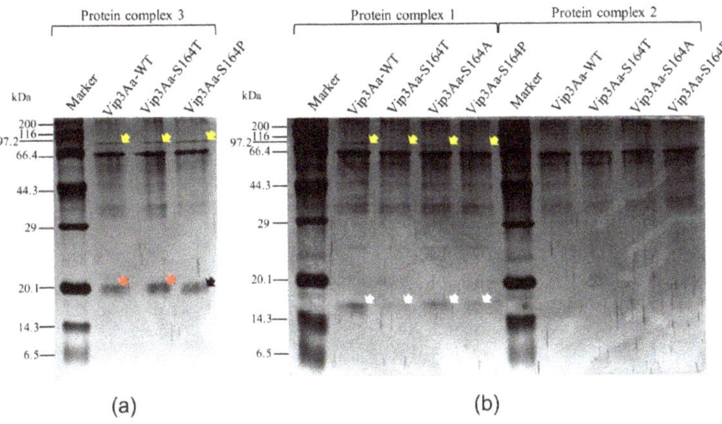

Figure 3. Separation of peptides from protein complexes of tryptic Vip3Aa proteins. Major protein bands representing different protein complexes in Figure 2a were sliced and separated in an SDS-PAGE gel. (**a**) peptides separated from the protein complex 3 in Figure 2a; (**b**) peptides separated from the protein complexes 1 and 2 respectively in Figure 2a. The yellow, red, black and white arrows indicate the bands of 95 kDa, 19 kDa, 17 kDa and 15 kDa fragments respectively.

2.5. Identification of Tryptic Fragments from the 15, 17 and 19 kDa Protein Fragments by Peptide Fingerprinting

The 15 and 19 kDa fragments, isolated from protein complexes 1 and 3 of Vip3Aa-WT and the 17 kDa fragment from protein complexes 3 of Vip3Aa-S164P were analyzed by nano LC-MS/MS to identify the protein fragments. The identified peptides derived from the 15, 17 and 19 kDa fragments were mapped to the amino acid sequence from D32 to K195 of Vip3Aa protein, located at the N terminal region (Figure 4).

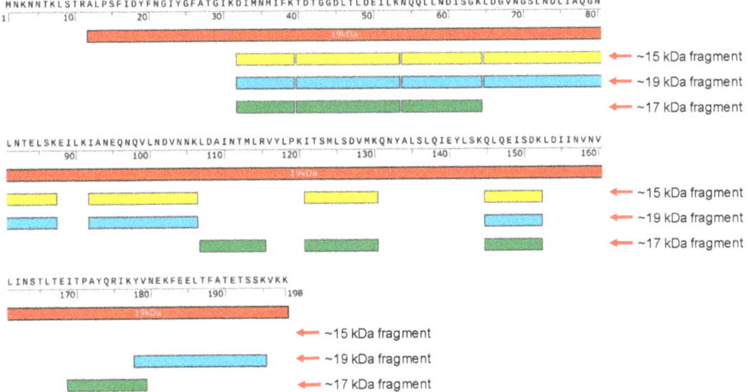

Figure 4. Schematic representation of the 15 kDa, 17 kDa and 19 kDa fragments from Vip3Aa protein. The first 198 amino acids at the N terminus of Vip3Aa were presented. The red box indicates amino acids corresponding to the N terminal 19 kDa fragment of Vip3Af. The yellow, blue and green boxes represent LC-MS/MS identified peptides from the 15 kDa, 19 kDa and 17 kDa fragments respectively.

2.6. Correlation of Toxicity of Vip3Aa Protein with the Formation of the Protein Complex 3 Composed of 19 kDa and 65 kDa Peptides

Trypsin-processed Vip3Aa-WT, Vip3Aa-S164T, Vip3Aa-S164A and Vip3Aa-S164P were fed to the neonates of *S. litura*, respectively, to assay their insecticidal activity. After 96 h, 100% mortality was observed by feeding *S. litura* larvae with 5 µg/mL and 50 µg/mL of trypsin-processed Vip3Aa-WT or Vip3Aa-S164T. In contrast, neither trypsin-processed Vip3Aa-S164A nor Vip3Aa-S164P showed significant toxicity to the larvae of *S. litura* (Figure 5). The insecticidal activity of trypsin treated Vip3Aa proteins was consistent to that of Vip3Aa protoxins (Table 1) and correlated with the formation of the protein complex 3 composed of the 19 kDa and the 65 kDa peptides (Figures 2a and 3a)

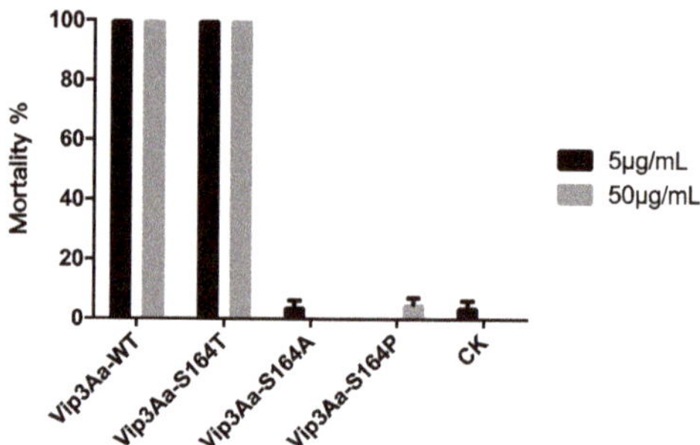

Figure 5. Mortality of *S. litura* larvae fed with trypsin-processed Vip3Aa proteins. After tryptic processing, Vip3Aa-WT, Vip3Aa-S164T, Vip3Aa-S164A and Vip3Aa-S164P were respectively fed to neonates of *S. litura* for 96 h to test their insecticidal activity. Error bars indicate the standard error of mortality among five replications.

3. Discussion

Previous studies have indicated that proteolytic processing of Vip3A proteins in insect midgut is a key step to exert the insecticidal activity [3,13]. In insect midgut, Vip3A proteins are processed by midgut proteases to produce a 62–66 kDa protease resistant toxic core from the C-terminal part of Vip3A. However, a recent study indicated that deletion of the first 198 residues at the N-terminus outside the ~65 kDa fragment region could lead to a complete loss of insecticidal activity and the resulting Vip3Aa fragments became sensitive to trypsin degradation [28]. Current studies have also shown that with treatment of Vip3A by trypsin, a 19~20 kDa peptide from the N-terminal region could bind with the C-terminal 62~65 kDa peptide, leading to the formation of a ~360 kDa homo-tetramer, which was tolerant to degradation in the protease-rich environment [29]. This 19~20 kDa peptide was proposed to play a functional role in protecting the 62~65 kDa peptide from proteolytic degradation and is necessary for the toxicity of the toxin in insects [30]. The K152 to E168, included in the N terminus 19~20 kDa peptide of Vip3Aa, were predicted to be a loop structure by the three-dimensional structure modeling software LOMETS (https://zhanglab.ccmb.med.umich.edu/LOMETS/). Significant decrease of toxicity of Vip3Af in *Spodoptera frugiperda* and *Agrotis segetum* was observed after substitution of T167 or E168 by Ala [14,20]. S164 was considered to be a polar amino acid located at the C terminus of the K152-E168 loop. Both K152 (carrying a basic polar side chain) and D154 (carrying an acidic polar side chain) were predicted to be at the N terminus of the K152-E168 loop. Consequently, in this study, we chose K152, D154 and S164 as our targets to analyze potential effects on the toxicity of Vip3Aa after substitution of these three amino acids respectively by Ala. Results of bioassay showed that only substitutions at S164 affect the toxicity of Vip3Aa in *S. litura*. Three main protein complexes were observed in protease treated Vip3Aa-WT and its three S164 mutants by native PAGE (Figure 2a,b). The protein complexes were composed of protein fragments of 19 and 65 kDa, 17 and 65 kDa, 15 and 65 kDa, or a single 65 kDa peptide only, respectively (Figure 3). The 95 kDa protein band was observed in protein complexes 1 and 3 but not protein complex 2 from each Vip3Aa protein. We interpret that the 95 kDa fragment is complexed with a 15 kDa peptide in protein complex 1 or a 19 kDa peptide in protein complex 3 with a 65 kDa peptide (Figure 3). LC-MS/MS analysis of the peptides between 15~19 KDa from the complexes indicated that the 15 kDa, 17 kDa and 19 kDa peptides were all from the N terminus of Vip3Aa protein, which corresponds to the previous reported domain I in Vip3Af protein [21]. Vip3Aa protoxin could be processed in vivo by the midgut proteases of *S. litura*. Toxicity of the Vip3Aa toxin, pretreated by midgut proteases, in *S. litura* should be similar to that of Vip3Aa protoxins. In order to build the relationship between the toxicity and the presence of the protein complex 3, mortality of *S. litura* fed with trypsin-processed Vip3Aa toxin was calculated and compared to the bioassay results by feeding *S. litura* with Vip3Aa protoxins (Table 1). Bioassay results showed that the trypsin-processed Vip3Aa-WT and Vip3Aa-S164T, in which the protein complex 3 was formed (composed of 19 and 65 kDa peptides), had significant toxicity in larvae of *S. litura* while the trypsin-processed Vip3Aa-S164A and Vip3Aa-S164P did not form the protein complex 3 with the 19 and 65 kDa peptides and completely lost the toxicity (Figure 5). These results are corresponding to the results of bioassay using Vip3Aa protoxins (Table 1). The toxicity of Vip3Af against *S. frugiperda* and *A. segetum* has been suggested to correlate with the transient formation of a tetramer composed of 20 kDa and 62 kDa peptides before the final processing to smaller fragments [14,21]. In this study, 100% mortality was observed in *S. litura* larvae fed with tryptic Vip3Aa-WT or Vip3Aa-S164T, while <10% mortality was observed in *S. litura* larvae fed with tryptic Vip3Aa-S164A or Vip3Aa-S164P (Figure 5). The protein complex 3 (composed of the 19 and 65 kDa fragments but not the 17 and 65 kDa fragments) could only be observed in tryptic Vip3Aa-WT or Vip3Aa-S164T (Figure 3). Consequently, formation of the protein complex 3 from Vip3Aa-WT and Vip3Aa-S164T was directly correlated to the toxicity of Vip3Aa in *S. litura*. A protein complex showing closed molecular weight to the protein complex 3 in Vip3Aa-WT and Vip3Aa-S164T was observed from tryptic Vip3Aa-S164P (Figure 2a) but not observed from midgut proteases-processed Vip3Aa-S164P (Figure 2b). Composition of this protein complex was identified to be the 17 kDa and the 65 kDa fragments (Figure 3a), different from the complex 3 from Vip3Aa-WT and Vip3Aa-S164T (Figure 4). This protein complex was degraded after treatment of Vip3Aa-S164P with midgut proteases of *S. litura*, while the protein complex 3 from Vip3Aa-WT and Vip3Aa-S164T remained stable (Figure 2b). Both the 17 kDa and the 19 kDa peptides were identified from the N terminal 199 amino acid residues of Vip3Aa proteins (Figure 4). These results indicated that the complete 19 kDa fragment is essential for the stability of the protein complex 3 which correlated to the insecticidal activity of Vip3Aa toxin.

The oligomer of Vip3A formed after proteolytic processing was observed through gel filtration chromatography analysis [14,21,29,31]. In this study, three protein complexes were observed from Vip3Aa-WT and its three mutants by native PAGE (Figure 2). Molecular weights of protein complexes 1 and 2 observed in the native SDS-PAGE gel were predicted to be ~240 kDa and ~200 kDa (Figure 2c). It is interesting that the protein complex 3 from tryptic Vip3Aa-WT could not be observed in the native SDS-PAGE. Previous studies identified a further degradation of Vip3A proteins due to the introduction of the secondary cleavage sites after treating with SDS contained SDS-PAGE sample buffer [17]. We speculate that protein complex 3 is easy to be disassembled in the native SDS-PAGE. Due to the disappearance of the less abundant protein complex 3 in the native SDS-PAGE gel, its molecular weight could only be predicted to be >240 kDa. From the toxin Vip3Af, a ~360 kDa homo-tetramer composed of 20 kDa and 62 kDa peptides has been proposed to correlate with the toxicity of Vip3Af toxin [14,21]. In this study, Vip3Aa-WT and Vip3Aa-S164T were observed to form the complex 3 composed of the 19 and 65 kDa peptides (Figure 3a) and have toxicity (Figure 5). It is possible that the protein complex 3 from Vip3Aa-WT and Vip3Aa-S164T corresponds to the previously reported 360 kDa homo-tetramer from Vip3Af, which was predicted to be formed by four 85–90 kDa protein complexes, each of them was composed of a 19 kDa peptide and a 65 kDa peptide [14,29]. SDS-PAGE analysis of trypsin- or midgut proteases-processed Vip3Aa proteins showed nearly the same protein fragment patterns below 20 kDa from Vip3Aa-WT and three S164 mutants (Figure 1). However, the protein complex 3 composed of the 19 kDa and 65 kDa fragments could only be observed from Vip3Aa-WT and Vip3Aa-S164T

(Figure 2a,b). This is because the 15 kDa, 17 kDa and 19 kDa fragments were all identified from the N terminus of Vip3Aa (Figure 4). The S164 is critical for the formation of the protein complex 3, composed of 19 kDa and 65 kDa peptides from Vip3Aa proteins.

In conclusion, the present study identified a > 240 kDa protein complex composed of the 19 kDa and 65 kDa fragments from Vip3Aa after proteolytic processing. The formation of this protein complex was determined to correlate with the toxicity of Vip3Aa in *S. litura* larvae. The S164 in Vip3Aa is critical for the formation of the >240 kDa protein complex and consequently the insecticidal activity. The results from this study provided new information on the insecticidal mechanism of Vip3Aa toxins.

4. Materials and Methods

4.1. Site Directed Mutation on the vip3Aa Gene

The *vip3Aa* gene (NCBI Accession No. AF500478.2) was cloned from plasmid of Bt WB7, a native strain isolated from soil collected in Wuyi mountain (Fujian, China) [32], by the use of primer pairs P-3Aa-F and P-3Aa-R (Table 2). The pGEX-KG vector [33] was used for the heterologous expression of the *vip3Aa* gene in *E. coli* BL21 (DE3). To substitute nucleotides coding for a single amino acid residue in Vip3Aa protein, primer pairs containing the site substitutions in *vip3Aa* gene was carried out by PCR using the pGEX-Vip3Aa as the template. Primers for the site directed mutation were listed in Table 2. To replace the S164 by an Ala in Vip3Aa, primers P-3Aa-F and P-164A-R were used as the primer pair for the 1st round PCR to obtain the N-terminal part of *vip3Aa* gene. P-164A-F and P-3Aa-R were used as the primer pair for the 2nd round PCR to obtain the C-terminal part of *vip3Aa* gene. Both PCR products were diluted 1000 folds in water and used as the template for the 3rd round PCR using P-3Aa-F and P-3Aa-R as the primer pair to obtain the full-length *vip3Aa* gene with the codons coding for S164 replaced by codons coding for A164. PCR reactions were performed using the iProof™ High-Fidelity Master Mix DNA polymerase (Bio-RAD, Hercules, CA, USA). Other site substituted mutants of *vip3Aa* gene were generated according to the same procedure by the use of corresponding primers (Table 2). The final products from the 3rd round PCR were purified and digested with restricted enzymes of *Nco* I and *Sac* I. Digested products were purified and ligated with pGEX-KG plasmid linearized with *Nco* I and *Sac* I. Plasmids carrying mutated *vip3Aa* gene were transformed into *E. coli* BL21 (DE3) cells for protein preparations.

Table 2. Primers for the site substitutions in *vip3Aa* gene.

Primers	Sequence 5′–3′
P-3Aa-F	CATGCCATGGACATGAACAAGAATAATACTAAAT
P-3Aa-R	CGAGCTCTTACTTAATAGAGACATCGT
P-164P-R	TTCAGTAAGTGTaggGTTAATAAGTACA
P-164P-F	ATGTACTTATTAACcctACACTTACTG
P-164T-R	TTCAGTAAGTGTggtGTTAATAAGTACA
P-164T-F	GTACTTATTAACaccACACTTACTGAAA
P-164A-F	ACTTATTAACgcgACACTTACTG
P-164A-R	AGTAAGTGTcgcGTTAATAAGTA
P-152A-F	GATTTCTGATgcgTTGGATATTA
P-152A-R	ATAATATCCAAcgcATCAGAAAT
P-154A-F	TGATAAGTTGgcgATTATTAATG
P-154A-R	ATTAATAATcgcCAACTTATCAG

Underlined sequences indicate the restricted enzyme sites of *Nco* I and *Sac* I. Lower case sequences indicate the mutant nucleotides in each primer.

4.2. Expression and Purification of Vip3Aa Proteins

To prepare Vip3Aa proteins, 250 μL of overnight culture of BL21 cells carrying a plasmid of pGEX-Vip3Aa were inoculated to 250 mL of LB in a 1 L flask. The bacterial cultures were incubated at 37 °C and shaken at 150 rpm until OD_{600} reached 0.5. Protein expression was induced by addition of 0.8 mM IPTG (Isopropyl β-D-thiogalactoside) to the cultures, followed by incubation at 16 °C for 24 h. The *E. coli* cells were then harvested by centrifugation at 14,000× *g* for 1 min and the cell pellets were resuspended and washed in GST binding buffer (PBS, pH 7.3). The cell suspension was sonicated with a sonicator (VC-50, Sonics & Materials Inc. Danbury, CT, USA), followed by centrifugation at 21,000× *g* for 10 min to pellet the cell debris. The supernatant containing soluble GST-Vip3Aa proteins was loaded onto a Glutathione Sepharose column. Purification of GST fusion proteins and removing of GST tag by thrombin followed the standard purification procedure described by manufacturer of Glutathione Sepharose 4B (GE healthcare, Chicago, IL, USA). All purification steps were conducted at 4 °C or on ice. The purified GST-Vip3Aa fusions or thrombin treated Vip3Aa were quantified by the Bradford method [34].

4.3. Insects Rearing and Bioassays

An inbred colony of *S. litura* reared in the laboratory for over 3 years (~30 generations) was used in this study. The *S. litura* colony was maintained on a soybean-based artificial diet at 27°C with 50% humidity and a photoperiod 16 h of light and 8 h of darkness.

Bioassays were conducted by diet overlay method [35]. GST tag of Vip3Aa-WT was removed while other Vip3Aa mutants were prepared as GST-Vip3Aa fusions for bioassays. Briefly, the Vip3Aa-WT toxins or GST-Vip3Aa fusions in a series of dilutions were prepared in water. A 200 μL aliquot of diluted toxin was overlaid on the surface (~7 cm^2) of diet in each cup (30 mL plastic cup containing ~5 mL diet). Each concentration included 5 replications. Ten neonates were placed into each cup. Cups were covered with lids and kept in the rearing room at 27 °C, 50% humidity and a photoperiod of 16:8 (light:dark) for at least 96 h. Mortality of larvae in each cup was recorded every 24 h. Cups contain diet overlaid by 100 μg/mL or 250 μg/mL of GST tag protein diluted in water were prepared as negative controls. Probit analysis of the bioassay data was carried out using the POLO program [36] to estimate the LC_{50} and 95% confidence limits (CL). For bioassays using tryptic Vip3Aa proteins, Glutathione Sepharose carrying GST-Vip3Aa fusions were digested with 10 mg/mL trypsin. Tryptic Vip3Aa proteins were quantified by the Bradford method and diluted to the concentration of 5 μg/mL and 50 μg/mL for the diet overlay bioassays described above. Mortality of *S. litura* larvae fed with tryptic Vip3Aa proteins were recorded in 96 h and analyzed by Prism (version 8.2.0, GraphPad Software, San Diego, CA, USA).

4.4. In Vitro Proteolytic Processing of Vip3Aa Proteins

To prepare midgut proteases of *S. litura* larvae, mid-fifth-instar larvae of *S. litura* were immobilized on ice for several minutes and dissected to isolate the complete midgut without loss of its contents. Midgut homogenates were prepared by thorough homogenization of 5 midguts in a 1.5 mL microcentrifuge tube. Grounded tissues were centrifugated at 16,000× *g* for 10 min at 4 °C. The supernatant was collected and distributed into small aliquots, snap frozen in liquid nitrogen and then stored at −80 °C until use. The protein concentration of midgut protease preparations was measured using the Bradford method.

Affinity-purified Vip3Aa proteins were subjected to in vitro proteolytic processing with trypsin (Sigma-Aldrich Inc. St. Louis, MO, USA) or midgut proteases of *S. litura*. Vip3Aa fusions were incubated with 10 mg/mL of trypsin or 400 μg/mL midgut protease preparation in Tris-HCl buffer (20 mM Tris-HCl, 0.15 M NaCl, 5 mM EDTA, pH 8.6) at the ratio of 120:100 (trypsin:Vip3Aa, *w/w*) for the trypsin treatment and 40:100 (midgut protease:Vip3Aa, *w/w*) for the midgut protease treatment. In vitro digestion was carried out at 30 °C for 6 h.

4.5. Analysis of Vip3Aa Proteins by the Native Gel and SDS-PAGE Gel

Protease treated Vip3Aa proteins were immediately analyzed by native PAGE or SDS-PAGE. To analyze protein complexes of Vip3Aa by native PAGE, trypsin- or midgut proteases-processed Vip3Aa proteins were mixed with the 2 × native PAGE sample buffer (0.5 M Tris-HCl pH 6.8, 5% bromophenol blue, 30% glycerol) and separated by the electrophoretic analysis on a 6% native PAGE gel. To analyze denatured Vip3Aa proteins by SDS-PAGE, the processed Vip3Aa proteins or protein complexes excised from the native PAGE gel were mixed with 5 × SDS-PAGE sample buffer (0.2 M Tris-HCl pH 6.8, 1 M sucrose, 5 mM EDTA, 0.1% bromophenol blue, 10% SDS and 5% β-mercaptoethanol), heated at 99 °C for 10 min and centrifugated at 15,000× g for 5 min. The supernatant was loaded to a 10% SDS-PAGE gel for the electrophoretic analysis. To estimate the molecular weight of protein complexes of Vip3Aa proteins by native SDS-PAGE, trypsin-processed Vip3Aa proteins were mixed with 5 × SDS-PAGE gel sample buffer with the absence of β-mercaptoethanol and loaded to a 6% SDS-PAGE gel for the electrophoretic analysis.

4.6. Identification of Trypsin-Processed Fragments

Protein bands of the 19 kDa fragment and the 15 kDa fragment contained by the protein complexes of trypsin treated wild-type Vip3Aa and a protein band of the 17 kDa fragment contained by the protein complex from trypsin treated Vip3Aa-S164P were excised from the SDS-PAGE gel after staining with Coomassie blue and detained with 30% acetonitrile and 100 mM NH_4HCO_3. The gel slices were dried in a vacuum centrifuge, and the proteins were reduced in-gel with dithiothreitol (10 mM DTT and 100 mM NH_4HCO_3) for 30 min at 56 °C, then alkylated with iodoacetamide (200 mM IAA and 100 mM NH_4HCO_3) in dark at room temperature for 30 min. Gel slices were briefly rinsed with 100 mM NH_4HCO_3 and acetonitrile, respectively, followed by digestion with 12.5 ng/μL trypsin in 25 mM NH_4HCO_3 overnight. The peptides were extracted three times with 60% acetonitrile and 0.1% trifluoroacetic acid. The extracts were pooled and dried completely in a vacuum centrifuge. The peptide mass and sequence were determined by Liquid Chromatography (LC)—Electrospray Ionization (ESI) Tandem mass spectrometry (MS/MS) in a Q Exactive mass spectrometer which was coupled to Easy nLC (Proxeon Biosystems, Thermo Fisher Scientific, Shanghai, China). The MS data were analyzed using Max Quant (version 1.6.4.0, Max Planck Institute of Biochemistry, Munich, Germany) by searching the data against the amino acid sequence of Vip3Aa, and the intensity of sequenced peptide in the target protein was calculated.

Author Contributions: E.S. and P.W. conceived and designed research. E.S., A.Z., Y.Y., Y.W. and X.J. conducted experiments. E.S., A.Z., L.S., and Z.H. analyzed data. E.S., P.W., X.G. and Z.H. wrote the manuscript. All authors have read and agreed to the published version of the manuscript.

Funding: This research was funded by National Natural Science Foundation of China (31772539); National Key R&D Program of China (2017YFE0121700); FAFU Science Fund for Distinguished Young Scholars (XJQ201819); Science and Technology Innovation Fund of FAFU (CXZX2018040).

Acknowledgments: We thank Forestry college of Fujian Agriculture and Forestry University for providing us with an insect breeding room for the breeding of *S. litura* species.

Conflicts of Interest: The authors declare no conflict of interest.

References

1. Christou, P.; Capell, T.; Kohli, A.; Gatehouse, J.A.; Gatehouse, A.M. Recent developments and future prospects in insect pest control in transgenic crops. *Trends Plant Sci.* **2006**, *11*, 302–308. [CrossRef] [PubMed]
2. Pradhan, S.; Chakraborty, A.; Sikdar, N.; Chakraborty, S.; Bhattacharyya, J.; Mitra, J.; Manna, A.; Dutta Gupta, S.; Sen, S.K. Marker-free transgenic rice expressing the vegetative insecticidal protein (Vip) of *Bacillus thuringiensis* shows broad insecticidal properties. *Planta* **2016**, *244*, 789–804. [CrossRef] [PubMed]
3. Chakroun, M.; Banyuls, N.; Bel, Y.; Escriche, B.; Ferré, J. Bacterial vegetative insecticidal proteins (Vip) from entomopathogenic bacteria. *Microbiol. Mol. Biol. Rev.* **2016**, *80*, 329–350. [CrossRef] [PubMed]

4. Geng, J.; Jiang, J.; Shu, C.; Wang, Z.; Song, F.; Geng, L.; Duan, J.; Zhang, J. *Bacillus thuringiensis* Vip1 Functions as a Receptor of Vip2 Toxin for Binary Insecticidal Activity against *Holotrichia parallela*. *Toxins* **2019**, *11*, 440. [CrossRef] [PubMed]

5. Bi, Y.; Zhang, Y.; Shu, C.; Crickmore, N.; Wang, Q.; Du, L.; Song, F.; Zhang, J. Genomic sequencing identifies novel *Bacillus thuringiensis* Vip1/Vip2 binary and Cry8 toxins that have high toxicity to Scarabaeoidea larvae. *Appl. Microbiol. Biotechnol.* **2015**, *99*, 753–760. [CrossRef]

6. Palma, L.; Muñoz, D.; Berry, C.; Murillo, J.; Caballero, P. *Bacillus thuringiensis* toxins: An overview of their biocidal activity. *Toxins* **2014**, *6*, 3296–3325. [CrossRef]

7. Palma, L.; de Escudero, I.R.; Maeztu, M.; Caballero, P.; Muñoz, D. Screening of vip genes from a Spanish *Bacillus thuringiensis* collection and characterization of two Vip3 proteins highly toxic to five lepidopteran crop pests. *Biol. Control* **2013**, *66*, 141–149. [CrossRef]

8. Sena, J.A.; Hernández-Rodríguez, C.S.; Ferré, J. Interaction of *Bacillus thuringiensis* Cry1 and Vip3A proteins with *Spodoptera frugiperda* midgut binding sites. *Appl. Environ. Microbiol.* **2009**, *75*, 2236–2237. [CrossRef]

9. Lee, M.K.; Miles, P.; Chen, J.-S. Brush border membrane binding properties of *Bacillus thuringiensis* Vip3A toxin to *Heliothis virescens* and *Helicoverpa zea* midguts. *Biochem. Biophys. Res. Commun.* **2006**, *339*, 1043–1047. [CrossRef]

10. Gouffon, C.V.; Van Vliet, A.; Van Rie, J.; Jansens, S.; Jurat-Fuentes, J. Binding sites for *Bacillus thuringiensis* Cry2Ae toxin on heliothine brush border membrane vesicles are not shared with Cry1A, Cry1F, or Vip3A toxin. *Appl. Environ. Microbiol.* **2011**, *77*, 3182–3188. [CrossRef]

11. Liu, J.-G.; Yang, A.-Z.; Shen, X.-H.; Hua, B.-G.; Shi, G.-L. Specific binding of activated Vip3Aa10 to *Helicoverpa armigera* brush border membrane vesicles results in pore formation. *J. Invertebr. Pathol.* **2011**, *108*, 92–97. [CrossRef] [PubMed]

12. Kurtz, R.W. A review of Vip3A mode of action and effects on Bt Cry protein-resistant colonies of lepidopteran larvae. *Southwest. Entomol.* **2010**, *35*, 391–394. [CrossRef]

13. Lee, M.K.; Walters, F.S.; Hart, H.; Palekar, N.; Chen, J.-S. The Mode of Action of the *Bacillus thuringiensis* Vegetative Insecticidal Protein Vip3A Differs from That of Cry1Ab δ-Endotoxin. *Appl. Environ. Microbiol.* **2003**, *69*, 4648–4657. [CrossRef] [PubMed]

14. Banyuls, N.; Hernández-Rodríguez, C.; Van Rie, J.; Ferré, J. Critical amino acids for the insecticidal activity of Vip3Af from *Bacillus thuringiensis*: Inference on structural aspects. *Sci. Rep.* **2018**, *8*, 1–14. [CrossRef] [PubMed]

15. Sellami, S.; Jemli, S.; Abdelmalek, N.; Cherif, M.; Abdelkefi-Mesrati, L.; Tounsi, S.; Jamoussi, K. A novel Vip3Aa16-Cry1Ac chimera toxin: Enhancement of toxicity against *Ephestia kuehniella*, structural study and molecular docking. *Int. J. Biol. Macromol.* **2018**, *117*, 752–761. [CrossRef]

16. Zheng, M.; Evdokimov, A.G.; Moshiri, F.; Lowder, C.; Haas, J. Crystal structure of a Vip3B family insecticidal protein reveals a new fold and a unique tetrameric assembly. *Protein Sci.* **2019**, *29*, 824–829. [CrossRef]

17. Yu, C.G.; Mullins, M.A.; Warren, G.W.; Koziel, M.G.; Estruch, J.J. The *Bacillus thuringiensis* vegetative insecticidal protein Vip3A lyses midgut epithelium cells of susceptible insects. *Appl. Environ. Microbiol.* **1997**, *63*, 532–536. [CrossRef]

18. Banyuls, N.; Hernández-Martínez, P.; Quan, Y.; Ferré, J. Artefactual band patterns by SDS-PAGE of the Vip3Af protein in the presence of proteases mask the extremely high stability of this protein. *Int. J. Biol. Macromol.* **2018**, *120*, 59–65. [CrossRef]

19. Hamadou-Charfi, D.B.; Boukedi, H.; Abdelkefi-Mesrati, L.; Tounsi, S.; Jaoua, S. *Agrotis segetum* midgut putative receptor of *Bacillus thuringiensis* vegetative insecticidal protein Vip3Aa16 differs from that of Cry1Ac toxin. *J. Invertebr. Pathol.* **2013**, *114*, 139–143. [CrossRef]

20. Bel, Y.; Banyuls, N.; Chakroun, M.; Escriche, B.; Ferré, J. Insights into the structure of the Vip3Aa insecticidal protein by protease digestion analysis. *Toxins* **2017**, *9*, 131. [CrossRef]

21. Quan, Y.; Ferré, J. Structural Domains of the *Bacillus thuringiensis* Vip3Af Protein Unraveled by Tryptic Digestion of Alanine Mutants. *Toxins* **2019**, *11*, 368. [CrossRef] [PubMed]

22. Hernández-Martínez, P.; Hernández-Rodríguez, C.S.; Van Rie, J.; Escriche, B.; Ferré, J. Insecticidal activity of Vip3Aa, Vip3Ad, Vip3Ae, and Vip3Af from *Bacillus thuringiensis* against lepidopteran corn pests. *J. Invertebr. Pathol.* **2013**, *113*, 78–81. [CrossRef] [PubMed]

23. Chakroun, M.; Ferré, J. In vivo and in vitro binding of Vip3Aa to *Spodoptera frugiperda* midgut and characterization of binding sites by 125I radiolabeling. *Appl. Environ. Microbiol.* **2014**, *80*, 6258–6265. [CrossRef]

24. Gayen, S.; Hossain, M.A.; Sen, S.K. Identification of the bioactive core component of the insecticidal Vip3A toxin peptide of *Bacillus thuringiensis*. *J. Plant Biochem. Biotechnol.* **2012**, *21*, 128–135. [CrossRef]

25. Qin, H.; Ye, Z.; Huang, S.; Ding, J.; Luo, R. The correlation of the different host plants with preference level, life duration and survival rate of *Spodoptera litura* Fabricius. *Chin. J. Eco-Agric.* **2004**, *12*, 40–42.

26. Qiong, L.; Cao, G.; Zhang, L.; Liang, G.; Gao, X.; Zhang, Y.; Guo, Y. The binding characterization of Cry insecticidal proteins to the brush border membrane vesicles of *Helicoverpa armigera*, *Spodoptera exigua*, *Spodoptera litura* and *Agrotis ipsilon*. *J. Integr. Agric.* **2013**, *12*, 1598–1605.

27. Song, F.; Lin, Y.; Chen, C.; Shao, E.; Guan, X.; Huang, Z. Insecticidal activity and histopathological effects of Vip3Aa protein from *Bacillus thuringiensis* on *Spodoptera litura*. *J. Microbiol. Biotechnol.* **2016**, *26*, 1774–1780. [CrossRef] [PubMed]

28. Li, C.; Xu, N.; Huang, X.; Wang, W.; Cheng, J.; Wu, K.; Shen, Z. *Bacillus thuringiensis* Vip3 mutant proteins: Insecticidal activity and trypsin sensitivity. *Biocontrol Sci. Technol.* **2007**, *17*, 699–708. [CrossRef]

29. Kunthic, T.; Surya, W.; Promdonkoy, B.; Torres, J.; Boonserm, P. Conditions for homogeneous preparation of stable monomeric and oligomeric forms of activated Vip3A toxin from *Bacillus thuringiensis*. *Eur. Biophys. J.* **2017**, *46*, 257–264. [CrossRef]

30. Palma, L.; Scott, D.J.; Harris, G.; Din, S.-U.; Williams, T.L.; Roberts, O.J.; Young, M.T.; Caballero, P.; Berry, C. The Vip3Ag4 insecticidal protoxin from *Bacillus thuringiensis* adopts a tetrameric configuration that is maintained on proteolysis. *Toxins* **2017**, *9*, 165. [CrossRef]

31. Zack, M.D.; Sopko, M.S.; Frey, M.L.; Wang, X.; Tan, S.Y.; Arruda, J.M.; Letherer, T.T.; Narva, K.E. Functional characterization of Vip3Ab1 and Vip3Bc1: Two novel insecticidal proteins with differential activity against lepidopteran pests. *Sci. Rep.* **2017**, *7*, 1–12. [CrossRef] [PubMed]

32. Cai, W.; Sha, L.; Zhou, J.; Huang, Z.; Guan, X. Functional analysis of active site residues of *Bacillus thuringiensis* WB7 chitinase by site-directed mutagenesis. *World J. Microbiol. Biotechnol.* **2009**, *25*, 2147–2155. [CrossRef]

33. Guan, K.L.; Dixon, J.E. Eukaryotic proteins expressed in Escherichia coli: An improved thrombin cleavage and purification procedure of fusion proteins with glutathione S-transferase. *Anal. Biochem.* **1991**, *192*, 262. [CrossRef]

34. Bradford, M.M. A rapid and sensitive method for the quantitation of microgram quantities of protein utilizing the principle of protein-dye binding. *Anal. Biochem.* **1976**, *72*, 248–254. [CrossRef]

35. Kain, W.C.; Zhao, J.-Z.; Janmaat, A.F.; Myers, J.; Shelton, A.M.; Wang, P. Inheritance of resistance to *Bacillus thuringiensis* Cry1Ac toxin in a greenhouse-derived strain of cabbage looper (Lepidoptera: Noctuidae). *J. Econ. Entomol.* **2004**, *97*, 2073–2078. [CrossRef]

36. Software, L. *POLO-PC: A User's Guide to Probit or Logit Analysis*; LeOra Software: Berkeley, CA, USA, 1987.

Article

Mitochondria and Lysosomes Participate in Vip3Aa-Induced *Spodoptera frugiperda* Sf9 Cell Apoptosis

Xiaoyue Hou [1], Lu Han [1], Baoju An [1], Yanli Zhang [1], Zhanglei Cao [1], Yunda Zhan [1], Xia Cai [1], Bing Yan [1] and Jun Cai [1,2,3,*]

[1] Department of Microbiology, College of Life Sciences, Nankai University, Tianjin 300071, China; xiaoyuehou@mail.nankai.edu.cn (X.H.); luhan0325@mail.nankai.edu.cn (L.H.); 1120190491@mail.nankai.edu.cn (B.A.); Yanlizhang@mail.nankai.edu.cn (Y.Z.); caozhanglei@mail.nankai.edu.cn (Z.C.); yundazhan@mail.nankai.edu.cn (Y.Z.); xiacai@mail.nankai.edu.cn (X.C.); iceyan@nankai.edu.cn (B.Y.)
[2] Key Laboratory of Molecular Microbiology and Technology, Ministry of Education, Tianjin 300071, China
[3] Tianjin Key Laboratory of Microbial Functional Genomics, Tianjin 300071, China
* Correspondence: caijun@nankai.edu.cn

Received: 20 December 2019; Accepted: 12 February 2020; Published: 13 February 2020

Abstract: Vip3Aa, a soluble protein produced by certain *Bacillus thuringiensis* strains, is capable of inducing apoptosis in Sf9 cells. However, the apoptosis mechanism triggered by Vip3Aa is unclear. In this study, we found that Vip3Aa induces mitochondrial dysfunction, as evidenced by signs of collapse of mitochondrial membrane potential, accumulation of reactive oxygen species, release of cytochrome c, and caspase-9 and -3 activation. Meanwhile, our results indicated that Vip3Aa reduces the ability of lysosomes in Sf9 cells to retain acridine orange. Moreover, pretreatment with Z-Phe-Tyr-CHO (a cathepsin L inhibitor) or pepstatin (a cathepsin D inhibitor) increased Sf9 cell viability, reduced cytochrome c release, and decreased caspase-9 and -3 activity. In conclusion, our findings suggested that Vip3Aa promotes Sf9 cell apoptosis by mitochondrial dysfunction, and lysosomes also play a vital role in the action of Vip3Aa.

Keywords: Vip3Aa; lysosome; mitochondria; apoptosis; Sf9 cells

Key Contribution: Vip3Aa-induced apoptosis involves mitochondria and lysosomes.

---

1. Introduction

Vip3Aa is a protein produced by *Bacillus thuringiensis* (*Bt*) during vegetative growth. It can bind to the brush border membrane vesicles (BBMV) specifically in susceptible and non-susceptible insects [1–3]. Moreover, the brush border membrane binding sites of Vip3Aa are different from those of insecticidal crystal proteins (ICPs), and Vip3Aa could extend its activity to pests non-susceptible to ICPs. Consequently, it is widely accepted that Vip3Aa can not only broaden the insecticidal spectrum, it may also delay the resistance development in insects [3–5]. Thus, Vip3Aa is considered a second-generation insecticidal toxin and has been used in genetically modified crops, such as *Bt* cotton and *Bt* corn products [6].

The pore-forming model is generally accepted to explain the virulence of ICPs [7] and Lee et al. [3] corroborated that the Vip3 proteins share a similar mode of action. In short, the Vip3 proteins (protoxins) are ingested by the insect and activated to the active form (act-Vip3A) by the midgut proteases. After that, the act-Vip3A binds to its receptor on the BBMV and exerts toxicity to the midgut cells, eventually leading to the death of the pests. Additionally, Kunthic et al. [8] found that the pH could regulate the properties of the tetramer made by the act-Vip3Aa, which further supported the pore-forming model and suggested that the pH could regulate the post-binding events such as membrane insertion or pore formation. Regarding the binding sites of the Vip3Aa, recent research has found some proteins interacting with Vip3Aa that are closely related to cell toxicity in *Spodoptera frugiperda* cells, such as S2, SR-C, and FGFR [5,9,10]. Additionally, Jiang et al. [9] found that the toxicity of Vip3Aa to Sf9 cells correlated with its endocytosis mediated by Sf-SR-C and that internalization is essential for Vip3Aa to exert its toxic effects.

Bel et al. [11] showed Vip3Aa provoked a wide transcriptional response in *Spodoptera exigua* larvae. The upregulated genes were involved in innate immune response and pathogen response, while the downregulated ones were mainly related to metabolism. However, genes related to the action of ICPs were found to be slightly overexpressed. Crava et al. [12] further indicated that Vip3Aa upregulated genes coding for antimicrobial peptides and lysozymes in *S. exigua* midgut. Ayra-Pardo et al. [13] reported a transcriptomic study, showing that the decreased translation rate could be an important adaptation for Vip3Aa resistance in *Heliothis virescens*. Hernández-Martinez et al. [14] suggested Vip3Aa could activate different insect response pathways that trigger the regulation of some genes, APN shedding, and apoptotic cell death. These results suggest that there are other mechanisms that are participating in cell death apart from the pore-forming model. Jiang et al. [15] observed that the Vip3Aa-treated Sf9 cells had some apoptosis characteristics, such as DNA breakage, mitochondrial membrane potential ($\Delta\Psi m$) collapse, and Sf-caspase-1 activation. Hernández-Martinez et al. [14] confirmed that there was apoptosis occurrence in midgut epithelial cells when *S. exigua* larvae were treated with Vip3Aa and Vip3Ca. However, how Vip3Aa induces apoptosis is unclear and further experiments will be needed to determine the underlying mechanism.

Apoptosis is indispensable to the homeostasis and development of organisms [16]. Bcl-2 family proteins are crucial regulators of cell survival and cell death. They are divided into anti- and pro-apoptotic proteins. After apoptotic stimulation, Bax, a pro-apoptosis protein, can transfer to mitochondria, resulting in mitochondrial membrane permeability increase and cytochrome c release. The mitochondrion, a highly sensitive organelle, plays a critical role in apoptosis. Increased mitochondrial membrane permeabilization may represent the point of no return of the lethal stressors-induced signal [17]. Cytochrome c normally localizes in the inner mitochondrial membrane through weak electrostatic interactions with acidic phospholipids. When mitochondria permeability increases, it releases to the cytoplasm and subsequently activates the apoptotic cascades. Anti-apoptotic proteins, such as Bcl-2 and Bcl-XL, inhibit apoptosis by locally preventing $\Delta\Psi m$ loss [17,18]. Environmental stimuli may contribute to mitochondrial injury, which causes $\Delta\Psi m$ collapse, oxidative stress, resulting in increased cellular ROS, changed Bcl-2 family protein levels, and apoptosis factor release [19–21].

In this paper, we try to further explore the mechanism of Vip3Aa-induced apoptosis and probe the signaling pathways and molecules involved in Vip3Aa-induced cell death.

2. Results

2.1. The Effects of Vip3Aa on Sf9 Cell Viability and the Subcellular Localization of Vip3Aa in Sf9 Cells

Sf9 cells were exposed to Vip3Aa (10, 20, 30, 40, or 50 µg/mL) for different times (24, 48, 60, and 72 h). Cell viability of Sf9 cells was assessed by the CCK-8 assay, by measuring the amount of orange–yellow formazan that is directly proportional to the number of living cells. As illustrated in Figure 1, when Sf9 cells were exposed to the same Vip3Aa concentration, the cell viability decreased as the time of treatment prolonged. If the Vip3Aa-treated time was the same, cell viability decreased with the increase of Vip3Aa concentration. Vip3Aa (final concentration, 40 µg/mL) treatment for 48 h reduced the cell viability of Sf9 cells to about 50%. Thus, the final concentration of Vip3Aa used in the following experiments was 40 µg/mL.

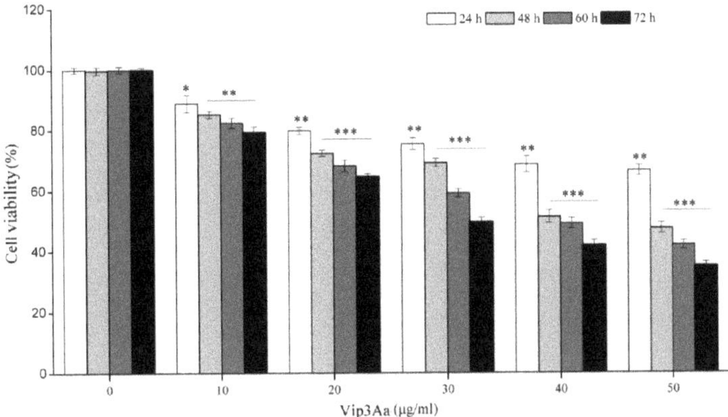

Figure 1. Viability impacts of Vip3Aa on Sf9 cells. The Sf9 cells were exposed to different concentrations of Vip3Aa for 24, 48, 60, and 72 h, respectively. Significant tests from the corresponding controls (without Vip3Aa treatment) are indexed via * $p < 0.05$, ** $p < 0.01$, and *** $p < 0.001$.

Jiang et al. [9] revealed that Vip3Aa could enter into the Sf9 cells via endocytosis. Since the lysosome is the endpoint of endocytosis, we further explored the subcellular localization of Vip3Aa in Sf9 cells. Sf9 cells were exposed to Vip3Aa-RFP for different times (2, 4, and 6 h). As shown in Figure 2, there was abundant co-localization of Vip3Aa and lysosomes from 4 h after Vip3Aa treatment. These results suggested that lysosomes might be involved in the action of Vip3Aa.

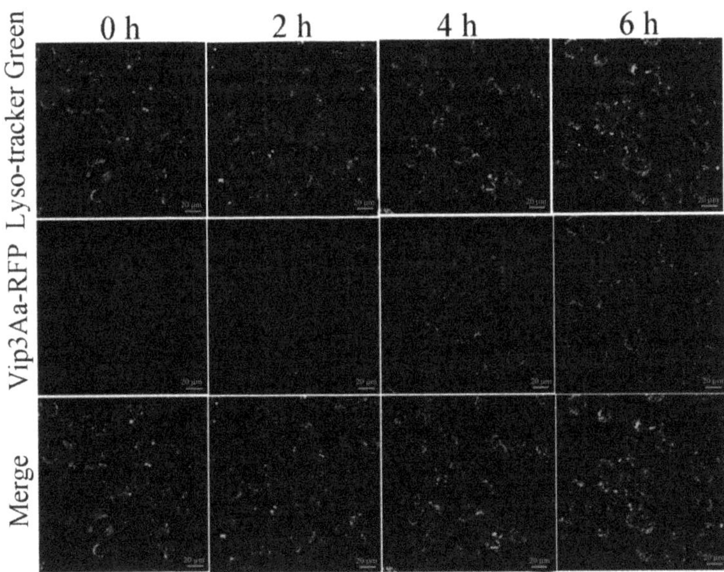

Figure 2. Co-localization of Vip3Aa and lysosomes in Sf9 cells. Cells were treated with Vip3Aa-RFP for 0, 2, 4, and 6 h, respectively, and were stained with fluorescent probe LysoSensor™ Green DND-189 at 28 °C for 45 min. Then, the cells were observed under a confocal laser scanning microscope. Scale bar, 20 μm.

2.2. Vip3Aa Impairs Mitochondrial Function and Induces Cytochrome c Release

Mitochondria are the meeting point of many apoptotic signals, so we examined the mitochondrial ultrastructure (red arrows) in Sf9 cells by transmission electron microscope (TEM) (Figure 3A). The Vip3Aa-untreated cells showed a normal ultrastructure with an intact cristae structure. However, the number of twisted and swollen mitochondria increased from 12 to 24 h after Vip3Aa treatment. After 36 h, the outer membranes of most mitochondria were intact, but the cristae structures were disrupted.

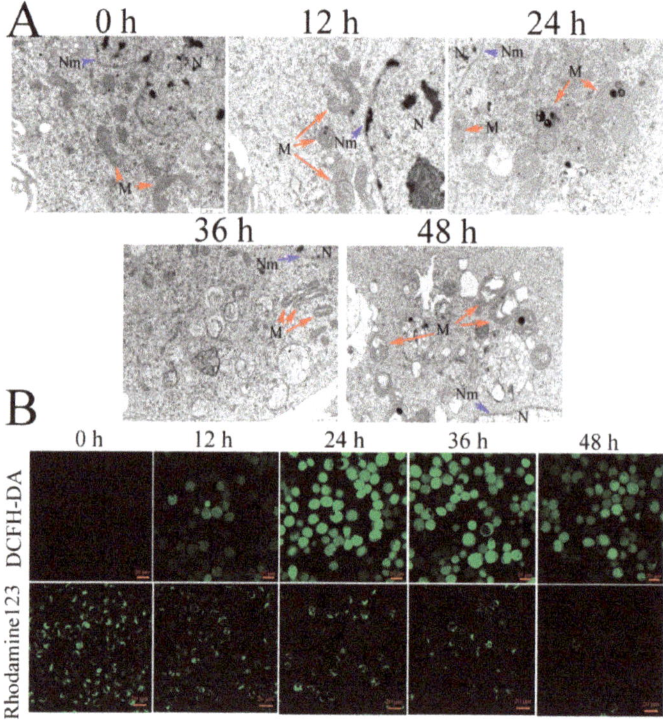

Figure 3. Effects of Vip3Aa on mitochondria in Sf9 cells. (**A**) Representative photographs of mitochondria ultrastructure in Sf9 cells after exposure to Vip3Aa, obtained by TEM. N, nucleus. Nm, nuclear membrane (blue arrows). M, mitochondria (red arrows). Magnification, 30000 ×. Scale bar, 1 μm. (**B**) Effects of Vip3Aa on ROS production and mitochondrial membrane potential ($\Delta\Psi m$) in Sf9 cells, which were determined by the fluorescent probe DCFH-DA and Rhodamine123, respectively. Scale bar, 20 μm.

Mitochondria are not only the main source of endogenous ROS but also the "absorption bank" of ROS. Mitochondria play an essential role in regulating ROS metabolism. In turn, ROS also impacts the function of mitochondria [22]. We then explored the impact of Vip3Aa on the ROS in Sf9 cells. For Vip3Aa-treated cells, ROS levels increased within 12 h, peaked at 24 h, and then decreased (Figure 3B).

Mitochondrial membrane potential ($\Delta\Psi m$) plays a key role in mitochondria function [23]. Rhodamine 123 was used to detect $\Delta\Psi m$. Results showed a significant decrease in $\Delta\Psi m$ appeared firstly at 24 h after Vip3Aa treatment, and the fluorescence intensity reduced to a lower level at 48 h (Figure 3B).

To determine the influence of Vip3Aa on cytochrome c distribution, we evaluated the cytochrome c content in the cytosol and mitochondria via Western blotting. As indicated in Figure 4A and

Figure S1D, the cytochrome c content in the cytoplasm increased, while that in the mitochondria decreased significantly. Subsequently, this phenomenon became more apparent.

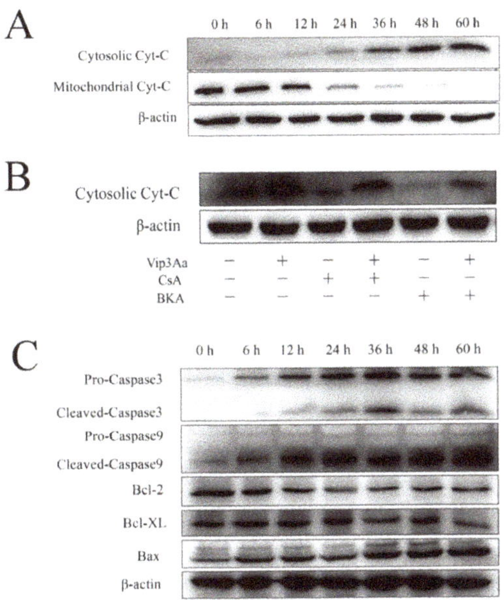

Figure 4. Subcellular distribution of cytochrome c and the levels of mitochondria-associated proteins in Sf9 cells after Vip3Aa treatment. (**A**) Subcellular distribution of cytochrome c after Vip3Aa treatment. Cytochrome c distribution in cytoplasm and mitochondria was detected by Western blotting. (**B**) Cytochrome c distribution in Sf9 cells after different treatments. CsA (5 µM) or BKA (10 µM) pretreated the Sf9 cells for 2 h before Vip3Aa treatment. (**C**) The mitochondria-associated proteins levels in Sf9 cells after Vip3Aa treatment. The original pictures of Western blotting (with protein maker) are shown in Supplementary Materials (Figure S1).

To further explore the way of cytochrome c release, mitochondria permeability transition pore (mPTP) inhibitors, CsA and BKA, were used. Results showed that both inhibitors prevented cytochrome c release partly, while BKA exerted a stronger inhibitory effect than CsA did (Figure 4B). These results suggested that an mPTP-dependent mechanism was involved in cytochrome c release in Vip3Aa-treated Sf9 cells.

2.3. Effects of Vip3Aa on the Mitochondria-Associated Proteins Levels

The protein levels of the Bcl-2 family are related to mitochondrial function. As shown in Figure 4C and Figure S1E, Bax expression increased, while Bcl-2 and Bcl-XL decreased with the extension of Vip3Aa treatment time.

In the mitochondrial pathway, cytochrome c release induces the formation of apoptotic protein complexes, which convert pro-caspase-9 into active caspase 9. Subsequently, caspase-9 will further activate caspase-3 and leads to cell apoptosis. So, we investigated whether caspase activation was involved in the Vip3Aa-induced cell death. Results suggested that the levels of cleaved-caspase-9 and cleaved-caspase-3 increased in different degrees with the extension of Vip3Aa treatment time (Figure 4C and Figure S1F). Additionally, when cytochrome c in the cytosol increased significantly, the protein level of cleaved-caspase increased accordingly. These results indicated that Vip3Aa induced dysregulation of mitochondria-associated proteins and subsequently led to the activation of caspases.

2.4. Effects of Vip3Aa on Lysosome Morphological and Physicochemical Property

Several lines of evidence suggest that the lysosomal pathway contributes to apoptosis. To explore the impact of Vip3Aa on lysosomes, we observed the lysosomal ultrastructure in Sf9 cells by TEM (Figure 5A). The control cells showed a few lysosomes and the cytoplasm was homogeneous. However, the type of lysosomes in the Vip3Aa-treated cells became diverse, and some lysosomes increased distinctly in volume. Meanwhile, we measured the ability of lysosomes to retain acridine orange (AO). As shown in Figure 5B, the fluorescence signals for Sf9 cells were mostly kept in the Q2 region (normal cells), while the percentage of the Q3 region (cells with weak lysosomes) increased from 6.07% to 29.16% with the prolongation of Vip3Aa treatment time. These data showed that Vip3Aa increased the proportion of cells with abnormal lysosomes, and the ability of these lysosomes to keep AO was poor.

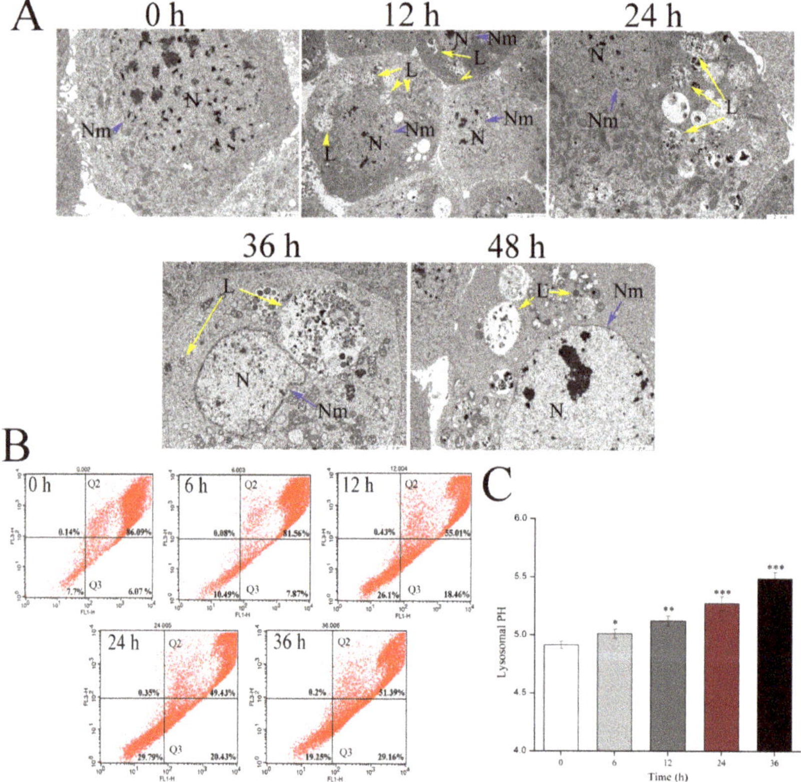

Figure 5. Effects of Vip3Aa on lysosomes in Sf9 cells. (**A**) Representative photographs of lysosomes ultrastructure in Sf9 cells after exposure to Vip3Aa, obtained by TEM. N, nucleus. Nm, nuclear membrane (blue arrows). L, lysosomes (yellow arrows). Magnification, 0 h, 24 h, 36 h, and 48 h, 10000 ×; 12 h, 5000 ×. (**B**) The physicochemical property of lysosomes was detected by acridine orange (AO) staining in Sf9 cells. (**C**) The lysosomal pH in Sf9 cells was detected using LysoSensor Yellow/Blue DND-160. Significant tests from the corresponding controls (without Vip3Aa treatment) are indicated by * $p < 0.05$, ** $p < 0.01$, and ** *$p < 0.001$.

Additionally, we also measured lysosomal pH to further study the impact of Vip3Aa on lysosomes. The lysosome pH value of the Vip3Aa-untreated cells was estimated at 4.91, whereas the lysosomal pH increased at 5.48 after Sf9 cells were exposed to Vip3Aa for 36 h (Figure 5C).

2.5. The Relationship between Sf9 Cell Cathepsins and Vip3Aa-Induced Apoptosis and Cytotoxicity

Lysosomes could be involved in apoptosis via lysosomal proteases, especially cathepsins. Therefore, we measured the mRNA level of cathepsins B, L, and D, which are the significant proteins in the lysosome function. The results indicated that the expression levels of cathepsins (L and D) increased differently depending on the cathepsins analyzed (Figure S2). The mRNA level of cathepsin L and cathepsin D peaked at 36 and 6 h, respectively. However, the expression level of cathepsin B had little change after the cells exposed to Vip3Aa in all the times analyzed.

Meanwhile, we detected the effects of cathepsins (B, L, and D) on Vip3Aa-induced apoptosis and toxicity using the inhibitors CA-074me (a cathepsin B inhibitor), Z-Phe-Tyr-CHO (a cathepsin L inhibitor) and pepstatin (a cathepsin D inhibitor). As illustrated in Figure 6A, the percentage of late apoptotic cells was 0.1%, which rose to 12.56% after 48 h of Vip3Aa treatment. Z-Phe-Tyr-CHO and pepstatin reduced the percentage of late apoptotic cells from 12.56% to 1.7% and 1.44%, respectively. However, CA-074me, a cathepsin B inhibitor, had a little impact on the late apoptotic rate. Compared with Vip3Aa used alone, when Z-Phe-Tyr-CHO was used, there was a little effect on the proportion of early apoptotic cells, but the proportion of late apoptotic cells decreased significantly. However, when pepstatin was used, the proportion of early and late apoptotic cells all decreased significantly. These results suggested that cathepsin D contributes to Vip3Aa-induced apoptosis more than cathepsin L, and cathepsin D plays a more important role in apoptotic signal transduction and enhancement.

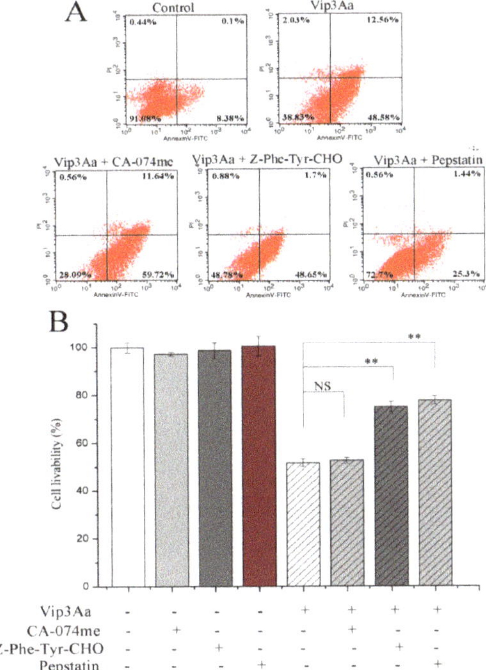

Figure 6. Effects of cathepsin inhibitors on Vip3Aa-treated Sf9 cells. Sf9 cells were pretreated with the inhibitors, CA-074me (10 μM), Z-Phe-Tyr-CHO (10 μM), or pepstatin (15 μM) 2 h before Vip3Aa was added. (**A**) The apoptotic rate of Vip3Aa-treated cells incubated with or without inhibitors. The apoptotic rate was evaluated by Annexin V-FITC/PI stains. (**B**) The cell viability of Vip3Aa-treated cells incubated with or without inhibitors. The cell viability was measured by a CCK-8 assay. Significant tests from the corresponding controls (without Vip3Aa treatment) are indicated by NS, not significant, ** $p < 0.01$.

We also detected the effect of inhibitors on cell livability. The results showed that Z-Phe-Tyr-CHO and pepstatin increased cell livability from 51.3% to 75.1% and 77.8%, respectively (Figure 6B). However, CA-074me had little effect on cell viability. These results suggested that cathepsins (L and D) play a critical role in Vip3Aa-induced cell death rather than cathepsin B.

2.6. Effects of Inhibition of Cathepsins (L and D) on Cytochrome c Distribution and Caspase-9 and -3 Activity

Cytochrome c plays a vital role in apoptosis when the mitochondrial pathway is the executor. Thus, we investigated whether the cathepsin (L and D) inhibitors, Z-Phe-Tyr-CHO and pepstatin, could impact the release of cytochrome c. As shown in Figure 7A, the cathepsin (L and D) inhibitors, especially pepstatin, could reduce cytochrome c release. These results indicated that the function of lysosomes affected Vip3Aa-induced mitochondrial dysfunction.

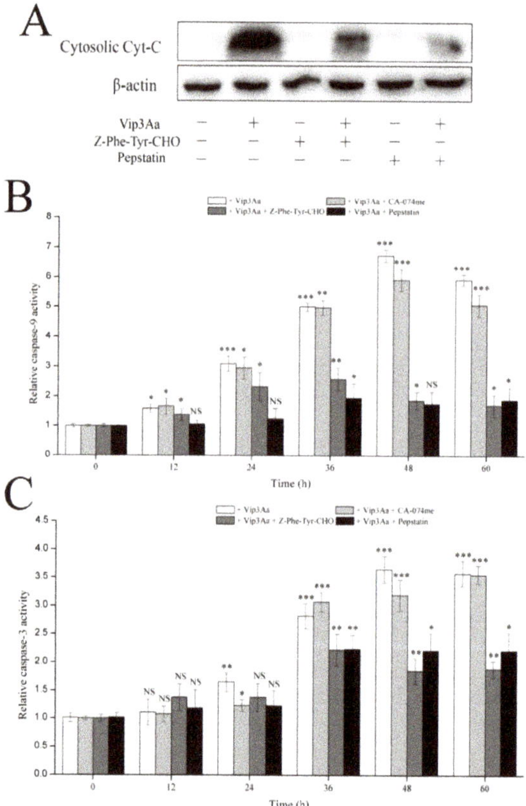

Figure 7. Effects of cathepsins inhibitors on caspases activity and cytochrome c release. (**A**) Impacts of cathepsin (L and D) inhibitors on cytochrome c distribution. The original pictures of Western blotting (with protein maker) are shown in Supplementary Materials (Figure S3). (**B**) Impacts of cathepsin inhibitors on caspase-9 activity. (**C**) Impacts of cathepsin inhibitors on caspase-3 activity. Significant tests from the corresponding controls (without Vip3Aa treatment) and densitometry of the protein bands are indicated by NS, not significant, * $p < 0.05$, ** $p < 0.01$, and *** $p < 0.001$.

Vip3Aa induced apoptosis in Sf9 cells in a caspase-dependent mode [15]. Mitochondrial membrane permeation and the release of cytochrome c from mitochondria activate caspase-9. Caspase-9 further activates caspase-3 and induces apoptosis. To further confirm the influence of lysosomes on Vip3Aa-induced mitochondrial pathway, we detected the caspase-9 and -3 activity without or with the cathepsin inhibitors. As shown in Figure 7B,C, the activity of caspase-9 and -3 increased from 24 h, peaked at 48 h after Vip3Aa treatment without cathepsin inhibitor. However, the caspase-9 and -3 activity decreased significantly after Vip3Aa treatment for 36 h with Z-Phe-Tyr-CHO or pepstatin. As expected, CA-074me had a little impact on caspase activity, especially on caspase-3. These findings indicated that lysosomes are involved in Vip3Aa-induced apoptosis and that cathepsins (L and D) have a vital impact on the Vip3Aa-induced mitochondrial pathway.

3. Discussion

Vip3Aa is a potent toxin against lepidopteran pests, especially to some pests of Noctuidae which are insensitive to ICPs. Recently, studies have shown that Vip3Aa could exert cytotoxicity by triggering apoptosis of insect cells and tissues besides formatting pores [3,5,14,15]. However, the specific mechanism of apoptosis induced by Vip3Aa remains unclear. Hence, we dissected the mechanism of mitochondrial pathway in Vip3Aa-induced apoptosis and found the lysosomes play a crucial role in Vip3Aa-induced apoptosis. The action mechanism of Vip3Aa found in Sf9 cells may also occur in insect intestinal epithelial cells.

Apoptosis includes two important pathways: the extrinsic pathway and the intrinsic pathway, mediated by the death receptor and mitochondria, respectively [24]. Jiang et al. [15] found that mitochondrial membrane potential decreased in Vip3Aa-treated Sf9 cells. We confirmed that Vip3Aa reduced Sf9 cell viability and caused mitochondria morphological alterations, which included swelling and disrupted cristae structure. In our study, we also found that Vip3Aa-induced apoptosis was mediated by mitochondrial dysfunction caused by loss of $\Delta\Psi m$, which subsequently led to cytochrome c release and caspase-9 and -3 activation. Bcl-2 family proteins and caspases are involved in programmed cell death by regulating the protein levels. Moreover, the trends of capase-3 activity were consistent with those of caspase-9 in Vip3Aa-treated Sf9 cells. These results supported that the intrinsic mitochondrial pathway is involved in Vip3Aa-induced apoptosis in Sf9 cells.

Studies have indicated that the extrinsic pathway can be triggered by activating the death receptor on the cell membrane [25]. Additionally, the receptor-mediated pathway contains two types of mechanisms. In type I cells, the extrinsic apoptotic pathway leads to the activation of caspase-8, which directly activates effector caspases (caspase-3), causing apoptosis [26]. Nevertheless, in type II cells, the two apoptotic pathways, i.e., extrinsic pathway and intrinsic pathway, can be linked by caspase-8, which can cleave non-activated Bid protein into truncated Bid (tBid) [27]. tBid could activate Bax, resulting in cytochrome c release and caspase-3 activation [28]. To explore whether the death receptor pathway involves Vp3Aa-induced apoptosis, we also detected the activity of caspase-8 (Figure S4). The caspase-8 activity increased a little bit from 12 to 36 h, but it was lower than that of untreated cells from 48 h. This suggested that caspase-8 might not contribute to the activation of caspase-3. Moreover, the two Vip3Aa receptors SR-C and FGFR did not contain the death domain [9,10]. Jiang et al. [9] revealed the toxicity of Vip3Aa to Sf9 cell correlates with its endocytosis mediated by Sf-SR-C and the internalization is essential for Vip3Aa to exert its toxic effects. On this basis, we further showed that internalized Vip3Aa impacted the features of lysosomes. These results suggested that Vip3Aa-induced apoptosis might involve the internalization of Vip3Aa and the denaturation of lysosomes rather than the death receptor-mediated pathway.

In this study, we found that the abundant colocalization of Vip3Aa and lysosome in Sf9 cells and Vip3Aa had a distinct effect on lysosome morphological and physicochemical properties. Thus, we thought the lysosomes contain the Vip3Aa, and the deformed lysosomes might be the consequence of Vip3Aa action. Duve et al. [29] put forward that lysosomes play a role in apoptosis in 1966. A new death theory, the lysosome–mitochondria axis, mainly emphasized that hydrolytic enzymes were released to the cytosol from the lysosome when lysosomal membranes were permeabilized, resulting in mitochondrial dysfunction, cytochrome c release, and caspase activation. Many studies had reported that cathepsins could be involved in the signaling of apoptosis. Cathepsin D can activate Bax and the active form of Bax translocates to the mitochondria, leading to the opening of transition pores on the mitochondrial membrane, which cause apoptosis factors such as cytochrome c to be released [16]. Another study indicated that cathepsin L acts as a death signal integrator and cytosolic cathepsin L regulated the cytochrome c release and caspase-3 activity in cervical cancer cells [30]. In this study, the results (Figures 6 and 7) showed that cathepsin (L and D) inhibitors could protect Sf9 cells from Vip3Aa and suppress cytochrome c release and inhibit the caspase-9 and -3 activity, suggesting that cathepsins (L and D) played a significant role in Vip3Aa-induced cell death. Moreover, some studies found that cathepsin B associated with programmed cell death of the fat body cells in the process of silkworm metamorphosis [31]. However, there was a little effect of cathepsin B on the Vip3Aa-treated Sf9 cells. As for the role of cathepsins (L and D), they may be released to the cytoplasm and could cleave Bid to tBid, and the latter triggers the mitochondrial outer membrane permeabilization, resulting in mitochondria dysfunction. On the other hand, cathepsins (L and D) may contribute to activating Vip3Aa in lysosomes. Many studies indicated that cathepsins were involved in the physiological reaction of insects, but the exact mechanism is unclear. In this study, lysosomes were firstly found to be involved in the process of Vip3Aa-induced apoptosis. However, the mechanism needs further investigation.

We found the caspase inhibitor (Z-VAD-FMK) could not protect all the Sf9 cells from Vip3Aa (Figure S5). This result suggested that some other apoptosis-independent cell death mechanisms, such as pore-forming, might be involved in cell death caused by Vip3Aa [3,8]. Some microbial toxins, such as aerolysin produced by *Aeromonas hydrophila* and α-toxin generated by *Staphylococcus aureus*, could contribute to pore-forming and apoptosis in their target cells [32]. Similarly, Vip3Aa may cause insect cell death through two mechanisms at the same time.

In conclusion, in Sf9 cells, we showed that the mitochondria pathway serves as the executor in Vip3Aa-induced apoptosis, while lysosomes are involved in Vip3Aa-induced mitochondrial dysfunction and apoptosis. Our findings can provide a venue for promoting the knowledge of Vip3Aa action.

4. Materials and Methods

4.1. Cell Culture and Reagents

The Sf9 cells were cultured in Sf-900 II SFM medium (Gibco, 10902088, Grand Island, NY, USA) supplemented with 6% FBS (GIBCO, Grand Island, NY, USA), at 28 °C. RIPA buffer (#9806S), and antibodies against Bax (#2772), Bcl-2 (#15071), caspase-9 (#9508), cytochrome c (#11940), and β-actin (#8457) were obtained from Cell Signaling Technology (Beverly, MA, USA). Acridine orange (#A8120) was purchased from Solarbio Life Science (Beijing, China). Antibody against Bcl-XL (#abs131907) was purchased from Absin Bioscience (Shanghai, China). Antibody against Caspase-3 (#A2156) was purchased from ABclonal (Wuhan, China). Goat anti-mouse IgG-HRP conjugate (#sc-2005) or anti-rabbit (#sc-2357) was purchased from Santa Cruz (Santa Cruz, TX, USA). DCFH-DA (#S0033) and Rhodamine123 (#C2007) were purchased from Beyotime Biotechnology (Nanjing, China). LysoSensor™ Green DND-189 (#40767ES50) was purchased from Yeasen Biotechnology (Shanghai, China).

4.2. Vip3Aa Purification

pET-28a (+) vector was used to construct a recombinant expression plasmid. The BL21 (DE3) strains transferred with pET28a-Vip3Aa were cultured to OD600 0.8–1.0, and IPTG (0.5 mM) were used to induce the protein expression at 16 °C for 12–16 h. Then, the cells were collected, broken by ultrasonication, and purified using Ni Sepharose™ affinity column. The Vip3Aa was dialyzed in a buffer containing 25 mM Tris-Hcl (pH 7.4) and 150 mM NaCl at 4 °C. The result of purified Vip3Aa was shown in Figure S6. The concentration of Vip3Aa was measured via the protein-dye method of Bradford. BSA was used as a standard protein. The full-length Vip3Aa was used directly in Sf9 cells.

4.3. Cell Viability Assay

The cell viability was detected using the CCK-8 Counting Kit (Dojindo, Kumamoto, Japan). Cell suspensions (100 µL, 2.5×10^5 cells/mL) were pipetted into a 96-well plate and incubated overnight at

28 °C. Then, Vip3Aa was added into the suspensions. The cells were exposed to Vip3Aa for 24, 48, 60, and 72 h. The final concentration of Vip3Aa was 10, 20, 30, 40, and 50 µg/mL. Sf-900 II SFM medium and cell suspensions without Vip3Aa were used as blank group and control group, respectively. Then, CCK-8 reagent (10 µL) was added and incubated in darkness for 2–4 h at 28 °C. The results were monitored at 450 nm using a microplate reader (PerkinElmer, Boston, MA, USA). The experiments were performed six times. Cell viability was the ratio of absorbance of Vip3Aa-treated group/control group.

4.4. Vip3Aa Subcellular Localization in Sf9 Cells

The cells were exposed to Vip3Aa-RFP for 0, 2, 4, and 6 h. Then, the cells were incubated with Sf-900 II SFM medium containing 1 µM LysoSensor™ Green DND-189 at 28 °C in the darkness for 45 min. The cells were washed with phosphate-buffered saline (PBS) (pH 7.4) three times and imaged with a Zeiss LSM710 fluorescence microscope.

4.5. Transmission Electron Microscopy (TEM)

Cell suspensions (5×10^5–1×10^6 cells/mL) were incubated overnight in 25 cm^2 flasks. The cultures were exposed to Vip3Aa (final concentration, 40 µg/mL) for 12, 24, 36, and 48 h, respectively. Transmission electron microscope (TEM) was used to observe and record the ultrastructure of Sf9 cells. The cells were fixed in 2% paraformaldehyde and 2.5% glutaraldehyde in phosphate-buffered saline (PBS, pH 7.4) for 2 h after washing with PBS three times. Then, the fixed cells were treated with 1% osmic acid (OsO_4) at 25 °C for 1 h after washing with PBS. The cell samples were dehydrated in different concentration ethanol solutions, soaked, and embedded in EPON812. Ultrathin (60 nm) sections were cut and counterstained with lead citrate and uranyl acetate. The sections were observed with TEM (JEOL-1200EX).

4.6. Measurement of Intracellular ROS and Mitochondrial Membrane Potential ($\Delta\Psi m$)

DCFH-DA and Rhodamine 123 were utilized to measure intracellular ROS [33] and $\Delta\Psi m$, respectively. The cells were exposed to Vip3Aa (final concentration, 40 µg/mL) for different times. Then, the cells were incubated with Sf-900 II SFM medium containing 10 mM of DCFH-DA or 50 nM Rhodamine 123 at 28 °C in the darkness for 30 min. The cells were washed with PBS (pH 7.4) three times and imaged with a Zeiss LSM710 fluorescence microscope.

4.7. Total Protein and Cytosolic Protein Extraction

The cells were exposed to Vip3Aa (final concentration, 40 µg/mL) for different times. The cells were lysed in 350 µL RIPA buffer with 1 mM PMSF and incubated on ice for 15 min after washing with PBS (pH 7.4) three times. The suspension was centrifuged at 12,000× g for 15 min. Then, the supernatant, which was the total protein extraction, was collected carefully.

Cells were washed and collected by centrifugation at 200 × g for 5 min. The cells were resuspended with 500 µL isotonic buffer (IB, 10 mM HEPES, 200 mM mannitol, 1 mM EGTA, 70 mM sucrose). The cell suspension was centrifuged at 3000× g for 5 min. The collected cells were resuspended in 500 µL IB with 20 mM NaF, 20 mM Na$_3$VO$_4$ and 1 mM PMSF. Then, 26-G needles were used to homogenize the cell suspension, which was passed through 14 times and stood on ice for 5 min. The suspension was centrifuged at 4 °C, 10,000× g for 15 min. Centrifugation sediment contains lysosomes and mitochondria. Then, the supernatant was diverted to a fresh cold centrifuge tube and centrifuged at 4 °C, 14,000× g for 30 min in an ultracentrifuge. The supernatant, which was the cytosolic protein extraction, was collected carefully.

4.8. Quantitative Real-Time PCR

Trizol reagent (Invitrogen, Carlsbad, CA, USA) was used to extract total RNA. Chloroform and isopropanol were used to isolate RNA. The primers used in this study are listed in Table 1.

A Primescript^{TM} RT reagent kit with gDNA Eraser (TakaRa, Dalian, China) was utilized to reverse-transcribe RNA. The quantitative real-time PCR was performed with SYBR® Premix Ex Taq™ (TakaRa, Dalian, China) in an ABI Prism 7900HT Real-Time PCR System (Applied Biosystems, Carlsbad, CA, USA). *GAPDH* was used as the control for normalization by the $2^{-\Delta\Delta Ct}$ method [34].

Table 1. Primers used in this study.

Primers	Primer Sequence
qCathepsin B-F	5′-GAAGTGAGGGACCAAGGAT-3′
qCathepsin B-R	5′-TCTGCGGAGAAGTGGAAAT-3′
qCathepsin L-F	5′-CAGGGTGATGAGGAGAAGC-3′
qCathepsin L-R	5′-TCGGTGGACGAGCAGTT-3′
qCathepsin D-F	5′-CAGGGGCTGGTGAAGCCA-3′
qCathepsin D-R	5′-CACGTACGTGAAGTTGCC-3′
qGAPDH-F	5′-GTGCCCAGCAGAACATCAT-3′
qGAPDH-R	5′-GGAACACGGAAAGCCATAC-3′

4.9. Western Blotting Analysis

A BCA Protein Assay kit (TIANGEN, Beijing, China) was used to test the concentrations of protein samples. Next, 12% SDS-PAGE electrophoresis was utilized to separate the target proteins, which were transferred onto the PVDF membrane. Primary antibodies were anti-caspase-3 (1:1000), anti-caspase-9 (1:500), anti-Bcl-XL (1:500), anti-Bcl-2 (1:500), anti-Bax (1:500), anti-cytochrome c (1:500), anti-cathepsin L (1:500), and anti-β-actin (1:500). Mouse anti-rabbit IgG-HRP (1:1000) and goat anti-mouse IgG-HRP (1:1000) were the secondary antibodies. Finally, the PVDF membranes were visualized using Immobilon Western Chemiluminescent HRP Substrate (Millipore, Milan, Italy).

4.10. Acridine Orange (AO) Staining Analysis

The AO staining analysis was performed by flow cytometry (Becton Dickinson, USA). The Sf9 cells were treated with Vip3Aa (final concentration, 40 μg/mL) for different times. Then, the cells were incubated with Sf-900 II SFM medium containing 5 μg/mL AO at 28 °C in the darkness for 10 min. The stained cells were used to analyze the fluorescence distribution (FL1-H/FL3-H) after washing with PBS (pH 7.4) three times. After adjusting the fluorescence compensation of the channels, the number of recorded cells was 10,000.

4.11. Lysosomal pH Assay

The LysoSensor Yellow/Blue DND-160 (Life Technologies, Carlsbad, CA), a lysosomal pH indicator, was used to measure the Sf9 cells lysosomal pH. Cell suspensions (100 μL, 2.5×10^5 cells/mL) were pipetted into a 96-well black plate. All the cells were incubated with Sf-900 II SFM medium containing 5 μM fluorescent probe at 28 °C in the darkness for 5 min. Then, the cells were washed and cultured in an MES calibration buffer (1.2 mM $MgSO_4$, 115 mM KCl, 5 mM NaCl, 25 mM MES, pH 3.5–6.0) containing 10 μM monensin and 10 μM nigericin. The fluorescence value (Ex340 nm/Em540 nm and Ex380 nm/Em540 nm) was monitored by a microplate reader (PerkinElmer, Boston, MA, USA). The pH calibration curve was generated using ratios of the two light emission intensities and the corresponding pH value. To find the effect of Vip3Aa on lysosomal pH, the Vip3Aa-treated cells were incubated with Sf-900 II SFM medium containing 5 μM fluorescent probe at 28 °C in the darkness for 5 min, washed, and resuspended in MES buffer (pH 7.0) and detected using a microplate reader (PerkinElmer, Boston, MA, USA). The lysosomal pH was estimated using the ratios and the pH calibration curve. Sf-900 II SFM medium and cell suspensions without Vip3Aa were used as blank group and control group, respectively.

4.12. Apoptosis Assay

Sf9 cells were treated with Vip3Aa (final concentration, 40 μg/mL) for 48 h with/without cathepsins inhibitor for 2 h. We evaluated the proportion of apoptotic cells using the FITC annexin V apoptosis detection kit I (BD Biosciences, USA). Cells without Vip3Aa-treated were used as a control group. After washing twice with PBS (100 × g, 5 min), cells incubated with 1 × binding buffer containing FITC annexin V at 28 °C in the darkness for 30 min. Then, 1 × binding buffer containing propidium iodide was added to each sample. After incubating at 28 °C in the darkness for 5 min, the cells were monitored with a flow cytometer (Becton Dickinson, Franklin Lakes, NJ, USA).

4.13. Caspase Activity Analysis

Sf9 cells were treated with Vip3Aa (final concentration, 40 μg/mL) for different time with/without cathepsins inhibitor for 2 h. Caspase-Glo® assay kit (Promega, Madison, WI, USA) was utilized to determine caspase activity. Cell suspensions (100 μL, 2.5×10^5 cells/mL) were pipetted into a 96-well white plate and incubated overnight at 28 °C. Then, Vip3Aa was added into the suspensions. The cells were exposed to Vip3Aa for 12, 24, 36, 48, 60, and 72 h. Sf-900 II SFM medium and cell suspensions without Vip3Aa were used as blank group and control group, respectively. Caspase-Glo® Reagent was prepared according to the protocol and all the operations should be performed in the darkness. Equilibrate the reagent and plates to room temperature. Caspase-Glo® Reagent (100 μL) was added to the plates containing cells in Sf-900 II SFM medium. A plate shaker was used to mix the plates containing cells and reagent at 300–500 rpm for 0.5–2 min. Then, the plates were incubated at room temperature for 2 h. Finally, the luminescence of each plate was detected using a microplate reader (PerkinElmer, Boston, MA, USA).

4.14. Statistical Analysis

The results were obtained from at least three independent experiments. The densitometry values were evaluated by the software Image J. Origin 8.0 (OriginLab, Northampton, MA, USA) was used to draw the graphs. The significance was tested by one-way analysis of variance utilizing Student t test. If p-value ≤0.05, the results were considered significant.

Supplementary Materials: The following are available online at http://www.mdpi.com/2072-6651/12/2/116/s1, Figure S1: Impacts of Vip3Aa in Sf9 cells on the cytochrome c distribution and the level of mitochondria-associated proteins. (A) Cytochrome c distribution in mitochondria and cytosol was detected by Western blotting. (B) Cytochrome c distribution in Sf9 cells after various treatments. CsA (5 μM) or BKA (10 μM) was pretreated the Sf9 cells for 2 h before adding Vip3Aa. (C) Mitochondria-associated protein levels in Sf9 cells were tested using Western blotting. (D, E, and F) Densitometry analysis of (A) and (C). Figure S2: Effect of Vip3Aa on cathepsins mRNA level in Sf9 cells at different time. Figure S3: Effects of cathepsin (L and D) inhibitors on cytochrome c distribution. (A) cytochrome c distribution in cytosol was detected by Western blotting. (B) Densitometry analysis of (A). Figure S4: Effects of Vip3Aa on caspase-8 activity. Significant tests from the corresponding controls (without Vip3Aa treatment) are indicated by NS, not significant, *$p < 0.05$, **$p < 0.01$, ***$p < 0.001$. Figure S5: Effects of caspase-3 inhibitor on Sf9 cell viability. Significant tests from the corresponding controls (without Vip3Aa treatment) are indicated by * $p < 0.05$, ** $p < 0.01$, *** $p < 0.001$. Figure S6: Result of purified Vip3Aa detected by SDS-PAGE electrophoresis analysis. M, protein maker 26619.

Author Contributions: Data curation, X.H.; funding acquisition, J.C.; investigation, X.H., L.H., B.A., Y.Z. (Yanli Zhang), Z.C., Y.Z. (Yunda Zhan), X.C., and B.Y.; methodology, X.H. and L.H.; project administration, J.C.; supervision, J.C.; writing—original draft, X.H.; writing—review and editing, J.C. All authors have read and agreed to the published version of the manuscript.

Funding: This study was funded by grants from the National Key R&D Program of China (No. 2017YFD0200400), and the National Natural Science Foundation of China (No. 31670081 and 31371979).

Conflicts of Interest: The authors declare no conflict of interest.

References

1. Chakroun, M.; Ferré, J. In vivo and in vitro binding of Vip3Aa to *Spodoptera frugiperda* midgut and characterization of binding sites by (125)I radiolabeling. *Appl. Environ. Microbiol.* **2014**, *80*, 6258–6265. [CrossRef]
2. Liu, J.G.; Yang, A.Z.; Shen, X.H.; Hua, B.G.; Shi, G.L. Specific binding of activated Vip3Aa10 to *Helicoverpa armigera* brush border membrane vesicles results in pore formation. *J. Invertebr. Pathol.* **2011**, *108*, 92–97. [CrossRef]
3. Lee, M.K.; Walters, F.S.; Hope, H.; Narendra, P.; Jeng-Shong, C. The mode of action of the *Bacillus thuringiensis* vegetative insecticidal protein Vip3A differs from that of Cry1Ab delta-endotoxin. *Appl. Environ. Microbiol.* **2003**, *69*, 4648–4657. [CrossRef]
4. Hernández-Martínez, P.; Hernández-Rodríguez, C.S.; Rie, J.V.; Escriche, B.; Ferré, J. Insecticidal activity of Vip3Aa, Vip3Ad, Vip3Ae, and Vip3Af from *Bacillus thuringiensis* against lepidopteran corn pests. *J. Invertebr. Pathol.* **2013**, *113*, 78–81. [CrossRef]
5. Gatikrushna, S.; Bindiya, S.; Nathilal, S.; Rakesh, S.; Bhatnagar, R.K. Interaction of *Bacillus thuringiensis* vegetative insecticidal protein with ribosomal S2 protein triggers larvicidal activity in *Spodoptera frugiperda*. *Appl. Environ. Microbiol.* **2010**, *76*, 7202–7209. [CrossRef]
6. Chakroun, M.; Banyuls, N.; Bel, Y.; Escriche, B.; Ferré, J. Bacterial vegetative insecticidal proteins (Vip) from entomopathogenic bacteria. *Microbiol. Mol. Biol. Rev.* **2016**, *80*, 329–350. [CrossRef] [PubMed]
7. Palma, L.; Muñoz, D.; Berry, C.; Murillo, J.; Caballero, P. *Bacillus thuringiensis* Toxins: An Overview of Their Biocidal Activity. *Toxins* **2014**, *6*, 3296–3325. [CrossRef] [PubMed]
8. Kunthic, T.; Watanabe, H.; Kawano, R.; Tanaka, Y.; Promdonkoy, B.; Yao, M.; Boonserm, P. pH regulates pore formation of a protease activated Vip3Aa from *Bacillus thuringiensis*. *Biochim. Biophys. Acta* **2017**, *1859*, 2234–2241. [CrossRef]
9. Jiang, K.; Hou, X.Y.; Tan, T.T.; Cao, Z.L.; Mei, S.Q.; Yan, B.; Chang, J.; Han, L.; Zhao, D.; Cai, J. Scavenger receptor-C acts as a receptor for *Bacillus thuringiensis* vegetative insecticidal protein Vip3Aa and mediates the internalization of Vip3Aa via endocytosis. *PLoS Pathog.* **2018**, *14*, e1007347. [CrossRef]
10. Jiang, K.; Hou, X.; Han, L.; Tan, T.; Cai, J. Fibroblast growth factor receptor, a novel receptor for vegetative insecticidal protein Vip3Aa. *Toxins* **2018**, *10*, 546. [CrossRef]
11. Bel, Y.; Jakubowska, A.K.; Costa, J.; Herrero, S.; Escriche, B. Comprehensive analysis of gene expression profiles of the beet armyworm *Spodoptera exigua* larvae challenged with *Bacillus thuringiensis* Vip3Aa toxin. *PloS ONE* **2013**, *8*, e81927. [CrossRef] [PubMed]
12. Crava, C.M.; Jakubowska, A.K.; Escriche, B.; Herrero, S.; Bel, Y. Dissimilar Regulation of Antimicrobial Proteins in the Midgut of *Spodoptera exigua* Larvae Challenged with *Bacillus thuringiensis* Toxins or Baculovirus. *PLoS ONE* **2015**, *10*, e0125991. [CrossRef] [PubMed]
13. Ayra-Pardo, C.; Ochagavía, M.E.; Raymond, B.; Gulzar, A.; Rodríguez-Cabrera, L.; Rodríguez de la Noval, C.; Morán Bertot, I.; Terauchi, R.; Yoshida, K.; Matsumura, H.; et al. HT-SuperSAGE of the gut tissue of a Vip3Aa-resistant *Heliothis virescens* (Lepidoptera: Noctuidae) strain provides insights into the basis of resistance. *Insect Sci.* **2019**, *26*, 479–498. [CrossRef] [PubMed]
14. Hernández-Martínez, P.; Gomis-Cebolla, J.; Ferré, J.; Escriche, B. Changes in gene expression and apoptotic response in *Spodoptera exigua* larvae exposed to sublethal concentrations of Vip3 insecticidal proteins. *Sci. Rep.* **2017**, *7*, 16245. [CrossRef]
15. Jiang, K.; Mei, S.; Wang, T.; Pan, J.; Chen, Y.; Cai, J. Vip3Aa induces apoptosis in cultured *Spodoptera frugiperda* (Sf9) cells. *Toxicon* **2016**, *120*, 49–56. [CrossRef]
16. Guang, Y.; Shaopeng, W.; Laifu, Z.; Xu, D.; Wenli, Z.; Liping, J.; Chengyan, G.; Xiance, S.; Xiaofang, L.; Min, C. 6-Gingerol induces apoptosis through lysosomal-mitochondrial axis in human hepatoma G2 cells. *Phytother. Res.* **2012**, *26*, 1667–1673. [CrossRef]
17. Kroemer, G.; Reed, J.C. Mitochondrial control of cell death. *Nat. Med.* **2001**, *1*, 513–519. [CrossRef]
18. Kroemer, G. The proto-oncogene Bcl-2 and its role in regulating apoptosis. *Nat. Med.* **1997**, *3*, 614–620. [CrossRef]
19. Bouchier-Hayes, L.; Lartigue, L.; Newmeyer, D.D. Mitochondria: Pharmacological manipulation of cell death. *J. Clin. Investig.* **2005**, *115*, 2640–2647. [CrossRef]
20. Nick, L. Mitochondrial disease: Powerhouse of disease. *Nature* **2006**, *440*, 600–602. [CrossRef]

21. Surico, D.; Farruggio, S.; Marotta, P.; Raina, G.; Mary, D.; Surico, N.; Vacca, G.; Grossini, E. Human chorionic gonadotropin protects vascular endothelial cells from oxidative stress by apoptosis inhibition, cell survival signalling activation and mitochondrial function protection. *Cell. Physiol. Biochem.* **2015**, *36*, 2108–2120. [CrossRef] [PubMed]

22. Starkov, A.A. The role of mitochondria in reactive oxygen species metabolism and signaling. *Ann. N. Y. Acad. Sci.* **2008**, *1147*, 37–52. [CrossRef] [PubMed]

23. Green, D.R. Apoptotic pathways: Ten minutes to dead. *Cell* **2005**, *121*, 671–674. [CrossRef]

24. Lee, Y.; Gustafsson, Å.B. Role of apoptosis in cardiovascular disease. *Apoptosis* **2009**, *14*, 536–548. [CrossRef]

25. alluzzi, L.; Vitale, I.; Abrams, J.M.; Alnemri, E.S.; Baehrecke, E.H.; Blagosklonny, M.V.; Dawson, T.M.; Dawson, V.L.; El-Deiry, W.S.; Fulda, S. Molecular definitions of cell death subroutines: Recommendations of the Nomenclature Committee on Cell Death. *Cell Death Differ.* **2012**, *19*, 107–120. [CrossRef]

26. Tait, S.W.G.; Green, D.R. Mitochondria and cell death: Outer membrane permeabilization and beyond. *Nat. Rev. Mol. Cell Biol.* **2010**, *11*, 621–632. [CrossRef]

27. Ichim, G.; Tait, S.W. A fate worse than death: Apoptosis as an oncogenic process. *Nat. Rev. Cancer* **2016**, *16*, 539–548. [CrossRef]

28. Kantari, C.; Walczak, H. Caspase-8 and bid: Caught in the act between death receptors and mitochondria. *BBA-Mol. Cell Res.* **2011**, *1813*, 558–563. [CrossRef]

29. Duve, C.D.; Wattiaux, R. Functions of lysosomes. *Annu. Rev. Physiol.* **1966**, *28*, 435–492. [CrossRef]

30. Keng-Fu, H.; Chao-Liang, W.; Soon-Cen, H.; Ching-Ming, W.; Jenn-Ren, H.; Yi-Te, Y.; Yu-Hung, C.; Ai-Li, S.; Cheng-Yang, C. Cathepsin L mediates resveratrol-induced autophagy and apoptotic cell death in cervical cancer cells. *Autophagy* **2009**, *5*, 451–460. [CrossRef]

31. Lee, K.S.; Bo, Y.K.; Choo, Y.M.; Yoon, H.J.; Kang, P.D.; Woo, S.D.; Sohn, H.D.; Roh, J.Y.; Zhong, Z.G.; Je, Y.H. Expression profile of cathepsin B in the fat body of *Bombyx mori* during metamorphosis. *Comp. Biochem. Phys. A* **2009**, *154*, 188–194. [CrossRef] [PubMed]

32. Diep, D.B.; Nelson, K.L.; Lawrence, T.S.; Sellman, B.R.; Tweten, R.K.; Buckley, J.T. Expression and properties of an aerolysin—Clostridium septicum alpha toxin hybrid protein. *Mol. Microbiol.* **1999**, *31*, 785–794. [CrossRef] [PubMed]

33. Lee, Y.J.; Lim, S.S.; Baek, B.J.; An, J.M.; Nam, H.S.; Woo, K.M.; Cho, M.K.; Kim, S.H.; Lee, S.H. Nickel(II)-induced nasal epithelial toxicity and oxidative mitochondrial damage. *Environ. Toxicol. Pharmacol.* **2016**, *42*, 76–84. [CrossRef] [PubMed]

34. Livak, K.J.; Schmittgen, T.D. Analysis of relative gene expression data using real-time quantitative PCR and the $2^{-\Delta\Delta CT}$ method. *Methods* **2001**, *25*, 402–408. [CrossRef]

MDPI

St. Alban-Anlage 66

4052 Basel

Switzerland

Tel. +41 61 683 77 34

Fax +41 61 302 89 18

www.mdpi.com

Toxins Editorial Office

E-mail: toxins@mdpi.com

www.mdpi.com/journal/toxins